An Introduction to Neuroendocrinology

Second Edition

How does the brain regulate sexual behavior, or control our body weight? How do we cope with stress? Addressing these questions and many more besides, this thoroughly revised new edition reflects the significant advances that have been made in the study of neuroendocrinology over the last 20 years.

The text examines the importance of the hypothalamus in regulating hormone secretion from the endocrine glands, describing novel sites of hormone release, including bone, heart, skeletal muscle, and liver. The role of steroid hormone, neurotransmitter and peptide receptors, and the molecular responses of target tissues, is integrated into the discussion of the neuroendocrine brain, especially through changes in gene expression. Particular attention is attached to neuropeptides, including their profound influence on behavior.

Complete with new full-color figures throughout, along with review and essay questions for each chapter, this is an ideal resource for undergraduate and graduate students of neuroscience, psychology, biology, and physiology.

Michael Wilkinson has 40 years of experience in teaching neuroscience and neuroendocrinology to undergraduate and graduate students as a Professor in the Department of Obstetrics and Gynaecology and IWK Health Centre, Dalhousie University, Halifax, Canada. His research laboratory has focused on neurodevelopmental aspects of female reproduction with a specific interest in the neuroendocrine regulation of hypothalamic function, including the impact of sex hormones on sleep.

Richard E. Brown is a University Research Professor in the Department of Psychology and Neuroscience at Dalhousie University. He has taught courses on hormones and behavior, measuring behavior, and the neurobiology of learning and memory for more than 35 years. His research is on mouse models of Alzheimer's Disease, Fragile X Syndrome, ADHD, and other neurological disorders. He is currently examining the age-related hormonal changes in transgenic Alzheimer's mice.

An Introduction to

Neuroendocrinology

Second Edition

Michael Wilkinson

Professor of Obstetrics & Gynaecology
Professor of Physiology & Biophysics
Dalhousie University Faculty of Medicine, Halifax, Nova Scotia, Canada

Richard E. Brown

Professor of Psychology
Dalhousie University, Halifax, Nova Scotia, Canada

CAMBRIDGE
UNIVERSITY PRESS

CAMBRIDGE
UNIVERSITY PRESS

University Printing House, Cambridge CB2 8BS, United Kingdom

Cambridge University Press is part of the University of Cambridge.

It furthers the University's mission by disseminating knowledge in the pursuit of education, learning and research at the highest international levels of excellence.

www.cambridge.org
Information on this title: www.cambridge.org/9780521806473

First published 1994
Second edition 2015
Reprinted 2022

Printed in the United Kingdom by TJ Books Limited, Padstow Cornwall

A catalogue record for this publication is available from the British Library

Library of Congress Cataloguing in Publication data
Wilkinson, Michael, 1943–, author.
An introduction to neuroendocrinology / Michael Wilkinson, Richard E. Brown. – Second edition.
  p. ; cm.
Richard E. Brown's name appears first in the previous edition.
Includes bibliographical references and index.
ISBN 978-0-521-80647-3 (hardback) – ISBN 978-0-521-01476-2 (paperback)
I. Brown, Richard E., author.   II. Title.
[DNLM: 1. Neuroendocrinology – methods.   2. Endocrine
Glands.   3. Neuropeptides.   4. Neurosecretory Systems.   5. Peptide Hormones.
6. Receptors, Neurotransmitter. WL 105]
QP356.4
612.8–dc23

                                                                          2014041712

ISBN 978-0-521-80647-3 Hardback
ISBN 978-0-521-01476-2 Paperback

This book is dedicated, first, to the more than 2,000 Dalhousie University students who were enrolled in the "Hormones and Behavior" undergraduate course and who were the original inspiration for writing the book. Many of them provided critical comments on early drafts of the first edition.

Second, one of us (M. W.) acknowledges the mentorship of the late Professor Kurt B. Ruf, a neuroendocrinologist and friend.

CONTENTS

Preface to the second edition | xiii
Acknowledgements | xv
List of abbreviations | xvi

1 Classification of chemical messengers | 1
1.1 Hormones, the brain and behavior | 1
1.2 The body's three communication systems | 2
1.3 Methods of communication between cells | 5
1.4 Types of chemical messenger | 7
1.5 Neuropeptides and neuromodulators | 14
1.6 Summary | 16

2 The endocrine glands and their hormones | 19
2.1 The endocrine glands | 19
2.2 The hormones of the endocrine glands | 19
2.3 Summary | 40

3 The pituitary gland and its hormones | 45
3.1 The pituitary gland | 45
3.2 The hormones of the pituitary gland | 49
3.3 Pituitary hormones in the brain | 54
3.4 Summary | 54

4 The hypothalamic hormones | 57
4.1 Functions of the hypothalamus | 57
4.2 Hypothalamic neurosecretory cells | 58
4.3 Hypothalamic hypophysiotropic hormones | 62
4.4 Complexities of hypothalamic-pituitary interactions | 71
4.5 Summary | 73

5 Neurotransmitters | 78
5.1 The neuron and the synapse | 78
5.2 Categories of neurotransmitters | 80
5.3 Neurotransmitter biosynthesis and storage | 88
5.4 Release of neurotransmitters | 92
5.5 Receptors for neurotransmitters | 94

5.6   Inactivation of neurotransmitters | 97
5.7   Neurotransmitter pathways | 97
5.8   Drugs influencing neurotransmitters, and their receptors, in the
      nervous system | 102
5.9   Can nutrients modify neurotransmitter levels and behavior? | 107
5.10 Divisions of the nervous system | 109
5.11 Summary | 114

6   Neurotransmitter and neuropeptide control of hypothalamic,
    pituitary and other hormones | 120
6.1   The cascade of chemical messengers | 120
6.2   Neural control of hypothalamic neurosecretory cells | 120
6.3   Neurotransmitter regulation of anterior pituitary hormone
      secretion | 124
6.4   Neurotransmitter regulation of neurohypophyseal hormone
      secretion | 133
6.5   Electrophysiology of neurosecretory cells | 135
6.6   Neurotransmitter regulation of other endocrine glands | 135
6.7   Complications in the study of neurotransmitter control of
      hypothalamic hormone release | 138
6.8   Neuroendocrine correlates of psychiatric disorders and
      psychotropic drug treatment of these disorders | 141
6.9   Glial cells and the regulation of hormone release | 142
6.10 Summary | 150

7   Regulation of hormone synthesis, storage, release, transport
    and deactivation | 157
7.1   The chemical structure of hormones | 157
7.2   Hormone synthesis | 159
7.3   Storage and intracellular transport of hormones | 162
7.4   Hormone release | 163
7.5   Hormone transport | 164
7.6   Deactivation of hormones | 166
7.7   Methodology for neuroendocrine research | 166
7.8   Summary | 168

8   Regulation of hormone levels in the bloodstream | 170
8.1   Analysis of hormone levels | 170
8.2   Mechanisms regulating hormone levels | 174
8.3   Hormonal modulation of neurotransmitter release | 183
8.4   The cascade of chemical messengers revisited | 184

8.5   When hormone regulatory mechanisms fail | 186
8.6   Summary | 187

9   Steroid and thyroid hormone receptors | 192
9.1   The intracellular receptor superfamily | 193
9.2   How are steroid hormone target cells identified? | 196
9.3   How are steroid hormone target cells differentiated from
      non-target cells? | 198
9.4   Genomic and non-genomic actions of steroid hormones | 198
9.5   Measurement and regulation of hormone receptor numbers | 205
9.6   Gonadal steroid hormone target cells in the brain | 206
9.7   Adrenal steroid target cells in the brain | 210
9.8   Steroid hormone-induced changes in neurotransmitter release | 214
9.9   Functions of steroid hormone modulation of nerve cells | 215
9.10 Thyroid hormone receptors in the brain | 223
9.11 Summary | 225

10 Receptors for peptide hormones, neuropeptides
    and neurotransmitters | 236
10.1 Membrane receptors | 236
10.2 Signal transduction by G proteins | 240
10.3 Second messenger systems | 242
10.4 Interactions in second messenger systems | 248
10.5 Signal amplification | 249
10.6 Second messengers in the brain and neuroendocrine system | 249
10.7 Comparison of neurotransmitter/neuropeptide and steroid hormone
      actions at their target cells | 252
10.8 Summary | 253

11 Neuropeptides I: classification, synthesis and co-localization
    with classical neurotransmitters | 257
11.1 Classification of neuropeptides | 257
11.2 Synthesis, storage, release and deactivation of neuropeptides | 259
11.3 Exploring the relationships among neuropeptides, neurotransmitters
      and hormones | 261
11.4 Coexistence (co-localization) of neurotransmitters
      and neuropeptides | 266
11.5 Localization of neuropeptide cell bodies and pathways in the
      brain | 269
11.6 Neuropeptide receptors and second messenger systems | 272
11.7 Neuropeptides and the blood-brain barrier | 275
11.8 Summary | 278

12   Neuropeptides II: function | 286
     12.1 Neurotransmitter and neuromodulator actions of neuropeptides: a
          dichotomy or a continuum? | 286
     12.2 Neurotransmitter actions of neuropeptides | 287
     12.3 Neuromodulator actions of neuropeptides | 293
     12.4 Regulatory effects of neuropeptides on the neuroendocrine
          system | 299
     12.5 Kisspeptin and GnRH as hypothalamic regulators of fertility | 300
     12.6 Neuropeptides and the regulation of food intake and body
          weight | 302
     12.7 Visceral, cognitive and behavioral effects of neuropeptides | 313
     12.8 Summary | 332

13   Cytokines and the interaction between the neuroendocrine
     and immune systems | 351
     13.1 The cells of the immune system | 351
     13.2 The thymus gland and its hormones | 354
     13.3 Cytokines: the messengers of the immune system | 356
     13.4 The functions of cytokines in the immune and hematopoietic
          systems | 360
     13.5 Effects of cytokines and other immunomodulators on the brain and
          neuroendocrine system | 364
     13.6 Neural and endocrine regulation of the immune system | 374
     13.7 Hypothalamic integration of the neuroendocrine and immune
          systems | 384
     13.8 Summary | 387

14   Methods for the study of behavioral neuroendocrinology | 400
     14.1 Behavioral bioassays | 400
     14.2 Correlational studies of hormonal and behavioral changes | 403
     14.3 Experimental studies I: behavioral responses to neuroendocrine
          manipulation | 407
     14.4 Experimental studies II: neuroendocrine responses to
          environmental, behavioral and cognitive stimuli | 416
     14.5 Neural and genomic mechanisms mediating neuroendocrine-
          behavior interactions | 424
     14.6 Confounding variables in behavioral neuroendocrinology
          research | 434
     14.7 Summary | 444

15   An overview of behavioral neuroendocrinology: present, past
      and future | 458
      15.1 The aim of this book | 458
      15.2 The history of endocrinology and behavioral
            neuroendocrinology | 460
      15.3 The future of behavioral neuroendocrinology | 460

      Index | 469

PREFACE TO THE SECOND EDITION

In this second edition of *An Introduction to Neuroendocrinology*, we have rewritten and greatly extended the original content. The revised text includes entirely new reference lists and a complete new set of illustrations. The book reflects the many advances that have occurred in the study of neuroendocrinology during the past 20 years. Nevertheless, and although the text is based largely on modern references, our primary aim is to provide an introductory description of mammalian neuroendocrine control systems. Several books are available that cover this topical and clinically relevant field, but, although valuable, these tend to be advanced texts of the edited, multi-author type. Our book is designed to provide the basic principles necessary to understand how the brain controls, and responds to, the endocrine hormones. It will be suitable for a variety of different students and especially those who might not have been previously exposed to a focused course in neuroendocrinology. Thus, students in psychology, biology and science should be able to master much of the basic material. However, the book is also highly appropriate for honors students and first-year graduate students in physiology, anatomy, neuroscience and medicine. This book is therefore designed for students in two levels of classes: introductory classes, in which all of the material will be new to the student, and more advanced classes, in which the students will be familiar with many of the terms and concepts through courses in biology, physiology, psychology or neuroscience, but who have not studied neuroendocrinology as an integrated discipline.

This book offers an overall outline of the neuroendocrine system and will provide the vocabulary necessary to understand the interaction between hormones and the brain. In addition, we provide a concise description of those topics that must underpin any attempt to learn, and to teach, neuroendocrinology. For example, there are chapters on basic neuroscience (neurotransmitters and neuropeptides), the physiology of the endocrine glands (hormones), receptors and receptor signaling mechanisms (e.g. G proteins; nuclear receptors), hormone assay and gene expression techniques (e.g. ELISA; *in situ* hybridization) and a description of the immune system, with particular emphasis on the integration of immune and neuroendocrine pathways. This basic information is also essential to understand the profound effects of hormones on behavior, described in Chapter 14. Once this material is mastered, the study of how hormones influence developmental neural processes and behavior will be easier. Moreover, we have included throughout the book references to the clinical relevance of many topics; for example, the influence of neuropeptides in the control of body weight and obesity. However, this book focuses primarily on the neural actions of hormones, and many of the peripheral physiological actions of hormones, such as regulation of metabolism, water balance, growth, and the regulation of calcium, sodium and potassium levels, which are the focus of traditional endocrinology texts, are referred to only in reference to their importance in the neuroendocrine system.

The introductory (second- or third-year undergraduate) student can be expected to follow the material in this book at the level presented. To help in this, review/study questions are given at the end of each chapter. These should be treated as practice examination questions and answered after each chapter is completed. For further detailed information on the topics covered in each chapter, all students can consult selected references provided in the text. Additional references under "Further reading" are also included at the end of each chapter and these will be particularly useful to the more advanced student. The book will be especially relevant for more advanced (honors and graduate) students who can use this book as an introductory account of the subject matter covered in each chapter. These students may then take advantage of the many references cited in each chapter to provide current and relevant information on each topic. The essay questions at the end of each chapter also serve to provide topics for discussion, analysis and directed research papers for the advanced student.

ACKNOWLEDGEMENTS

The authors are indebted to friends and colleagues who offered generous and invaluable assistance in the writing of this book. Paul Wilkinson, Ms. Alex Pincock and Ms. Diane Wilkinson created several figures; Alex Pincock and Dr. Jim Pincock carefully read, and made useful suggestions for improvement of, several early chapters. Special thanks are due to Diane Wilkinson, who typed all the tables and assisted in compiling the extensive reference lists. The following scientists unselfishly provided illustrations from their published material: Dr. O. Almeida, Dr. A. Armario, Dr. R. Bridges, Dr. R. Goyal, Dr. L. De Groot, Dr. L. Hale, Dr. J. Herman, Drs. T. Horvath and M. Dietrich, Ms. A. Rain, Dr. T. Smith, Dr. J. Ström, Dr. J. Wakerley, Dr. A. Winokur and Dr. S. Winters. As far as we are aware, all sources of the illustrations used have been acknowledged. Permission to use previously published figures was obtained either from the original authors or via *RightsLink* (Copyright Clearance Centre).

Finally, thanks are due to Megan Waddington of Cambridge University Press for her patience in awaiting the delivery of this manuscript.

ABBREVIATIONS

IIIv	third ventricle	CART	cocaine- and amphetamine-regulated transcript
2-AG	2-arachidinoyl glycerol		
5-HIAA	5-hydroxyindoleacetic acid	cGMP	cyclic guanosine monophosphate
5-HT	5-hydroxytryptamine (serotonin)	CB1	cannabinoid receptor 1
5-HTP	5-hydroxytryptophan	CBG	corticosteroid binding globulin (transcortin)
6-OHDA	6-hydroxy-dopamine		
AC	adenyl cyclase	CCK	cholecystokinin
ACh	acetylcholine	CCK-KO	CCK knockouts
ACTH	adrenocorticotropic hormone	CGRP	calcitonin gene related peptide
ADH	antidiuretic hormone (vasopressin)	ChAT	choline acetyltransferase
ADHD	attention deficit hyperactivity disorder	CL	centrolateral thalamus
AEA	anandamide	Cl⁻	chloride ion
AgRP	agouti-related protein	CLIP	corticotropin-like intermediate lobe peptide
AH	anterior hypothalamus		
AHA	anterior hypothalamic area	CM	centromedial thalamus
AMPA	α-amino-3-hydroxy-5-methyl-4-isoxazole propionic acid	CNS	central nervous system
		COMT	catechol o-methyl transferase
AMYG	amygdala	CP	caudate/putamen
ANP	atrial natriuretic peptide	CREB	cAMP responsive element binding protein
ANS	autonomic nervous system		
AP	area postrema	CRF	corticotropin-releasing factor (also called CRH)
APC	antigen presenting cell		
APUD	amine precursor uptake and decarboxylation	CRH	corticotropin-releasing hormone (also called CRF)
AR	androgen receptor	CSF	cerebrospinal fluid
ARC	arcuate nucleus	CVO	circumventricular organs
AT	angiotensin	D	diestrus
ATP	adenosine triphosphate	D2R	dopamine 2 receptor
AVP	arginine vasopressin	D3	diestrus 3
AVPV	anteroventral periventricular nucleus	DA	dopamine
β2-AR	β2-adrenergic receptor	DAG	diacylglycerol
β-END	β-endorphin	DBD	DNA binding domain
β-Gal-ir	β-Galactosidase immunoreactivity	DBH	dopamine beta-hydroxylase
BBB	blood-brain barrier	DG	dentate gyrus
BDNF	brain-derived neurotrophic factor	DHEA	dehydroepiandrosterone
BLA	basolateral amygdala	DHT	dihydrotestosterone
BNP	B-type natriuretic peptide	dlSON	dorsolateral supraoptic nucleus
Ca²⁺	calcium ion	DMN	dorsomedial hypothalamic nucleus
CAH	congenital adrenal hyperplasia	DMT	dimethyltryptamine
cAMP	cyclic adenosine monophosphate	DNA	deoxyribonucleic acid

DNES	Diffuse Neuroendocrine System	GnIH	gonadotropin inhibitory hormone
DYN	dynorphin	GnRH	gonadotropin-releasing hormone
E	estradiol	GPR54	G-protein-coupled receptor 54
EDC	endocrine disrupting chemicals	GR	glucocorticoid receptor
EGF	epidermal growth factor	GRE	glucocorticoid response element
EGL	external granule cell layer	G_S	stimulatory G protein
EL	ejaculation latency	GTF	general transcription factor
ELISA	enzyme-linked immunosorbent assay	GTP	guanosine triphosphate
		HBD	hormone binding domain
ENK	enkephalin	HCG	human chorionic gonadotropin
ENS	enteric nervous system	HCS	human chorionic somatomammotropin
EOP	endogenous opioid peptide		
EPO	erythropoietin	HDC	histidine decarboxylase
ER	endoplasmic reticulum	HFD	high fat diet
ER	estrogen receptor	HGP	hepatic glucose production
ERE	estrogen response element	H-P-A	hypothalamic-pituitary-adrenal
FGF	fibroblast growth factor	HPL	human placental lactogen
fMRI	functional magnetic resonance imaging	HPLC	high performance liquid chromatography
FS	folliculostellate	HRE	hormone response element
FSH	follicle-stimulating hormone	HRT	hormone replacement therapy
FSH-RH	follicle-stimulating hormone-releasing hormone	HSP	heat shock protein
		HVA	homovanillic acid
FX	fornix	ICo	nucleus intercollicularis
G	granule cells	IF	intromission frequency
G-CSF	granulocyte colony stimulating factor	IFNγ	interferon γ
		I_g	immunoglobulin
GABA	gamma-aminobutyric acid	IGF	insulin-like growth factor; somatomedin
GABA-T	GABA transaminase		
GAD	glutamic acid decarboxylase	IGFBP	insulin-like growth factor binding protein
GDNF	glial-derived neurotrophic factor		
GDP	guanosine diphosphate	IGL	internal granule cell layer
GFP	green fluorescent protein	III	inter-intromission interval
GH	growth hormone	IL	interleukin
GHRH	growth hormone releasing hormone	IL	intromission latency
GH-RIH	growth hormone release inhibiting hormone (see SOM)	lMAN	lateral magnocellular nucleus of the anterior nidopallium
GI	gastrointestinal	IP3	inositol triphosphate
G_i	inhibitory G protein	iR	ion channel
GIP	gastrin inhibitory peptide	IRS-1	insulin receptor substrate 1
GLP-1	glucagon-like peptide-1	JAK	janus kinase
GLP-2	glucagon-like peptide-2	K^+	potassium ion
Glu	glutamate	K_P	kisspeptin
GM-CSF	granulocyte-macrophage colony stimulating factor	LH	luteinizing hormone (also lateral hypothalamus)

LHRH	luteinizing hormone releasing hormone	NMDA	N-methyl-D-aspartate
LPH	lipotropic hormone (also β-lipotropin)	NO	nitric oxide
LSD	lysergic acid diethylamide	NOS	nitric oxide synthase
M	muscarinic	NP	neurophysin
MAO	monoamine oxidase	NPY	neuropeptide Y
MBH	mediobasal hypothalamus	NSF	N-ethylmaleimide sensitive factor
MC	melanocortin	NT	neurotransmitter
M-CSF	macrophage colony stimulating factor	NTD	amino terminal domain
MD	dorsomedial thalamus	NTS	nucleus tractus solitarius
ME	median eminence	nXIIts	tracheosyringeal portion of the nucleus
MET	metestrus		hypoglossus
mf	mossy fibers	OB	olfactory bulb
MF	mount frequency	OT	oxytocin
mGluR	metabotropic glutamate receptor	ORL1	opioid receptor-like receptor
MHC	major histocompatibility complex	OTR	oxytocin receptor
MHPG	3-methoxy-4-hydroxyphenylglycol	OVLT	organum vasculosum of the lamina(e)
mIU	milli international units		terminalis
ML	mount latency	OXM	oxyntomodulin
ML	molecular layer	OXY	oxytocin
MMGB	medial geniculate body	P	progesterone (also Purkinje cells)
MOE	main olfactory epithelium	PACAP	pituitary adenylate cyclase-activating
MPOA	medial preoptic area		polypeptide
mR	metabotropic membrane receptor	PC	proprotein convertase
MR	mineralocorticoid receptor	PCP	phencyclidine
MRF	midbrain reticular formation	PCR	polymerase chain reaction
MRI	magnetic resonance imaging	pCREB	phosphorylated CREB
mRNA	messenger ribonucleic acid	PEI	post-ejaculatory interval
α-MSH	α-melanocyte-stimulating hormone	PeN	anterior periventricular nucleus
MSH-RF	melanocyte-stimulating hormone – releasing factor	PENK	preproenkephalin
		PET	positron emission tomography
MSH-RH	melanocyte-stimulating hormone – releasing hormone	pf	parallel fibers
		PFA	perifornical area
MSH-RIF	melanocyte-stimulating hormone – release-inhibiting factor	PGE2	prostaglandin E2
		PH	posterior hypothalamus
MSH-RIH	melanocyte-stimulating hormone – release-inhibiting hormone	PI3K	phosphoinositide 3 kinase
		PIF	prolactin releasing inhibiting factor
MT	melatonin	PIP2	phosphatidylinositol diphosphate
MUA	multiple unit activity	PIR	piriform cortex
NA	noradrenaline (also norepinephrine, NE)	PKA	protein kinase A
Na$^+$	sodium ion	PL	placental lactogen
NE	norepinephrine (also noradrenaline, NA)	PLC	phospholipase C
		PNS	parasympathetic nervous system
NGF	nerve growth factor	POA	preoptic area
NK	natural killer cell	POL	RNA polymerase
NKT	natural killer T cell	POMC	pro-opiomelanocortin

PP	pancreatic polypeptide		TIDA	tuberoinfundibular DA
PR	progesterone receptor		TNFα	tumor necrosis factor α
PRF	prolactin releasing factor		TR	thyroid hormone receptors
PRH	prolactin-releasing hormone		TRF	thyrotropin (TSH) releasing factor (also TRH)
PRL	prolactin			
PRO	proestrus		TRH	thyroid hormone releasing hormone
PrRP	prolactin-releasing peptide		trk	tyrosine receptor kinase
PTH	parathyroid hormone		T_S	suppressor T cell
PTSD	post-traumatic stress disorder		TSH	thyroid-stimulating hormone
PV	periventricular nucleus		TSHR	TSH receptor
PVN	paraventricular nucleus		TSH-RH	thyroid-stimulating hormone-releasing hormone (TRH)
PYY	peptide YY			
RA	robust nucleus of the arcopallium		VEGF	vascular endothelial growth factor
RER	rough endoplasmic reticulum		VIP	vasoactive intestinal polypeptide
RSP	retrosplenial cortex		vmSON	ventromedial supraoptic nucleus
SC	subcutaneous		VMH	ventromedial hypothalamic nucleus
SCN	suprachiasmatic nucleus		VMN	ventromedial nucleus of hypothalamus
SDN	sexually dimorphic nucleus			
SEM	standard error of the mean		VNO	vomeronasal organ
SHBG	sex hormone binding globulin		VP	vasopressin
SNAP	soluble SNF attachment proteins		WAT	white adipose tissue (fat)
SNARE	SNAP receptor protein			
SNB	spinal nucleus of the bulbocavernosus			
SNS	sympathetic nervous system			
SOCS	suppressor of cytokine signaling			
SOM	somatostatin			
SON	supraoptic nucleus			
SP	Substance P			
SS	somatosensory cortex			
SST	somatostatin receptor			
STAT	signal transducer and activator of transcription / signal transduction and transcription			
T	testosterone			
T3	triiodothyronine			
T4	thyroxine			
TBG	thyroid hormone binding globulin			
T_C	cytotoxic T cell			
TF5	thymosin fraction 5			
TGFα	transforming growth factor α			
TGFβ1	transforming growth factor β1			
TH	tyrosine hydroxylase			
T_H	helper T cell			
THC	tetrahydrocannabinol			

Classification of chemical messengers

1.1 Hormones, the brain and behavior

Neuroendocrinology is the study of how the brain controls the endocrine systems that keep us alive and able to reproduce. However, an essential and critical characteristic of this neural control of the endocrine systems is that endocrine hormones in turn have profound effects on brain function through feedback systems. Research on hormones and the brain is intensive and covers many fields: from cell and molecular biology and genetics to anatomy, physiology, pharmacology, biochemistry, medicine, psychiatry and psychology. This book will examine the interactions between hormones, the brain and behavior. Thus, the primary focus will be on how the endocrine and nervous systems affect each other to produce an integrated functional neuroendocrine system that influences physiological and behavioral responses. As preliminary background reading, students are referred to any modern text on Human Physiology (see "Further reading" at the end of this chapter).

When you hear the term "hormone," for example *steroid hormone*, you think of the endocrine glands and how their secretions influence physiological responses in the body, but this is only part of the picture. Many of the endocrine glands (although not all of them) are influenced by the pituitary gland, the so-called "master gland," and the pituitary is itself controlled by various hormones secreted from the hypothalamus, a part of the brain situated directly above the pituitary gland. The release of hypothalamic hormones is in turn regulated by neurotransmitters released from nerve cells (neurons) in the brain. Some neurotransmitters released within the brain also control behavior, and the secretion of neurotransmitters from specific nerve cells can be modulated by the level of specific endocrine hormones in the circulation. This is called hormone feedback. Thus, neurotransmitter release influences both hormones and behavior and, in turn, hormones regulate the release of neurotransmitters. This interaction between hormones, the brain and behavior involves a wide variety of chemical messengers which are described in this chapter.

This chapter provides an introduction to the chemical messengers found in the neuroendocrine system. Later chapters describe the endocrine glands and their hormones (Chapter 2), the pituitary gland and its hormones (Chapter 3) and the regulation of the pituitary gland by hypothalamic hormones (Chapter 4). Chapter 5 outlines the role of neurotransmitters in communicating between nerve cells and Chapter 6 discusses neurotransmitter control of hypothalamic, pituitary and other hormones. The regulation of hormone synthesis, transport, storage, release and deactivation is described in Chapter 7. Hormones from the endocrine glands, pituitary gland and hypothalamus influence each other through feedback mechanisms, which are described in Chapter 8. Hormones act on target cells in the body and the brain that have specific hormone recognition sites (*receptors*). The nature of steroid and thyroid hormone receptors is discussed in Chapter 9 and the receptors for peptide hormones and neurotransmitters, which function by activating intracellular second messenger signals in their target cells, are described in Chapter 10. In the brain, hormones influence the release of both neurotransmitters and hypothalamic hormones by their action on neural target cells. The brain is also influenced by a number of newly discovered substances called *neuropeptides*, which are introduced in Chapter 11. Neuropeptides are important because they can act as *neurotransmitters* to modify neural activity or as *neuro-modulators* to influence the synthesis, storage, release and action of other neurotransmitters in modifying brain function (Chapter 12). The cells of the immune system also produce chemical messengers called cytokines that interact with the neural and endocrine systems as described in Chapter 13. When hormones, neuropeptides or cytokines alter the synthesis and release of neurotransmitters in the brain, one result is a change in behavior. Methods for the study of hormones and behavior are discussed in Chapter 14, and current developments in behavioral neuroendocrinology, as well as a historical overview, are given in Chapter 15.

The neuroendocrine system, therefore, involves a network of hormone-brain-behavior interactions and an example is depicted in Figure 1.1 (Hyman 2009). This figure illustrates how adrenal steroid hormones (*glucocorticoids*) are involved in our response to stress. The perception of an environmental stimulus such as a light, odour, sound or touch occurs through the sense organs and their neural connections to the brain. These stimuli can be interpreted as physical stressors, sexual stimuli, etc. by the cerebral cortex and other brain areas that influence the neuroendocrine system. Two different responses then occur. There is a rapid neuromuscular response, resulting in an immediate behavioral change: for example, you see a truck coming and you jump out of the way. This is accompanied by complex neuroendocrine changes. Your hypothalamic-pituitary-adrenal response to the oncoming truck involves an immediate (seconds to minutes) release of many different hormones which circulate through the bloodstream to stimulate their target cells in the heart, adrenal glands, liver, skeletal muscles, adipose tissue and, of course, the brain. When the target cell is stimulated, it undergoes a physiological change caused by the hormonal action. The hormones released into circulation then exert feedback to the hypothalamus and pituitary gland, to alter further hormone release. Finally, when the brain is a target for hormonal action, the result may be a behavioral as well as a physiological change.

1.2 The body's three communication systems

The body has three different communication systems: the nervous system, the endocrine system and the immune system, each of which uses its own types of chemical messenger. Nerve cells communicate through the release of neurotransmitters; endocrine glands secrete hormones, and the immune system operates through the release of cytokines. These three systems are not independent; each one interacts with the other two, as outlined in Figure 1.2 (Glaser and Kiecolt-Glaser 2005).

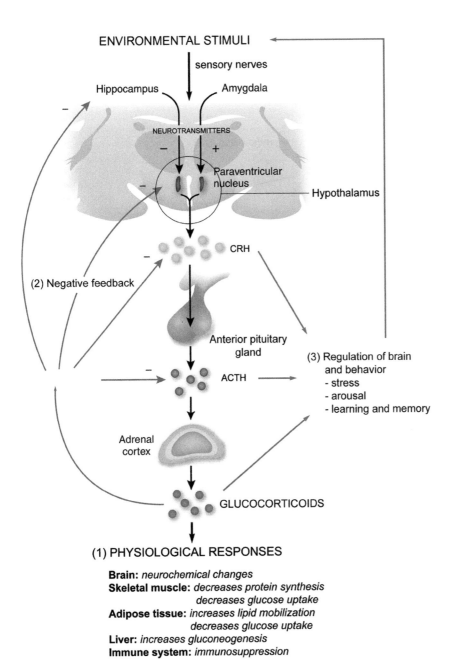

Figure 1.1 The interaction between hormones and the brain
Environmental stimuli influence the brain through sensory nerves and the brain regulates behavior and hormone secretion through the release of neurotransmitters that stimulate nerve impulses. The hormones released from the hypothalamus, pituitary gland and other endocrine glands when the neuroendocrine system is activated stimulate: (1) physiological responses in target cells in the brain and body; (2) feedback regulation of hypothalamic and pituitary hormone release; and (3) brain and behavioral responses through their action on neurotransmitter and neuropeptide release from neurons in the brain. The example used here is the hypothalamic-pituitary-adrenal response to an environmental stressor.

Abbreviations: ACTH, adrenocorticotropic hormone; CRH, corticotropin-releasing hormone. Reproduced with permission (Hyman 2009).

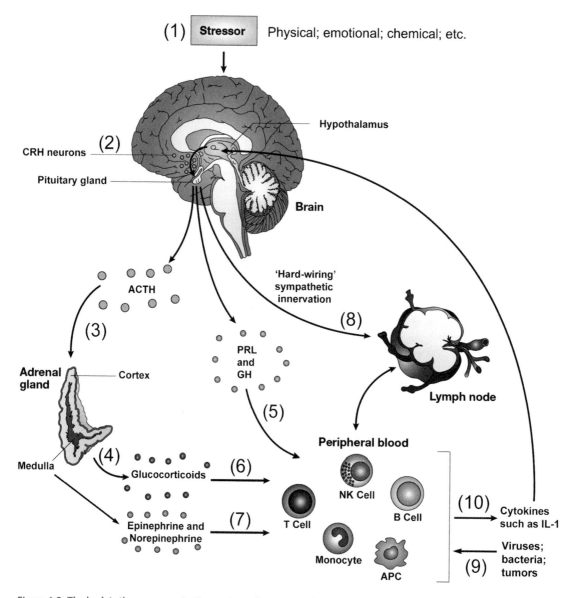

Figure 1.2 The body's three communication systems do not act independently
The brain and nervous system influence the neuroendocrine and immune systems, which also influence each other and the brain. This example shows that cognitive stimuli (stressors (1)) activate the neuroendocrine system through the brain and nervous system (e.g. through hypothalamic activation of CRH neurons (2)). The secretion of CRH from hypothalamic neurons stimulates the anterior pituitary gland to release ACTH (3), which in turn acts on the adrenal cortex to release glucocorticoids (4). Pituitary hormones such as prolactin (PRL) and GH (5), as well as glucocorticoids (6), also influence cells of the immune system.

Autonomic neural activity is also regulated as a result of stressors (1). For example, the sympathetic branch of the autonomic nervous system induces release of epinephrine and norepinephrine from the adrenal medulla (7). Other branches of the autonomic system stimulate cells of the immune system to release cytokines (8). Note that non-cognitive stimuli such as viruses and bacteria can directly activate cells of the immune system (9) and the resulting release of cytokines (10) activates the neuroendocrine system.

Abbreviations: ACTH, adrenocorticotropic hormone; APC, antigen-presenting cell; CRH, corticotropin-releasing hormone; GH, growth hormone; IL-1, interleukin-1; NK, natural killer; PRL, prolactin. Reproduced with permission (Glaser and Kiecolt-Glaser 2005).

Because these systems interact, they are often referred to as the neuroendocrine, neuroimmune or neuroimmunoendocrine systems. To designate the influence of these systems on behavior, the terms *psychoneuroendocrinology* (Smythies 1976) and *psychoneuroimmunology* (http://en.wikipedia.org/wiki/Psychoneuroimmunology) have been coined. These important fields of science have specific journals devoted to them.

As shown in Figure 1.2, the nervous system controls the release of hormones that can influence the release of cytokines from the immune system. In turn, hormones and other chemical messengers modulate the activity of both the nervous system and the immune system. Likewise, the immune system can modulate both neural activity and the release of hormones by the release of cytokines. While cognitive-sensory stimuli influence neural, immune and endocrine activity through the brain and nervous system, non-cognitive stimuli, such as bacteria and viruses, influence these systems through their action on the immune system.

1.3 Methods of communication between cells

As shown in Figure 1.3, hormones and other chemical signals may communicate with their target cells through *endocrine, paracrine, autocrine* and *neuroendocrine* mechanisms. These are compared with neurochemical signaling between neurons, sometimes called *neurocrine*. A special case is when the signal is not released from the cell but interacts with receptors *inside* the cell; this is termed *intracrine* communication.

1.3.1 Endocrine communication
Endocrine cells release their hormones into the bloodstream and these hormones then travel via the circulation to distant target cells. For example, thyroid-stimulating hormone (TSH) is released from the pituitary gland and travels through the bloodstream to stimulate its target cells in the thyroid gland (see Figure 6.6).

1.3.2 Paracrine communication
Endocrine cells also release hormones that act on adjacent cells. These hormones may diffuse from one cell to the next, or go into the bloodstream, but travel only a very short distance. Paracrine secretion is, therefore, a localized hormone action. This happens, for example, in the ovaries. In order to produce the sex hormone estradiol, granulosa cells must first take up androgen which is released from the adjacent thecal cells. The androgen, for example, androstenedione, is then converted to estradiol (see Widmaier *et al.* 2010). The target cell is located immediately adjacent to the hormone-secreting cell, resulting in a localized chemical communication within a particular tissue or organ. Paracrine secretion is also important in the immune system and nervous system (see section 1.3.3).

1.3.3 Neurocrine communication
A special type of *paracrine* communication is that between cells in the nervous system. Here, nerve cells (neurons) secrete neurotransmitters, such as acetylcholine, that travel

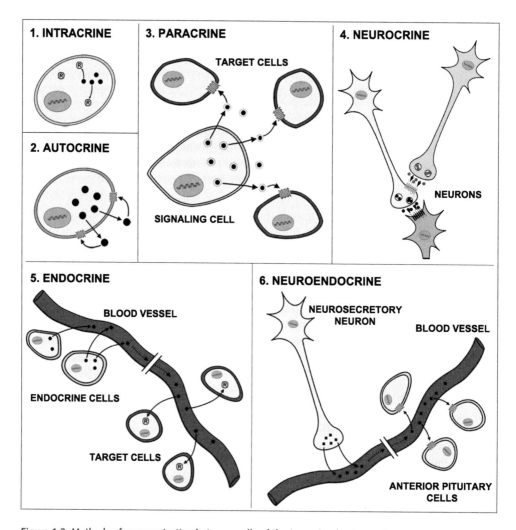

Figure 1.3 Methods of communication between cells of the (neuro)endocrine system
In *intracrine* communication (1) hormonal messengers, either synthesized in the same cell or imported from the bloodstream, act at intracellular receptors. In *autocrine* communication (2), hormones act on the cells that release them. In *paracrine* communication (3), hormones act on adjacent cells such as occurs in the testes, gastrointestinal tract and brain. A special form of paracrine signaling, sometimes called *neurocrine* (4), is the close-range signaling that occurs between nerve cells (neurons). *Endocrine* communication (5) occurs when hormones are released into the bloodstream and act on cells at distant sites throughout the body. *Neuroendocrine* communication (6) occurs when neuropeptides are released from presynaptic cells into the synapse and act on receptors of postsynaptic cells in the central and peripheral nervous systems. Neuroendocrine secretion also involves the release of neurohormones from neurosecretory neurons in the hypothalamus and adrenal medulla. Copyright P. M. H. Wilkinson.

ultra-short distances across a synapse to either stimulate or inhibit other postsynaptic neurons (see Chapter 5). Another example is when neuropeptides are secreted from neurons in the brain. These behave as neuromodulators by regulating the sensitivity of other cells to stimulation (see Chapter 11).

1.3.4 Autocrine communication

This is a modified form of paracrine secretion in which a cell releases a hormone or neurotransmitter that then has a direct feedback effect on the secretory cell itself. This is referred to as autocrine action. A specific example of autocrine communication would be a neurotransmitter acting presynaptically to modify its own secretion.

1.3.5 Neuroendocrine communication

Neuroendocrine (neurosecretory) cells are neurons that release peptide hormones either into the peripheral circulatory system, so that they can stimulate distant target cells (e.g. the release of oxytocin by the posterior pituitary to stimulate targets cells in the uterus – see Figure 6.10) or into the hypothalamic portal vessels to induce the release of pituitary hormones (e.g. gonadotropin-releasing hormone [GnRH] released from the hypothalamus to stimulate the release of luteinizing hormone from the anterior pituitary – see Figure 6.7).

1.3.6 Intracrine signaling

Some hormones, such as the steroid hormones estradiol and testosterone, are biosynthesized in the gonads, transported in the bloodstream, and act via receptors that are inside the target cells (see Chapter 9). However, some tissues, such as vagina, bone and prostate, possess estradiol and testosterone receptors, but are also able to biosynthesize these steroids within the cell without them being secreted. Thus, they can bind to receptors in the same cell and exert so-called *intracrine* signaling (Labrie *et al.* 2001). Other examples include growth factors, such as fibroblast growth factor (FGF), which are produced inside cells and bind directly to intracellular receptors, *without actually leaving the cell*, perhaps as a mechanism for control of cell proliferation.

1.4 Types of chemical messenger

The classification of chemical messengers is a constantly changing endeavor as new substances and new functions for known substances are continuously being discovered.

1.4.1 Phytohormones

Phytohormones are chemical messengers such as auxins, kinins, gibberellins and other growth regulators produced by the higher plants. Although this book is not directly concerned with the actions of phytohormones, they should interest us for two reasons. First, many phytohormones are similar to known mammalian hormones and neurotransmitters and may thus be important in understanding the evolution of the neuroendocrine system. Second, some phytohormones are used as drugs that can influence the human neuroendocrine system. Thus, plant substances such as muscarine, nicotine and morphine stimulate highly specific receptors on mammalian target cells, while atropine, ergocornine and strychnine block target cell receptors, preventing their response to hormones and neurotransmitters. In some cases, the same chemical may be found in both plants and

Table 1.1 **Types of chemical messenger**
Phytohormones: plant hormones – kinins, auxins, gibberellins, etc.
"True" hormones: these are (a) chemical messengers which are (b) synthesized in ductless (endocrine) glands and (c) secreted into the bloodstream. They (d) act on specific target cell receptors and (e) exert specific physiological (biochemical) regulatory actions in the target cells. They can be steroid hormones (e.g. estradiol and testosterone) or peptide hormones (e.g. insulin or growth hormone).
Neurohormones: hormones that are released by hypothalamic neurosecretory cells (neurons) via the posterior pituitary into the circulation (e.g. oxytocin and vasopressin) or via the portal system, into the anterior pituitary (the so-called hypothalamic releasing/inhibiting hormones).
Neurotransmitters: these are released by presynaptic nerve cells into a synapse (e.g. acetylcholine, dopamine, norepinephrine, etc.), where they stimulate receptors on postsynaptic nerve cells.
Pheromones: these are (a) volatile chemical messengers that are (b) synthesized in exocrine (duct) glands and (c) secreted into the environment. They (d) act on other individuals, usually of the same species, through olfactory (smell) or gustatory (taste) receptors and (e) alter behavior (releaser effects) or the neuroendocrine system (primer effects).
Parahormones: hormone-like substances which are not necessarily produced in endocrine glands (e.g. histamine, prostaglandins, leukotrienes, vitamin D). Prostaglandins and leukotrienes, for example, are generated locally during a tissue inflammatory response and act in a paracrine fashion.
Prohormones: these can be (a) large peptide molecules that may be processed into single or multiple hormones (e.g. pituitary beta-lipotropin is converted to β-endorphin (β-END) and adrenocorticotropic hormone (ACTH), or (b) steroid hormones converted to other bioactive steroids (e.g. testosterone to estradiol).
Growth factors: hormone-like substances which promote growth of body or brain tissue; e.g. nerve growth factor (NGF) or epidermal growth factor (EGF).
Cytokines. hormone-like factors released from lymphocytes, macrophages and other cells of the immune system that regulate the activity of cells of the immune system (e.g. interferon-γ and the interleukins).
Adipokines: one of the largest endocrine glands in the body is our fat tissue, which secretes a large number of hormone-like substances, such as leptin. Leptin regulates appetite and body weight. Some adipokines, such as interleukins, are also cytokines.
Vitamins: chemicals that regulate metabolism, growth and development in the body. Vitamin D for example, is synthesized in the body and has many hormone-like properties. If not produced in sufficient quantities, it must be taken as a dietary supplement in order to maintain bone strength.

animals. For example, abscisic acid – a phytohormone that causes leaves and fruit to fall from trees – is also found in human granulocytes, perhaps acting as part of the immune system (Minorsky 2002).

1.4.2 Hormones

A hormone is defined as: (a) a chemical messenger which is biologically effective in minute quantities (nanomolar 10^{-9} M, or picomolar 10^{-12} M); (b) synthesized in a ductless or endocrine gland; (c) secreted into the circulatory system, and transported through the body in the blood to (d) act on receptors on specific target cells located at a distance from the site of synthesis by (e) exerting a specific physiological or biochemical regulatory action on the target cell.

Chemical messenger

Hormones are chemical messengers that regulate the physiological actions of their target cells, but not all physiological regulators are hormones. For example, there are a number of non-hormonal chemicals, such as carbon dioxide, nitric oxide, glucose (blood sugar), histamine and the prostaglandins, which also regulate the physiological actions of their target cells.

Hormones are effective in minute quantities, although the physiological concentrations vary, depending on the hormone. Normally, the hypothalamic-releasing hormones (e.g. thyrotropin-releasing hormone, TRH; corticotropin-releasing hormone, CRH) are secreted in very small quantities (femtograms, 10^{-15} g). Pituitary hormones are released in greater quantities (picograms, 10^{-12} g) and gonadal, adrenal and thyroid hormones are released in much larger quantities (nanograms, 10^{-9} g). Modern assay technology has enabled these hormones to be detected and quantified on a routine clinical basis.

Biosynthesis in an endocrine gland

Although hormones are synthesized in endocrine glands, some hormone-like chemicals are produced in other locations. The production of angiotensin I, for example, occurs in the bloodstream, not in an endocrine gland. Neurohormones (e.g. hypothalamic-releasing hormones, oxytocin and vasopressin) are synthesized in neurosecretory cells which are modified neurons. Growth factors such as somatomedin and nerve growth factor act to promote tissue growth, but are not synthesized in endocrine glands. Likewise, the lymphokines, which have hormone-like activity, are synthesized by lymphocytes.

Some hormones are synthesized in a number of locations. Insulin, for example, is synthesized in both the pancreatic islets and in the brain and somatostatin is produced in both the pancreas and the brain. Estradiol is synthesized in the ovaries, testes, adrenal cortex, placenta, brain and by tumor cells. Peptides, such as somatostatin, are called hormones when they are secreted from endocrine glands, but if they are produced by neurons in the brain, they are called neuropeptides. Finally, some hormones such as adrenocorticotropin (ACTH) are secreted from the pituitary gland, from lung cancer cells and from lymphocytes and other cells of the mammalian and non-mammalian immune systems. As we will see in Chapter 12, hormones such as ACTH can also be synthesized in the brain and act as neuropeptides.

Secreted into the bloodstream

The traditional definition of a hormone is that it is secreted into the bloodstream and transported to its target cells through the circulatory system. But as we have seen already, many chemical messengers are not secreted into the bloodstream; that is, hormones can activate the cell adjacent to the one that releases them (*paracrine* action) or even the same cell that releases them (*autocrine* action). Neurotransmitters and neuropeptides are secreted from neurons into a synapse (i.e. the junction between two nerve cells). Neurohormones, neuropeptides and neurotransmitters may also be transported by the cerebrospinal fluid (CSF), as well as by the circulatory system, and the small quantities which enter the blood are quickly degraded. Pheromones are released into the air by one individual to act on another individual.

Act on specific target cells located at a distance from the site of synthesis

Although hormones are defined as acting on specific target cells, some hormones act on several different cell types in the body, rather than on a single and specific cell type. For example, growth hormone affects a variety of cells – such as liver, muscle and fat tissue – so it is hard to say that there is a specific target cell for this hormone. Likewise, glucocorticoids such as cortisol act on virtually every type of cell in the body.

Hormones are also defined as acting at target cells located at a distance from the site of synthesis, but how large is this distance? Hypothalamic hormones travel only a short distance down the portal venous system to the anterior pituitary gland. Paracrine hormones influence the very next cell. Autocrine hormones act on the same cell that secretes them. On the other hand, pituitary hormones such as luteinizing hormone (LH) influence the gonads, so they travel a long distance through the bloodstream. In turn, gonadal hormones such as testosterone influence the brain by traveling through the whole circulatory system before reaching their target cells.

Specific physiological actions

Although hormones are defined as having a specific action at their target cell, this action may vary according to the type of target cell stimulated. For example, prolactin stimulates the production of milk in the breast, and this is an example of a specific hormonal function. Estradiol, on the other hand, can increase tissue growth, such as in bone, or in the uterus during pregnancy, and it can increase the sensitivity of the pituitary to stimulation by gonadotropin-releasing hormone (GnRH) during the menstrual cycle. Moreover, some hormones can interact with multiple receptor types such that they have different functions in different target cells. For example, norepinephrine (sometimes called noradrenaline) released from the adrenal medulla can bind to either α-adrenergic or β-adrenergic receptors on target cells in muscle tissue and the physiological action stimulated may differ depending on which receptor is activated (see Chapter 10). Heart rate is increased via β-adrenergic receptor stimulation, whereas bronchoconstriction is mediated by α-adrenergic receptors. In contrast, many different hormones, no matter what their target cell, induce the same signaling response; that is, an increase in intracellular cyclic adenosine monophosphate (cAMP) production.

1.4.3 Neurohormones

A neurohormone differs from a "true" hormone because it is synthesized and released from a neurosecretory cell, which is a modified neuron. Two of the best studied neuro-hormones are oxytocin and vasopressin. These are hormones manufactured in the brain by hypothalamic neurosecretory cells, although they are stored and released from nerve terminals located in the posterior pituitary gland. The releasing hormones of the hypothalamus, such as corticotropin-releasing hormone (CRH) and gonadotropin-releasing hormone (GnRH), are also neurohormones. Chapters 3 and 4 will describe the production of neurohormones in neurosecretory cells and the important interactions between the hypothalamus and the pituitary gland.

Now is a convenient time to bring up the problem that many hormones have more than one name. Vasopressin, for example, is also known as *antidiuretic hormone* (ADH) and adrenaline is also known as *epinephrine*. This confusion occurs because many hormones were discovered independently by different researchers and, since they each thought that they had found something new, they gave them different names. Depending on the textbook used, both names remain in use to describe the same chemical. We will point out many examples of this in the next few chapters.

1.4.4 Neurotransmitters

Neurotransmitters differ from "true" hormones because they are synthesized by nerve cells rather than endocrine glands. They are also released into the synapse between nerve cells rather than into the bloodstream and therefore they are not neurohormones. Some examples of neurotransmitters are glutamate, acetylcholine, dopamine and norepinephrine. Neurotransmitters facilitate communication between nerve cells in the brain, and in the central and autonomic nervous systems (CNS and ANS, respectively) (Chapter 5). The definition of a neurotransmitter, like that of a hormone, is currently undergoing revisions as new discoveries force a reformulation of many concepts in chemical communication. For example, two unusual but important neurochemical messengers are *nitric oxide*, a gas, and the so-called *endocannabinoids* that are endogenous cannabis-like signaling molecules. This is discussed in Chapter 11.

1.4.5 Pheromones

Pheromones differ from "true" hormones in many respects. Pheromones are: (a) chemical messengers which are (b) produced in exocrine or ducted glands. (c) They are secreted to the outside environment rather than into the bloodstream. (d) They act on other individuals, usually of the same species, rather than on other cells within the secreting individual by (e) stimulating their olfactory (smell) or gustatory (taste) receptors. When pheromones were first discovered they were called "external hormones" (ectohormones) because they act like hormones, but on a different animal from the one that produces them. The secretion of pheromones in mammals is regulated by a wide variety of hormones.

There is a complex terminology to describe the various actions of pheromones in both vertebrates and invertebrates, but this need not concern us here. Pheromones have two general effects on the receiving individual: "releaser effects" which result in rapid behavioral changes (i.e. they "release" behavior), or "primer effects," which alter the neuroendocrine system, resulting in a later behavioral change (i.e. they "prime" the behavioral change). It normally takes about 48 hours for the "primer effect" of a pheromone to cause a change of behavior, although faster responses have been detected (Grimm 2014).

Pheromones are detected by the main olfactory receptors or the special *vomeronasal organ*, both of which stimulate the olfactory nerves causing neurotransmitter release in the olfactory bulb and other brain areas that influence the neuroendocrine system. This results in the release of hypothalamic, pituitary and gonadal hormones that stimulate physiological and behavioral changes in the receiver animal (Murata *et al.* 2014). Thus, there is a complex series of neuroendocrine events that must occur before a pheromone affects behavior through its primer action on the neuroendocrine system. Little more will be said about pheromones in this book.

1.4.6 Parahormones

"Parahormone" is a term that is not often used, but can be defined as a hormone-like chemical messenger which is generally not produced in an endocrine gland, but has all the other characteristics of a "true" hormone. Probably the best example is that of the prostaglandins. Prostaglandins are not stored in any particular tissue, but are generally formed rapidly when needed in response to, for example, tissue trauma. They are very potent substances and have profound autocrine/paracrine effects on blood pressure, the inflammatory response and in bronchodilation. Their sensitization of nerve endings, causing pain, is prevented by drugs such as aspirin.

1.4.7 Prohormones

Prohormones are hormone precursors. They tend to be large molecules, often peptides, that are cleaved to form hormones, or molecules which are in other ways modified to form hormones. The best-described prohormone is pro-opiomelanocortin (POMC), a giant peptide molecule which, when split into different segments, produces a number of different hormones (see Figure 3.4). One segment of the prohormone becomes β-lipotropin, which in turn is the prohormone for the endogenous opioid, β-endorphin. β-endorphin functions both as a hormone and a neurotransmitter (see Chapter 3). Other segments of POMC are cleaved to produce the pituitary hormones melanocyte-stimulating hormone (MSH) and adrenocorticotropic hormone (ACTH). In contrast, a small-molecule, non-peptide example of a prohormone is testosterone. This hormone is the precursor of two other hormones: estradiol and dihydrotestosterone (DHT). The significance of the conversion of testosterone to estradiol and DHT is discussed in Chapter 9.

Another example of a prohormone is angiotensinogen. Angiotensinogen is released by the liver into the bloodstream where it is converted to angiotensin I by the enzyme renin. Thus, the "hormone" angiotensin I is made not in an endocrine gland, but in the blood. Angiotensin I is then converted to angiotensin II in the lungs, and angiotensin II acts like a hormone to stimulate adrenal cells to produce aldosterone (see Chapter 2). Because the synthesis of angiotensin I occurs in the bloodstream rather than in an endocrine gland, angiotensin I is not a true hormone, even though it has some characteristic hormonal actions.

1.4.8 Growth factors

Growth factors are chemical messengers that are synthesized in various types of cells and act to stimulate tissue growth or maintain cell survival. A large number of growth factors are now known, but three of the most familiar are nerve growth factor (NGF), which is synthesized in innervated structures such as the heart and brain; epidermal growth factor (EGF), which is synthesized in the salivary glands; and fibroblast growth factor (FGF), which is synthesized throughout the neural and endocrine systems. Growth factors tend to act locally, in an autocrine/paracrine fashion, although there is now evidence that some may also be intracrine messengers. Growth factors in the brain are often called *neurotrophins* and these include NGF and BDNF (brain-derived neuro-trophic factor).

1.4.9 Cytokines

Cytokines are a superfamily of peptidergic signaling molecules secreted by cells of the immune system, such as lymphocytes and macrophages, although they are also released from a wide variety of other cell types, including neurons and glial cells. Nevertheless, we tend to think of them as critical modulators of the immune system, defending the body from bacteria and viruses by producing an inflammatory response. For example, interleukins are released by lymphocytes and macrophages, whereas interferons are derived from leukocytes and fibroblasts. New cytokines continue to be discovered and already include numerous interleukins (37 in all!), interferons, tumor necrosis factor and transforming growth factor.

Cytokines are usually thought of as *paracrine* or sometimes *autocrine* messengers acting within the immune system. However, they do have what looks like an *endocrine* role when we consider that they can reach the brain from the general circulation (see Figure 1.2). For example, we know that when we suffer from a bacterial infection we experience a number of effects that are mediated by the brain; e.g. fever, anorexia, disturbed sleep, and these are caused by cytokines affecting the brain. In addition, the effects of cytokines specifically on the hypothalamus cause disturbances in the endocrine system, including activation of the stress response.

Because neurotransmitters, hormones and neuropeptides can influence the immune system, they are also considered to be immunomodulators. The brain can regulate the immune system through direct nerve pathways to the thymus gland, through the release of neurotransmitters or through its control over the release of hormones. For example, the cytokine-mediated release of glucocorticoids (Figure 1.2) serves to inhibit the immune system and to reduce the immune response. This is why glucocorticoids are so powerful in the treatment of immune disorders such as rheumatoid arthritis. Figure 1.2 summarizes the mechanisms through which the neuroendocrine and immune systems are co-regulated. This interdependency is covered in more detail in Chapter 13.

1.4.10 Adipokines

The extensive and ongoing studies into the causes and possible treatments of obesity have revealed the existence of a completely new class of hormones that are secreted from fat (adipose) tissue. These *adipokines* circulate in the blood like the hormones discussed in the present chapter. This discovery means that body fat represents one of the largest endocrine organs in the body. In addition, we now know that there may be at least 50 different adipokines released by fat tissue (see Chapter 2). Fat tissue also secretes cytokines such as interleukin-6 (IL-6). Leptin is the best known of all the adipokines and this hormone is responsible for the maintenance of body weight by instructing the brain to regulate food intake and energy expenditure (see Chapter 12). Figure 1.4 illustrates how an obese, *leptin-deficient* child (i.e. born with an abnormal leptin gene) can be treated with leptin to attain a normal body weight (Broberger 2005; Farooqi 2011). Unfortunately, leptin does not reverse obesity in children, and adults, who are overweight because of their poor diet, and the effects of other adipokines on obesity need to be studied.

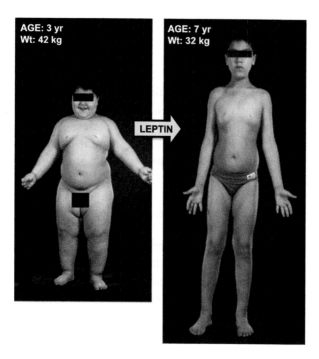

Figure 1.4 Leptin treatment of an obese child

Left panel, a 3-year-old boy with congenital leptin deficiency (body weight: 42 kg). On the right, the same boy, after four years of daily injections of recombinant leptin. Leptin induced a striking decrease in body fat (body weight: 32 kg). Reproduced with permission (Farooqi 2011).

1.4.11 Vitamins

Most vitamins, such as the B vitamins, do not qualify as hormones because they are not produced in the body, but are consumed as nutrients in food, though there are suggestions that food itself should be considered as a hormone (Ryan and Seeley 2013). On the other hand, vitamins A, D and K are formed in the body and have hormone-like actions. Vitamin A is formed in the body from carotene; vitamin D in the skin through the action of sunlight; and vitamin K by bacteria in the intestines. For example, vitamin D is released from the skin into the blood, and then acts at target cells in the intestines, kidney and bone to help regulate calcium levels in the body (see Chapter 2).

1.5 Neuropeptides and neuromodulators

The categorization of chemical messengers as shown in Table 1.1 is an extensive one. As mentioned above, there are problems in determining what is a "true" hormone and what isn't. It could be that the use of the word "hormone" as a classification is no longer necessary, as long as we understand the general rule that cells communicate via chemical messengers in several ways (intracrine, paracrine, endocrine, etc.). Of great interest is the fact that some hormones synthesized in endocrine glands and in the gastrointestinal tract are also produced in the brain and may act as neurotransmitters (see Chapter 5). For example, somatostatin and glucagon, normally found in the pancreas, exert paracrine control of insulin secretion, but are also found in the brain, along with their receptors (see Chapter 12). These chemicals are usually called *neuropeptides*, because they are made up of a string of amino acids. Somatostatin, for example, consists of 14 amino acids.

Table 1.2 Neuroactive substances: neurotransmitters, neuromodulators and neuropeptides

Neuroactive substance: A neurotransmitter or other chemical (e.g. a neuromodulator) that alters nerve cell activity

Neurotransmitter	Neuromodulator
A chemical messenger	A chemical messenger
Alters neural activity	Alters neural activity
Released from a presynaptic cell	Released from neural and non-neural cells (glial cells, neuroendocrine cells, endocrine glands)
Acts via the synapse on the receptors of the postsynaptic cell	Acts non-synaptically on both the presynaptic and postsynaptic cell receptors to alter synthesis, storage, release and re-uptake of the neurotransmitter
Can be monoamine, indoleamine or catecholamine	Can be a steroid or neuropeptide, hormone or non-hormonal peptide

Neuropeptide: A hormonal or non-hormonal peptide that acts as a neuromodulator

In general, when these neuropeptides act in the brain, they are called *neuromodulators* rather than neurotransmitters. Because this might be confusing, the definitions of substances that have neuroactivity listed in Table 1.2 will be used in this book. *Neuroactive substance* is defined as any chemical messenger that regulates the activity of neurons. They can be neurotransmitters or neuromodulators.

1.5.1 Neuropeptides

As noted above, neuropeptides can be hormones operating within the endocrine system (such as somatostatin) or peptides that are synthesized and act inside the brain. Other hormones, such as insulin, that have important effects in the periphery – such as regulating our glucose levels – may also reach the brain and act as a neuropeptide via insulin receptors on neurons. Leptin is another example of this type of neuropeptide; it is released from fat cells and travels to the brain to regulate food intake via leptin receptors. The reverse of this is also well described; that is, some peptide hormones, such as GnRH, are synthesized in the brain and are secreted by the brain into the peripheral circulation to control pituitary secretion of gonadotropins. However, GnRH is also released in the brain, and the spinal cord, to modify neural activity. Thus, it behaves as a peptide hormone and also as a neuropeptide. Neuropeptides are discussed in detail in Chapters 11 and 12.

1.5.2 Neuromodulators

A neuromodulator is a chemical messenger, usually a neuropeptide, that is released by brain cells to act on neurons to modulate their response to a neurotransmitter. They are a special kind of neurotransmitter, but instead of being released from a nerve terminal into a synapse, they may be secreted from varicosities (i.e. release sites located on axons away from the synapse) and act over long distances. One mechanism through which neuromodulators influence neural activity is by altering the permeability of the nerve cell membrane to ions such as sodium or calcium: this changes the way in which the incoming neurotransmitter signal affects the neuron (see Chapter 12). Examples of non-peptide neuromodulators are the sex hormones estradiol and testosterone, which act as neuromodulators to regulate the neural control of reproduction and also have profound effects on behavior. The concept of neuromodulation has become central to our understanding of the action of the neuroendocrine system.

1.6 Summary

This chapter introduced the concept of hormone, brain and behavioral interactions through the body's three communication systems: the nervous, endocrine and immune systems. Some of the mechanisms through which the brain interacts with and regulates the secretions of the neuroendocrine and neuroimmune systems are shown in Figures 1.2 and 1.5.

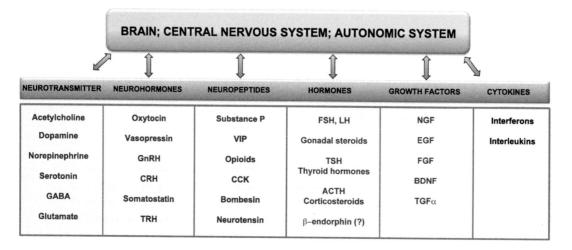

Figure 1.5 Reciprocal interactions of brain, central nervous system and ANS with chemical messengers of the neuroendocrine and neuroimmune systems

The figure shows that the brain, the central nervous system and the ANS control the release of a variety of messengers, including neurotransmitters, neuropeptides, neurohormones, etc. The arrows indicate that all of these also influence brain function, either by feedback mechanisms (such as corticosteroids) or by local synthesis and release (such as neuropeptides). *Note* that although β-endorphin is secreted from the anterior pituitary in a hormone-like fashion, it has at present no known function as a hormone. On the other hand, it is a well-described neuropeptide acting within the brain and spinal cord.

Abbreviations: ACTH, adrenocorticotropic hormone; BDNF, brain-derived neurotrophic factor; CCK, cholecystokinin; CRH, corticotropin-releasing hormone; EGF, epidermal growth factor; FGF, fibroblast growth factor; FSH, follicle-stimulating hormone; GABA, gamma-aminobutyric acid; GnRH, gonadotropin-releasing hormone; LH, luteinizing hormone; NGF, nerve growth factor; TGFα, transforming growth factor α; TRH, thyrotropin-releasing hormone; TSH, thyroid-stimulating hormone; VIP, vasoactive intestinal polypeptide.

The types of chemical messengers in these three systems are described, with reference to the definition of "true" hormones, and the distinctions between hormones, neurohormones, neurotransmitters, pheromones, parahormones and prohormones are discussed. Neuroregulators are defined as chemical messengers which regulate the activity of a nerve cell, and the distinction is made between neuromodulators, neuropeptides and neurotransmitters. The differences between endocrine, intracrine, autocrine, paracrine, neurocrine and neuroendocrine communication are discussed and the mechanisms for the neuroendocrine influence on the immune system are outlined. The concept of an immunomodulator is discussed and the cytokines (peptides produced by the cells of the immune system) are defined. Because neurotransmitters, hormones and neuropeptides can regulate the immune system, they are also considered to be immunomodulators. Finally, a new class of hormone, the adipokines, are introduced as messengers released from fat cells to interact with the brain as neuropeptides. All of these systems will be described in more detail in subsequent chapters.

FURTHER READING

Widmaier, E. P., Raff, H. and Strang, K. T. (2010). *Vander's Human Physiology: The Mechanisms of Body Function*, 12th edn. (New York: McGraw-Hill).

REVIEW QUESTIONS

1.1   What is the difference between a prohormone and a phytohormone?

1.2   What is the difference between a neurohormone and a neurotransmitter?

1.3   What is a prohormone?

1.4   Which hormone could be considered a prohormone for estradiol?

1.5   What is the difference between a "true" hormone and a "pheromone"?

1.6   Why is a neurohormone different from a "true" hormone?

1.7   What is the difference between paracrine and endocrine communication?

1.8   What is the difference between neurocrine and neuroendocrine communication?

1.9   What is a parahormone?

1.10   What is the difference between a neurotransmitter and a neuromodulator?

1.11   What are cytokines?

ESSAY QUESTIONS

1.1   Define the term "neuroendocrine" and describe the components of the neuroendocrine system. How does the term "psychoneuroendocrine" relate to this system?

1.2   Describe the "common receptor mechanisms" which might act to integrate the neural, endocrine and immune systems.

1.3   Discuss the differences between endocrine, autocrine and paracrine communication.

1.4   Discuss the problems in the classical definition of a "hormone."

1.5 Discuss the problems in differentiating a true hormone from neurohormones, parahormones and prohormones.

1.6 Discuss the problems in differentiating between neurotransmitters, neuropeptides and neuromodulators.

1.7 Discuss the similarities and differences between neurohormones and neurotransmitters.

1.8 Why might cytokines be called "immunotransmitters"?

REFERENCES

Broberger, C. (2005). "Brain regulation of food intake and appetite: molecules and networks," *J Intern Med* 258, 301–327.

Farooqi, I. S. (2011). "Genetic, molecular and physiological insights into human obesity," *Europ J Clin Invest* 41, 451–455.

Glaser, R. and Kiecolt-Glaser, J. K. (2005). "Stress-induced immune dysfunction: implications for health," *Nat Rev Immunol* 5, 243–251.

Grimm, D. (2014). "Male scent may compromise biomedical studies," *Science* 344, 461.

Hyman, S. E. (2009). "How adversity gets under the skin," *Nat Neurosci* 12, 241–243.

Labrie, F., Luu-The, V., Labrie, C. and Simard, J. (2001). "DHEA and its transformation into androgens and estrogens in peripheral target tissues: intracrinology," *Front Neuroendocrinol* 22, 185–212.

Minorsky, P. V. (2002). "Abscisic acid: a universal signalling factor?" *Plant Physiol* 128, 788–789.

Murata, K., Tamogami, S., Itou, M., Ohkubo, Y., Wakabayashi, Y., Watanabe, H., Okamura, H., Takeuchi, Y. and Mori, Y. (2014). "Identification of an olfactory signal molecule that activates the central regulator of reproduction in goats," *Curr Biol* 24, 681–686.

Nussey, S. S. and Whitehead, S. A. (2001). *Endocrinology: An Integrated Approach* (Oxford: Bios Scientific Publishers), www.ncbi.nlm.nih.gov/books/NBK20.

Ryan, K. K. and Seeley, R. J. (2013). "Food as a hormone," *Science* 339, 918–919.

Smythies, J. R. (1976). "Perspectives in psychoneuroendocrinology," *Psychoneuroendocr* 1, 317–319.

Widmaier, E. P., Raff, H. and Strang, K. T. (2010). *Vander's Human Physiology: The Mechanisms of Body Function*, 12th edn. (New York: McGraw-Hill).

Wikipedia definition of "psychoneuroimmunology", http://en.wikipedia.org/wiki/Psychoneuroimmunology.

The endocrine glands and their hormones

2.1 The endocrine glands

The location of the human endocrine glands is shown in Figure 2.1. The *pineal gland* is a small gland lying deep between the cerebral cortex and the cerebellum at the posterior end of the third ventricle in the middle of the brain. The *hypothalamus* exerts some degree of control over most of the endocrine glands through the release of neurohormones, neuropeptides and neurotransmitters. The *pituitary gland* hangs from the bottom of the hypothalamus at the base of the brain and sits in a small cavity of bone above the roof of the mouth.

The *thyroid gland* is located in the neck and the small *parathyroid glands* are embedded in the surface of the thyroid. In the chest is the *thymus gland*, which is very important for the production of T lymphocytes that play a critical role in the immune response. The *heart* and *lungs* also act as endocrine glands that secrete hormones. The *gastrointestinal (GI) tract*, consisting of the stomach and intestines, is also an important source of hormones. The *liver* secretes several hormones such as somatomedin (also called IGF-1), important for growth. The *adrenal glands* are complex endocrine glands situated on top of the kidneys. The *pancreas* secretes hormones involved in regulating blood sugar levels and the *kidney* also produces hormone-like chemicals. The *testes* and *ovaries* produce gonadal hormones, or sex hormones, which, in addition to the maintenance of fertility and sex characteristics, have important effects on behavior. During pregnancy, the *placenta* acts as an endocrine gland. The endocrine glands occur in similar locations in all vertebrates. A large endocrine gland is fat (*adipose tissue*) which can be found beneath the skin (subcutaneous), in the abdominal cavity surrounding the heart and GI tract (see Figure 2.1), and within tissues such as liver and muscle. Fat secretes a variety of hormones called *adipokines*. Finally, two of the largest tissues in the body – *skeletal muscle* and *bone* – secrete factors that act in an endocrine fashion.

2.2 The hormones of the endocrine glands

Each endocrine gland secretes one or more hormones. Table 2.1 summarizes the main hormones produced by each gland and some of the functions of these hormones.

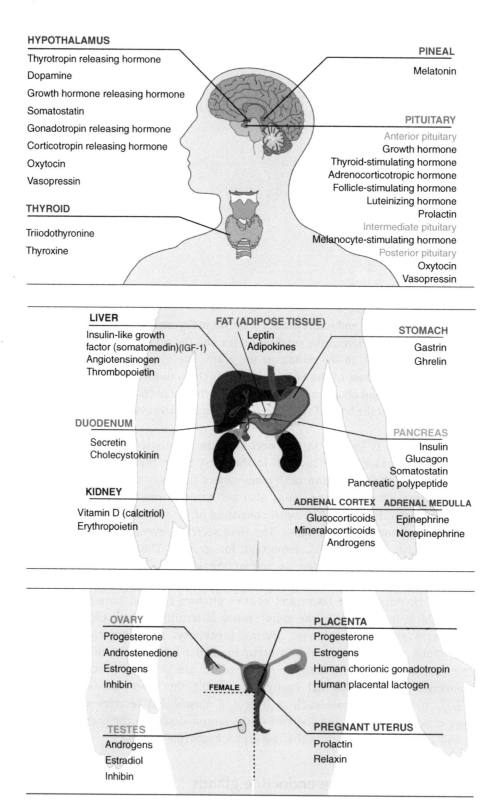

HYPOTHALAMUS

Thyrotropin releasing hormone
Dopamine
Growth hormone releasing hormone
Somatostatin
Gonadotropin releasing hormone
Corticotropin releasing hormone
Oxytocin
Vasopressin

THYROID

Triiodothyronine
Thyroxine

PINEAL

Melatonin

PITUITARY

Anterior pituitary
Growth hormone
Thyroid-stimulating hormone
Adrenocorticotropic hormone
Follicle-stimulating hormone
Luteinizing hormone
Prolactin
Intermediate pituitary
Melanocyte-stimulating hormone
Posterior pituitary
Oxytocin
Vasopressin

LIVER

Insulin-like growth
factor (somatomedin)(IGF-1)
Angiotensinogen
Thrombopoietin

FAT (ADIPOSE TISSUE)

Leptin
Adipokines

STOMACH

Gastrin
Ghrelin

DUODENUM

Secretin
Cholecystokinin

PANCREAS

Insulin
Glucagon
Somatostatin
Pancreatic polypeptide

KIDNEY

Vitamin D (calcitriol)
Erythropoietin

ADRENAL CORTEX

Glucocorticoids
Mineralocorticoids
Androgens

ADRENAL MEDULLA

Epinephrine
Norepinephrine

OVARY

Progesterone
Androstenedione
Estrogens
Inhibin

FEMALE

PLACENTA

Progesterone
Estrogens
Human chorionic gonadotropin
Human placental lactogen

TESTES

Androgens
Estradiol
Inhibin

PREGNANT UTERUS

Prolactin
Relaxin

Figure 2.1 Anatomical location and representative hormones of the major endocrine organs
A more complete list of hormones is provided in Table 2.1. Public Domain images from *Ladyofhats*:
http://en.wikipedia.org/wiki/File:Endocrine_central_nervous_en.svg; and http://en.wikipedia.org/wiki/File:
Endocrine_Alimentary_system_en.svg#file.

Table 2.1 Endocrine glands and their hormones

Gland	Hormone(s)	Action
Pineal	Melatonin	Seasonal breeding (e.g. sheep) Skin pigmentation Gonadal function Sleep/jet lag?
Hypothalamus	Multiple hormones	See Chapter 4
Pituitary gland	Many hormones	See Chapter 3
Thyroid	Triiodothyronine (T3) and Thyroxine (T4)	Regulate metabolic rate Brain development Bone growth
	Calcitonin	Reduces blood calcium levels
Parathyroid	Parathyroid hormone (PTH)	Elevates blood calcium level
Thymus	Thymosins	Regulate immune system development and response to pathogens; release of cytokines
Heart	Atrial natriuretic peptide (ANP)	Reduction of blood pressure through vasodilation and action on kidney
	B-type natriuretic peptide (BNP)	As for ANP, but blood levels also used as a marker for congestive heart failure
Gastrointestinal (GI) system	Multiple hormones	See details in Table 2.2
Pancreas (Islets of Langerhans)	Insulin (β cells), glucagon (α cells), somatostatin (δ cells), and pancreatic polypeptide (F cells).	Insulin lowers and glucagon increases glucose levels; somatostatin inhibits release of both of these
Adrenal cortex	Mineralocorticoid (aldosterone) Glucocorticoid (cortisol, corticosterone) Sex steroids: estradiol and testosterone	Sodium retention in kidney Stress; anti-inflammatory Puberty; sexual differentiation; Secondary sexual characteristics
Adrenal medulla	Epinephrine (adrenaline) Norepinephrine (noradrenaline)	Epinephrine secretion is stress-related; increases heart rate and blood glucose Norepinephrine increases blood pressure; constricts blood vessels
Kidney	Vitamin D (calcitriol)	Acts with PTH to maintain blood calcium levels; important for bone and immune system
	Erythropoietin	Acts on bone marrow to produce red blood cells (erythrocytes)
Testis:		
Leydig cells	Androgens (testosterone and dihydrotestosterone)	Male sexual characteristics, including sex behavior and sperm production
Sertoli cells	Inhibin and activin	Inhibin exerts inhibitory control over FSH secretion; activin may oppose inhibin's effects

Table 2.1 (Cont.)

Gland	Hormone(s)	Action
Ovary:		
Granulosa cells	Estradiol (an estrogen)	Female sex characteristics; e.g. breast formation; uterine function and menstrual cycle
Luteal cells	Progesterone, Inhibin, relaxin	Uterine function; important for pregnancy Inhibin regulates FSH secretion; relaxin only present in pregnancy and acts on uterus
Placenta	Many hormones; see Table 2.3 for details	Maintenance of pregnancy
Fat (adipose tissue)	Adipokines (e.g. leptin, adiponectin and resistin)	Control of body weight/energy balance; role in insulin signaling and diabetes
Muscle	Myostatin, interleukins, irisin	Control of fat mass; control of insulin secretion
Bone	Osteocalcin	Insulin secretion↑ and testosterone levels ↑
Liver	Fetuin-A	Insulin resistance↑ Type 2 diabetes↑
	FGF-21	Insulin resistance ↓ (mouse) Insulin resistance ↑ (human) Inhibition of reproduction
	Betatrophin	β-cell proliferation ↑

2.2.1 The pineal gland

Melatonin is the main hormone secreted by the pineal gland and secretion is normally high during the night. It has no clear biological role in humans, but is under investigation for its possible use in improving sleep patterns, in ameliorating jet lag (see also section 8.2.1; Figure 8.5) and in the management of depression, epilepsy, Alzheimer's Disease, diabetes, obesity, migraine, and cancer (Singh and Jadhav 2014). In some mammals, melatonin mediates reproductive activity in response to changes in environmental light cycles. For example, it is critical for the correct timing of puberty in seasonal breeders such as sheep. In non-mammalian vertebrates, such as frogs, melatonin causes lightening of pigment coloration.

2.2.2 The hypothalamus and pituitary gland

The hypothalamus secretes a number of hormones, neuropeptides and neurotransmitters (Chapter 4), many of which regulate the secretion of at least nine hormones from the pituitary gland (Chapter 3).

2.2.3 The thyroid gland

The thyroid gland produces three main hormones. Two of these, *triiodothyronine* (T3) and *thyroxine* (T4), are quite similar in structure and depend on dietary iodine for their synthesis. T4 is much more prevalent in the blood (98 percent) than T3 (2 percent), but T4 is converted to T3 in all target tissues. T4 is therefore a *prohormone* for T3. These hormones regulate body metabolism, and are also important in bone growth and in the development and maturation of the brain and nervous system. Congenital lack of thyroid hormones (hypothyroidism) leads to severely reduced brain development, a syndrome originally called *cretinism*, but which is now termed "congenital hypothyroidism." The third thyroid hormone, **calcitonin**, reduces blood calcium levels.

2.2.4 The parathyroid glands

The parathyroids are small glands, typically four in humans, embedded in the posterior surface of the thyroid gland (see Figure 2.2). *Parathyroid hormone* (PTH) is a large polypeptide released in response to a fall in blood calcium levels. In order to maintain blood calcium levels, PTH causes release of calcium from bone and at the same time induces calcium to be reabsorbed into the blood through the kidneys and the gut. PTH also works alongside vitamin D in the maintenance of optimal calcium levels in blood (Figure 2.2; and

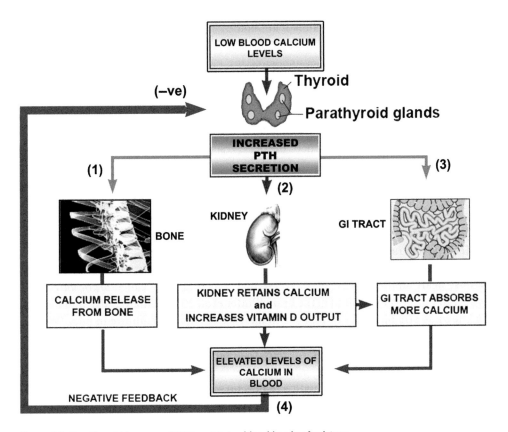

Figure 2.2 Parathyroid hormone (PTH) maintains blood levels of calcium

PTH secretion is *increased* when plasma calcium levels fall. This elevated PTH output exerts effects on bone, kidney and gut to restore calcium levels: (1) PTH induces calcium release from bone; (2) PTH increases reabsorption of calcium and at the same time increases vitamin D production in the kidneys; (3) vitamin D then aids in the reabsorption of calcium in the gut; (4) when blood levels of calcium increase, negative feedback of Ca^{2+} inhibits further secretion of PTH.

see section 2.2.10) (Widmaier *et al.* 2010). This figure is an excellent illustration of the principle of *hormonal feedback*, many other examples of which will be given later. This is also a good time to point out that hormones often occur in pairs, which have antagonistic actions. One hormone will stimulate a response and the other will inhibit it. Thus, calcitonin lowers blood calcium levels and parathyroid hormone raises blood calcium levels in an effort to maintain calcium levels within the normal range. Many pairs of hormones have such opposing effects.

2.2.5 The thymus gland

The thymus gland, located in the upper chest cavity, produces a family of polypeptide hormones called *thymosins* (Goldstein and Badamchian 2004). Thymosins (e.g. thymosin α1 and thymosin β3) stimulate the production and differentiation of lymphocytes in the immune system, and increase the secretion of cytokines. The T cells, which are involved in cellular immunity, depend on the thymus gland for their development. Thymosins are essential for the conversion of pre-thymic cells into T lymphocytes and for the development of immunocompetence, the ability to respond to antigenic stimulation. The thymus is also important for the neonatal production of antibodies. The thymus gland is one of the points of interaction between the endocrine, neural and immune systems. The thymosins act locally within the thymus as autocrine/paracrine messengers, and also circulate in the blood as endocrine signals.

2.2.6 The heart and lungs

As well as its function as a pump for the blood, the heart is an endocrine gland. Muscle cells (cardiomyocytes) in the atrium of the heart respond to an increase in blood pressure by secreting a hormone called *atrial natriuretic peptide* (ANP). ANP opposes the increase in blood pressure by causing (a) vasodilation and (b) a reduced reabsorption of water and sodium ions through its action on the kidneys. ANP also acts as a neuropeptide in the brain, where it regulates salt and water intake, heart rate and vasopressin secretion. A second natriuretic peptide released as a hormone from the heart ventricles is called BNP (*B-type natriuretic peptide*). BNP seems to work in the same way as ANP, on the same receptors, but it is interesting because high blood levels of BNP are used to diagnose heart failure (Ritchie *et al.* 2009). BNP was originally called *brain natriuretic peptide*, since it was first found in the brain (Potter *et al.* 2006). Finally, ANP and BNP are also known to stimulate brown fat cells to increase energy expenditure in mice (Whittle and Vidal-Puig 2012).

The lungs also contain groups of cells that produce neuropeptides. These include cholecystokinin and peptide YY, peptides that are more usually linked with the GI system (see below). They may regulate growth and development of the airways, but could also be involved in some pathologies (van Lommel *et al.* 1999). For example, lung cancers secrete peptides such as ACTH, the hormone normally responsible for regulating the stress system (see Chapter 3).

2.2.7 Gastrointestinal hormones

Tissues of the gastrointestinal (GI) tract (stomach and intestines) secrete over 20 hormones (Figure 2.3). Many of these hormones function in the GI tract to regulate

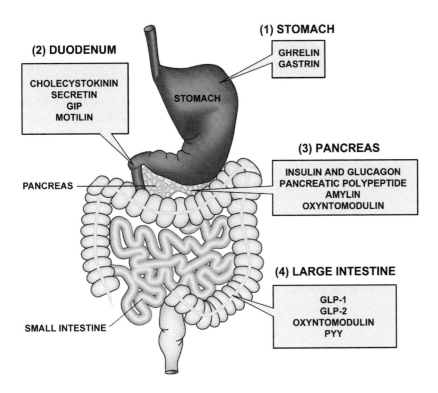

(1) STOMACH

GHRELIN
GASTRIN

(2) DUODENUM

CHOLECYSTOKININ
SECRETIN
GIP
MOTILIN

STOMACH

(3) PANCREAS

INSULIN AND GLUCAGON
PANCREATIC POLYPEPTIDE
AMYLIN
OXYNTOMODULIN

PANCREAS

(4) LARGE INTESTINE

GLP-1
GLP-2
OXYNTOMODULIN
PYY

SMALL INTESTINE

Figure 2.3 Location and function of some GI tract hormones

The gastrointestinal tract releases many hormones, including: (1) ghrelin and gastrin from the *stomach*; (2) cholecystokinin, secretin, GIP and motilin from the *duodenum/small intestine*; (3) insulin, glucagon, pancreatic polypeptide, amylin and oxyntomodulin from the *pancreas*; and (4) GLP-1, GLP-2, oxyntomodulin and PYY from the *large intestine*. These hormones can signal to other cells in the gut (paracrine and endocrine) and to the central nervous system (endocrine) to regulate gastrointestinal motility, hunger, satiation and energy balance. See also Table 2.2.

Abbreviations: GIP, gastrin inhibitory peptide; GLP-1, glucagon-like peptide-1; GLP-2, glucagon-like peptide-2; PYY, peptide YY. Reproduced with permission (Murphy and Bloom 2006).

contraction/relaxation of smooth muscle walls and sphincters, secretion of digestive enzymes, and secretion of fluids and electrolytes. It could be argued that hormones of the GI tract are not "true" hormones because they are not secreted from specific endocrine glands. However, since they are released from endocrine cells dispersed, singly or in groups, throughout the GI tract, they are correctly classed as hormones. As we have defined earlier (Chapter 1), some of these act as endocrine hormones at distant sites (e.g. cholecystokinin acts on the gall bladder and pancreas), whereas others act locally in a paracrine fashion (e.g. somatostatin is released from the stomach to regulate gastrin secretion from so-called G cells elsewhere in the stomach). Yet others are released in a neuroendocrine fashion; that is, NPY and VIP are released from neurons of the enteric nervous system (see Chapter 5; section 5.10.3) and form an integral part of the neuroendocrine system.

The three best-known GI hormones are gastrin, secretin and cholecystokinin.

Gastrin

G cells in the walls of the stomach produce gastrin in response to distension caused by eating a meal. Gastrin has many actions, including stimulation of hydrochloric acid secretion by the stomach, stimulation of pancreatic enzyme secretion, and increasing intestinal motility. Gastrin is thus an important hormone for digesting food.

Cholecystokinin belongs to the gastrin family of hormones and is released from cells in the walls of the duodenum by the presence of food, particularly fats and fatty acids. It has many functions in common with gastrin, including the stimulation of gall bladder contraction and pancreatic enzyme secretion (Figure 2.3), and it acts as a neuropeptide in the brain to reduce food intake.

Secretin

The passage of partially digested food from the stomach to the duodenum stimulates the release of secretin from the duodenal mucosa. Secretin stimulates the secretion of pancreatic bicarbonate and has many other functions in digestion, including the stimulation of hepatic bile flow and the potentiation of CCK-stimulated pancreatic enzyme secretion (Figure 2.3). Secretin was the first hormone to be discovered by Baylis and Starling (1902).

The gut-brain axis

The interactions between the gastrointestinal tract and the brain has been labeled "the gut-brain axis" (Romijn *et al.* 2008) and many of the GI hormones also act as neuropeptides in the brain to regulate energy homeostasis, hunger and appetite (Murphy and Bloom 2006; Sam *et al.* 2012; Chapter 12). CCK represents one example of these GI hormones that regulate energy homeostasis (Murphy and Bloom 2006). Others include *ghrelin, peptide YY* (PYY) and *glucagon-like peptides 1 and 2* (GLP-1 and GLP-2) and those listed in Table 2.2.

Ghrelin is a peptide hormone that is synthesized in the stomach. Ghrelin, sometimes called the "hunger hormone," is currently the only hormone known to increase appetite via the circulation. Ghrelin levels in the blood are increased by fasting and are reduced following meals. It is possible that blocking the effects of ghrelin might help to avoid, or prevent, obesity. But a much more useful application of its properties could be to stimulate appetite in people who need to gain weight, such as cancer patients and people suffering from anorexia nervosa (Hall *et al.* 2009). Like other GI neuropeptides, ghrelin is also biosynthesized in the brain and may regulate growth hormone secretion as well as appetite.

Peptide YY is secreted from the gut but in a manner opposite to that of ghrelin; that is, blood levels *increase* after food restriction and decrease after meals. Studies in humans and animals revealed that injection of PYY could reduce food intake, probably by acting on the brain, and therefore it might be therapeutically useful as an anti-obesity drug (Gardiner *et al.* 2008).

Glucagon-like peptide-1 is a GI peptide made in the same cells as PYY and is derived from a large *preproglucagon* molecule (see section 2.2.8, below, for **glucagon**). This large precursor peptide is processed to produce a group of smaller peptides that include glucagon, GLP-1 and GLP-2 (Gardner *et al.* 2008). As with PYY, blood levels of GLP-1

Table 2.2 Peptide hormones of the gut-brain axis

Peptide	Source	Function
1. Cholecystokinin	small intestine & stomach	stimulates vagus nerve to signal satiation
2. Ghrelin	stomach	increases hunger
	small intestine	stimulates feeding
	hypothalamus	stimulates GH release
3. PYY	large intestine	released after eating
		inhibits feeding after eating
		reduces food intake
4. GLP-1	large intestine	inhibits food intake
		increases insulin
GLP-2	brain and large intestine	inhibits food intake
5. Oxyntomodulin	large intestine	inhibits food intake
		increases energy expenditure
	pancreas	reduces food intake
6. Insulin	pancreas (β cells)	lowers glucose levels
7. Glucagon	pancreas (α cells)	increases glucose
8. PP	pancreas (F cells)	slows transit of food through the gut
		reduces feeding
9. Amylin	pancreas (β cells)	reduces food intake
10. Somatostatin	pancreas (δ cells)	slows gastric emptying
		inhibits insulin release
11. Neurotensin	GI tract	inhibits GI motility
12. VIP	GI tract neurons	relaxes gut smooth muscle reduces food intake
13. Substance P	GI tract (enteric) neurons and endocrine cells	increases gut motility
14. Somatostatin	GI tract (and pancreas, above) stomach	inhibits motility
15. Bombesin	GI tract neurons	inhibits gastrin secretion
16. NPY	GI tract neurons brain	inhibits gut motility stimulates appetite
17. GIP	duodenum	increases insulin secretion

Abbreviations: GH, growth hormone; GI, gastrointestinal; GIP, gastrin inhibitory peptide; GLP-1, glucagon-like peptide-1; GLP-2, glucagon-like peptide-2; NPY, neuropeptide Y; PP, pancreatic polypeptide; PYY, peptide YY; VIP, vasoactive intestinal polypeptide.

are rapidly increased following a meal and GLP-1 stimulates insulin secretion from the pancreas. An important clinical question is whether GLP-1 can be used as an appetite suppressor to control body weight. Clinical trials with a synthetic analog of GLP-1 (*Exenatide*) suggest this is possible, especially in the treatment of Type 2 diabetes (Bradley *et al.* 2010). In contrast, GLP-2 does not affect appetite, but exerts several important effects in the intestine, including growth, motility and blood flow.

Other peptides, not included in Figure 2.3, but listed in Table 2.2, are also secreted by endocrine cells in the GI tract. These include vasoactive intestinal polypeptide (VIP), substance P, somatostatin, bombesin, neurotensin and neuropeptide Y (NPY). These GI hormones are also synthesized in the brain and function as neuropeptides (see Chapters 11 and 12) (Murphy and Bloom 2006; Romijn *et al.* 2008; Gardiner *et al.* 2008; Sam *et al.* 2012).

Many gastrointestinal peptides also act, together with neurotransmitters, as chemical signals in the enteric nervous system. This is a branch of the autonomic nervous system (ANS) which innervates the GI system with sensory neurons, motor neurons and interneurons (see Chapter 5; Benarroch 2007).

2.2.8 The pancreas

The endocrine cells of the pancreas consist of small clusters of cells (*Islets of Langerhans*) dispersed throughout the pancreas. The islets are surrounded by pancreatic exocrine cells that secrete the pancreatic juices essential for the digestion of food. The islets themselves consist of four cell types. The *β cells* of the Islets of Langerhans secrete *insulin* in response to increased blood glucose levels which result from carbohydrate intake. Insulin lowers glucose levels in the blood by increasing glucose uptake in fat cells, liver cells or muscle cells where it is stored as glycogen or utilized as an energy source (see Figure 8.6). The *α cells* of the Islets of Langerhans secrete the hormone glucagon which opposes the influence of insulin by increasing blood glucose levels (Figure 8.6). Glucagon does this by stimulating the conversion of glycogen to glucose in the liver. Glucagon is formed from the same prohormone as GLP-1 (see section 2.2.7), but in the pancreas only glucagon is processed. Figure 2.4 illustrates how α cells and β cells are juxtaposed within a human pancreatic Islet (Unger and Orci 2010). In this way, insulin exerts a *paracrine* inhibition of glucagon secretion.

Two other hormones are made in the Islets of Langerhans: *somatostatin* and *pancreatic polypeptide* (see http://en.wikipedia.org/wiki/Islets_of_Langerhans). Somatostatin (SOM) is made in the *δ cells* and inhibits the secretion of both glucagon and insulin through a paracrine mechanism. Somatostatin is also made in the hypothalamus and inhibits growth hormone secretion (see Chapter 4). Pancreatic polypeptide (PP) is synthesized in *F cells* of the Islets of Langerhans, and like PYY and GLP-1, blood levels of PP are increased following food intake. Some evidence suggests that PP could be used clinically to control

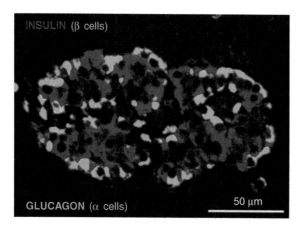

Figure 2.4 Location of insulin and glucagon cells in a human Islet of Langerhans

Microscopic image of a normal human islet shows the extensive juxtaposition of insulin and glucagon-containing cells. Note that clusters of insulin-containing cells (red; β cells) are surrounded by glucagon-containing cells (green; α cells). The scale bar is 50 microns. Reproduced with permission (Unger and Orci 2010).

body weight, but an understanding of how all of these GI hormones interact to control energy homeostasis, appetite and satiation is necessary for understanding the regulation of body weight and the problem of obesity.

2.2.9 The adrenal glands

The adrenal glands sit on top of the kidneys (Figure 2.5). Each adrenal is really two glands, consisting of a medulla surrounded by a cortex. The cortex is further divided into three different layers, each consisting of a different endocrine cell type.

The adrenal cortex

The adrenal cortex is a true endocrine gland that secretes three categories of steroid hormones: mineralocorticoids, glucocorticoids and sex steroids (Figure 2.5). *Aldosterone* is the primary mineralocorticoid produced by the outer layer of the cortex. Aldosterone secretion is stimulated when sodium ions (Na^+) in the blood are decreased. Aldosterone acts to increase the reabsorption of sodium ions in the kidneys, salivary glands and sweat glands.

The synthesis and release of the glucocorticoids (*cortisol* in humans, and *corticosterone* in rats) is stimulated by adrenocorticotropic hormone (ACTH), released from the anterior pituitary gland (Chapter 3). Glucocorticoids are released in response to stress and function to stimulate glycogen and glucose synthesis in the liver (thus increasing blood sugar levels), increase carbohydrate and fat metabolism, activate the neural stress response and suppress immune system activity (see Figures 1.1 and 1.2). Glucocorticoids inhibit inflammatory and allergic reactions and inhibit the production of lymphocytes by the immune system. This is why synthetic glucocorticoids are used for treatment of arthritis and as anti-transplant rejection drugs. Excessive glucocorticoid secretion during chronic stress affects a wide variety of tissues (Figure 2.6; see also Kadmiel and Cidlowski 2013) resulting, for example, in inhibition of growth and reproduction, immunosuppression, insulin resistance (diabetes), and increased deposition of fat. Chronic high levels of glucocorticoids can also damage neurons and lead to neuron death via apoptosis (Charmandari *et al.* 2005; and see section 9.9.2).

The adrenal cortex also produces small amounts of the *sex steroids* testosterone and estradiol and the androgen precursor molecules dehydroepiandrosterone (DHEA) and androstenedione that can be converted to testosterone in other tissues. The adrenal sex steroids, and their precursors, may influence sexual differentiation and the bodily changes that occur at puberty.

The adrenal medulla

The adrenal medulla is completely surrounded by the adrenal cortex (Figure 2.5) and resembles brain tissue more than an endocrine gland; that is, the cells behave like neurons and the mode of communication is *neuroendocrine* (see Chapter 1). Secretion of hormones from the adrenal medulla into the bloodstream is controlled by neurons of the sympathetic branch of the ANS. The two hormones released from the adrenal medulla are *epinephrine* (adrenaline) and *norepinephrine* (noradrenaline). These two chemicals are also produced in the brain, where they act as neurotransmitters. Epinephrine is released from the adrenal

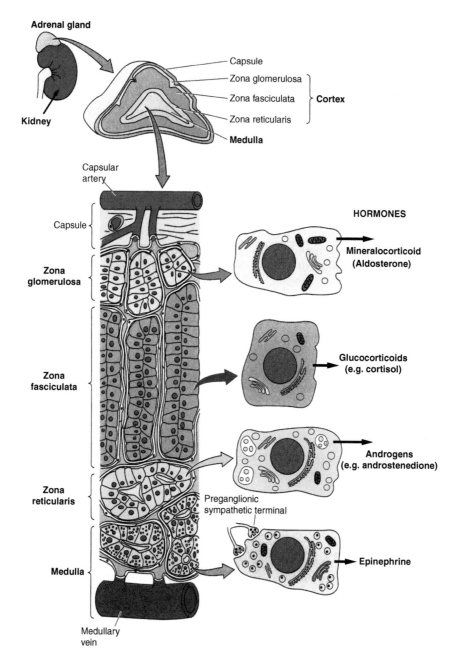

Figure 2.5 Diagrammatic cross section of an adrenal gland illustrating the three cortical layers and the medulla

The adrenal is really two glands: a cortex surrounding the medulla. Each cortical zone is responsible for a specific steroid secretion: *aldosterone* from the zona glomerulosa; *glucocorticoids* from the zona fasciculata; and *androgens* from the zona reticularis. *Epinephrine and norepinephrine* are secreted from the adrenal medulla. Reproduced with permission (Boron and Boulpaep 2005).

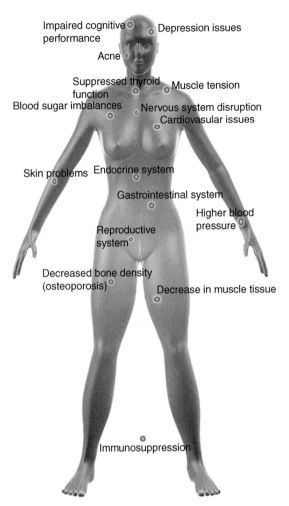

Impaired cognitive performance

Depression issues

Acne

Suppressed thyroid function

Muscle tension

Blood sugar imbalances

Nervous system disruption

Cardiovasular issues

Skin problems

Endocrine system

Gastrointestinal system

Higher blood pressure

Reproductive system

Decreased bone density (osteoporosis)

Decrease in muscle tissue

Immunosuppression

Figure 2.6 Adverse effects of prolonged cortisol secretion
Chronic stress has negative effects on many physiological systems, shown here. In addition, there is an increase in deposition of abdominal fat, which is associated with more health problems than fat deposited elsewhere. Some of the health problems associated with increased stomach fat are heart attacks and strokes. Reproduced with permission (Alesandra Rain: http://pointofreturn.com/relax.html).

medulla following stress due to environmental extremes such as cold, physical exertion or fear and acts to increase heart rate and blood glucose levels (via the liver), thus increasing the amount of work the muscles can do. Norepinephrine increases blood pressure and constricts blood vessels. Under chronic high stress, there is hyperactivity of the adrenal medulla as well as the adrenal cortex and, as a result, high levels of epinephrine and norepinephrine are secreted.

2.2.10 The kidneys

The kidneys are responsible for the secretion of at least four signaling molecules. Two of these – renin and kallikrein – are enzymes necessary for the production of the signaling molecules angiotensin and bradykinin, respectively, in the bloodstream and are therefore not true hormones. *Vitamin D* and *erythropoietin* are true hormones. Under the influence of sunlight, the skin synthesizes two prohormones (vitamins D3 and D2) that are in turn metabolized by the liver and then the kidneys to form the biologically active 1,25-dihydroxyvitamin D (also known as *calcitriol*). PTH is necessary for the final step in the production of calcitriol by the kidney tubules (Figure 2.2). Thus, PTH and calcitriol, acting together, maintain blood calcium levels. Calcitriol also regulates calcium absorption from the gut and assists in maintaining bone mineralization. An adequate level of calcitriol, supplemented by vitamin D tablets, may be necessary for good health in terms of bone strength, immune response and cancer prevention (Bikle 2010; Manson *et al.* 2011).

Erythropoietin (EPO) is synthesized in the interstitial cells of the kidney and acts as a hormone to stimulate bone marrow to produce red blood cells (erythrocytes) that carry oxygen to all tissues. Recombinant (genetically engineered) EPO has been used to treat patients with kidney failure and also those suffering from AIDS. EPO may also be useful in preventing neural apoptosis (cell death) after brain injury (Xiong *et al.* 2011). However, EPO has become an illegal drug used by athletes in their attempts to increase performance by blood doping (Jelkmann and Lundby 2011).

2.2.11 The gonads

The gonads (testes and ovaries) secrete three major classes of steroid hormones, referred to as sex steroids: androgens (e.g. testosterone), estrogens (e.g. estradiol) and progestins (e.g. progesterone).

The testes

The male gonads (testes) produce *androgens* from the Leydig (interstitial) cells. The primary androgen is testosterone, but there are other androgens, such as dihydrotestosterone and androstenedione. Dihydrotestosterone is formed from testosterone and is an important hormone in its own right. Some target tissues, such as the prostate, respond to testosterone by first converting it to dihydrotestosterone. Thus, testosterone, in this case, is a prohormone. Testosterone also acts as a prohormone when it is converted to estradiol in the brain in order to induce sexual differentiation. Testosterone is important for masculinization of the genitalia during sexual differentiation, for the control of sperm production, the development of male secondary sexual characteristics at puberty, and for the activation of sexual, aggressive and other behaviors in adulthood. In addition to testosterone, the testes also secrete several peptide-signaling molecules. The Sertoli cells produce *inhibin*, a peptide hormone that helps to inhibit the secretion of follicle-stimulating hormone (FSH) from the pituitary gland (see Chapter 3). Another testicular peptide, *activin*, appears to stimulate the secretion of FSH by opposing the inhibitory effect of inhibin (Strauss and Barbieri 2004).

The ovaries

The ovaries are the female gonads, and these produce two major classes of hormones, the *estrogens* and *progestins*. The primary estrogen is *estradiol*, but there are a variety of other less active estrogens, including estrone and estriol. There are synthetic versions of these hormones, such as ethinyl estradiol and medroxyprogesterone acetate, that are used in oral contraceptive pills. Estradiol is produced in the granulosa cells of the ovarian follicle and is important at puberty for the development of female secondary

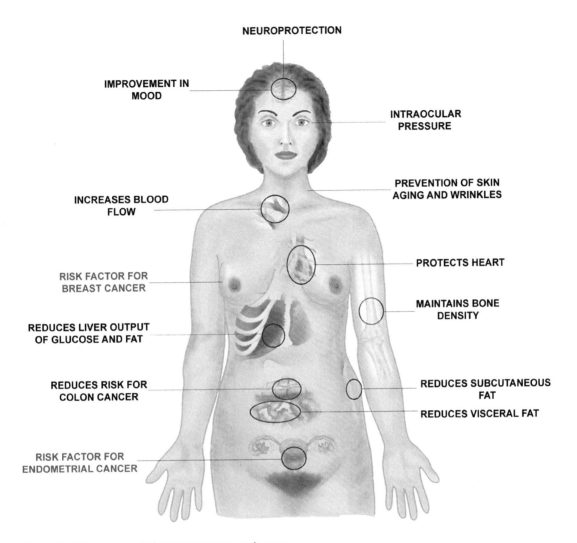

Figure 2.7 Effects of estradiol in various tissues and organs
Beneficial effects of estradiol are numerous. For example, this female sex hormone appears to exert neuroprotection and to reduce mood swings in post-menopausal women. Estradiol lowers intraocular pressure and may have cardioprotective actions. Estradiol also prevents and reverses osteoporosis, increases skin collagen production, and reduces the depth of wrinkles. It is possible it may also reduce the incidence of colon cancer. In contrast, however, estradiol can stimulate the growth of uterine and breast tumors. Copyright Alex Pincock.

sex characteristics and for subsequent menstrual cycles and pregnancy. Estradiol also has widespread influences on mood, body temperature, skin texture and fat distribution, and serves to protect the heart and bone density (Figure 2.7) (see Gruber *et al.* 2002; Khosla *et al.* 2012). Estradiol also influences sexual, parental and other behaviors in the female.

Progesterone is produced in the corpus luteum of the ovary and is important for uterine, vaginal and mammary gland growth. During the menstrual cycle, progesterone's primary role is to prepare the uterus for possible implantation of a fertilized egg (Widmaier *et al.* 2010). In maintaining pregnancy, progesterone inhibits the menstrual or reproductive cycle and inhibits the sexual behavior associated with it in rats, mice and other mammals. Estradiol and progesterone usually act synergistically since most actions of progesterone require prior estradiol priming of the target cells. For example, breast development at puberty requires estradiol to prime the cells and then progesterone causes cell differentiation and growth. As noted above, progesterone acts on the estradiol-primed uterus to prepare for pregnancy. Progesterone also has a variety of influences on behavior.

The ovaries also secrete the peptides inhibin and activin that regulate FSH secretion from the pituitary gland. As in the male, inhibin reduces FSH secretion and activin stimulates FSH release. Inhibin is also detectable in blood during pregnancy and abnormal high levels are useful as a clinical diagnosis for pre-eclampsia (a pregnancy-associated increase in blood pressure) and Down's Syndrome. During pregnancy, the ovary secretes another peptide, *relaxin*, a hormone that acts to prepare and maintain the uterine lining during pregnancy.

2.2.12 Placental hormones

During pregnancy, the placenta is a major endocrine organ and the first hormone to be produced, at the very beginning of pregnancy, is *human chorionic gonadotropin* (HCG). HCG is released immediately from the embryo implantation site in the uterine wall and its detection is routinely used as a pregnancy test. HCG stimulates the corpus luteum of the mother's ovary to keep progesterone secretion at levels high enough to maintain the uterine lining so that placental development can proceed. If progesterone levels drop, the pregnancy is aborted. HCG stimulates progesterone release for only a certain period of time, after which the placenta begins to produce its own progesterone to maintain the pregnancy. Once the placenta develops, the fetus, placenta and mother form an integrated *maternal-feto-placental* unit that produces a large number of hormones critical for the maintenance of pregnancy (Table 2.2; Strauss and Barbieri 2004). Another critical hormone, unique to human pregnancy, is *human placental lactogen* (HPL), also called *human chorionic somatomammotropin* (HCS). HPL functions in a manner similar to growth hormone and prolactin and stimulates the mammary glands to differentiate and to begin to manufacture and secrete milk. HPL is not secreted until pregnancy is well established. The placenta also produces steroid hormones (estradiol and progesterone), pituitary-like hormones (LH, FSH, GH, prolactin), neuropeptides (CRH, TRH, cholecystokinin, somatostatin, VIP) and biogenic amines such as epinephrine and serotonin (Table 2.3; Strauss and Barbieri 2004). Towards the end of pregnancy, the placenta begins to secrete relaxin to prepare the birth canal for parturition. Relaxin increases the flexibility of the cervix ("cervical ripening") to facilitate childbirth.

2.2.13 Hormones of fat tissue: the adipokines

There are two main types of fat (adipose) tissue. White adipose tissue acts as an endocrine organ that secretes a family of hormones called *adipokines* (Figure 2.8; Henry and Clarke 2008). More than 50 adipokines have now been

Table 2.3 **Neuropeptides, peptide hormones, steroid hormones and catecholamines produced by the placenta**

Neuropeptides	Pituitary-like hormones	Steroid hormones	Monoamines and adrenal-like peptides
CRH	ACTH	Progesterone	Epinephrine
TRH	TSH	Estradiol	Norepinephrine
GnRH	GH	Estrone	Dopamine
Melatonin	PL	Estriol	Serotonin
Cholecystokinin	Inhibin	Estetrol	Adrenomedullin
Met-enkephalin	LH	Allopregnanolone	
Dynorphin	FSH	Pregnenolone	
Neurotensin	β-endorphin	5α-dihydroprogesterone	
VIP	Prolactin		
Galanin	Oxytocin		
Somatostatin	Leptin		
CGRP	Activin		
Neuropeptide-Y	Follistatin		
Substance P			
Endothelin			
ANP			
Renin			
Angiotensin			
Urocortin			

Abbreviations: ACTH, adrenocorticotropic hormone; ANP, atrial natriuretic peptide; CRH, corticotropin-releasing hormone; CGRP, calcitonin gene related peptide; FSH, follicle-stimulating hormone; GnRH, gonadotropin-releasing hormone; GH, growth hormone; LH, luteinizing hormone; PL, placental lactogen; TRH, thyroid hormone releasing hormone; TSH, thyroid-stimulating hormone; VIP, vasoactive intestinal polypeptide.

described (Dhillo *et al.* 2006). Brown adipose tissue is a heat-generating tissue that is especially important in human infants, although it is now known to be present in adult humans as well and can be activated under certain conditions by the hormone irisin (see section 2.2.14; Frühbeck *et al.* 2009).

The primary hormones in white adipose tissue are leptin and resistin. *Leptin* is secreted from fat cells and acts to reduce food intake and prevent an increase in body fat. It does this by stimulating leptin receptors in the hypothalamus that cause a reduction in appetite and increased energy expenditure (Figure 2.9; Morton *et al.* 2006).

Figures 1.4, 12.15 and 12.16 demonstrate what happens when leptin is missing; that is, the leptin gene does not work properly. Fortunately for people like this, leptin is available for their treatment. But what about the millions of people who are obese through overeating and lack of exercise? It turns out that such people paradoxically have *elevated* levels of circulating leptin, but leptin does not work in the same way as in thin people. The reason for this is that the receptors in the hypothalamus are somehow desensitized (called "*leptin resistance*") and do not respond to leptin feedback. One of the

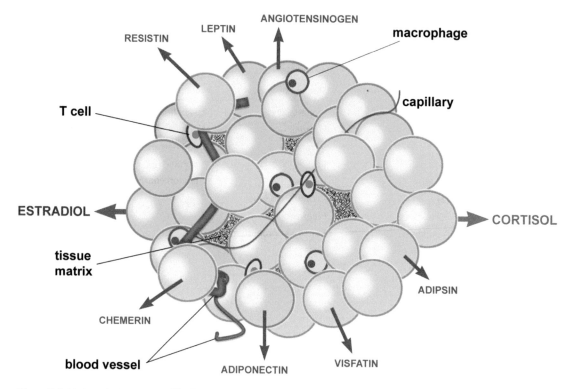

Figure 2.8 Various factors secreted by fat cells
Adipocytes can synthesize and secrete many factors (adipokines), including leptin, adiponectin, angiotensinogen, visfatin, adipsin and resistin. They can also synthesize and release steroids such as estradiol and cortisol. The cellular matrix is supportive tissue, predominantly fibroblasts. Immune cells such as macrophages and T cells are responsible for the release of cytokines such as interleukins (IL-1 and IL-6), tumor necrosis factor α (TNF α), and prostaglandin E2.

great challenges of endocrine research is to understand the causes of leptin resistance in obesity (Chicurel 2000; Morton *et al.* 2006; Myers *et al.* 2012). A more detailed description of the neuroendocrinology of leptin is provided in section 12.6.3. *Resistin* is a peptide hormone secreted by adipose tissue that is involved in insulin resistance, obesity, diabetes and hypertension (Schwartz and Lazar 2011). Resistin and adiponectin, another adipokine, modify insulin signaling and may regulate the onset of diabetes (Ahima and Lazar 2008).

Many other hormones, cytokines, and additional chemical signals are secreted from adipose tissue (Figure 2.8). These include the adipokines adiponectin, adipsin and visfatin; the cytokines IL-1, IL-6 and TNFα; prostaglandins; and steroid hormones estradiol and cortisol (Henry and Clarke 2008). Much has yet to be discovered about the role of adipokines in the control of energy balance.

2.2.14 Hormones of skeletal muscle: the myokines

Skeletal muscle secretes multiple factors that have autocrine, paracrine and endocrine effects (Pedersen and Febbraio 2012). They communicate with other organs such as

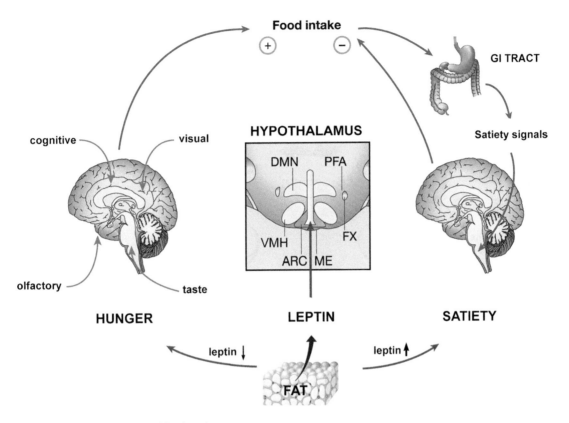

Figure 2.9 Leptin and control of food intake
Leptin circulates in proportion to the mass of body fat. It reaches the brain from the blood stream and acts on neuronal pathways that regulate food intake. Leptin diminishes perception of food reward – food palatability – and enhances the response to satiety signals that inhibit feeding and lead to meal termination. Weight loss lowers leptin levels and this in turn increases the rewarding properties of food while at the same time diminishing satiety. This effectively increases food intake.

Abbreviations: ARC, arcuate nucleus; DMN, dorsomedial hypothalamic nucleus; FX, fornix; ME, median eminence; PFA, perifornical area; VMH, ventromedial hypothalamic nucleus.

Reproduced with permission (Morton *et al.* 2006).

adipose tissue, pancreas, liver, and brain. Blood levels of *interleukin-6* (IL-6), for example, are increased up to 100-fold during exercise (see section 13.3.2 for discussion of interleukins in the immune system). The hormonal effects of IL-6 become obvious in mice in which the IL-6 gene is missing (IL-6 knockouts). Such mice develop mature-onset obesity, suggesting that IL-6 has an important effect on fat cells (Wallenius *et al.* 2002). *Myostatin* also modulates fat tissue function such that mice lacking myostatin have a greatly *reduced* fat mass (McPherron and Lee 2002). These authors theorize that blocking the action of myostatin, for example with an antagonist, might be effective in preventing the development of obesity (see also LeBrasseur 2012). Another muscle hormone, *irisin*, also regulates fat tissue and, like IL-6, is secreted from skeletal muscle during exercise in both animals and humans (Villarroya 2012; Kelly 2012; Bostrum *et al.* 2012). As mentioned in section 2.2.13, irisin stimulates the production of brown adipocytes

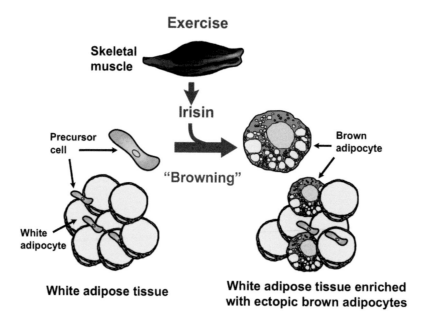

Figure 2.10 "Browning" effect of irisin in white fat tissue
Irisin is secreted from skeletal muscle during exercise in mice and humans. Irisin induces the conversion (differentiation) of precursor cells, contained within the white fat depot, into brown fat cells. Reproduced with permission (Villarroya 2012).

within white fat depots, a process called "browning" (Figure 2.10; Villarroya 2012). This promotes energy expenditure through thermogenesis, a characteristic of brown, but not white, adipose tissue.

Bostrum *et al.* (2012) showed that an elevation in irisin levels in obese mice caused an increase in browning of fat tissue, increased energy expenditure, plus weight loss and improvements in the diabetic state. Since physical activity is known to expend calories and to provide benefits in terms of combating human obesity, is irisin the hormone responsible? Blood levels of irisin are increased by exercise in humans (Bostrum *et al.* 2012; Huh *et al.* 2012), but evidence for an effect of irisin on browning and weight loss in humans has yet to be determined (Kelly 2012).

Is it possible that other types of muscle also secrete hormones? As mentioned already in section 2.2.6, heart (cardiac) muscle secretes two hormones, ANP and BNP, and, like irisin, they induce brown fat cells to increase thermogenesis (Whittle and Vidal-Puig 2012; Bordicchia *et al.* 2012). This suggests that tissues involved in high physical activity, such as skeletal and cardiac muscle, send endocrine signals to fat cells.

2.2.15 Bone as an endocrine organ
Bone is normally regarded as an inert tissue that provides a rigid structural scaffold for the body (Chapter 13 will describe how the *bone marrow* is vital for the function of the immune system). Nevertheless, and as noted already, bone is a hormonal target tissue. For example: (1) parathyroid hormone (PTH) induces the release of calcium ions from bone into the blood (see Figure 2.2); (2) estradiol is responsible for maintenance of bone mass in men and

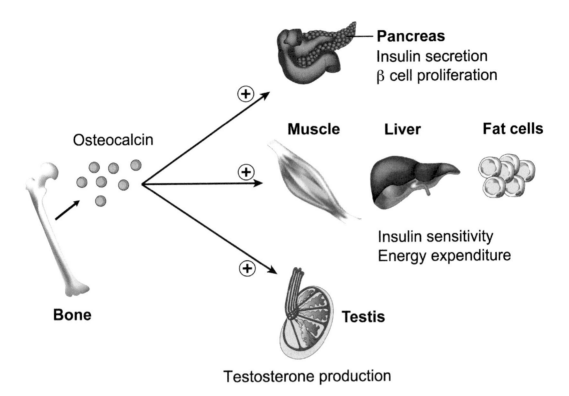

Figure 2.11 Osteocalcin is a bone-derived multifunctional hormone
Osteocalcin is secreted from osteoblast cells in the skeleton and targets the β cells of the pancreas. Proliferation of β cells is increased and insulin secretion is stimulated. Osteocalcin also stimulates energy expenditure in muscle and at the same time increases the insulin sensitivity of muscle, fat and liver. In addition, it promotes male fertility by stimulating testosterone output by testicular cells and increasing sperm count. Reproduced with permission (Karsenty and Ferron 2012).

women (see Figure 2.7) (Khosla *et al.* 2012); and (3) IGF-1, secreted from the liver under the influence of growth hormone, targets bone in the regulation of growth (see section 3.2.2 and Figure 6.9). However, bone, one of the largest tissues in the body, is also an active endocrine "gland" in its own right. The major hormone secreted by bone is called *osteocalcin*, which is secreted from osteoblasts, the primary bone-forming cells. Figure 2.11 illustrates that osteocalcin has four main effects: (1) it increases testosterone production by the testes; (2) it stimulates proliferation of pancreatic β cells that produce insulin; (3) it increases insulin release from the β cells; and (4) osteocalcin increases insulin sensitivity of muscle, liver and fat cells and stimulates energy expenditure by muscle (Karsenty and Oury 2012; Guntur and Rosen 2012).

    These effects of osteocalcin were discovered in mice whose osteocalcin gene was inactivated (*osteocalcin knockouts*). Such mice were obese and poor breeders, but in addition were hyperglycemic (high glucose levels), hypoinsulinemic (low insulin levels) and insulin resistant in liver, fat and muscle (i.e. unresponsive to insulin). The absence of osteocalcin therefore induces a diabetic state that can be reversed by treatment with osteocalcin. The poor breeding performance is due to reduced testosterone output from the testes as well as low sperm counts, also reversed by osteocalcin. *In summary*, the skeleton secretes osteocalcin from osteoblasts to regulate glucose homeostasis, energy expenditure and fertility. This indicates that bone should be considered a classic endocrine organ.

2.2.16 Liver as an endocrine organ: the hepatokines

The liver is another major organ that secretes several factors that act as endocrine or paracrine signals. Three of these *hepatokines* – fetuin-A, fibroblast growth factor-21 (FGF-21), and betatrophin – will be briefly described here. *Fetuin-A* is a glycoprotein (a large peptide attached to carbohydrate residues) secreted from the liver under conditions of chronic over-nutrition and high glucose levels (Stefan and Häring 2013). Elevated glucose induces the liver to secrete fetuin-A, which reduces insulin receptor signaling on both skeletal muscle (an endocrine effect) and liver (a paracrine effect). Desensitization of the insulin receptor (insulin resistance) prevents these tissues from taking up glucose, thus inducing a diabetic state (see section 2.2.8). This is confirmed in animal studies where mutant mice that lack the fetuin-A gene had improved insulin signaling. In contrast, in mice, *FGF-21* is secreted from the liver and induces *increased* insulin sensitivity and muscular uptake of glucose. In humans, however, insulin sensitivity may be reduced by FGF-21 (Stefan and Häring 2013). In mice, it also acts as a starvation indicator (increased FGF-21 levels) and may be responsible for fasting-induced inhibition of reproduction through an effect on the hypothalamus (Bass 2013; Yates 2013). *Betatrophin* is a liver hormone that induces proliferation of pancreatic β-cells (see section 2.2.8) and a higher output of insulin in mice (Lickert 2013; Raghow 2013).

2.3 Summary

This chapter provides a concise overview of endocrine glands and tissues, their location in the body, and the hormones they produce. You should become familiar with each of these tissues and be able to identify the principal hormones produced by each and the primary functions of each hormone as outlined in Table 2.1. In Chapter 3, you will become familiar with the stimuli that regulate the release of hormones. Thus, understanding of material in future chapters depends on knowing the information in this chapter.

FURTHER READING

Gardner, D. G. and Shoback, D. (2011). *Greenspan's Basic and Clinical Endocrinology*, 9th edn. (New York: McGraw-Hill).

Griffin, J. E. and Ojeda, S. R. (2011). *Textbook of Endocrine Physiology*, 6th edn. (New York: Oxford University Press).

Gruber, C. J., Tschugguel, W., Schneeberger, C. and Huber, J. C. (2002). "Production and actions of estrogens," *New Eng J Med* 346, 340–352.

Jones, B. J., Tan, T. and Bloom, S. R. (2012). "Glucagon in stress and energy homeostasis," *Endocr* 153, 1049–1054.

Ogawa, T. and de Bold, A. J. (2014). "The heart as an endocrine organ," *Endocr Connections* 3, R31–R44.

Petraglia, F., Imperatore, A. and Challis, J. R. (2010). "Neuroendocrine mechanisms in pregnancy and parturition," *Endocr Rev* 31, 783–816.

Reis, F. M., Florio, P., Cobellis, L., Luisi, S., Severi, F. M., Bocchi, C. *et al.* (2001). "Human placenta as a source of neuroendocrine factors," *Biol Neonate* 79, 150–156.

Singh, M. and Jadhav, H. R. (2014). "Melatonin: functions and ligands," *Drug Discov Today* 19, 1410–1418.

REVIEW QUESTIONS

2.1   Which endocrine glands secrete the following hormones?
    (a) calcitonin
    (b) melatonin
    (c) glucagon
    (d) gastrin

2.2   Each of the following glands secretes two primary hormones. Name these hormones.
    (a) duodenum
    (b) ovary

2.3   Name the two parts of the adrenal gland and two hormones secreted from each part.

2.4   Name five steroid hormones.

2.5   Name two hormones produced in the following glands:
    (a) testis
    (b) stomach
    (c) thyroid
    (d) pancreas (beta cells)

2.6   Which endocrine gland is important for the development of the immune system?

2.7   Which placental hormone stimulates the ovaries to keep producing progesterone in the early stages of pregnancy?

2.8   Which endocrine glands secrete the following hormones?
    (a) human placental lactogen
    (b) thymosin
    (c) aldosterone
    (d) epinephrine

2.9   Name two hormones that are also neurotransmitters.

2.10 As well as the gonads, the sex steroids are produced in which other endocrine gland?

2.11 Name two hormones secreted from fat cells.

ESSAY QUESTIONS

2.1 Discuss the concept of opposing actions between pairs of hormones such as calcitonin and parathyroid hormone or insulin and glucagon.

2.2 Discuss the relationship of the thymus gland with the immune system.

2.3 Discuss the gastrointestinal hormones with regard to the definition of a "true" hormone given in Chapter 1. Do they meet the criteria or not?

2.4 How is it that the sex steroids are produced in both the gonads and the adrenal cortex?

2.5 Why are epinephrine and norepinephrine classed both as hormones and as neurotransmitters?

2.6 Discuss the different roles of the pineal gland in amphibia and mammals.

2.7 Describe the changes in the ovarian follicle that regulate the timing of estradiol and progesterone secretion.

2.8 Discuss the placental hormones, the timing of their secretion, and their functions.

2.9 Discuss "leptin resistance" and its importance in obesity.

REFERENCES

Ahima, R. S. and Lazar, M. A. (2008). "Adipokines and the peripheral and neural control of energy balance," *Mol Endocrinol* 22, 1023–1031.

Bass, J. (2013). "Forever (FGF) 21," *Nat Med* 19, 1090–1092.

Benarroch, E. E. (2007). "Enteric nervous system: functional organization and neurologic implications," *Neurol* 69, 1953–1957.

Bikle, D. D. (2010). "Vitamin D: newly discovered actions require reconsideration of physiologic requirements," *Trends Endocrinol Metab* 21, 375–384.

Bordicchia, M., Liu, D., Amri, E.-Z., Ailhaud, G., Dessi-Fulgheri, P., Zhang, C. *et al.* (2012). "Cardiac natriuretic peptides act via p38 MAPK to induce the brown fat thermogenic program in mouse and human adipocytes," *J Clin Invest* 122, 1022–1036.

Boron, W. F. and Boulpaep, E. L. (2005). *Medical Physiology*, updated edn. (Philadelphia, PA: Elsevier Saunders), p. 1011.

Bostrum, P., Wu, J., Jedrychowski, M. P., Korde, L. Y., Ye, L., Lo, J. C. *et al.* (2012). "A PGC1-α-dependent myokine that drives brown-fat-like development of white fat and thermogenesis," *Nature* 481, 463–468.

Bradley, D. P., Kulstad, R. and Schoeller, D. A. (2010). "Exenatide and weight loss," *Nutrition* 26, 243–249.

Charmandari, E., Tsigos, C. and Chrousos, G. (2005). "Endocrinology of the stress response," *Annu Rev Physiol* 67, 259–284.

Chicurel, M. (2000). "Whatever happened to leptin?" *Nature* 404, 538–540.

Dhillo, W. S., Murphy, K. G. and Bloom, S. (2006). "Endocrinology: The next 60 years," *J Endocrinol* 190, 7–10.

Frühbeck, G., Becerril, S., Sainz, N., Garrastachu, P. and Garcia-Velloso, M. J. (2009). "BAT: A new target for human obesity?" *Trends Pharmacol Sci* 30, 387–396.

Gardiner, J. V., Jayasena, C. N. and Bloom, S. R. (2008). "Gut hormones: A weight off your mind," *J Neuroendocrinol* 20, 834–841.

Goldstein, A. L. and Badamchian, M. (2004). "Thymosins: Chemistry and biological properties in health and disease," *Expert Opin Biol Ther* 4, 559–573.

Guntur, A. R. and Rosen, C. J. (2012). "Bone as an endocrine organ," *Endocr Pract* 18, 758–762.

Hall, J., Roberts, R. and Vora, N. (2009). "Energy homoeostasis: The roles of adipose tissue-derived hormones, peptide YY and Ghrelin," *Obes Facts* 2, 117–125.

Henry, B. A. and Clarke, I. J. (2008). "Adipose tissue hormones and the regulation of food intake," *J Neuroendocrinol* 20, 842–849.

Huh, J. Y., Panagiotou, G., Mougios, V., Brinkoetter, M., Vamvini, M. T., Schneider, B. E. *et al.* (2012). "FNDG5 and irisin in humans: I. Predictors of circulating concentrations in serum and plasma and II. mRNA expression and circulating concentrations in response to weight loss and exercise," *Metabolism* 61, 1725–1738.

Jelkmann, W. and Lundby, C. (2011). "Blood doping and its detection," *Blood* 118, 2395–2404.

Kadmiel, M. and Cidlowski, J. A. (2013). "Glucocorticoid receptor signaling in health and disease," *Trends Pharmacol Sci* 34, 518–530.

Karsenty, G. and Ferron, M. (2012). "The contribution of bone to whole-organism physiology," *Nature* 481, 314–320.

Karsenty, G. and Oury, F. (2012). "Biology without walls: The novel endocrinology of bone," *Ann Rev Physiol* 74, 87–105.

Kelly, D. P. (2012). "Irisin, light my fire," *Science* 336, 42–43.

Khosla, S., Oursler, M. J. and Monroe, D. G. (2012). "Estrogen and the skeleton," *Trends Endocr Metab* 23, 576–581.

LeBrasseur, N. K. (2012). "Building muscle, browning fat and preventing obesity by inhibiting myostatin," *Diabetologia* 55, 13–17.

Lickert, H. (2013). "Betatrophin fuels β cell proliferation: First step toward regenerative therapy?" *Cell Metab* 18, 5–6.

Manson, J. E., Mayne, S. T. and Clinton, S. K. (2011). "Vitamin D and prevention of cancer – ready for prime time?" *N Engl J Med* 364, 1385–1387.

McPherron, A. C. and Lee, S.-J. (2002). "Suppression of body fat accumulation in myostatin-deficient mice," *J Clin Invest* 109, 595–601.

Morton, G. J., Cummings, D. E., Baskin, D. G., Barsh, G. S. and Schwartz, M. W. (2006). "Central nervous system control of food intake and body weight," *Nature* 443, 289–295.

Murphy, K. G. and Bloom, S. R. (2006). "Gut hormones and the regulation of energy homeostasis," *Nature* 444, 854–859.

Myers, M. G., Heymsfield, S. B., Haft, C., Kahn, B. B., Laughlin, M., Leibel, R. L. *et al.* (2012). "Challenges and opportunities of defining clinical leptin resistance," *Cell Metab* 15, 150–156.

Pedersen, B. K. and Febbraio, M. A. (2012). "Muscles, exercise and obesity: Skeletal muscle as a secretory organ," *Nat Rev Endocr* 8, 457–465.

Potter, L. R., Abbey-Hosch, S. and Dickey, D. M. (2006). "Natriuretic peptides, their receptors, and cyclic guanosine monophosphate-dependent signaling functions," *Endocr Revs* 27, 47–72.

Raghow, R. (2013). "Betatrophin: A liver-derived hormone for pancreatic β-cell proliferation," *World J Diabet* 4, 234–237.

Ritchie, R. H., Rosenkranz, A. C. and Kaye, D. M. (2009). "B-type natriuretic peptide: endogenous regulator of myocardial structure, biomarker and therapeutic target," *Curr Mol Med* 9, 814–825.

Romijn, J. A., Corssmit, E. P., Havekes, L. M. and Pijl, H. (2008). "Gut-brain axis," *Curr Opin Clin Nutr Metab Care* 11, 518–521.

Sam, A. H., Troke, R. C., Tan, T. M. and Bewick, G. A. (2012). "The role of the gut-brain axis in modulating food intake," *Neuropharmacol* 63, 46–56.

Schwartz, D. R. and Lazar, M. A. (2011). "Human resistin: Found in translation from mouse to man," *Trends Endocrinol Metab* 22, 259–265.

Stefan, N. and Häring, H.-U. (2013). "The role of hepatokines in metabolism," *Nat Rev Endocr* 9, 144–152.

Strauss, J. F. and Barbieri, R.L. (2004). *Yen and Jaffe's Reproductive Endocrinology*, 5th edn. (Philadelphia, PA: Elsevier Saunders).

Unger, R. H. and Orci, L. (2010). "Paracrinology of islets and the paracrinopathy of diabetes," *Proc Natl Acad Sci USA* 107, 16009–16012.

Van Lommel, A., Bolle, T., Fannes, W. and Lauweryns, J. M. (1999). "The pulmonary neuroendocrine system: The past decade," *Arch Histol Cytol* 62, 1–16.

Villarroya, F. (2012). "Irisin, turning up the heat," *Cell Metab* 15, 277–278.

Wallenius, V., Wallenius, K., Ahren, B., Rudling, M., Carlsten, H., Dickson, S. L. *et al.* (2002). "Interleukin-6 deficient mice develop mature-onset obesity," *Nat Med* 8, 75–79.

Whittle, A. J. and Vidal-Puig, A. (2012). "NPs – heart hormones that regulate brown fat," *J Clin Invest* 122, 804–807.

Widmaier, E. P., Raff, H. and Strang, K. T. (2010). *Vander's Human Physiology: The Mechanisms of Body Function*, 12th edn. (New York: McGraw-Hill).

Xiong, T., Qu, Y., Mu, D. and Ferriero, D. (2011). "Erythropoietin for neonatal brain injury: Opportunity and challenge," *Int J Dev Neurosci* 29, 583–591.

Yates, D. (2013). "Signalling starvation," *Nat Rev Neurosci* 14, 670–671.

The pituitary gland and its hormones 3

3.1 The pituitary gland

The pituitary gland, which is also called the *hypophysis*, is attached to the hypothalamus at the base of the brain (Figure 3.1). Secretion of the hormones of the pituitary gland is regulated by the hypothalamus and it is through the hypothalamic-pituitary connection that external and internal stimuli can influence the release of the pituitary hormones, thus producing the neural-endocrine interaction. The pituitary has been called the body's "master gland" because its hormonal secretions stimulate a variety of endocrine glands to synthesize and secrete their own hormones. However, it is really the hypothalamus that is the master gland, because it controls the pituitary.

The pituitary gland consists of two primary organs: the anterior pituitary (*adenohypophysis* or *pars distalis*) which is a true endocrine gland, and the posterior pituitary (*neurohypophysis*) which is formed from neural tissue and is an extension of the hypothalamus (Figure 3.1). The pituitary gland is attached to the hypothalamus by the pituitary (hypophyseal) stalk. Further details of the anatomy and physiology of the pituitary gland can be found elsewhere (Norman and Litwack 1997; Amar and Weiss 2003; Boron and Boulpaep 2005; Gardner and Shoback 2011).

3.1.1 The neurohypophysis (posterior pituitary)

The neurohypophysis consists of neural tissue and contains the nerve terminals (about 100,000) of axons whose cell bodies are located in the paraventricular nucleus (PVN) and supraoptic nucleus (SON) of the hypothalamus. The axons of these large magnocellular neurosecretory cells project down from the hypothalamus through the part of the pituitary stalk called the *infundibulum* and terminate in the posterior pituitary gland (Figure 3.2). The neurosecretory cells of the PVN and SON manufacture the hormones oxytocin and vasopressin (also called antidiuretic hormone, ADH), which are transported down the axons and stored in nerve terminals in the posterior pituitary. The axon terminals in the posterior pituitary are surrounded by supporting cells called pituicytes. The posterior pituitary gland receives its blood supply from the inferior hypophyseal artery and when the hypothalamic PVN and SON neurons are stimulated oxytocin and/or vasopressin are secreted into the bloodstream and travel to their target cells throughout the body.

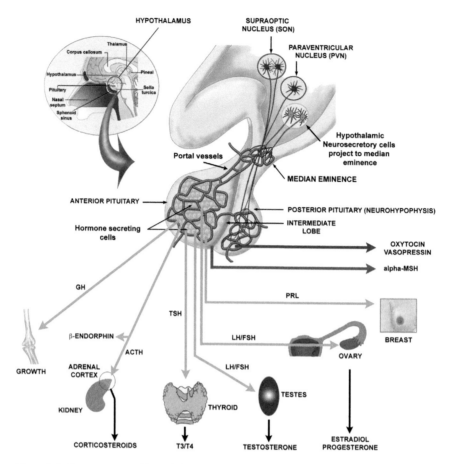

Figure 3.1 Neuroanatomy of hypothalamic-pituitary unit
The pituitary and hypothalamus are interconnected through neural and vascular links. Neurons in the para-ventricular nucleus (PVN) and supraoptic nucleus (SON) project axons down into the posterior pituitary where they release oxytocin and vasopressin. α-MSH is secreted from the intermediate lobe of the pituitary. In contrast, neurosecretory neurons responsible for secretion of releasing/inhibiting hormones project only to the median eminence. The anterior pituitary is connected to the median eminence by a portal system of blood vessels. These blood vessels convey the releasing/inhibiting hormones to the anterior pituitary where they regulate secretion of seven hormones. The target organs respond to stimulation with these hormones by releasing steroid and thyroid hormones.

    Abbreviations can be found in Table 3.1. Copyright Alex Pincock; inset reproduced with permission (Heaney and Melmed 2004).

3.1.2 The adenohypophysis (anterior pituitary)

The adenohypophysis consists of two distinct parts: the anterior pituitary and the intermediate pituitary. The adenohypophysis is attached to the hypothalamus by that part of the pituitary stalk that contains the *hypophyseal portal system* of blood vessels (Figure 3.1). The nerve endings of the neurosecretory cells of the hypothalamus (described in Chapter 4) terminate at the median eminence, where their hormones are released into the hypophyseal portal system, through which they are carried to the

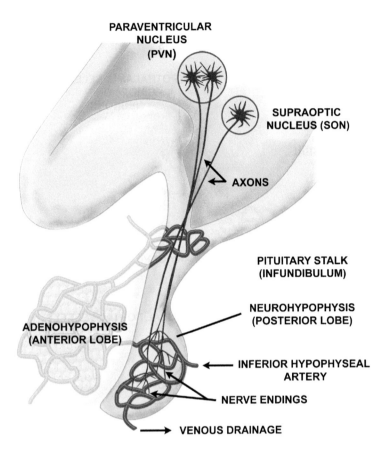

PARAVENTRICULAR NUCLEUS (PVN)

SUPRAOPTIC NUCLEUS (SON)

AXONS

PITUITARY STALK (INFUNDIBULUM)

NEUROHYPOPHYSIS (POSTERIOR LOBE)

ADENOHYPOPHYSIS (ANTERIOR LOBE)

INFERIOR HYPOPHYSEAL ARTERY

NERVE ENDINGS

VENOUS DRAINAGE

Figure 3.2 Anatomy of the neurohypophysis
The neurohypophysis (posterior lobe; pars nervosa) contains the nerve endings of the neurosecretory cells whose cell bodies are found in the supraoptic (SON) and paraventricular nuclei (PVN) of the hypothalamus. The axons of these SON and PVN neurons project through the infundibulum to the posterior lobe, where they release their hormones (oxytocin and vasopressin) from nerve endings into a capillary system fed by the inferior hypophyseal artery. Hormones reach the peripheral circulation via venous drainage from the gland. Copyright Alex Pincock.

anterior pituitary. The pituitary stalk thus contains both nerve axons, traveling to the posterior pituitary, and blood vessels that connect the hypothalamus and anterior pituitary gland. This vascular connection between the hypothalamus and the anterior pituitary consists of the superior hypophyseal artery delivering blood to the median eminence of the hypothalamus, where it forms a series of tiny blood vessels (capillaries), called the "primary plexus," into which the hypothalamic hormones are released. These hormones then travel through the hypophyseal portal veins to the secondary plexus, another series of capillaries in the anterior pituitary. Here, the hypothalamic hormones stimulate pituitary cells to release their hormones into the secondary plexus, from which they enter the general circulation (Figure 3.3).

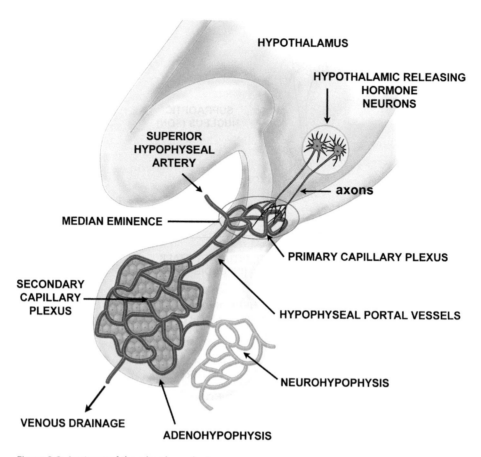

Figure 3.3 Anatomy of the adenohypophysis

Hypothalamic releasing/inhibiting hormone neurons are located in various parts of the hypothalamus. For example, gonadotropin-releasing hormone (GnRH) neurons are found in the arcuate nucleus in rats. These neurosecretory cells project their axons to the primary capillary plexus in the median eminence of the hypothalamus, where they secrete releasing/inhibiting hormones. The hormones are then transported through the hypophyseal portal vessels to the secondary capillary plexus in the anterior pituitary, where they can bind to specific receptors on pituitary cells to stimulate release of hormones such as growth hormone (GH) and luteinizing hormone (LH). Copyright Alex Pincock.

3.1.3 The intermediate lobe of the pituitary

The intermediate lobe of the pituitary gland varies in size among the vertebrates, being large in amphibians and in lower vertebrates, but is not present in all mammals. Rodents have a distinct intermediate lobe containing melanotrophs, but adult humans do not. However, it is well developed in the human fetus during fetal growth, but becomes indistinct after birth, as the melanotrophs become distributed throughout the anterior lobe (Mauri *et al.* 1993; Saland 2001). The melanotroph cells of the intermediate lobe are innervated by hypothalamic neurons and actively synthesize and release neuropeptides derived from the pro-opiomelanocortin gene (POMC, as described below).

Table 3.1 **The hormones of the pituitary gland**

Posterior pituitary (neurohypophysis or pars nervosa)

Oxytocin – stimulates uterine muscle contractions and milk ejection from the mammary glands via contraction of myoepithelial cells.

Vasopressin (antidiuretic hormone, ADH) – elevates blood pressure and promotes reabsorption of water by the kidneys.

Neurophysins – not hormones, but carrier proteins released along with oxytocin and vasopressin.

Anterior pituitary (adenohypophysis or pars distalis)

Growth hormone (GH; somatotropin or somatotropic hormone) – promotes protein synthesis, reduction of fat mass, and carbohydrate metabolism; direct stimulatory effect on muscle mass; increases bone growth by first stimulating somatomedin (IGF-1) release from the liver.

Adrenocorticotropic hormone (ACTH) – stimulates glucocorticoid secretion from the adrenal cortex.

Thyroid-stimulating hormone (TSH; thyrotropin) stimulates thyroxine (T4) and triiodothyronine (T3) secretion from the thyroid gland.

Prolactin (PRL) – initiates milk production in the mammary glands and has many other functions, including stimulation of the gonads.

Gonadotropic hormones

Follicle-stimulating hormone (FSH) – stimulates growth of the primary follicle and estradiol secretion from the ovary in females; sperm production and inhibin secretion in the testis of males. FSH secretion is also regulated locally by two factors produced in folliculostellate cells (i.e. follistatin and activin).

Luteinizing hormone (LH) – stimulates ovulation, formation of the corpus luteum and progesterone secretion in females; stimulates Leydig (interstitial) cells to secrete androgens (testosterone) in males.

β-endorphin – true function unknown.

Intermediate pituitary (pars intermedia)

α-melanocyte-stimulating hormone (α-MSH) – stimulates melanophores to darken skin color in amphibia. Some evidence for a similar effect in human, e.g. in Addison's disease.

β-endorphin – acts as a neuromodulator in the brain to regulate neurotransmitter release, and possibly as a circulating analgesic. Also secreted from anterior pituitary.

3.2 The hormones of the pituitary gland

3.2.1 Neurohypophysis (posterior pituitary)

The two hormones secreted from the posterior pituitary, oxytocin and vasopressin, are strictly speaking hypothalamic or neurosecretory hormones (Table 3.1). They are manufactured in the neurosecretory cells of the paraventricular nucleus (PVN) and supraoptic nucleus (SON) of the hypothalamus and transported down the axons of these neurosecretory cells, via the infundibulum (pituitary stalk), to the posterior pituitary. Here, they are stored in nerve terminals and then released into the venous circulation (see Figure 6.10). There are specific oxytocin-producing neurons and vasopressin-producing neurons in both the SON and PVN of the hypothalamus.

Oxytocin has two primary functions: it promotes uterine contractions during parturition (childbirth) and it stimulates milk ejection, or letdown, from the mammary glands during lactation. Oxytocin also has a number of neuropeptide functions in the brain (see Chapter 11 and section 12.7.2). *Vasopressin* (antidiuretic hormone, ADH) acts to raise blood pressure and to promote water reabsorption in the kidneys – that is, it acts as an antidiuretic. As a neuropeptide, vasopressin may enhance memory (see Chapter 12). As well as these hormones, the posterior pituitary releases two large proteins called *neurophysins* that function as carrier proteins for oxytocin and vasopressin (see Chapter 7). A good description of the posterior pituitary and its hormones is provided by Nussey and Whitehead (2001).

3.2.2 Adenohypophysis

The adenohypophysis secretes six hormones from the anterior pituitary and two hormones from cells of the intermediate lobe, as summarized in Table 3.1. As shown in Figure 3.1, a seventh "hormone" is the opioid peptide β-endorphin, secreted from the anterior lobe into the bloodstream along with ACTH, but at present it has no known hormonal actions. The following section provides a brief outline of the functions of these hormones. More detailed descriptions of the hormones of the anterior pituitary can be found in most textbooks on endocrinology, but a good online source is Nussey and Whitehead (2001).

Growth hormone (GH)

Growth hormone is also known as somatotropin or somatotropic hormone. The suffix – *tropin* – refers to a substance that has a stimulating effect on its target organ; thus, somatotropin is a body- (soma-) stimulating hormone. Growth hormone is the most abundant hormone of the anterior pituitary and is produced in somatotroph cells. GH has effects in almost all body cells: bone, muscle, brain, heart, fat, etc. Growth hormone has direct effects on some cells, such as muscle, and therefore it increases muscle mass. However, its effect on bone growth, and therefore height, occurs indirectly by stimulating the release of *somatomedin*, a peptide growth factor (also known as insulin-like growth factor; IGF-1) from the liver (see Figure 6.9). GH also stimulates the liver to increase glucose output.

Prolactin (PRL)

Prolactin is produced in lactotroph (or mammotroph) cells in the anterior pituitary. Prolactin is essential for initiating milk production in the mammary glands and also has many functions related to growth, osmoregulation, fat and carbohydrate metabolism, reproduction, and parental behavior. In many of these actions, prolactin interacts with other hormones, including estradiol, progesterone and oxytocin.

Adrenocorticotropic hormone (ACTH)

ACTH is produced in the corticotroph cells of the anterior pituitary and acts to stimulate the synthesis and release of glucocorticoid hormones (cortisol, corticosterone, etc.) in the adrenal cortex. It does this with a distinct rhythm so that levels of ACTH, and cortisol, are high in the early morning (see Figure 6.5). ACTH is also involved in regulating the immune system (Figure 1.1; and Chapter 13).

Thyroid-stimulating hormone (TSH)

TSH, also known as thyrotropin or thyrotropic hormone, is produced in thyrotroph cells of the anterior pituitary. TSH stimulates the synthesis and release of thyroxine (T4) and triiodothyronine (T3) from the thyroid gland.

The gonadotropic hormones

The gonad-stimulating or gonadotropic hormones, follicle-stimulating hormone (FSH), and luteinizing hormone (LH) are produced in the gonadotroph cells of the anterior pituitary.

Follicle-stimulating hormone (FSH)

Follicle-stimulating hormone has a similar function in both sexes: it promotes the development of the gametes and the secretion of gonadal hormones. In the female, FSH stimulates the growth of the primary follicle in the ovary, promoting development of the ovum, and the secretion of the female sex hormone estradiol (Widmaier *et al.* 2010). In the male, FSH stimulates sperm production (spermatogenesis) and the secretion of the hormone inhibin by acting on the Sertoli cells of the testis.

Luteinizing hormone (LH)

In the female, luteinizing hormone stimulates ovulation by rupturing the follicle and releasing the ovum. The residual cells of the follicle then form the progesterone-secreting luteal cells (corpora lutea) in the ovary. In the male, luteinizing hormone stimulates the Leydig cells (also called interstitial cells) to secrete androgens such as testosterone.

β-endorphin

As well as the hormones discussed above, the anterior pituitary also produces an opioid peptide, β-endorphin. β-endorphin and ACTH are derived from the same large pre-propeptide (pro-opiomelanocortin – POMC, as outlined in Figure 3.4). POMC is a large polypeptide (a *propeptide*) that is synthesized in the anterior and intermediate lobes of the pituitary, as well as in the brain. The POMC gene is expressed in each of these tissues, but its products are different depending upon cell-specific enzymes that cleave the peptide into smaller pieces. Figure 3.4 shows that it is broken down into active hormones by enzymes called *proprotein convertases* (PCs) (Hook *et al.* 2008). As shown in Figure 3.4, the POMC molecule contains the sequences for many pituitary peptides, including ACTH, α-MSH, β-lipotropin (LPH), and β-endorphin (Bicknell 2008; see also Figure 11.1).

First, ACTH and β-LPH are cleaved off from the propeptide and this occurs in both the anterior and intermediate pituitary (see section 3.3.2, below). ACTH and β-LPH are both secreted by the anterior pituitary, but all of the ACTH in the intermediate lobe is converted to α-MSH. In the anterior pituitary, all of the β-LPH is converted to β-endorphin and γ-lipotropin. Thus, the anterior pituitary co-releases ACTH, β-endorphin, and β-LPH from the same corticotroph cells. In terms of hormone-like effects, β-endorphin and γ-LPH have no known function when released into the bloodstream, but β-endorphin has a wide range of neuropeptide functions in analgesia, learning and memory, psychiatric diseases, feeding, thermoregulation, blood pressure regulation, and reproductive behavior (see Chapters 11

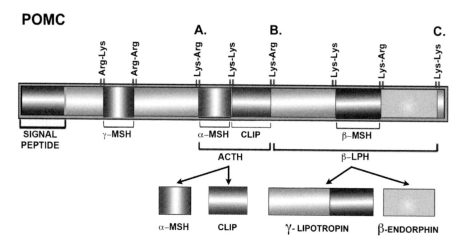

Figure 3.4 Cleavage of pro-opiomelanocortin (POMC) peptide
The large POMC polypeptide, encoded by the *POMC* gene, is proteolytically cleaved into smaller fragments by proprotein convertases PC1 and PC2. The cleavage sites are pairs of amino acids as shown, either arginine-lysine (Arg-Lys), or Arg-Arg, or Lys-Lys junctions. These junctions are recognized by PCs. In the example shown here, POMC in pituitary corticotrophs is first cleaved by PC1 at the sites labeled A, B and C. This step produces the hormones ACTH and β-lipotropin. Further cleavage takes place under the influence of PC2, this time yielding hormones α-MSH, γ-lipotropin and β-endorphin, and a peptide fragment called CLIP (corticotropin-like intermediate lobe peptide) of no known function.

Abbreviations: ACTH, adrenocorticotropic hormone; LPH, lipotropic hormone; MSH, melanocyte-stimulating hormone. See also Figure 11.1.

and 12). One popular suggestion for a role for *circulating* β-endorphin, released from the anterior pituitary, is in the so-called "Runner's High." It is generally assumed that β-endorphin, released from the anterior pituitary during exercise, somehow reaches the brain from the circulation. Since β-endorphin is a large molecule that is unlikely to cross the blood-brain barrier, it is much more likely that the source of β-endorphin that may be inducing the "high" is within the brain itself.

3.2.3 The intermediate lobe (pars intermedia)

As already stated above, the POMC pre-propeptide is also processed in the intermediate lobe of the pituitary to produce the hormones α-MSH and β-endorphin. Note that the pre-propeptide also contains the sequences for β-MSH and γ-MSH, although these are not produced in pituitary tissue (Figure 3.4). In amphibians, α-MSH acts on the melanophores to darken their skin to match background color. α-MSH can also affect pigmentation in mammals, including humans (Millington 2006). The secretion of α-MSH from the intermediate lobe of the pituitary may occur in the human fetus, although it has not been detected in normal healthy adults. In patients with Addison's disease, however, where cortisol levels are low and ACTH secretion high, there is a pronounced darkening of the skin that might be caused by α-MSH secreted along with ACTH. An important *non-pituitary* source of α-MSH is the hypothalamus, where α-MSH plays a critical role in influencing feeding behavior (see Chapter 11). α-MSH may also act to modulate the immune system (Catania *et al.* 2000; see Chapter 13).

Products of folliculostellate cells: *VEGF, FGF, IL-6, NO, FOLLISTATIN, ACTIVIN, OTHERS*

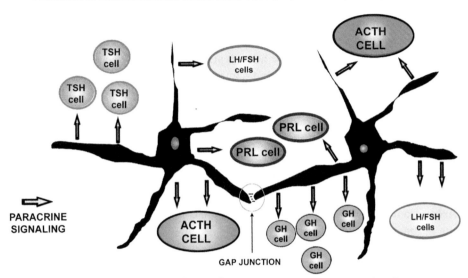

Regulators of folliculostellate cells: *cortisol, estradiol, somatostatin, others*

Figure 3.5 Paracrine relationship between folliculostellate cells and hormone-producing cells
of the anterior pituitary gland
This diagram shows how star-shaped folliculostellate (FS) cells are strategically interspersed between pituitary
hormone cells, allowing them to communicate in a paracrine fashion. FS cells modify the secretion of anterior
pituitary hormones by releasing paracrine signals such as FGF and NO. FS cells themselves are regulated by
incoming steroid signals such as cortisol and estradiol. Also shown is an example of a gap junction between
FS cells that permits widespread coordinated activation within the gland.
   Abbreviations: ACTH, adrenocorticotropic hormone; FGF, fibroblast growth factor; FSH, follicle-stimulating
hormone; GH, growth hormone; IL-6, interleukin 6; LH, luteinizing hormone; NO, nitric oxide; PRL, prolactin;
TSH, thyroid-stimulating hormone; VEGF, vascular endothelial growth factor. Figure generously provided by
Dr. S. J. Winters.

3.2.4 Folliculostellate cells of the anterior pituitary

So far, we have described the hormones secreted from five anterior pituitary cell types: LH
and FSH (gonadotrophs), PRL (lactotrophs), GH (somatotrophs), ACTH and β-endorphin
(corticotrophs), and TSH (thyrotrophs). In addition, there is a complex network of star-
shaped, so-called *folliculostellate (FS) cells*, which wires together, via paracrine signaling,
the whole anterior pituitary gland and which is probably responsible for extensive information transfer between all the cells of the anterior lobe (Figure 3.5) (Devnath and Inoue
2008; Winters and Moore 2007). The close association of FS cells with the endocrine cells
allows a variety of paracrine regulators to influence hormonal secretion. These paracrine
factors include growth factors (e.g. FGF, VEGF), cytokines (e.g. IL-6, LIF), and nitric oxide
(NO). Two additional factors released from FS cells, and known to specifically regulate FSH
secretion, are called *activin* and *follistatin*. FS cells are themselves controlled by some of the
same feedback signals that influence the hormonal cells; for example, cortisol, estradiol
and somatostatin. Thus, FS cells are key components of the anterior pituitary in the
regulation of the endocrine system.

3.3 Pituitary hormones in the brain

Some of the peptide hormones synthesized in the pituitary gland may also be produced as neuropeptides in neurons of the central nervous system, to function as neurotransmitters/neuromodulators. The functions of these neuropeptides, such as oxytocin, ACTH, α-MSH, and β-endorphin, are discussed in detail in Chapter 12.

3.4 Summary

The pituitary gland consists of the adenohypophysis (anterior), intermediate lobe, and neurohypophysis (posterior lobe), which are connected to the hypothalamus by the hypophyseal stalk. Axons from the PVN and SON of the hypothalamus project through the stalk to terminate in the posterior pituitary, which consists of axons and nerve endings supported by glial cells called pituicytes. The anterior and intermediate lobes of the pituitary consist of endocrine tissue, which receive hormonal stimulation from the hypothalamus through the blood vessels of the hypophyseal portal system. The pituitary gland produces nine hormones: six from the anterior pituitary, two from the posterior pituitary, and one from the intermediate lobe. Two further peptides, β-endorphin and γ-lipotropin, are secreted from the intermediate and anterior pituitary. The hormones of the posterior pituitary (oxytocin and vasopressin) are more correctly termed hypothalamic, or neurosecretory, hormones, and they are stored and released from the nerve terminals in the posterior pituitary. The posterior pituitary also releases the neurophysins which act as carrier proteins for oxytocin and vasopressin. Some of the hormones of the anterior pituitary, including TSH, ACTH, LH, FSH and GH, stimulate the release of hormones from other endocrine glands, such as the thyroid, adrenal cortex, gonads and liver, respectively. Other pituitary hormones (PRL, GH, α-MSH) act directly on non-endocrine target cells in the brain and body, as does GH (muscle, fat). All of the pituitary hormones, including β-endorphin, also function as central neuropeptides (neuromodulators). Finally, the discovery of the folliculostellate cell network that wires together all of the anterior pituitary cell types has added a new level of regulation for the secretion of pituitary hormones.

REVIEW QUESTIONS

3.1   What is the hypophysis?

3.2   Which two pituitary hormones are neurosecretory hormones?

3.3   Describe the connections between the hypothalamus and the neurohypophysis (posterior pituitary).

3.4   Which pituitary hormone, or hormones, serves the following functions?
     (a)   stimulate ovulation
     (b)   stimulates corticosteroid secretion

(c) stimulates milk secretion from the breast

(d) stimulates uterine contractions at childbirth

3.5 Name the six hormones of the anterior pituitary.

3.6 The six hormones referred to above are secreted from five cell types (e.g. corticotrophs). What is the name of the cell type that links activity in all of these cells?

3.7 What does "tropic" mean?

3.8 Which pituitary hormones have the following functions?
(a) cause skin color changes in amphibia
(b) stimulate T4 secretion

3.9 Which two pituitary hormones are gonadotropins?

3.10 Give the latin names for the three lobes of the pituitary gland.

3.11 Name the hormone released by the intermediate pituitary gland.

3.12 Which three pituitary hormones are synthesized from the prohormone/propeptide POMC?

3.13 Which enzymes are responsible for cleaving POMC into smaller peptides?

ESSAY QUESTIONS

3.1 Discuss the anatomy of the pituitary stalk and its importance for hypothalamic-pituitary connections.

3.2 How does it come about that the pituitary gland is made up of both endocrine and neural tissue? Discuss the embryological development of the pituitary gland.

3.3 Discuss the functions of oxytocin in male and female mammals.

3.4 Compare and contrast the functions of prolactin in humans, fish and birds.

3.5 Discuss the relationship between ACTH, α-MSH and β-endorphin.

3.6 Discuss the role of the somatomedins in mediating the growth-promoting functions of growth hormone.

3.7 Outline the role of anterior pituitary folliculostellate cells.

REFERENCES

Amar, A. P. and Weiss, M. H. (2003). "Pituitary anatomy and physiology," *Neurosurg Clin N Am* 14, 11–23.

Bicknell, A. B. (2008). "The tissue-specific processing of pro-opiomelanocortin," *J Neuroendocrinol* 20, 692–699.

Boron, W. F. and Boulpaep, E. L. (2005). *Medical Physiology*, updated edn. (Philadelphia, PA: Elsevier Saunders), p. 1011.

Catania, A., Airaghi, L., Colombo, G. and Lipton, J. M. (2000). "α-Melanocyte-stimulating hormone in normal human physiology and disease," *Trends Endocr Metab* 11, 304–308.

Devnath, S. and Inoue, K. (2008). "An insight to pituitary folliculo-stellate cells," *J Neuroendocrinol* 20, 687–691.

Gardner, D. G. and Shoback, D. (2011). *Greenspan's Basic and Clinical Endocrinology*, 9th edn. (New York: McGraw-Hill).

Heaney, A. P. and Melmed, S. (2004). "Molecular targets in pituitary tumours," *Nature Revs Cancer* 4, 285–295.

Hook, V., Funkelstein, L., Lu, D., Bark, S., Wegrzyn, J. and Hwang, S.-R. (2008). "Proteases for processing proneuropeptides into peptide neurotransmitters and hormones," *Ann Revs Pharmacol Toxicol* 48, 393–423.

Mauri, A., Volpe, A., Martellotta, M. C., Barra, V., Piu, U., Angioni, G. *et al.* (1993). "Alpha-Melanocyte-stimulating hormone during human perinatal life," *J Clin Endocrinol Metab* 77, 113–117.

Millington, G. W. M. (2006). "Proopiomelanocortin (POMC): the cutaneous roles of its melanocortin products and receptors," *Clin Exp Dermatol* 31, 407–412.

Norman, A. W. and Litwack, G. (1997). *Hormones*, 2nd edn. (San Diego, CA: Academic Press).

Nussey, S. S. and Whitehead, S. A. (2001). *Endocrinology: An Integrated Approach* (Oxford: Bios Scientific Publishers), www.ncbi.nlm.nih.gov/books/NBK20.

Saland, L. C. (2001). "The mammalian pituitary intermediate lobe: An update on innervation and regulation," *Brain Res Bull* 54, 587–593.

Widmaier, E. P., Raff, H. and Strang, K. T. (2010). *Vander's Human Physiology: The Mechanisms of Body Function*, 12th edn. (New York: McGraw-Hill).

Winters, S. J. and Moore, J. P. (2007). "Paracrine control of gonadotrophs," *Semin Reprod Med* 25, 379–387.

The hypothalamic hormones

4

. . . .

Chapters 2 and 3 surveyed the hormones of the endocrine and pituitary glands. This chapter outlines the functions of the hypothalamus, and the hypothalamic neurosecretory cells and examines the role of the hypothalamus in controlling the release of pituitary hormones.

4.1 Functions of the hypothalamus

The hypothalamus is located at the base of the forebrain, below the thalamus (see Figure 3.1), and is divided into two halves, along the midline, by the third ventricle, which is filled with cerebrospinal fluid (CSF). As shown in a coronal (frontal) section in Figure 4.1, the hypothalamus contains many groups, or nuclei, of nerve cell bodies. The medial basal hypothalamus, consisting of the VMH, ARC and median eminence, is often referred to as the "endocrine hypothalamus" because of its neuroendocrine functions. For students interested in further details, a description of the anatomy of the hypothalamus can be found elsewhere (Norris 2007; Page 2006; Squire *et al.* 2008).

It is beyond the scope of this book to consider in detail the many and complex roles of the hypothalamus in maintaining normal bodily functions. But bear in mind that this brain center exerts an amazing diversity of critical controls, including growth, reproduction, temperature control, metabolism and body weight, emotional behavior (anger, fear, euphoria), motivational arousal (hunger, thirst, aggression and sexual arousal), circadian rhythms, stress and fluid balance. It contains multiple internal connections between neurons, but in addition receives neural information from other brain regions such as amygdala, hippocampus and spinal cord. The hypothalamus is well supplied with blood vessels and is therefore the recipient of essential information from the bloodstream, such as temperature and hormone levels. Thus, we can appreciate the importance of the hypothalamus in integrating and responding to all of this information by modifying its output of neural and neuroendocrine signaling.

These functions of the hypothalamus can be "localized" to particular nuclei, although any boundaries, such as those outlined in Figure 4.1, should be regarded only as approximate guides. These nuclei, and associated functions, are itemized in Table 4.1.

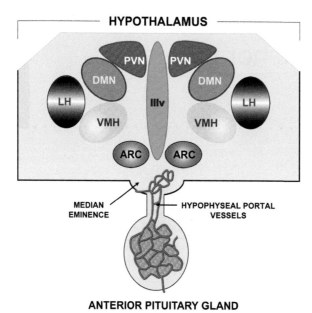

HYPOTHALAMUS

PVN   PVN

DMN   DMN

LH   IIIv   LH

VMH   VMH

ARC   ARC

MEDIAN
EMINENCE

HYPOPHYSEAL PORTAL
VESSELS

ANTERIOR PITUITARY GLAND

Figure 4.1 Neuroanatomical location of hypothalamic nuclei
Illustrates a coronal (frontal) view of the hypothalamus that emphasizes the paired nature of major nuclei either side of the third ventricle (IIIv). Compare this with Figure 3.3, where connections to the pituitary are shown in more detail.

Abbreviations: ARC, arcuate nucleus; DMN, dorsomedial hypothalamic nucleus; LH, lateral hypothalamus; PVN, paraventricular nucleus; VMH, ventro-medial hypothalamic nucleus.

4.2 Hypothalamic neurosecretory cells

Neurosecretory cells are neurons which, rather than secreting a neurotransmitter into a synapse, release a peptide hormone into the blood for neuroendocrine communication (see Figure 1.3). In the case of the releasing hormones, this is the hypophyseal portal circulation, or in the case of vasopressin/oxytocin neurons, this is directly into the peripheral circulation. There are two groups of hypothalamic neurosecretory cells: the *magnocellular* and *parvicellular* systems.

The magnocellular system

The large magnocellular neurosecretory cells are located in the paraventricular (PVN) and supraoptic nuclei (SON). The paraventricular nucleus consists of two types of neuron: one produces the hormone oxytocin and the other vasopressin (antidiuretic hormone, or ADH). Similarly, the supraoptic nucleus produces both oxytocin and vasopressin in separate neurons. These hormones are released from the nerve terminals of the axons of the magnocellular neurosecretory cells located in the posterior pituitary as is described in Chapter 3 (Figure 3.1). Details of the organization of the magnocellular neurosecretory system are given by Swanson and Sawchenko (1983).

The parvicellular system

The small parvicellular neurosecretory cells are found in the PVN (CRH cells), preoptic area (GnRH cells) and the arcuate nucleus (GnRH, GHRH cells), and project to the median eminence (Figure 3.1). The parvicellular neurosecretory cells terminating at the median eminence release their hypothalamic hormones into the hypophyseal portal system. These

Table 4.1 Major functions of hypothalamic nuclei

Preoptic area (POA) and anterior hypothalamus (AH)
Synthesis, and secretion, of GnRH (also called LHRH) in rodents
Temperature regulation; e.g. via vasodilation in response to heat; also responds to immune challenge by increasing body temperature (fever)
Regulates male sexual behavior and female parental behavior

Suprachiasmatic nucleus (SCN)
Biological clock, regulates rhythmic release of glucocorticoids, melatonin and other hormones
Regulates sleep-wake and other body rhythms, and seasonal reproduction

Periventricular nucleus (PV)
Synthesis of somatostatin (SOM)

Supraoptic nucleus (SON)
Synthesis of vasopressin (ADH), oxytocin and their carrier proteins (neurophysins) from magnocellular neurons
Regulation of thirst and fluid balance (ADH); milk let-down during suckling and regulates childbirth (oxytocin)

Paraventricular nucleus (PVN)
Synthesis of oxytocin and vasopressin from magnocellular neurons
Synthesis of CRH in the parvicellular division
Synthesis of TRH
Sympathetic control of heart and blood pressure

Lateral hypothalamus (LH)
Source of neuropeptide orexin (also called hypocretin), involved in regulation of sleep/wakefulness and in eating/energy balance

Dorsomedial hypothalamic nucleus (DMN)
Regulation of circadian rhythms
Control of feeding, drinking
Cardiovascular response to stress
Control of aggression

Ventromedial hypothalamic nucleus (VMH)
Controls digestive system, functions via cholecystokinin
Detects blood glucose levels (glucoreceptors) and regulates food intake and energy balance
Regulates female sexual behavior
Regulation of cardiovascular function

Posterior hypothalamus (PH)
Temperature regulation
Regulates sympathetic nervous system and visceral functions
Regulates fear response
Influences sleep and arousal
Involvement in "cluster" headaches

Arcuate nucleus (ARC) and median eminence
Synthesis of GnRH (in humans and primates) and GHRH
GnRH nerve terminals release GnRH into the portal system
Dopamine, released into portal veins from the tuberoinfundibular dopaminergic neurons, inhibits prolactin secretion from pituitary
POMC neurons synthesize β-endorphin and α-MSH
POMC neurons regulate NPY neurons in ARC to control body weight
Reproductive system also regulated by kisspeptins, encoded by the *kiss1* gene

Table 4.1 **(Cont.)**

Other functions of the hypothalamus
Controls heart rate and blood pressure
Controls respiration
Controls "emotions" – anger, fear, euphoria
Regulates calcium balance
Influences the immune system through the thymus gland
Influences the release of pancreatic and other gut hormones

hormones modulate the release of anterior pituitary hormones and are referred to as the *hypophysiotropic hormones.*

The history of this field, culminating in the award of a Nobel prize, followed considerable controversy about the concept of neurosecretion. G. W. Harris postulated that the hypothalamus controlled the release of anterior pituitary hormones by the release of neurohormones (hypothalamic releasing or inhibiting hormones) into the hypophyseal portal veins. A historical review of this research is given by Raisman (1997). After Harris's death, Guillemin and Schally were awarded the 1977 Nobel Prize in Physiology and Medicine for isolating several of the hypothalamic-releasing hormones and proving that Harris's theory was correct. Harris (1972) set three criteria for the definition of hypothalamic releasing/inhibiting hormones: (1) the hormone is present in the median eminence of the hypothalamus; (2) it is present in higher levels in the hypophyseal portal blood than in the rest of the circulatory system; and (3) the level of the hormone in the hypophyseal portal blood is correlated with the secretory rate of particular anterior pituitary hormones.

In a typical neurosecretory cell (Figure 4.2), neurohormones/neuropeptides are synthesized as peptide precursors (*pre-propeptides/propeptides*) and packaged into neurosecretory vesicles in the cell body (see also Figure 7.1). These secretory vesicles are then transported down the nerve axon and stored at the neuron terminal. During this transport, the propeptide is enzymatically cleaved and converted to a smaller neuropeptide (e.g. a releasing hormone such as GnRH), which is then ready for release. When the cell fires an action potential the nerve terminal is depolarized (see Chapter 5) and this releases the neurohormone into the circulation. The release of hypothalamic releasing/inhibiting hormones is stimulated by neurotransmitters released from other neurons (Chapter 6). Secretion of releasing hormones is also regulated by feedback from hormones, neuropeptides and other chemical messengers (see Chapters 8 and 11).

The concept of the neuroendocrine transducer

Because hypothalamic neurosecretory cells are stimulated by neurotransmitters (e.g. dopamine, serotonin and norepinephrine) to release their hormones (e.g. GnRH), they are essentially converting neural information into hormonal output. This has been termed "neuroendocrine transduction" (Wurtman 1980). As originally postulated, this

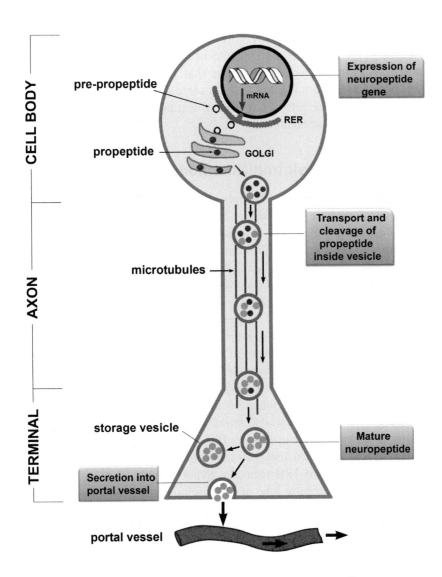

Figure 4.2 Synthesis and processing of a peptide in a neurosecretory cell
Pre-propeptides leave the rough endoplasmic reticulum (RER), are converted to propeptides, and are then consigned to secretory vesicles via the Golgi apparatus. During vesicular transport down the axon, the propeptides continue to be processed into smaller active peptides by convertase enzymes. The mature peptide (e.g. GnRH) is stored in vesicles in the nerve terminal until it is secreted into the portal system.
Abbreviations: mRNA, messenger ribonucleic acid; RER, rough endoplasmic reticulum.

term was applied to the pineal gland which converts information from the light-dark cycle into the secretion of the hormone melatonin (see section 8.2; Figure 8.5). The concept can thus be applied to hypothalamic neurosecretory neurons that are influenced by a variety of external stimuli such as light, taste, sound, fear, anger, hunger, depression and sexual arousal. These stimuli all influence neurotransmission that in turn modulates neurosecretory activity (see, e.g., Figure 1.1). Four examples of neuroendocrine transducers are: (1) the magnocellular neurons of the PVN and SON that release oxytocin/vasopressin; (2) the parvicellular neurons that secrete releasing/inhibiting hormones into the hypophyseal portal system; (3) the adrenal medulla that responds to sympathetic stimulation with the release of catecholamines; and (4) the pineal gland that receives adrenergic input to release melatonin.

4.3 Hypothalamic hypophysiotropic hormones

There are six well-established hypothalamic *hypophysiotropic hormones* that regulate the release of the anterior pituitary hormones (Table 4.2). These are also called hypothalamic releasing and inhibiting hormones. A concise description of the structures and functions of these hormones can be found elsewhere (Squire *et al.* 2008).

Hypothalamic hormones are sometimes called *factors* rather than hormones. There is no hard and fast rule for this, and this book will employ the term "hormone."

As shown in Table 4.2, there are five identified hypothalamic peptide hormones (TRH, CRH, GnRH, GHRH and somatostatin). The major control of prolactin secretion is not a peptide, but the neurotransmitter dopamine which inhibits prolactin secretion. A prolactin-releasing hormone may be present in the hypothalamus, although it does not seem to act in the same way as the other releasing hormones; that is, there is no evidence that it is released into the pituitary portal system (Onaka *et al.* 2010). Nevertheless, it is able to stimulate PRL release from the pituitary and may function as a versatile releasing hormone since it also stimulates LH, ACTH and vasopressin. Under some circumstances in the human female, TRH will also induce PRL secretion. For example, in hypothyroidism, when TSH secretion is low, PRL secretion is increased because of the enhanced release of TRH from the hypothalamus. The regulation of α-MSH secretion from the intermediate lobe of the pituitary gland may also involve a neurotransmitter, rather than a releasing hormone stimulation, mediated secretion (see section 4.3.6, below). Hypothalamic releasing/inhibiting hormones are secreted from neuroendocrine cells in a number of different hypothalamic nuclei (Table 4.3).

Table 4.2 Hypothalamic hypophysiotropic hormones

Releasing hormones

Thyrotropin-releasing hormone (TRH) = TSH-RH

Corticotropin-releasing hormone (CRH)

Gonadotropin-releasing hormone (GnRH) = LHRH. Thought to release LH and FSH and therefore GnRH is a more accurate abbreviation.

Paired releasing and inhibiting hormones

Growth hormone releasing hormone (GHRH)

Growth hormone release inhibiting hormone (GH-RIH) = Somatostatin

Prolactin releasing factor (PRF)

Prolactin releasing inhibiting factor (PIF) (probably dopamine)

Melanocyte-stimulating hormone–releasing factor (MSH-RF) (may be CRH and serotonin)

Melanocyte-stimulating hormone–release-inhibiting factor (MSH-RIF) (probably dopamine and GABA)

Table 4.3 Location of the hypothalamic neurosecretory cells which synthesize the hypothalamic hormones

The magnocellular neurosecretory cells that synthesize the hormones released from the posterior pituitary

Oxytocin	Paraventricular (PVN) and Supraoptic (SON) nuclei
Vasopressin (ADH)	Paraventricular (PVN) and Supraoptic (SON) nuclei

The parvicellular neurosecretory cells that synthesize the hypothalamic hypophyseal hormones

TRH (TSH-RH)	Primarily in the PVN
CRH	Primarily in the PVN
GnRH (LH-RH)	POA-AH in rodents
	ARC (in humans and monkeys)
FSH-RH (?)	Most probably GnRH in ARC or POA
GHRH	Primarily in the ARC
Somatostatin (SOM)	Periventricular region of AH
PRH	Could be TRH or PrRP in the PVN/DMN
PIF (dopamine)	ARC (tuberoinfundibular dopaminergic neurons)
MSH-RH	Could be CRH or serotonin in PVN or ARC?
MSH-RIH (dopamine or GABA)	ARC (tuberofundibular dopaminergic neurons)

Abbreviations are as in Tables 4.1 and 4.2. FSH-RH, follicle-stimulating hormone-releasing hormone; PrRP, prolactin-releasing peptide.

The secretion of a single hypothalamic hormone may be regulated through a number of different neural pathways (see Chapter 6). A concise outline of the hypothalamic hormones and their sites of secretion can be found elsewhere (Squire *et al.* 2008; Boron and Boulpaep 2005).

4.3.1 Thyrotropin releasing hormone (TRH)

TRH is also known as TRF (releasing factor) or thyroid-stimulating hormone-releasing hormone (TSH-RH). TRH is released from TRH neuron nerve terminals in the median eminence into the pituitary portal system from which it stimulates the thyrotroph cells of the anterior pituitary to produce and release TSH. TRH also acts as a neuromodulator in the brain (see section 12.7.3). TRH is synthesized primarily in the paraventricular nucleus (PVN). Like most hypothalamic releasing/inhibiting hormones, TRH secretion from hypothalamic neurons is regulated by inhibitory neurotransmitters such as dopamine, and by opiates such as morphine, as well as by cold and food intake (see Figure 6.6). Another important regulator of TRH release is CART (*cocaine- and amphetamine-regulated transcript*), a brain peptide implicated in food intake (Fekete and Lechan 2006). In addition, there are environmental factors which stimulate the release of TRH, including acute cold exposure and when babies suckle at their mother's breast (see Figure 8.3; Sanchez *et al.* 2001). Sensitive molecular biology techniques have also localized TRH gene expression in cells in the suprachiasmatic nuclei (SCN; Fliers *et al.* 1998), suggesting that TRH may be important in the control of circadian rhythms, distinct from its involvement in TSH secretion from the pituitary (see Chapter 12).

4.3.2 Corticotropin-releasing hormone (CRH)

CRH stimulates the release of ACTH and β-endorphin from the corticotroph cells of the anterior pituitary. CRH is synthesized primarily in PVN neurons and is released from nerve terminals in the median eminence into the portal circulation in response to pain and stress. CRH secretion from nerve terminals is regulated by the negative feedback of glucocorticoids as well as a number of neurotransmitters and neuropeptides. β-endorphin stimulates CRH release, whereas GABA is inhibitory (Yamauchi *et al.* 1997; Kovacs *et al.* 2004). CRH secretion also shows a circadian rhythm, and in humans there is an increase in CRH/ACTH/cortisol release in the morning hours (see Figure 6.5). The CRH peptide is also synthesized and released elsewhere in the brain, where it acts as a neuromodulator in addition to its role in the regulation of ACTH secretion (see Chapter 12). Thus, CRH gene expression is detectable in a wide variety of brain sites, including the cerebral cortex, amygdala and lateral hypothalamus, in addition to the PVN (Alon *et al.* 2009).

4.3.3 Gonadotropin-releasing hormone (GnRH)

It is generally accepted that release of both gonadotropins, FSH and LH, from the gonadotroph cells of the anterior pituitary is regulated by a single hypothalamic-releasing hormone, GnRH. In contrast, there is some evidence that there may be two different gonadotropin-releasing hormones: LHRH and FSH-RH. It is beyond the scope of this book to fully discuss this issue, given that a single releasing hormone, GnRH, can account for all the release of both LH and FSH. Nonetheless, for those interested in this topic, the views of the late S. M. McCann may be compelling (McCann *et al.* 2001).

Location of GnRH neurons: regulation of GnRH release

Rodent and monkey brains have been carefully mapped to identify the neuroanatomical location of GnRH-positive neuron cell bodies (Spergel *et al.* 2001; Standish *et al.* 1987). Unlike the neurons responsible for the synthesis of vasopressin and oxytocin, which are tightly grouped together in discrete nuclei (i.e. SON and PVN), GnRH neurons are relatively few in number (approx. 600 to 800 per brain) and are scattered through several brain areas. Herbison and colleagues (2008) have estimated that as few as 80 GnRH neurons are required for normal fertility in males, and only 200 for reproductive function in females, indicating a high level of redundancy within the reproductive system. In the mouse, GnRH cell bodies occur in a continuum from the olfactory bulbs at the front of the brain to the preoptic area (POA) of the hypothalamus. It is the POA neurons that send axons to the median eminence where GnRH is released into the portal system (Spergel *et al.* 2001). In monkey and human brains, there are some GnRH cell bodies in the POA, but the primary neurons that secrete GnRH into the portal system are found in the arcuate nucleus (Standish *et al.* 1987).

GnRH release is regulated by many different neurotransmitters and neuropeptides, many of which interact with each other and with feedback from gonadal steroid hormones such as estradiol (female) or testosterone (male), to modulate GnRH release (Figure 4.3). This multiple control system enables a wide variety of external and internal stimuli to influence GnRH release. For example, GnRH secretion is modulated by neurons projecting from the suprachiasmatic nucleus, which regulates circadian rhythms; by neurons from the PVN, which process visceral afferent input and may activate stress-induced changes in GnRH

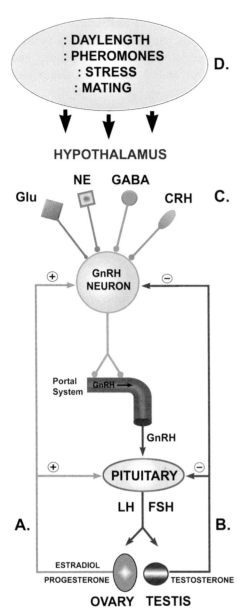

Figure 4.3 Control of GnRH secretion
Schematic diagram highlighting the role of GnRH neurons in the control of mammalian reproduction. Pink arrows (A) indicate the positive feedback effects of serum estradiol and progesterone on GnRH, LH and FSH release in the proestrous LH surge in female rodents (see also Figure 4.6). Note that this type of steroid feedback is normally negative (e.g. see Figure 4.10). Blue arrows (B) illustrate the negative feedback effects of testosterone on GnRH, LH and FSH release. Note that GnRH neurons are subjected to many neurochemical inputs (C), both stimulatory and inhibitory. A few examples are shown here. GnRH neurons are also sensitive to light, stress, mating, etc. (D).

Abbreviations: CRH, corticotropin-releasing hormone; FSH, follicle-stimulating hormone; GABA, γ-aminobutyric acid; Glu, glutamate; GnRH, gonadotropin-releasing hormone; LH, luteinizing hormone; NE, norepinephrine.

release; and by neural input from the olfactory and vomeronasal pathways, through which pheromones can influence GnRH release. Another neuropeptide, *kisspeptin*, produced by the *kiss1* gene, is an important regulator of GnRH secretion (see Chapter 6).

Extra-hypothalamic GnRH release

The GnRH peptide, and its receptor, has been localized to many brain regions in addition to those implicated in the hypothalamic-pituitary system (Skinner *et al.* 2009;

Albertson *et al.* 2008). As occurs in the CRH system (section 4.3.2), GnRH neurons release GnRH in other brain areas, including the hippocampus, amygdala, olfactory system, cerebellum and spinal cord, where it has neuropeptide/neuromodulator actions and may play a role in sexual behavior, motor control and Alzheimer's Disease (see Chapter 12).

It is possible that release of GnRH in non-hypothalamic regions occurs simultaneously with the release of GnRH into the portal system to facilitate female sexual receptivity (Pfaff *et al.* 1987; Sakuma 2002). Also, the high concentration of GnRH associated with the olfactory pathways could explain why pheromones have such potent "primer effects" on the neuroendocrine system. Further details on the functions of GnRH as a neuromodulator are discussed in Chapter 12. That GnRH may be a particularly versatile neuropeptide is also emphasized by reports of effects *outside* the brain (e.g. in heart, kidney, lymphocytes, adrenal and skin) (Skinner *et al.* 2009).

Tonic and pulsatile secretion of GnRH and generation of the ovulatory LH/FSH surge

As described above, GnRH is synthesized in neurons of the preoptic area in rodents, whereas in primates and humans the corresponding cell bodies are primarily located in the mediobasal hypothalamus (MBH) (Plant 2008). These neurosecretory cells extend their axons to the median eminence where they release GnRH into the pituitary portal system to regulate the "tonic," or basal, secretion of LH and FSH from the anterior pituitary. However, GnRH neurons release GnRH in an episodic, or *pulsatile*, fashion that results in a corresponding pulsatile secretion of LH and FSH from the anterior pituitary. Figure 4.4 illustrates the pattern of basal secretion of LH in an adult human male with normal levels of testosterone. The neurosecretory neurons that release GnRH fire in discrete bursts every 1.5 to 2 hours to secrete GnRH and stimulate the anterior pituitary to release LH in pulses. This pattern of LH secretion is also seen in human females and in rodents, sheep and monkeys of both sexes, but the *absence* of LH pulses

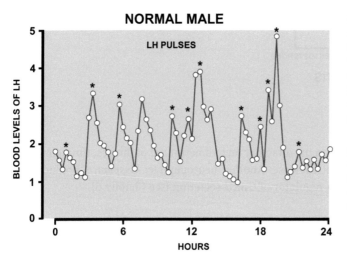

Figure 4.4 Pulsatile secretion of luteinizing hormone (LH) in a normal male
Blood levels of LH (mIU per ml) measured during a 24-hour period in a 36-year-old adult male with normal levels of testosterone.
Key: *Indicates significant pulses above the background. Pulses occur approximately every 2 to 3 hours. Reproduced with permission and redrawn (Besser and Thorner 2002).

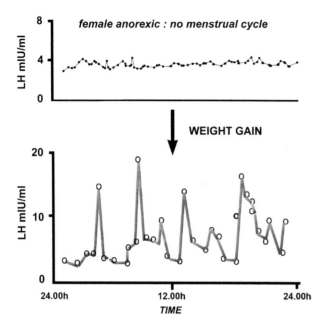

Figure 4.5 LH pulses in anorexia nervosa and following weight gain The *upper graph* shows a typical non-pulsatile pattern of LH secretion, over 24 hours, in a female patient with low body weight due to anorexia nervosa. Such a pattern of LH secretion causes amenorrhea (no menstrual cycles). The *lower graph* shows the dramatic effect of restoration of normal body weight; LH pulses are now greatly increased in frequency and magnitude. Graphs based on data from Boyar *et al.* (1974).

induces a state of infertility in both sexes. A good example of this is when young women starve themselves and become anorexic; their menstrual periods stop because the LH/FSH pulses disappear. Starvation inhibits GnRH release and regaining their correct body weight re-establishes pulsatile LH secretion and menstrual cyclicity in these women (Figure 4.5) (Boyar *et al.* 1974).

In order to induce an LH surge and ovulation during the female reproductive cycle, the GnRH neurons switch from their continuous pulsatile activity to generate the massive surge of GnRH that is necessary to induce female reproduction. This is achieved through the *positive feedback* effect of estradiol on GnRH neurons. GnRH neurons respond to estradiol by increasing the magnitude and frequency of GnRH pulses, causing the pituitary to release a surge of LH that induces ovulation (Figure 4.6). The ability of estradiol to generate a positive feedback on LH/FSH secretion represents an important sex difference in GnRH secretion, since neither estradiol nor testosterone has this effect in the male, where these steroid hormones have only a negative effect on LH/FSH secretion. The principles of hormonal feedback are discussed in more detail in sections 8.2.3 and 9.9.1.

4.3.4 Growth hormone releasing and inhibiting hormones

Growth hormone secretion from the somatotroph cells of the anterior pituitary is stimulated by growth hormone releasing hormone (GHRH; stimulatory) and inhibited by growth hormone release inhibiting hormone somatostatin (SOM; inhibitory). GHRH is released in pulses approximately every 1 to 3 hours from neurosecretory cells in the arcuate nucleus of the hypothalamus. GHRH cells are regulated by catecholaminergic and serotonergic neurotransmitters, as well as by the opioid neuropeptide enkephalin, VIP, TRH and NPY (Squire *et al.* 2008). GHRH release has a circadian rhythm with

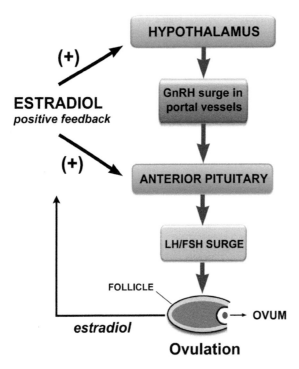

HYPOTHALAMUS

(+)

ESTRADIOL
positive feedback

**GnRH surge in
portal vessels**

(+)

ANTERIOR PITUITARY

LH/FSH SURGE

FOLLICLE

estradiol

→ OVUM

Ovulation

Figure 4.6 Positive feedback of estradiol on GnRH and luteinizing hormone (LH) and follicle-stimulating hormone (FSH) secretion
At the midpoint of the human menstrual cycle, high levels of estradiol (E) are released from the ovary. The normally negative feedback of E is transformed to a positive feedback that stimulates the release of large pulses of GnRH from the hypothalamus. This GnRH stimulus causes massive secretion of LH and, to a lesser extent, FSH, from the pituitary, which in turn causes release of the ovum from a preovulatory follicle (ovulation).
Abbreviations: FSH, follicle-stimulating hormone; GnRH, gonadotropin-releasing hormone; LH, luteinizing hormone.

increased release during sleep (Figure 4.7). Patients who suffer from *acromegaly* have high GH levels more or less throughout the day (Melmed 2009). In humans, GH secretion is also increased by sleep, stress, exercise, starvation and gonadal steroids such as estradiol and testosterone (Nussey and Whitehead 2001). These stimuli could presumably affect both GHRH (+ve) and SOM (–ve).

Somatostatin is synthesized in neurosecretory cells of the periventricular nuclei that, like GHRH cells, send axons to the median eminence. Somatostatin release is regulated by a number of neurotransmitters and neuropeptides, including dopamine, norepinephrine, GABA, neurotensin, ghrelin and substance P. The pattern of GH pulses from the anterior pituitary results from the interaction between the stimulatory effects of GHRH and the inhibitory effects of SOM. These effects are exerted at the level of the somatotroph cells via G-protein-coupled receptors and cellular signaling mechanisms such as cAMP (Figure 4.8) (for more details, see Chapter 10). The peptide *ghrelin* also releases GH from the pituitary (Castaneda *et al.* 2010). Ghrelin is produced in the stomach, enters the circulation, and is a powerful stimulus for GH release and for food intake (see Chapter 12 and Figure 4.8). It is possible that this peptide could be used therapeutically for the treatment of malnutrition and wasting (DeBoer 2011). Somatostatin gene expression is also found widely in the brain, suggesting that SOM is a neuropeptide that influences other neural pathways distinct from its role in GH secretion (see Chapter 12). Thus, SOM-expressing neurons are localized to, for example, cortex, amygdala, hippocampus, basal ganglia, midbrain, suprachiasmatic nucleus and retina (Mengod *et al.* 1992; Johnson *et al.* 2000).

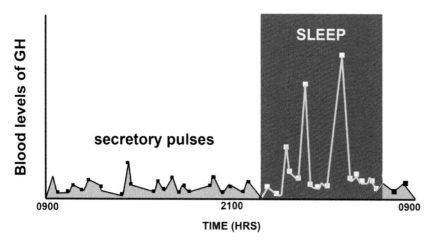

Figure 4.7 Sleep-related growth hormone secretion
GH levels in blood are shown. GH is secreted in pulses from the pituitary gland and reflects the pulsatile secretion of GHRH from hypothalamic neurons. GHRH/GH secretion is also under circadian control, with greatly increased release during sleep.

4.3.5 Prolactin releasing and inhibiting factors

Prolactin secretion from the lactotroph cells of the anterior pituitary is controlled primarily by dopamine, which acts as an inhibitory factor. The discovery of a prolactin-releasing hormone (PRH) has proved to be elusive, although a *prolactin-releasing peptide* (PrRP) has attracted attention because of its apparent role in energy homeostasis and stress responses (Onaka *et al.* 2010). PrRP does release PRL when it is injected directly into the brain, but nerve fibers that contain PrRP are not detectable in the median eminence where other releasing hormones are secreted into the portal vessels. On the other hand, PrRP-positive nerve fibers are found throughout the hypothalamus, including the PVN and DMN and in the anterior pituitary. Stimuli for PRL secretion include TRH, opioids, estradiol, lactation/suckling, oxytocin, VIP, sleep and stress.

Sex differences in the control of prolactin secretion

In rats, the hypothalamic control of prolactin secretion, like that of LH, differs in males and females (see section 4.3.3); that is, a surge of PRL occurs at proestrus in cycling female rats, whereas estradiol treatment is unable to reproduce this in males (Neill 1972). In humans, a circadian rhythm in PRL secretion is present in men and women, but the amplitude of PRL secretion is significantly higher in women (Waldstreicher *et al.* 1996). The sex difference in prolactin secretion appears to be the result of sex differences in the organization of the hypothalamic medial preoptic area due to masculinization of the brain by androgens during prenatal development (see Chapter 9) (McCarthy *et al.* 2009). Both men and women secrete basal levels of PRL in pulses, largely during sleep, with a frequency of about one per 90 minutes (Strauss and Barbieri 2004). However, sex differences in PRL secretion also occur

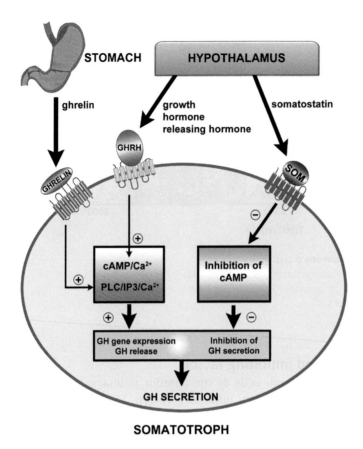

Figure 4.8 Regulation of growth hormone secretion

GH secretion and synthesis are induced by growth hormone releasing hormone (GHRH) and ghrelin (from the stomach) acting at G-protein-coupled receptors on pituitary somatotrophs. GHRH signals through the cAMP system to increase intracellular Ca^{2+} and GH secretion. GHRH also increases GH gene expression to produce more GH. Ghrelin also increases GH release, but acts via phospholipase C and IP3 to increase Ca^{2+} availability. Secretion of GH is moderated by the inhibitory effects of somatostatin (SOM), which acts at specific G-protein-coupled receptors to inhibit cAMP production.

Abbreviations: cAMP, cyclic adenosine monophosphate; GH, growth hormone; IP3, inositol triphosphate; PLC, phospholipase C.

because of sex differences in physiology. For example, women secrete PRL in pregnancy and during suckling of babies.

4.3.6 Melanocyte-stimulating hormone – releasing and inhibiting hormones

As noted in Chapter 3, α-MSH and β-endorphin are derived from the POMC propeptide (Figure 3.4) and are released from melanotroph cells of the intermediate lobe of the pituitary. It has been assumed that their secretion is controlled by melanocyte-stimulating hormone – releasing hormone (MSH-RH) and by melanocyte-stimulating hormone – release-inhibiting hormone (MSH-RIH). However, neither MSH-RH nor MSH-RIH has been identified, and recent evidence suggests that nerve fibers releasing dopamine and γ-aminobutyric acid (GABA) exert inhibitory control of α-MSH secretion, whereas CRH, serotonin and epinephrine may stimulate α-MSH release (Saland 2001). As discussed in Chapter 3, the intermediate lobe of the pituitary is not thought to be important in adult humans, although circulating levels of α-MSH have been detected in trauma patients (Todd *et al.* 2009).

4.4 Complexities of hypothalamic-pituitary interactions

The relationship between the endocrine hypothalamus and the pituitary gland involves many complexities, five of which are mentioned here.

1. *Hypothalamic hormones are multifunctional*; that is, they do not always have a one-to-one relationship with pituitary hormones. For example, while some influence only one pituitary hormone (e.g. GHRH releases GH), TRH can stimulate both prolactin and TSH release. Somatostatin primarily inhibits the release of GH but can also inhibit secretion of TSH and prolactin (Figure 4.9). Dopamine inhibits PRL, TSH and α-MSH secretion, and GnRH stimulates the secretion of both LH and FSH.

2. *Pituitary hormones may be transported back to the hypothalamus to act as neuromodulators of neuronal activity*. Figure 3.3 illustrates the hypophyseal stalk as carrying hormones in one direction, from the hypothalamus to the pituitary. However, there are three mechanisms by which pituitary hormones may be transported retrogradely back to the hypothalamus. Neurohypophyseal hormones, such as oxytocin, may be transported from the posterior pituitary to the hypothalamus by retrograde axonal transport (i.e. the reverse of that shown in Figure 5.7). Adenohypophyseal hormones, such as GH, may be carried to the hypothalamus by efferent portal vessels; that is, via blood flow from the pituitary towards the hypothalamus (Bergland and Page 1979;

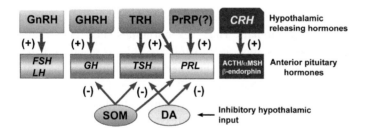

Figure 4.9 Summary of the hypothalamic releasing and inhibiting hormones that modulate the secretion of anterior pituitary hormones

Abbreviations: ACTH, adrenocorticotropic hormone; CRH, corticotropin-releasing hormone; DA, dopamine; FSH, follicle-stimulating hormone; GH, growth hormone; GHRH, growth hormone releasing hormone; GnRH, gonadotropin-releasing hormone; LH, luteinizing hormone; α-MSH, α-melanocyte-stimulating hormone; PRL, prolactin; PrRP, prolactin-releasing peptide; SOM, somatostatin; TRH, thyroid-stimulating hormone-releasing hormone; TSH, thyroid-stimulating hormone.

Page 2006). Pituitary hormones transported in this way may also be released into the CSF and influence hypothalamic nuclei situated adjacent to the third ventricle. β-endorphin and ACTH, for example, are detectable in CSF, while other adenohypophyseal hormones enter the CSF from peripheral blood through capillaries of the choroid plexus in the ventricles (Lenhard and Deftos 1982). The functional nature of CSF hormones, if any, remains to be determined.

3. *Not all hypothalamic hormones are secreted into the portal system.* Several releasing/inhibiting hormones are also present in regions other than the hypothalamus. They are released there either from hypothalamic neurons that project to these regions, or they are released from neurons that are present in these brain areas. When secreted to, or in, other brain regions, these hypothalamic hormones act as neuropeptides to modulate neural excitability, and thus influence neurotransmitter release and behavior (see section 12.7.3). For example, GnRH is involved in the control of female sexual behavior (Sakuma 2002).

4. *Release of pituitary hormones may be regulated by a number of neuropeptides and neurotransmitters, as well as by hypothalamic hypophysiotropic hormones.* Data from older papers suggested that various neurotransmitters and neuropeptides can have direct effects on the release of pituitary hormones from the pituitary gland (Ben-Jonathan *et al.* 1989; Weiner *et al.* 1988; Bennett and Whitehead 1983). Other evidence includes dopamine regulation of PRL secretion (Figure 4.9) and the stimulatory effect of vasopressin on ACTH secretion described in Chapter 6 (Figure 6.4). Although several neurotransmitters/neuropeptides have been implicated in regulating pituitary hormone secretion (e.g. oxytocin, VIP, enkephalins, substance P), probably the best-described example is GABA. There is evidence that GABA is present in hypophyseal portal blood (Mitchell *et al.* 1983) and GABA receptors are found in the pituitary gland (Bianchi *et al.* 2004). In addition, pituitary GABA receptors play a role in regulating GH secretion (Gamel-Didelon *et al.* 2002; End *et al.* 2005) and GABA influences TSH secretion (Tapia-Arancibia *et al.* 1987). Evidence that neurotransmitters/neuropeptides are *co-localized* with pituitary hormones in the pituitary gland is described in section 11.4.4.

5. *Hypothalamic hormones interact with other hormones.* The primary hormones that regulate secretion of hypophyseal hormones are those hormones released from target glands such as gonads and adrenals. Thus, sex hormones such as estradiol and testosterone, and adrenal steroids such as cortisol, are able to reach both the hypothalamus and pituitary gland from the peripheral circulation. This control system is called hormone feedback and will be covered in more detail in Chapter 8. However, a simple illustration of negative feedback is shown in Figure 4.10. The ability of the hypothalamus to release CRH into the hypophyseal portal system, and the subsequent secretion of ACTH, is regulated by the negative feedback of cortisol released from the adrenal cortex. Note that the negative feedback is also exerted at the level of the pituitary. Another example of negative feedback is the release of inhibin from the gonads that regulates FSH secretion from the pituitary (section 2.2.11).

NEGATIVE FEEDBACK

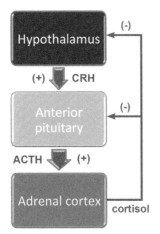

Figure 4.10 Example of a negative feedback signaling system
The stress response is mediated by the hypothalamus generating a surge of CRH release into the pituitary portal system. This results in increased ACTH output that drives adrenal cortical cells to release cortisol. High levels of cortisol exert negative feedback in order to curtail excessive release of CRH and ACTH. Similar feedback systems exist for other steroids such as estradiol and testosterone (see Chapter 8 for more details).
   Abbreviations as for Figure 4.9.

4.5 Summary

This chapter describes the functions of the hypothalamus with particular reference to the hypothalamic control of pituitary hormones. The hypothalamus contains nuclei that control the autonomic nervous system, temperature regulation, biological rhythms, emotional responses, motivational arousal and hormone secretion. Hypothalamic hormones are synthesized in neurosecretory (neuroendocrine) cells that convert neural input to hormonal output. The magnocellular neurosecretory cells of the SON and PVN produce the neurohypophyseal hormones oxytocin and vasopressin, which are released from the posterior pituitary. The parvicellular neurosecretory cells of the hypothalamus release hypophysiotropic hormones (releasing/inhibiting hormones) into the hypophyseal portal vessels in the median eminence. Two hypothalamic hormones, CRH and GnRH, act individually to stimulate the release of pituitary hormones. Others act in tandem to stimulate or inhibit pituitary hormones: GHRH (+ve) and SOM (–ve) for GH secretion; TRH and dopamine for PRL secretion; and CRH/dopamine for α-MSH secretion. There is some controversy over whether there is a FSH-RH separate from GnRH and there are a number of complications in the hypothalamic control of pituitary hormones, including sex differences, multiple effects of hypothalamic hormones on the pituitary, retrograde transport of pituitary hormones to the hypothalamus, the stimulation of pituitary hormones by neuropeptides, neurotransmitters/neuropeptides acting directly on the endocrine cells of the pituitary and feedback effects of hormones such as estradiol, testosterone and cortisol.

REVIEW QUESTIONS

4.1   Name the three areas of the medial basal hypothalamus that are referred to as "the endocrine hypothalamus."

4.2   What are the two types of neurosecretory cells in the hypothalamus and what are their functions?

4.3   What is a neuroendocrine transducer?

4.4   Name the seven hypothalamic hypophysiotrophic hormones and describe how they influence the release of the adenohypophyseal hormones.

4.5   How do the hormones of the hypothalamus reach the anterior pituitary?

4.6   What is the difference between tonic and cyclic GnRH secretion?

4.7   What two neurotransmitters act as prolactin inhibiting factors?

4.8   Which two pituitary hormones do each of the following hypothalamic hormones affect, and are these effects stimulatory or inhibitory?
   (a) TRH
   (b) somatostatin
   (c) GnRH

ESSAY QUESTIONS

4.1 Discuss the functional anatomy of the nuclei of the hypothalamus, with particular emphasis on describing "the endocrine hypothalamus."

4.2 Discuss the neuroanatomy and neural connections of the magnocellular neurosecretory cells.

4.3 Outline the history of the discovery of the hypothalamic hormones from 1935 to the present.

4.4 Discuss the mechanism of the neuroendocrine transducer using one of the four different neuroendocrine transducers as an example.

4.5 Discuss the evidence of the existence of one versus two GnRHs. Which theory do you believe?

4.6 Discuss the sex differences in the secretion of prolactin and the gonadotropins.

4.7 Discuss the extra-hypothalamic sources of GnRH and their possible functions.

4.8 Discuss some of the problems in trying to understand the hypothalamic control of the anterior pituitary gland.

4.9 Imagine that you discovered a new hypothalamic hormone which was the mysterious prolactin releasing factor and you named it prolactinotropin. How would you prove that this hormone was what you said it was? Describe the experiments necessary to demonstrate that you had really discovered a new hypothalamic hypophysiotropic hormone.

REFERENCES

Albertson, A. J., Navratil, A., Mignot, M., Dufourny, L., Cherrington, B. and Skinner, D. C. (2008). "Immunoreactive GnRH type I receptors in the mouse and sheep brain," *J Chem Neuroanat* 35, 326–333.

Alon, T., Zhou, L., Perez, C. A., Garfield, A. S., Friedman, J. M. and Heisler, L. K. (2009). "Transgenic mice expressing green fluorescent protein under the control of the corticotropin-releasing hormone promoter," *Endocr* 150, 5626–5632.

Ben-Jonathan, N., Arbogast, L. A. and Hyde, J. F. (1989). "Neuroendocrine regulation of prolactin release," *Progress in Neurobiology*, 33, 399–447.

Bennett, G. W. and Whitehead, S. A. (1983). *Mammalian Neuroendocrinology* (New York: Oxford University Press).

Bergland, R. M. and Page, R. B. (1979). "Pituitary-brain vascular relations: A new paradigm," *Science* 204, 18–24.

Besser, G. M. and Thorner, M. O. (2002). *Comprehensive Clinical Endocrinology*, 3rd edn. (St. Louis, MO: Mosby).

Bianchi, M. S., Catalano, P. N., Bonaventura, M. M., Silveyra, P., Bettler, B., Libertun, C. *et al.* (2004). "Effect of androgens on sexual differentiation of pituitary γ-aminobutyric acid receptor subunit GABA_B expression," *Neuroendocrinology* 80, 129–142.

Boron, W. F. and Boulpaep, E. L. (2005). *Medical Physiology,* updated edition (Philadelphia, PA: Elsevier Saunders).

Boyar, R. M., Katz, J., Finkelstein, J. W., Kapen, S., Weiner, H., Weitzman, E. D. *et al.* (1974). "Anorexia nervosa. Immaturity of the 24-hour luteinizing hormone secretory pattern," *N Engl J Med* 291, 861–865.

Castaneda, T. R., Tong, J., Datta, R., Culler, M. and Tschop, M. H. (2010). "Ghrelin in the regulation of body weight and metabolism," *Front Neuroendocr* 31, 44–60.

DeBoer, M. D. (2011). "Ghrelin and cachexia: Will treatment with GHSR-1a agonists make a difference for patients suffering from chronic wasting syndromes?" *Mol Cell Endocrinol* 340, 97–105.

End, K., Gamel-Didelon, K., Jung, H., Tolnay, M., Ludecke, D., Gratzl, M. *et al.* (2005). "Receptors and sites of synthesis and storage of gamma-aminobutyric acid in human pituitary glands and in growth hormone adenomas," *Am J Clin Pathol* 124, 550–558.

Fekete, C. and Lechan, R. M. (2006). "Neuroendocrine implications for the association between cocaine- and amphetamine-regulated transcript (CART) and hypophysiotropic thyrotropin-releasing hormone (TRH)," *Peptides* 27, 2012–2018.

Fliers, E., Wiersinga, W. M. and Swaab, D. F. (1998). "Physiological and pathophysiological aspects of thyrotropin-releasing hormone gene expression in the human hypothalamus," *Thyroid* 8, 921–928.

Gamel-Didelon, K., Corsi, C., Pepeu, G., Jung, H., Gratzl, M. and Mayerhofer, A. (2002). "An autocrine role for pituitary GABA: Activation of GABA-B receptors and regulation of growth hormone levels," *Neuroendocrinology* 76, 170–177.

Harris, G. W. (1972). "Humours and hormones," *J Endocrinol* 53, 2–23.

Herbison, A. E., Porteous, R., Pape, J. R., Mora, J. M. and Hurst, P. R. (2008). "Gonadotropin-releasing hormone neuron requirements for puberty, ovulation, and fertility," *Endocr* 149, 597–604.

Johnson, J., Rickman, D. W. and Brecha, N. C. (2000). "Somatostatin and somatostatin subtype 2A expression in the mammalian retina," *Microsc Res Tech* 50, 103–111.

Kovacs, K. J., Miklos, I. H. and Bali, B. (2004). "GABAergic mechanisms constraining the activity of the hypothalamo-pituitary-adrenocortical axis," *Ann NY Acad Sci* 1018, 466–476.

Lenhard, L. and Deftos, L. J. (1982). "Adenohypophyseal hormones in the CSF," *Neuroendocrinology* 34, 303–308.

McCann, S. M., Karanth, S., Mastronardi, C. A., Dees, W. L., Childs, G., Miller, B. *et al.* (2001). "Control of gonadotropin secretion by follicle-stimulating hormone-releasing factor, luteinizing hormone-releasing hormone, and leptin," *Arch Med Res* 32, 476–485.

McCarthy, M. M., Wright, C. L. and Schwarz, J. M. (2009). "New tricks by an old dogma: Mechanisms of the organizational/activational hypothesis of steroid-mediated sexual differentiation of brain and behavior," *Horm Behav* 55, 655–665.

Melmed, S. (2009). "Acromegaly pathogenesis and treatment," *J Clin Invest* 119, 3189–3202.

Mengod, G., Rigo, M., Savasta, M., Probst, A. and Palacios, J. M. (1992). "Regional distribution of neuropeptide somatostatin gene expression in the human brain," *Synapse* 12, 62–74.

Mitchell, R., Grieve, G., Dow, R. and Fink, G. (1983). "Endogenous GABA receptor ligands in hypophysial portal blood," *Neuroendocrinology* 37, 169–176.

Neill, J. D. (1972). "Sexual differences in the hypothalamic regulation of prolactin secretion," *Endocr* 90, 1154–1159.

Norris, D. O. (2007). *Vertebrate Endocrinology*, 4th edn. (San Diego: Academic Press).

Nussey, S. S. and Whitehead, S. A. (2001). *Endocrinology: An Integrated Approach* (Oxford: Bios Scientific Publishers), www.ncbi.nlm.nih.gov/books/NBK20.

Onaka, T., Takayanagi, Y. and Leng, G. (2010). "Metabolic and stress-related roles of prolactin-releasing peptide," *Trends Endocrinol Metab* 21, 287–293.

Page, R. B. (2006). "Anatomy of the hypothalamic-hypophyseal complex" in J. D. Neill (ed.), *Knobil and Neill's Physiology of Reproduction* (San Diego: Academic Press), vol. 1, pp. 1309–1414.

Pfaff, D. W., Jorgenson, K. and Kow, L. M. (1987). "Luteinizing hormone-releasing hormone in rat brain: Gene expression, role as neuromodulator, and functional effects," *Ann NY Acad Sci* 519, 323–333.

Plant, T. M. (2008). "Hypothalamic control of the pituitary-gonadal axis in higher primates: Key advances over the last two decades," *J Neuroendocrinol* 20, 719–726.

Raisman, G. (1997). "An urge to explain the incomprehensible: Geoffrey Harris and the discovery of the neural control of the pituitary gland," *Annu Rev Neurosci* 20, 533–566.

Sakuma, Y. (2002). "GnRH in the regulation of female rat sexual behavior," *Prog Brain Res* 141, 293–301.

Saland, L. C. (2001). "The mammalian pituitary intermediate lobe: An update on innervation and regulation," *Brain Res Bull* 54, 587–593.

Sanchez, E., Uribe, R. M., Corkidi, G., Zoeller, R. T., Cisneros, M., Zacarias, M. *et al*. (2001). "Differential responses of thyrotropin-releasing hormone (TRH) neurons to cold exposure or suckling indicate functional heterogeneity of the TRH system in the paraventricular nucleus of the rat hypothalamus," *Neuroendocrinology* 74, 407–422.

Skinner, D. C., Albertson, A. J., Navratil, A., Smith, A., Mignot, M., Talbott, H. *et al*. (2009). "Effects of gonadotrophin-releasing hormone outside the hypothalamic-pituitary-reproductive axis," *J Neuroendocrinol* 21, 282–292.

Spergel, D. J., Kruth, U., Shimshek, D. R., Sprengel, R. and Seeburg, P. H. (2001). "Using reporter genes to label selected neuronal populations in transgenic mice for gene promoter, anatomical, and physiological studies," *Prog Neurobiol* 63, 673–686.

Squire, L. R., Berg, D. E., Bloom, F. E., du Lac, S., Ghosh, A. and Spitzer, N. C. (2008). *Fundamental Neuroscience*, 3rd edn. (London: Academic Press).

Standish, L. J., Adams, L. A., Vician, L., Clifton, D. K. and Steiner, R. A. (1987). "Neuroanatomical localization of cells containing gonadotropin-releasing hormone messenger ribonucleic acid in the primate brain by in situ hybridization histochemistry," *Mol Endocrinol* 1, 371–376.

Strauss, J. F. and Barbieri, R. L. (2004). *Yen and Jaffe's Reproductive Endocrinology*, 5th edn. (Philadelphia, PA: Elsevier Saunders).

Swanson, L. W. and Sawchenko, P. E. (1983). "Hypothalamic integration: Organization of the paraventricular and supraoptic nuclei," *Annu Rev Neurosci* 6, 269–324.

Tapia-Arancibia, L., Roussel, J. P. and Astier, H. (1987). "Evidence for a dual effect of γ-aminobutyric acid on thyrotropin (TSH)-releasing hormone-induced TSH release from perifused rat pituitaries," *Endocrinology* 121, 980–986.

Todd, S. R., Kao, L. S., Catania, A., Mercer, D. W., Adams, S. D. and Moore, F. A. (2009). "α-melanocyte stimulating hormone in critically injured trauma patients," *J Trauma* 66, 465–469.

Waldstreicher, J., Duffy, J. F., Brown, E. N., Rogacz, S., Allan, J. S. and Czeisler, C. A. (1996). "Gender differences in the temporal organization of PRL secretion: Evidence for a sleep-independent circadian rhythm of circulating PRL levels – a clinical research center study," *J Clin Endocr Metab* 81, 1483–1487.

Weiner, R. I., Findell, P. R. and Kordon, C. (1988). "Role of classic and peptide neuromediators in the neuroendocrine regulation of LH and prolactin" in E. Knobil, J. D. Neill *et al.* (eds.), *The Physiology of Reproduction* (New York: Raven Press), vol. 1, pp. 1235–1281.

Wurtman, R. J. (1980). "The pineal as a neuroendocrine transducer," *Hosp Pract* 15, 82–86.

Yamauchi, N., Shibasaki, T., Wakabayashi, I. and Demura, H. (1997). "Brain β-endorphin and other opioids are involved in restraint stress-induced stimulation of the hypothalamic-pituitary-adrenal axis, the sympathetic nervous system, and the adrenal medulla in the rat," *Brain Res* 777, 140–146.

5 Neurotransmitters

In general terms, neurons communicate with each other through chemical messengers called neurotransmitters. Given the complexity of the brain, it should not be surprising that there are more than 100 known neurotransmitters (Purves *et al.* 2008). Neurotransmitters are synthesized in nerve cells, sometimes using precursors from the diet (e.g. tyrosine; see Figure 5.8), and are released into the synapse where they bind to specific receptors located on the postsynaptic cell. As discussed in the present chapter, this simple view conceals the many fascinating ways in which neurons communicate with each other. This chapter focuses on the different categories of neurotransmitters, the synthesis, storage, transport and release of neurotransmitters, their action at receptors and their deactivation. The influence of drugs on neurotransmitter function will also be discussed. Chapter 6 examines the specific effects of neurotransmitters in the neuroendocrine system, and Chapter 10 covers the actions of neurotransmitters at their receptors on postsynaptic cells.

5.1 The neuron and the synapse

A typical neuron is shown in Figure 5.1. Neurons possess a cell body, which contains the nucleus, and the characteristic dendrites plus an axon. The dendrites receive messages from other cells onto their spines and shafts, while the axon transmits information to other cells. Although each nerve cell has only one axon, this axon may have a number of branches and the nerve terminals at the end of each branch can form synapses with other neurons.

Nerve cells communicate with each other by the release of neurotransmitters from the nerve terminals of the axon into the synapse, the space that separates the presynaptic and postsynaptic cells. Neurotransmitters released into the synapse then bind to their receptors on the postsynaptic cell. As shown in Figure 5.1, synapses can form between the axon of the presynaptic cell and a number of different sites on the postsynaptic cell, including the shafts and/or spines of dendrites (axodendritic synapses), the cell body (axosomatic synapses) and the axons (axoaxonic synapses). As discussed in section 5.2, synapses can be either excitatory or inhibitory. At an excitatory synapse, the neurotransmitter released from the presynaptic cell excites the postsynaptic cell, changing the electrical potential of the cell membrane, and causing it to become depolarized and generate an action potential

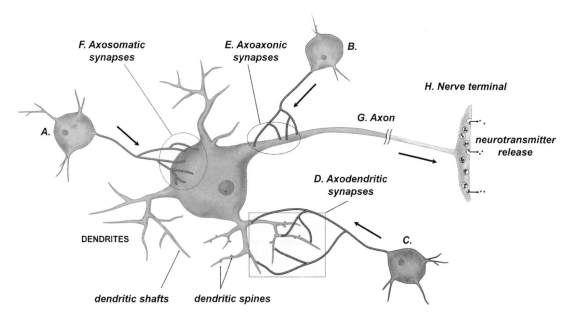

Figure 5.1 Shape of a typical neuron
Neurons have complex branches called dendrites that receive incoming information via synapses from other neurons (A, B and C). Synapses may be located on dendritic shafts or on dendritic spines (D – *axodendritic synapses*). Synapses are also found on axons (E – *axoaxonic synapses*) and on the cell body (soma; F – *axosomatic synapses*). Connections are made with other neurons via a single axon (G) and the nerve terminal (H) that releases a neurotransmitter into the synaptic space. Large black arrows indicate the direction of the action potentials (see also Figure 5.10). Copyright Alex Pincock.

down the axon. At inhibitory synapses, the neurotransmitter released from the presynaptic cell inhibits the postsynaptic cell from firing. Excitatory synapses are usually found on the dendrites, while inhibitory synapses occur on the cell body and axon terminals.

5.1.1 What is a neurotransmitter?

A neurotransmitter is a chemical messenger that conveys information from one neuron to another cell that may or may not be another neuron. For example, most neurons regulate other neurons, but some also control muscles and endocrine glands. Neurons can respond to their own neurotransmitters through autocrine signaling through autoreceptors, also called presynaptic receptors (see Figure 1.3). For a neurochemical to be considered as a neurotransmitter, it should meet certain criteria that distinguish "true" neurotransmitters from chemicals that also have neural activity. For example, smokers, and cocaine and heroin users, have high levels of nicotine and narcotics in their brains. Also, psychiatric patients depend on the ability of antidepressant or anti-anxiety medications to act in the brain. All of these substances interfere with normal neurotransmission in order to induce their pleasurable or therapeutic effects, but they are not neurotransmitters. In order to establish whether a chemical substance is a neurotransmitter, it has to meet a number of criteria. These criteria, listed in Table 5.1, apply to both *classical* neurotransmitters, such as acetylcholine, and to *neuropeptide* neurotransmitters, such as opioids or substance

> ### Table 5.1 Criteria for establishing that a neuroregulatory chemical is a neurotransmitter
>
> 1. The neurotransmitter is made in a neuron and therefore the precursors and synthetic enzymes must be present in that neuron. For example, as shown in Figure 5.8, the precursor tyrosine is enzymatically converted to dopamine. Thus, both tyrosine and tyrosine hydroxylase must be present in dopamine neurons.
>
> 2. The substance must be present, usually contained in vesicles, in the terminals of presynaptic neurons.
>
> 3. Neurotransmitters are unevenly distributed throughout the brain. They are found in specific areas and are associated with specific nerve pathways (see Figure 5.15). For example, brain regions associated with pain will often have opioid peptides as a neurotransmitter.
>
> 4. Stimulation of a neuron or neuron pathway should cause release of the substance from the nerve terminals of neurons in physiologically significant amounts.
>
> 5. Specific receptors that bind the substance should be localized in close proximity to the presynaptic neuron terminals. When the released neurotransmitter binds to these receptors, they should induce changes in postsynaptic neurons. For example, the substance should induce changes in postsynaptic membrane permeability leading to excitatory or inhibitory potentials in the postsynaptic cell.
>
> 6. Effects of direct application of the substance into the synapses in appropriate brain regions should reproduce effects identical to those produced by stimulating nerve pathways (as in #4).
>
> 7. Interventions at postsynaptic sites – for example, using specific receptor *antagonist* drugs – should block the effects of nerve stimulation (#4), whereas application of *agonist* drugs should mimic the actions of the neurotransmitter.
>
> 8. Specific mechanisms should exist which terminate the effects of the neurotransmitter. These mechanisms include enzymatic degradation or re-uptake into the presynaptic terminal (see Figure 5.14).

P (Squire *et al.* 2008). A more detailed description of the brain's neurotransmitter substances is provided by Nestler *et al.* (2009) and Iversen *et al.* (2009).

5.2 Categories of neurotransmitters

Based on the criteria presented in Table 5.1, a large number of substances can now be categorized as neurotransmitters. As summarized in Table 5.2, the major neurotransmitters belong to four general categories: amino acids, acetylcholine, monoamines and peptides (see Purves *et al.* 2008).

However, other substances acting in the brain defy such categorization; that is, they regulate the transfer of information between neurons, but they do not satisfy all of the criteria outlined in Table 5.1. These molecules include gases, such as nitric oxide, and endogenous cannabis-like substances called endocannabinoids. These unusual neurochemicals are described in section 5.2.5.

5.2.1 The amino acid neurotransmitters

The amino acid neurotransmitters are the most abundant neurotransmitters in the mammalian CNS and occur in neurons throughout the brain and spinal cord. For example, the excitatory neurotransmitter glutamate has been implicated in almost every physiological brain pathway and is involved in pathological processes as epilepsy, brain damage following stroke and addiction. Amino acid neurotransmitter receptors are actually ion channels

Table 5.2 Categories of neurotransmitters

Category	Neurotransmitter
A. Amino acid transmitters	
Excitatory	Aspartic acid
	Glutamic acid
Inhibitory	Gamma-aminobutyric acid (GABA), Glycine
B. Cholinergic neurotransmitter	Acetylcholine (ACh)
C. Monoamine neurotransmitters	
Catecholamines	
Dopaminergic	Dopamine (DA)
Adrenergic	Norepenephrine (NE; sometimes Noradrenaline)
	Epinephrine (or Adrenaline)
Indoleamine	Serotonin (5-HT)
Other	Histamine
D. Peptide neurotransmitters	Substance P
	Somatostatin
	Neurotensin
	Cholecystokinin (CCK)
	Enkephalins, β-endorphin (opioids)
E. Unusual neurotransmitters	Gases: nitric oxide and hydrogen sulphide
	Endogenous cannabis: Endocannabinoids, Zinc;
	D-serine
F. Putative neurotransmitters	Prostaglandins
	ATP (adenosine triphosphate), Adenosine

in the membrane of their postsynaptic target cells. Binding of the amino acids to these channels causes them to open, thus inducing rapid excitatory or inhibitory actions on these postsynaptic cells. For example, γ-aminobutyric acid (GABA) acts as an inhibitory neuro-transmitter by opening chloride ion channels via GABA$_A$ receptors (Figure 5.2). This has the effect of inhibiting electrophysiological activity in postsynaptic neurons. Behavioral effects of this inhibition are readily seen, and the effects of tranquilizers (e.g. Diazepam), anaesthetics (e.g. thiopental; propofol), barbiturates and alcohol are all mediated through GABA receptors (Belelli and Lambert 2005; Nestler *et al.* 2009). Another major amino acid inhibitory transmitter is glycine, which acts through a different ion channel that also regulates entry of chloride ions into the postsynaptic neuron. Its primary sites of action are in the spinal cord. Glycine is also implicated in excitatory neurotransmission (see Figure 5.3).

The excitatory amino acids glutamic acid (glutamate) and aspartic acid (aspartate) act as neurotransmitters in the brain and spinal cord (Nestler *et al.* 2009; Purves *et al.* 2008). Like GABA, they are ubiquitously localized in the brain and it has been estimated that approximately 50 percent of all brain synapses use glutamate as a neurotransmitter. In contrast to GABA, however, glutamate and aspartate receptors – some of which are also ion channels – induce excitation by opening postsynaptic channels for the entry of calcium

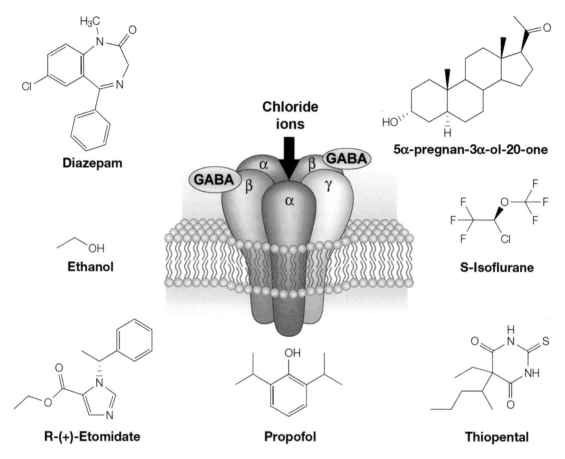

Figure 5.2 Structure of the inhibitory GABA_A receptor
This receptor is located in neuron cell membranes and is a chloride ion channel made up of five subunits (α, β and γ). Binding of GABA to the receptor opens the central channel to allow chloride ions to enter the neuron. The receptor complex also contains binding sites for drugs such as anaesthetics (thiopental, propofol, isoflurane), alcohol and anti-anxiety drugs such as diazepam. These drugs act to enhance the effects of GABA by facilitating chloride entry into the neuron. Neurosteroids such as 5α-pregnan-3α-ol-20-one can influence mood and behavior in various physiological and pathophysiological situations (see Chapter 9 for discussion of neurosteroids). Reproduced with permission (Belelli and Lambert 2005).

and sodium ions (Figure 5.3). Given that these excitatory neurotransmitters are found throughout the brain, it is not surprising that they have been implicated in a multitude of physiological processes. One of the most important, and most-studied, consequences of glutamate action is the induction of long-lasting changes in synaptic function (called synaptic plasticity) that forms the basis for neural changes underlying learning and memory.

5.2.2 Acetylcholine (Ach)

Acetylcholine is released from neurons of the *cholinergic* pathways. As well as acting widely in the brain and central nervous system, acetylcholine is the neurotransmitter used at the neuromuscular junction, in the parasympathetic branch of the autonomic nervous system (ANS) and in autonomic ganglia. Acetylcholine is important in attention and

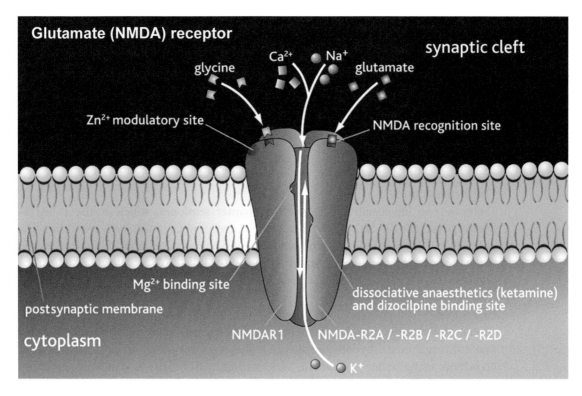

Figure 5.3 Structure of an excitatory glutamate receptor

Excitatory glutamate receptors are ion channels that admit calcium and sodium ions to induce depolarization of the target neuron. The example shown here is the NMDA (N-methyl-D-aspartate) subtype of glutamate receptor. A description of the complex nature of glutamate receptors is beyond the scope of this book and only the essential facts will be presented here. For example, the NMDA receptor consists of four subunits arranged around a central pore, usually NMDAR1 in combination with various NMDAR2 subunits (NMDA-R2A, -R2B, -R2C or -R2D). In addition to glutamate, the receptor complex is regulated by several other chemicals. Thus, glycine is a critical co-agonist that is necessary for glutamate to open the channel pore so that sodium and calcium ions can enter the neuron. Binding of zinc ions is also necessary (see also Figure 5.6), and magnesium ions act to block the channel until depolarization occurs. In addition, the receptor is the target of anaesthetic and hallucinatory drugs such as ketamine and dizocilpine. Copyright CNSforum, Lundbeck Institute: http://www.cnsforum.com.

memory processes and may be involved in diseases of aging such as Alzheimer's Disease (Nicholls *et al.* 2001). Acetylcholine also functions in motivated behaviors including aggression, sexual behavior and the regulation of thirst and drinking. ACh also binds to receptors that act as ion channels (nicotinic receptors), but can also operate through muscarinic receptors (see section 5.5).

5.2.3 The monoamine neurotransmitters

The monoamine neurotransmitters occur at a much lower concentration in the brain than the amino acid neurotransmitters, but nevertheless are extremely important as drug targets, for example in the treatment of depression, anxiety, Parkinson's disease and drug addiction (Nestler *et al.* 2009). When these neurotransmitters bind to their receptors, they activate a series of chemical changes involving second messenger systems in the cytoplasm of the cell (see Nestler *et al.* 2009).

The catecholamines

Dopamine (DA), norepinephrine (NE) (sometimes called noradrenaline (NA)) and epinephrine (or adrenaline) are synthesized and released by neurons of the *adrenergic* pathways. Their neuronal cell bodies are localized to a small number of brain nuclei, but their axons project widely throughout the brain (see section 5.7). For example, a prominent NE nucleus is found in the locus ceruleus, whereas dopamine neurons are found in the substantia nigra. NE is an important neurotransmitter in the sympathetic branch of the ANS (section 5.10). Catecholamines are important in the arousal of emotional and motivated behavior and in the regulation of the endocrine hypothalamus (see Chapter 6).

The indoleamines

There are two indoleamines, the neurotransmitter serotonin, which is also called 5-hydroxytryptamine (5-HT) and its close relative, melatonin, a hormone secreted from the pineal gland. The cell bodies of serotonin-secreting neurons are found almost exclusively in the Raphé region of the upper brainstem, but extend their axons widely throughout the brain (see Nestler *et al.* 2009; Iversen *et al.* 2009).

Histamine

Histamine neurons occur primarily in the posterior hypothalamus and nerve terminals are found throughout the hypothalamus, cerebral cortex and spinal cord. This widespread occurrence of histamine innervation suggests that it has many functions in the brain, including regulation of hormone release, learning and memory, and sleep-wake control (Haas *et al.* 2008).

5.2.4 Peptide neurotransmitters

The neuropeptides that act as neurotransmitters in the CNS were originally discovered in the gastrointestinal system. Cholecystokinin, vasoactive intestinal polypeptide (VIP), gastrin and bombesin occur in very low concentrations in many central neural pathways (see Chapters 11 and 12). Other neuropeptides, including substance P, neurotensin, somatostatin and the enkephalins, are produced in neurons and meet most of the criteria of neurotransmitter action given in Table 5.1. An important difference between neuropeptides and the classical neurotransmitters such as dopamine is the manner in which they are synthesized by neurons. As described in Chapter 4, neuropeptides are synthesized via large propeptides that are processed to yield smaller, active, peptides (see section 4.2 and Figure 3.4). The enkephalins, or *endogenous opioids*, for example, are synthesized first as large proenkephalins in neuron cell bodies in the brain and spinal cord (see Figure 11.1). The final products, enkephalins, are small molecules with only five amino acids that have neurotransmitter-like activity, particularly in the regulation of pain perception. Other neuropeptides act as neuromodulators; that is, they alter the way in which a neuron responds to another neurotransmitter, such as dopamine. This will be discussed in Chapter 11.

5.2.5 Non-classical neurotransmitters

All of the substances mentioned so far satisfy most of the criteria that classify them as neurotransmitters (see Table 5.1). However, other neurochemical entities such as the gas

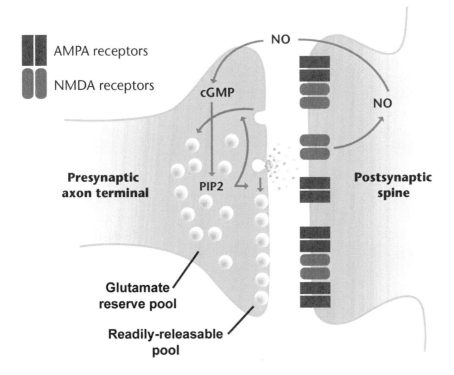

Figure 5.4 Nitric oxide (NO) is an unusual neurotransmitter

Release of glutamate from presynaptic axon terminals activates postsynaptic N-methyl-D-aspartate (NMDA) receptors (see Figure 5.3), which induce the production and release of NO. Another glutamate receptor subtype, called AMPA receptors, are also stimulated. NO acts as a retrograde messenger, enhancing cGMP and PIP2 production in the presynaptic terminal (see Chapter 9 for further details). The increase in PIP2 speeds up the rate at which vesicles are retrieved and readied for re-use in the readily-releasable pool. NO thus enhances glutamate neurotransmission.

Abbreviations: AMPA, α-amino-3-hydroxy-5-methyl-4-isoxazole propionic acid; cGMP, cyclic guanosine monophosphate; NO, nitric oxide; PIP2, phosphatidylinositol diphosphate. Reproduced with permission (Sullivan 2003).

nitric oxide (NO), and substances that behave like cannabis (*endocannabinoids*), are able to act as neurotransmitters even though they do not satisfy several of the criteria listed in Table 5.1. For example, NO is formed in neurons through the action of the neurotransmitter glutamate, which activates the enzyme nitric oxide synthase (NOS; Figure 5.4; Sullivan 2003) to convert the amino acid arginine to NO. Only 1 percent of brain neurons actually express this enzyme, but it is estimated that almost every neuron is exposed to NO (Squire *et al.* 2008), indicating the great importance of NO in brain function. Because NO is gaseous, it is released from neurons by diffusion; that is, it is not stored in vesicles and is not secreted. Further, although NO travels backward across the synapse (retrograde action) and acts on presynaptic neurons to modify secretion of glutamate, it also diffuses away from the neuron terminal to act on more distant neurons and glial cells. NO also differs from other neurotransmitters in that it does not act on specific receptors, but simply diffuses into other cells where it regulates signaling systems such as cyclic guanosine monophosphate (cGMP) (Baranano *et al.* 2001; see Vincent 2010).

Thus, at least three of the criteria in Table 5.1 are unsatisfied. Nevertheless, it is now accepted that NO acts as a neurotransmitter. Moreover, there is evidence that other gases, such as *carbon monoxide* and *hydrogen sulphide* (Olson 2011; Deng *et al.* 2014), may also act as neurotransmitters.

The psychoactive properties of cannabis have been known for more than a thousand years, but only recently was the search for brain cannabis receptors (CB1 receptors) successful. This knowledge, that the brain possessed receptors specifically for a plant chemical, led to the discovery of endogenous brain substances that mimicked the activity of tetrahydrocannabinol (THC). Such chemical messengers were christened *endocannabinoids*, and the primary member of this class is called *anandamide*. Like nitric oxide, anandamide is enzymatically produced only when needed (i.e. it is not stored) and is freely diffusible through cell membranes and is therefore not secreted. Its primary function seems to be as a feedback system onto presynaptic neurons. It is therefore a *retrograde messenger* that acts on presynaptic CB receptors to suppress neurotransmitter release (Figure 5.5;

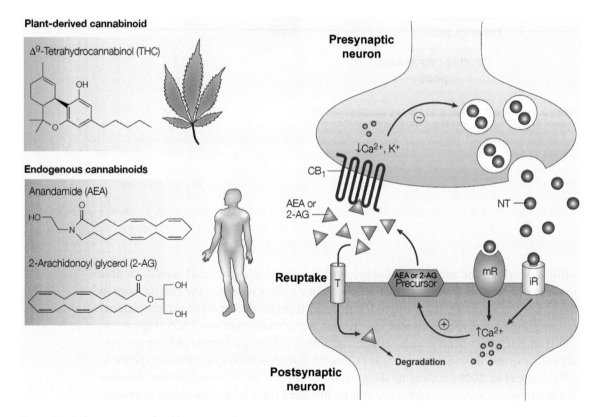

Figure 5.5 Endogenous cannabis-like compounds as neurotransmitters

Endocannabinoids, such as anandamide (AEA) or 2-arachidonoyl glycerol (2-AG), are formed from precursors in the postsynaptic cell membrane following neurotransmitter (NT) stimulation of ion channels (iR) or metabotropic membrane receptors (mR). Anandamide (or 2-AG) is released into the synapse where it binds to presynaptic, G-protein-coupled cannabinoid receptors (CB[1]). This results in an inhibition of NT release. Excess AEA is removed from the synapse by reuptake into the postsynaptic cell via a transporter (T), followed by enzymatic degradation. Reproduced with permission (Guzmán 2003).

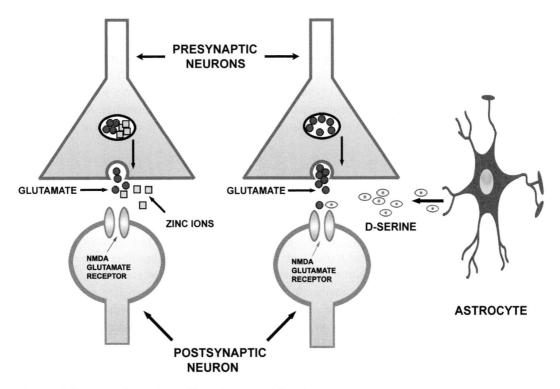

Figure 5.6 Two unusual neurotransmitters: zinc ions and D-serine
Zinc ions are co-localized and therefore co-released with glutamate and bind to glutamate receptors of the N-methyl-D-aspartate (NMDA) subtype (as shown in Figure 5.3). D-serine, on the other hand, is secreted from astrocytes and binds to the NMDA receptor, probably at the glycine site, to potentiate the action of glutamate (see Figure 5.3). The secretion of D-serine from astrocytes is also discussed in Chapter 6 (section 6.9.2).

Guzmán 2003; Christie and Vaughn 2001). CB receptors are distributed throughout the brain and this accounts for the many effects of cannabis on memory, mood, coordination, appetite, sleep and pain sensation. Since marijuana increases appetite, and a CB receptor *antagonist*, Rimonabant, reduces eating and body weight (Christensen *et al.* 2007; further reading see Ward and Raffa 2011), it is possible that CB receptors may also be implicated in obesity.

Two other unusual additions to this list of non-classical neurotransmitters are zinc ions and D-serine that act, with glutamate, as co-transmitters (Baranano *et al.* 2001). Zinc ions are stored along with glutamate in vesicles and act on neuronal NMDA postsynaptic receptors (Figure 5.6). D-serine is released from glial cells called astrocytes (glial cells will be covered in more detail in section 6.9), but acts synergistically with glutamate at glutamate receptors. D-serine probably binds to the glycine site shown in Figure 5.3. There is evidence that D-serine may be useful in enhancing memory (Otto 2011).

Finally, there are several other signaling molecules that are involved in intercellular communication in the brain that are not, strictly speaking, neurotransmitters. As noted in section 1.4.6, the *prostaglandins* and related compounds are made throughout the body including the CNS. Although prostaglandins are not stored in the manner of a neurotransmitter, they do have widespread signaling properties in the brain (Iversen *et al.* 2009). Similarly, *ATP* (adenosine triphosphate), and its breakdown product adenosine, acts like a neurotransmitter at specific receptor sites (Nestler *et al.* 2009; Iversen *et al.* 2009). Of course, ATP is present in all synaptic vesicles and acts as an energy source. But some nerve cells, termed *purinergic neurons*, also use ATP as a neurotransmitter. Adenosine, on the other hand, is only made outside the nerve terminal and is not therefore stored in vesicles.

5.3 Neurotransmitter biosynthesis and storage

There are significant differences between the synthesis of amino acid and monoamine neurotransmitters and the synthesis of neuropeptides. The enzymes necessary for the biosynthesis of amines and amino acids are produced on the ribosomes of the endoplasmic reticulum in the neuron cell body and stored in synaptic vesicles. The amino acid substrates, such as tyrosine (for dopamine) and tryptophan (for serotonin) are also stored in these synaptic vesicles. The synthesis of the neurotransmitters is completed inside the synaptic vesicles as they are transported down the axon to the nerve terminals. The precursors (*pre-propeptides*) of the neuropeptide neurotransmitters are also biosynthesized in the cell body and then stored in vesicles that contain enzymes (*convertases*) that further process the precursor peptides by cutting them into smaller active molecules (see section 5.2.4) as the vesicles are transported to the nerve terminals (Figure 5.7). Inside the nerve terminal, the synaptic vesicles exist in a storage pool before they are released into the synapse (Figure 5.7).

Synaptic vesicles perform a number of essential functions in the nerve cell: (1) they transport the neurotransmitter precursors/propeptides and biosynthetic enzymes from the cell body along the axon to the nerve terminal; (2) neurotransmitter synthesis is often completed within the vesicles; (3) the vesicles store the neurotransmitters until they are released; (4) vesicles protect the neurotransmitters from deactivation; (5) neurotransmitters are released from the nerve terminal by exocytosis when the vesicle fuses with the depolarized cell membrane in the presence of calcium ions (see section 5.4); and (6) vesicles are able to exert feedback control on the synthesis of neurotransmitter; i.e. when vesicles are depleted of neurotransmitter, for example after extended periods of secretion, enzymes such as tyrosine hydroxylase are stimulated to produce more norepinephrine (see section 5.3.2 and Figure 5.9). Conversely, when norepinephrine levels are restored to normal in the vesicles, the vesicle exerts a negative feedback effect on tyrosine hydroxylase to shut down further production of norepinephrine.

5.3.1 Synthesis and storage of particular neurotransmitters

Amino acid neurotransmitters

Glutamate and aspartate are synthesized from glucose and other precursors in neurons and glial cells. GABA is formed from its precursor, glutamic acid, by the enzyme glutamic acid

NEUROTRANSMITTERS   **NEUROPEPTIDES**

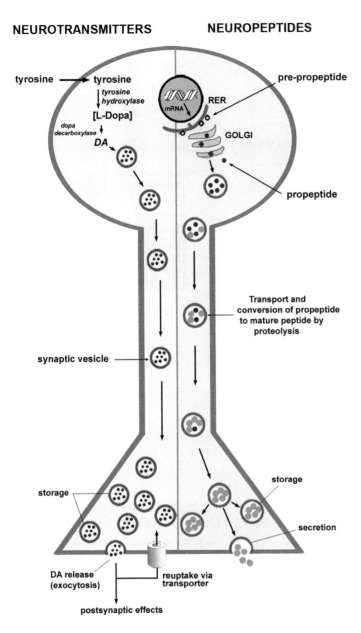

Figure 5.7 Comparison of neurotransmitter versus neuropeptide biosynthesis
Biosynthesis of neuropeptides is dependent on gene transcription and translation in the cell body, with subsequent modification to the final structure inside the vesicles. *In contrast*, classical neurotransmitters are dependent on uptake of dietary precursors such as tyrosine. As shown here, dopamine is made in the cytoplasm from the precursor tyrosine and then stored in vesicles for transport to the nerve terminal. In some neurons, norepinephrine is biosynthesized from dopamine inside vesicles. A further major difference between classical neurotransmitters and neuropeptides is that the latter cannot be reused after release; i.e. there is no reuptake system. For classical neurotransmitters, there is an efficient reuptake system that ensures a supply of fully stocked vesicles ready for further release (see also Figure 5.9).

Abbreviations: DA, dopamine; mRNA, messenger ribonucleic acid; RER, rough endoplasmic reticulum.

decarboxylase (GAD) in the nerve terminals. Details of the biosynthesis of the amino acid neurotransmitters can be found elsewhere (Nestler *et al.* 2009).

Acetylcholine

Acetylcholine is synthesized from choline, an amino acid taken up from the circulation, and acetyl coenzyme A through the action of the enzyme choline acetyltransferase (Nestler *et al.* 2009; Iversen *et al.* 2009). Choline acetyltransferase is synthesized in the cell body and transported down the axon to the nerve terminals where acetylcholine is synthesized.

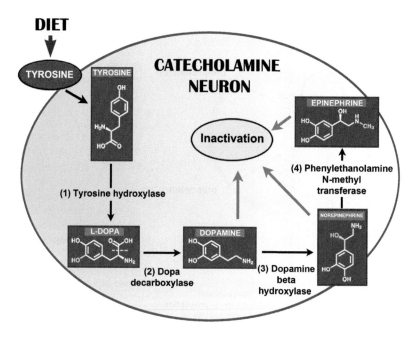

Figure 5.8 Biosynthesis of catecholamines
This figure illustrates the relatively simple structural changes that occur to convert tyrosine to three different catecholamines. Dopamine neurons convert tyrosine to L-Dopa through the enzymic action of tyrosine hydroxylase (1), which is then converted to dopamine by dopa decarboxylase (2). Norepinephrine neurons perform the same steps, as far as dopamine, but the enzyme dopamine beta hydroxylase (3) adds another hydroxyl group to yield norepinephrine. Finally, epinephrine neurons convert norepinephrine to epinephrine by adding a methyl group through the action of phenylethanolamine N-methyltransferase (4). All three catecholamines are inactivated by enzymes (described in section 5.6).

Monoamine (catecholamine) neurotransmitters

The synthesis of a typical catecholamine, dopamine, is shown in Figure 5.8 (see also Figure 5.7). The amino acid tyrosine is transported into the nerve cell from the blood and is converted to L-Dopa, through the action of the enzyme tyrosine hydroxylase. L-Dopa is then converted to dopamine in the cell body through the action of the enzyme dopa decarboxylase and dopamine is transported to the nerve terminal in the vesicles. Norepinephrine is biosynthesized in separate noradrenergic neurons by converting dopamine into norepinephrine inside the synaptic vesicle through the action of the enzyme dopa decarboxylase. Other neurons, using the enzyme phenylethanolamine N-methyl transferase, are able to convert norepinephrine to epinephrine (Figure 5.8). L-Dopa is used as a drug to treat Parkinson's disease because it readily enters the brain to act as a precursor for dopamine and facilitates dopamine synthesis.

Serotonin or 5-hydroxytryptamine (5-HT) is biosynthesized from the amino acid tryptophan using a similar pathway to the catecholamines; that is, tryptophan is transported from the blood into neurons and converted first to 5-hydroxytryptophan (5-HTP) by the enzyme tryptophan hydroxylase. 5-HTP is then converted to serotonin by the enzyme 5-HTP decarboxylase (Nestler *et al.* 2009; Iversen *et al.* 2009). These transformations take place inside vesicles as shown for catecholamines.

Neuropeptide neurotransmitters

The biosynthesis of neuropeptides is referred to in section 5.2.4, and will be discussed in more detail in Chapter 11.

5.3.2 Regulation of neurotransmitter synthesis and storage

Neurotransmitters are continually being synthesized, stored, released and deactivated. However, since it is crucial that the neuron terminal is not depleted of neurotransmitter, there are several ways by which an optimal level of releasable neurotransmitter is maintained. Using a norepinephrine neuron as an example, the fastest and most efficient way to restock the storage vesicles is to recapture norepinephrine that has already been released into the synapse. The neuron terminal achieves this through a reuptake mechanism, which acts like a vacuum cleaner, removing any free norepinephrine from the synaptic space and moving it back into the terminal where it is taken up again by the vesicles (Figures 5.9 and 5.14). Under conditions in which the neuron is continuously firing and releasing large amounts of norepinephrine, the reuptake may not be sufficient to restock the vesicles. In this case, either the empty vesicles signal tyrosine hydroxylase to make more norepinephrine, or, in extreme circumstances, the nucleus is signaled to increase expression of tyrosine hydroxylase mRNA (Iversen *et al.* 2009; Figure 5.9). When sufficient neurotransmitter has been stored in the vesicles, tyrosine hydroxylase is inhibited (*end product inhibition*) by a negative feedback loop. The synthesis of neurotransmitters can also be regulated by neural input from other cells (e.g. by entry of calcium ions or via cAMP), by hormones and by the availability of amino acids such as tyrosine. The latter suggests that our diet could, under certain circumstances, affect some aspects of brain function (see Richard *et al.* 2009; Fernstrom and Fernstrom 2007).

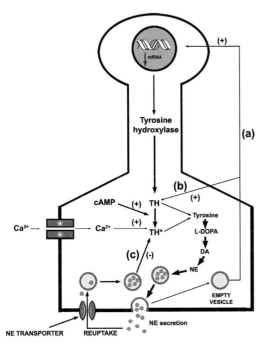

Figure 5.9 Maintenance of adequate levels of catecholamine in nerve terminals

This figure uses norepinephrine (NE) as an example, but the description is also appropriate for dopamine. Following high neuronal activity, and the release of large amounts of NE, optimal levels of NE are restored first by reuptake through the NE transporter, and second by increased synthesis of NE in the neuron terminal. For example, when vesicles are depleted of their NE contents, the empty vesicle sends two signals: (a) to the nucleus to activate the tyrosine hydroxylase (TH) gene and the synthesis of more TH; and (b) to activate existing TH within the terminal (i.e. TH to TH*). The reverse of this sequence is also true: (c) when vesicles are refilled, the activity of TH is switched off.

Abbreviations: cAMP, cyclic adenosine monophosphate; DA, dopamine.

5.4 Release of neurotransmitters

By definition, each neuron is both a post- and a presynaptic cell; that is, when a neuron is stimulated *postsynaptically* by neurotransmitters from another neuron, the neuron releases its neurotransmitter *presynaptically* into the next synapse. When these neurotransmitters stimulate their postsynaptic receptors at an excitatory synapse, they cause ion channels to open, and the nerve membrane is depolarized, that is, sodium (Na^+) flows in and potassium (K^+) moves out of the cell. This depolarization causes a measurable change in the electrical activity of the cell, called an action potential, which travels along the nerve cell membrane until it reaches the nerve terminal of the axon (Figure 5.10).

When depolarization of the nerve terminal occurs, the synaptic vesicles in which the neurotransmitters are stored fuse with the cell membrane and, under the influence of calcium ions, there is a change in the permeability of the cell membrane, releasing the

EXTRACELLULAR

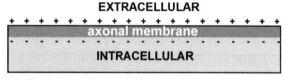

A. Resting membrane potential.

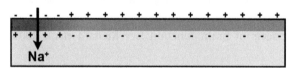

B. Action potential initiated by influx of Na+ ions: polarity of membrane reversed.

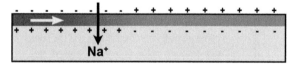

C. Polarization spreads with continued influx of Na+.

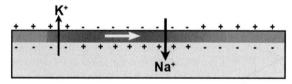

D. Action potential moves down axon and K+ ions are pumped out to restore polarization.

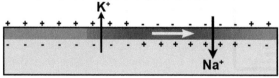

E. Action potential continues down axon.

Figure 5.10 Generation of an action potential

Propagation of a nerve impulse (action potential) along the axon coincides with a localized inflow of sodium (Na^+) ions followed by an outflow of potassium (K^+) ions through channels that are controlled by voltage changes across the axonal membrane. The cell membrane establishes a resting membrane potential by separating fluids that differ in their content of Na^+ and K^+ ions (A). The extracellular fluid has about ten times more Na^+ ions than K^+ ions, but in the intracellular fluid the ratio is the reverse. The electrical event that sends a nerve impulse traveling down the axon normally originates in the cell body. The action potential begins with a depolarization, or reduction in the negative potential, across the axonal membrane. This voltage shift opens some of the Na^+ channels. The inflow of Na^+ ions accelerates until the inner surface of the membrane shifts from a negative to a positive charge (B). This voltage reversal closes the Na^+ channel and opens the K^+ channel (C). The outflow of K^+ ions quickly restores the negative potential inside the axon and the action potential propagates itself down the axon (D and E). After a brief *refractory period*, during which Na^+ is pumped out of the cell and K^+ into the cell by the ion pumps, a second impulse can flow.

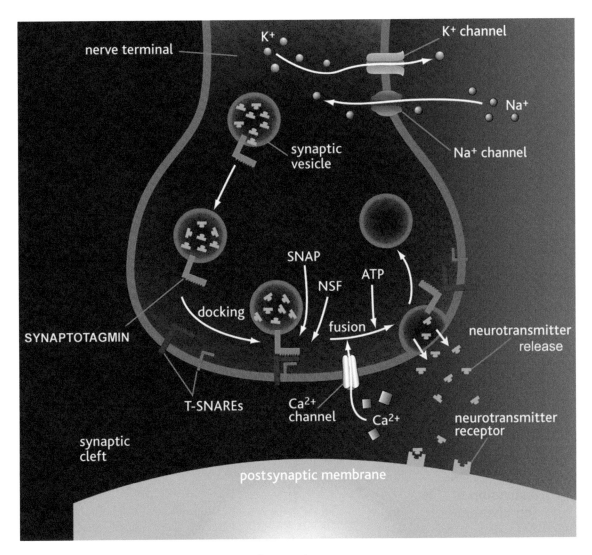

Figure 5.11 Mechanism for neurotransmitter release from vesicles
Neurotransmitters are released from the nerve terminal following fusion of the synaptic vesicle membrane with the neuronal membrane. This fusion is accomplished via specific proteins, including synaptotagmin, NSF, SNAP and SNARE that enable the fusion to occur. NSF is a fusion protein whereas SNAP is a NSF attachment protein; T-SNARE is a receptor for SNAP and synaptotagmin is a vesicle attachment protein for SNAP. Note that action potential propagation (via Na^+ and K^- ions; see Figure 5.10) induces Ca^{2+} influx at the presynaptic membrane, and membrane fusion takes place under the influence of synaptotagmin, SNARES, etc. Following neurotransmitter release, synaptic vesicle membrane components are recycled via an endocytic process.

Abbreviations: ATP, adenosine triphosphate; NSF, N-ethylmaleimide sensitive factor; SNAP, soluble SNF attachment proteins; SNARE, SNAP receptor protein. Copyright CNSforum, Lundbeck Institute: http://www.cnsforum.com.

neurotransmitters from the vesicle into the synapse. The release of the neurotransmitter into the synapse involves a number of complex biochemical actions via docking molecules (Figure 5.11) (see Burgoyne and Morgan 2003; Scheller 2013). After the cell fires, it returns to its resting potential until stimulated to fire again. The cell cannot, however, fire again instantaneously as the resting potential is reinstated through the action of the ion pump, which pumps Na$^+$ ions out of the cell and K$^+$ ions back into the cell. During the time that this ion pump takes to reinstate the resting potential, the cell cannot fire and is said to be in a refractory period.

5.5 Receptors for neurotransmitters

After the neurotransmitter is released into the synapse, it regulates postsynaptic neuronal functions by binding to a specific receptor protein or ion channel on the surface membrane of the postsynaptic cell (Figure 5.12).

Most neurotransmitters are too polar (water soluble) to enter their target cells (i.e. they will not cross the phospholipid cell membrane) and, therefore, they stimulate receptors located on the outside of the cell membrane. There are, however, exceptions, such as the endocannabinoids and NO (see section 5.2.5) that pass freely through cell membranes. Membrane receptors are of three general types: *ionotropic receptors*, which open ion channels and are stimulated primarily by the amino acid neurotransmitters such as

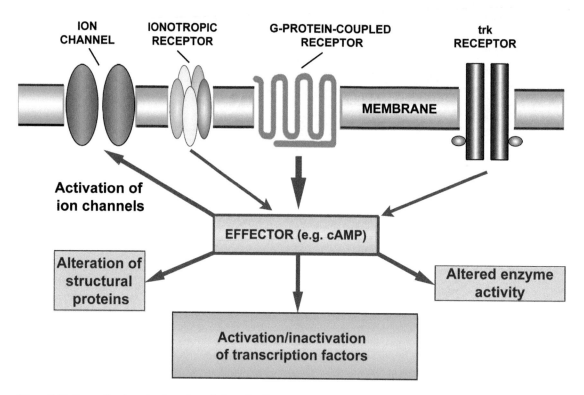

Figure 5.12 Generalized mechanisms for cellular activation

Neuronal activation is controlled through various receptors and ion channels. For example, the neurotransmitter norepinephrine binds to a metabotropic receptor protein (also called a G-protein-coupled receptor) that activates many cellular processes via an intracellular *effector* such as cAMP. Ionotropic receptors respond to neurotransmitters such as glutamate, and trk receptors bind hormones such as leptin. These signaling mechanisms are covered in more detail in Chapter 10.

Abbreviations: cAMP, cyclic adenosine monophosphate; trk, tyrosine receptor kinase.

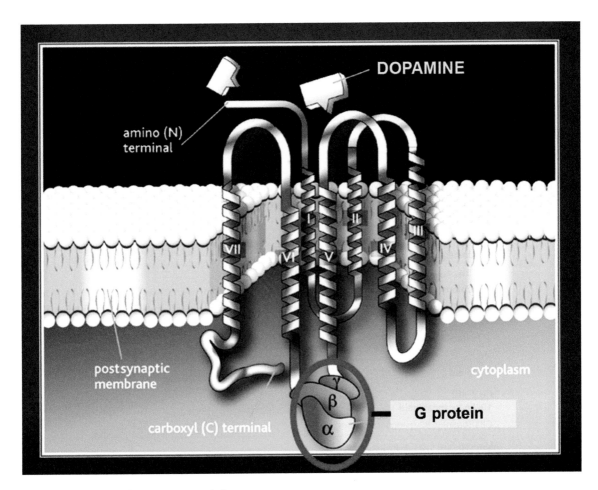

Figure 5.13 Structure of a G-protein-coupled receptor
Many neurotransmitters, including dopamine, bind to G-protein-coupled receptors. These membrane receptors that have seven transmembrane spanning α-helices. Dopamine binding to the extracellular portion of the receptor activates the G proteins, which initiate intracellular secondary messenger signaling pathways. A G protein is a complex of three subunits termed α, β and γ. The downstream effect will be either inhibitory or stimulatory, depending on the types of G protein linked to the receptor – dopamine D1, D5 receptors are linked to stimulatory G proteins (G_s), whereas dopamine D2, D3 and D4 are linked to inhibitory G proteins (G_i). Copyright CNSforum, Lundbeck Institute: http://www.cnsforum.com.

GABA (Figure 5.2), glutamate (Figure 5.3) and acetylcholine; *metabotropic receptors*, which activate intracellular second messenger systems such as cAMP to regulate metabolic changes within the postsynaptic cell, and are stimulated by catecholamine and peptide neurotransmitters. These receptors are also known as G-protein-coupled receptors (Figure 5.13). *A third type of receptor*, responsible for the activity of growth factors, neurotrophins (e.g. NGF) and adipokines (e.g. leptin), is called a *tyrosine receptor kinase* (trk receptor). These are single-stranded proteins that project through the cell membrane and that join together (dimerize) when they bind the ligand, such as leptin (see Figure 10.6). The mechanisms by which these three receptor types modify cell function are discussed in more detail in Chapter 10.

Table 5.3 Neurotransmitter receptor subtypes

1. GABAergic

$GABA_A$, $GABA_B$

2. Glutamatergic

NMDA, AMPA kainate, $mGluR_1$, $mGluR_2$, $mGluR_3$, $mGluR_4$, $mGluR_5$, $mGluR_6$

3. Glycinergic

Glycine

4. Cholinergic

Muscarinic: M_1, M_2, M_3, M_4, M_5

Nicotinic: muscle, neuronal (α-bungarotoxin-insensitive)

5. Dopaminergic

D_1, D_2, D_3, D_4, D_5

6. Adrenergic

α_{1A}, α_{1B}, α_{1C}, α_{1D}, α_{2A}, α_{2B}, α_{2C}, α_{2D}, β_1, β_2, β_3

7. Serotonergic

$5\text{-}HT_{1A}$, $5\text{-}HT_{1B}$, $5\text{-}HT_{1D}$, $5\text{-}HT_{1E}$, $5\text{-}HT_{1F}$, $5\text{-}HT_{2A}$, $5\text{-}HT_{2B}$, $5\text{-}HT_{2C}$, $5\text{-}HT_3$, $5\text{-}HT_4$, $5\text{-}HT_{5A}$, $5\text{-}HT_{5B}$, $5\text{-}HT_6$, $5\text{-}HT_7$

8. Histaminergic

H_1, H_2, H_3, H_4

9. Opioid

μ_1, μ_2, μ_3, δ_1, δ_2, κ_1, κ_2, κ_3

10. Endocannabinoid

CB1 (brain), CB2 (peripheral tissues)

Abbreviations: 5-HT, 5-hydoxytryptamine;
AMPA, α-amino-3-hydroxy-5-methyl-4-isoxazole propionic acid;
GABA, γ-aminobutyric acid; mGluR, metabotropic glutamate receptor.

Each neurotransmitter may have more than one type of receptor and these are called *receptor subtypes*. The study of such receptors – receptor neuropharmacology – is an enormous field. Table 5.3 provides a brief overview of the number of receptor subtypes for those neurotransmitters that we have discussed already. Receptor subtypes allow cells to respond in different ways to the same neurotransmitter. For example, the smooth muscle cells of blood vessels contract in response to norepinephrine binding to α1-adrenergic receptors, but they relax when NE binds to β2-adrenergic receptors (Purves *et al.* 2008). Dopamine receptor subtypes are also capable of eliciting quite different cellular responses; for example, dopamine D1 receptors stimulate the generation of cAMP, whereas D2 receptors inhibit cAMP formation. Knowledge of receptor subtypes is invaluable for the development of new drugs. For example, new anti-migraine drugs were developed that specifically bind to the $5\text{-}HT_{1B/1D}$ receptor subtype. The nervous system has now been carefully mapped for most receptor subtypes both in terms of the proteins themselves (Kuhar *et al.* 1986) and of the cellular location of their specific genes (e.g. *µ-opiate receptors*: Mansour *et al.* 1995; *glutamate receptor subtypes*: Van den Pol *et al.* 1994). Receptors for neuropeptides are discussed in Chapter 10.

5.6 Inactivation of neurotransmitters

Effective neurotransmission requires not only the secretion of the neurotransmitter from the presynaptic terminal, but also its efficient and rapid inactivation. This is important because continuous activation of postsynaptic receptors by a neurotransmitter can be deleterious to neurons. For example, uncontrolled stimulation of glutamate receptors can lead to neuronal death and seizures. To avoid this, neurotransmitters are quickly inactivated via several pathways. The most efficient way is to physically remove the neurotransmitter through reuptake into the presynaptic terminal via transporters. In addition, neurotransmitters can be enzymatically degraded (Nestler *et al.* 2009; Iversen *et al.* 2009). These mechanisms are illustrated for norepinephrine in Figure 5.14. Reuptake through the transporter, and repackaging in vesicles, is the most efficient way to remove norepinephrine from the synapse. This step also recycles unused neurotransmitter and reduces the need for further synthesis. Neuropeptides do not appear to undergo reuptake into presynaptic terminals.

An understanding of the importance of transporters in synaptic transmission led to the realization that drugs of abuse, such as cocaine and amphetamine, exert their effects by binding to dopamine transporters to prevent the reuptake and inactivation of dopamine. In other words, the stimulant effect of these drugs (the "rush") is caused by dopamine remaining in the synapse longer than normal. Another example is the mechanism of action of some antidepressant drugs, such as Prozac. This drug blocks the reuptake of serotonin and increases the level in the synapse. There are many other examples of the effects of blocking transporters with specific drugs.

Also shown in Figure 5.14 is the enzymatic degradation of catecholamines by two separate enzymes: monoamine oxidase (MAO) and catechol o-methyl transferase (COMT) (Figure 5.14). Other types of neurotransmitters have specific degrading enzymes and neuropeptide neurotransmitters are deactivated by peptidase enzymes. Thus, GABA is degraded by *GABA transaminase* (GABA-T), acetylcholine by *acetylcholinesterase* and histamine by *histaminase*. Because these enzymes exist in the synapse (see Figure 5.14), they can deactivate the neurotransmitters before they bind to postsynaptic receptors, as well as after. Enough neurotransmitter must therefore be released to stimulate the postsynaptic cell in spite of the active enzymic degradation process. In addition, any neurotransmitter that is not taken up into the vesicles in the presynaptic terminal is also deactivated (Figure 5.14).

5.7 Neurotransmitter pathways

Neurotransmitters occur in specific neuronal pathways in the human brain. Figure 5.15 shows examples of the pathways for dopamine, norepinephrine, acetylcholine, serotonin, GABA and glutamate.

Dopamine pathways are concentrated into three major systems (Figure 5.15A): (1) the substantia nigra projecting to the caudate nucleus/putamen (also called the striatum); (2) the ventral tegmentum projecting to the cerebral cortex and various limbic structures such as septum, hippocampus, amygdala and nucleus accumbens; and (3) the short tubero-infundibular sytem of the hypothalamus (Iversen *et al.* 2009; Nestler *et al.* 2009).

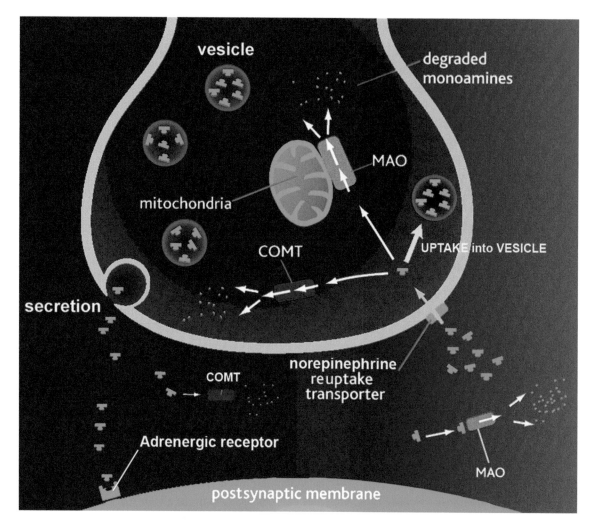

Figure 5.14 Deactivation mechanisms for catecholamines

The primary mechanism for inactivation of catecholamines is to remove them from the synapse by specific transporters that enable them to be repackaged in secretory vesicles. This step also recycles unused neurotransmitter and reduces the need for further synthesis. In addition, catecholamines are rapidly degraded by two separate enzymes: monoamine oxidase (MAO); and catechol o-methyl transferase (COMT). Note that the enzymes are present both inside and outside (in the synapse) the neuron terminal. All neurotransmitters have their own degrading enzymes, but not all have specific transporters.

Abbreviations: COMT, catechol o-methyl transferase; MAO, monoamine oxidase. Copyright CNSforum, Lundbeck Institute: www.cnsforum.com.

The substantia nigra-caudate nucleus dopamine pathway is critical for the integration of movement and the loss of dopamine cells in the substantia nigra causes the rigidity and tremor of patients with Parkinson's disease. Injection of L-Dopa, a dopamine precursor (see Figure 5.8) that is able to cross the blood-brain barrier, relieves many of these symptoms by enabling the synthesis of additional dopamine. The DA system is also part of a complex reward pathway as well as the target of drugs of abuse, such as amphetamine. It has also been implicated in the etiology of attention deficit hyperactivity disorder

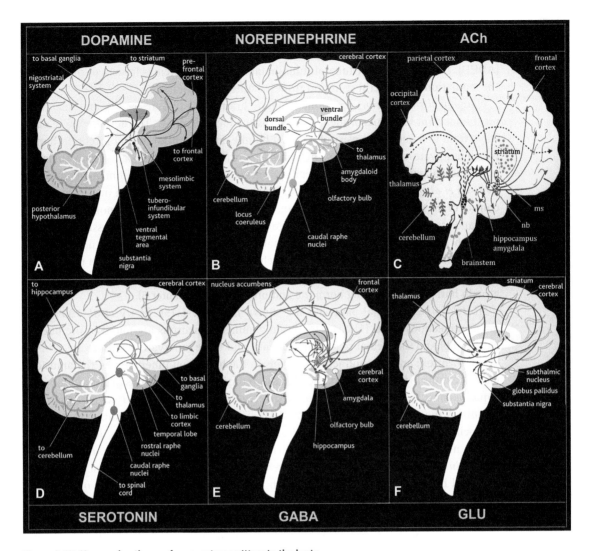

Figure 5.15 Neuronal pathways for neurotransmitters in the brain

This figure illustrates the location of neuronal cell bodies and projecting axons for six neurotransmitters: dopamine (A), norepinephrine (B), acetylcholine (ACh; C), serotonin (D), GABA (E) and glutamate (Glu; F). *Dopamine* is transmitted via three major pathways: (1) an extension from the substantia nigra to the caudate nucleus-putamen (neostriatum); (2) a pathway projects from the ventral tegmentum to the mesolimbic forebrain; and (3) a short pathway, known as the tuberoinfundibular system, is concerned with neuronal control of the hypothalamic-pituitary system. Many regions of the brain are supplied by the *norepinephrine (noradrenergic) systems* and noradrenergic neurons are found in the locus ceruleus and the caudal Raphé nucleus. The ascending nerves of the locus ceruleus project to the frontal cortex, thalamus, hypothalamus and limbic system. Norepinephrine is also transmitted from the locus ceruleus to the cerebellum. Nerves projecting from the caudal Raphé ascend to the amygdala and descend to the midbrain. *Acetylcholine* is present in two major pathways that project widely to different brain areas: basal-forebrain cholinergic neurons (red, including the nucleus basalis (nb) and medial septal nucleus (ms)) and pedunculopontine–lateral dorsal tegmental neurons (blue). Other cholinergic neurons include striatal interneurons (orange), cranial-nerve nuclei (green circles) and spinal-cord preganglionic and motor neurons (yellow). The principal centers for *serotonin* neurons are the rostral and caudal Raphé nuclei. From the rostral nuclei, axons ascend to the cerebral cortex, limbic regions and specifically to the basal ganglia. Serotonin nuclei in the caudal Raphé give rise to descending axons, some of which terminate in the medulla, while others descend into the spinal cord. *GABA* is the main inhibitory neurotransmitter in the central nervous system (CNS) and GABAergic inhibition is seen at all levels, including the hypothalamus, hippocampus, cerebral cortex and cerebellar cortex. In addition to the large well-established GABA pathways, GABA interneurons are highly abundant in the brain, with 50 percent of the inhibitory synapses in the brain being GABA mediated. *Glutamate* is the most abundant excitatory neurotransmitter and major connections between the thalamus and cortex are shown here. However, glutamate receptors are found extensively throughout the brain, particularly in the hypothalamus, indicating glutamate innervation of many brain areas. Panels A, B, D, E and F are copyright CNSforum, Lundbeck Institute: www.cnsforum.com. Panel C is reproduced with permission (Perry *et al.* 1999).

(ADHD). The hypothalamic tuberoinfundibular dopamine system is largely responsible for the regulation of prolactin secretion from the anterior pituitary.

Norepinephrine neurons are found in many brain areas (Figure 5.15B), and one of the most important locations of the cell bodies is the locus ceruleus. The noradrenergic pathways from this nucleus consist of a dorsal bundle, and a ventral medial forebrain bundle which innervate many structures, including the cortex, olfactory bulb, amygdala, thalamus and hypothalamus. Additional descending tracts innervate the cerebellum and spinal cord (Iversen *et al.* 2009; Nestler *et al.* 2009). Brain norepinephrine has been implicated in the control of the sleep-wake cycle, attention, feeding behavior and as a component of normal fear and stress responses. There is also good evidence for a role in depressive disorders. Through their projections to the hypothalamus and median eminence, noradrenergic neurons affect emotional and motivated behavior and regulate the secretion of hypothalamic releasing/inhibiting hormones that control secretion of pituitary hormones such as growth hormone, luteinizing hormone and prolactin (see Chapter 6).

Acetylcholine neuronal pathways are shown in Figure 5.15C (Perry *et al.* 1999). Two major projections exist: (1) basal forebrain cholinergic neurons are found in the nucleus basalis (nb) and the medial septal nucleus (ms) and these innervate extensive areas in the cerebral cortex; and (2) cholinergic neurons in the brainstem innervate the thalamus and motor neurons in the spinal cord. A third group consists of interneurons in the striatum. The hypothalamus is also a site of cholinergic regulation and Table 6.1 reveals that ACh stimulates the releasing hormones for ACTH, TSH and GH, and the neurons that secrete oxytocin and vasopressin. Other studies show that acetylcholine has a direct effect on GnRH neurons inducing secretion of the neuropeptide (Krsmanovic *et al.* 1998). *Serotonin* pathways (*serotoninergic* system; Figure 5.15D) extend almost exclusively from cell bodies in the Raphé nuclei, with axons ascending to most areas of the brain, including the cerebral cortex, the hippocampus, hypothalamus and amygdala. Axons from the caudal aspect of the Raphé nuclei descend to the cerebellum and spinal cord (Nestler *et al.* 2009). The widespread distribution of serotonin axons, and the multiple serotonin receptors (Table 5.3), indicate that this neurotransmitter is implicated in a wide variety of brain functions. These include sleep and wakefulness, depression, anxiety, effects of psychedelic drugs and migraine (Iversen *et al.* 2009). Serotonin neurons are also implicated in the regulation of anterior pituitary hormone secretion (see Table 6.1), but may also control posterior pituitary hormone (e.g. vasopressin) release (Jorgensen 2007).

Amino acid neurotransmitters are also key signaling molecules in the brain. As mentioned already, they consist of two distinct classes: excitatory neurotransmitters such as glutamate and aspartate, and inhibitory neurotransmitters such as GABA and glycine. *GABA* is the most abundant inhibitory neurotransmitter in the brain and spinal cord and many GABA neurons exist as local circuit inhibitory interneurons; that is, neurons with very short axons that release GABA to inhibit adjacent neurons. Some GABAergic neurons, however, have longer axons that project to many areas of the brain (Figure 5.15E). For example, cell bodies in the nucleus accumbens/substantia nigra innervate the hippocampus, thalamus and cortex. GABAergic neurons also regulate the release of hypothalamic and pituitary hormones (see Table 6.1). *Glutamate* is the principal excitatory neurotransmitter in the brain and some of the glutamate pathways are shown in Figure 5.15F. However, the figure gives only a limited view of the extent of glutamate neuron projections. This is because large amounts of glutamate are found in the brain that are not linked to a neurotransmitter function; that is, in addition to its role as a neurotransmitter, glutamate is a precursor in the biosynthesis of GABA, and glutamate also plays a critical role in neuronal metabolism (Iversen *et al.* 2009). For this reason, the location of *glutamate receptors* is more revealing of an extensive role for glutamate as a neurotransmitter in controlling brain function. The distribution of neurons that have glutamate receptor subtypes has been mapped using a technique called *in situ hybridization* (see section 9.2 for further details). For example, glutamate receptors of the NMDA subtype (see Figure 5.3) were localized in many brain areas, including thalamus, olfactory bulb, hippocampus and cortex (Watanabe *et al.* 1993), and in the hypothalamus (Eyigor *et al.* 2001). The latter findings are consistent with the important role for glutamate in the hypothalamic control of pituitary hormone secretion (Van den Pol *et al.* 1990; Brann 1995). For example, glutamate is a critical component of the hypothalamic regulation of fertility, through stimulatory effects on GnRH secretion, and is part of the neurochemical control of thyroid and adrenal hormone secretion via effects on TRH and CRH neurons (Iremonger *et al.* 2010; Wittman *et al.* 2005; Levy and Tasker 2012). Table 6.1 reveals that glutamate stimulates secretion of all pituitary hormones.

Glutamate has been implicated in regulating neural plasticity, especially as the basis for learning and memory (Watkins and Jane 2006), and both GABA and glutamate are associated with a number of psychiatric disorders. For example, abnormal GABA function is linked to epilepsy, Huntington's disease and sleep disorders, whereas glutamate is implicated in stroke, epilepsy and neurotoxicity (Iversen *et al.* 2009).

The gaseous neurotransmitters, such as *nitric oxide* (NO), are not stored in vesicles and diffuse away from their sites of biosynthesis. For this reason, it is difficult to localize specific brain pathways. However, neurons producing NO can be identified by localizing the enzyme *nitric oxide synthase* (NOS), which is responsible for the biosynthesis of NO. The rat brain has been mapped for NOS using immunohistochemistry, revealing NO-specific neurons through-out the brain, including cerebral cortex, olfactory bulbs, hippocampus, amygdala, cerebellum and substantia nigra (Rodrigo *et al.* 1994; Vincent 2010). The hypothalamus has the highest concentration of nitric oxide synthase in the brain, and NO is implicated in the release of most pituitary hormones (Prevot *et al.* 2000; Brann *et al.* 1997).

Endocannabinoids are also enzymatically produced only when needed (i.e. they are not stored in vesicles) and they freely diffuse through cell membranes rather than being secreted (section 5.2.5). The localization of endocannabinoid pathways is therefore a difficult problem. Endocannabinoids are biosynthesized in the brain, including the hypothalamus, and the distribution of endocannabinoid receptors (CB1; CB2 receptors are found in the immune system) in the brain provides a means to determine where endocannabinoids exert their effects. CB1 receptors are especially abundant in the cortex, hippocampus, amygdala, striatum and cerebellum (Breivogel and Sim-Selley 2009; Mackie 2005). CB1 receptors are found in the hypothalamus and endocannabinoid signaling is important in regulating the stress response (Hill and Tasker 2012) and energy balance. In the latter case, therapeutic targeting of CB1 receptors may be a way to combat obesity (de Kloet and Woods 2009). Hypothalamic and pituitary CB1 receptors also control the reproductive axis by reducing LH and PRL secretion and the secretion of oxytocin from the posterior pituitary (Maccarrone and Wenger 2005; Pagotto *et al.* 2006; Rettori *et al.* 2010).

5.8 Drugs influencing neurotransmitters, and their receptors, in the nervous system

Numerous drugs are available that influence neurotransmitter release and receptor activation. Much of what we know about the normal functioning of the nervous system has been achieved through the use of drugs designed to interact with specific neurotransmitter receptors and we will examine neurotransmitter receptors in more detail in Chapter 10. The knowledge obtained from using drugs as tools to explore brain function has resulted in treatments for a variety of nervous system disorders of mood, emotionality, motor function, neuroendocrine secretion and drug abuse (Nestler *et al.* 2009; Iversen *et al.* 2009).

Drugs can influence neurotransmitter action in at least three ways: (1) they can act as neuromodulators to influence the synthesis, storage, release and reuptake of the neurotransmitter by the presynaptic cell; (2) they can alter the activity of deactivating enzymes, such as MAO and COMT in the synapse, thus influencing synaptic concentrations of the neurotransmitter; or (3) they can mimic (or block) the effect of the neurotransmitter at the synapse by attaching directly to specific receptors on the postsynaptic cell. Examples of some of these drug actions are discussed here. Because drugs affect all areas of the brain involving a particular neurotransmitter system (see Figure 5.15), drugs developed to treat particular disorders may have a number of side effects, some of which involve the neuroendocrine system.

5.8.1 Drugs altering neurotransmitter synthesis, storage, release and reuptake

Extensive animal experiments revealed that neurotransmitter levels can be modified using various neuropharmacological techniques (see Table 5.4). For example, reserpine severely reduces norepinephrine levels in the brain by preventing its reuptake into vesicles, thus allowing degradation to take place (reuptake into vesicles shown in Figure 5.14). Drugs are also available for *reducing* neurotransmitter levels by blocking their synthesis. In those situations where neurotransmitter levels are required to be *increased*, for example, in Parkinson's disease, this has been achieved by treating patients with a precursor to dopamine, L-Dopa, which permits synthesis of additional dopamine in the striatum (see section 5.7). There is also evidence that neurotransmitter levels in the brain might be altered through components of our diet, such as tyrosine and tryptophan (see section 5.9). Neurotransmitter levels in the synapse can be increased by blocking the normal inactivation pathways. For example, drugs such as clorgyline block the degrading enzyme monoamine oxidase (MAO; see Figure 5.16), whereas entacapone blocks COMT (catechol o-methyl transferase), thereby elevating synaptic norepinephrine levels. Other drugs (e.g. Prozac) increase both serotonin and norepineprine levels by blocking their reuptake through their respective transporters (Figure 5.16). The effects of some of these drugs on postsynaptic cellular responses are outlined in Figure 5.17.

Table 5.4 lists examples of many drugs that affect neurotransmitter levels or activity, some of which are used clinically and could affect aspects of endocrine function.

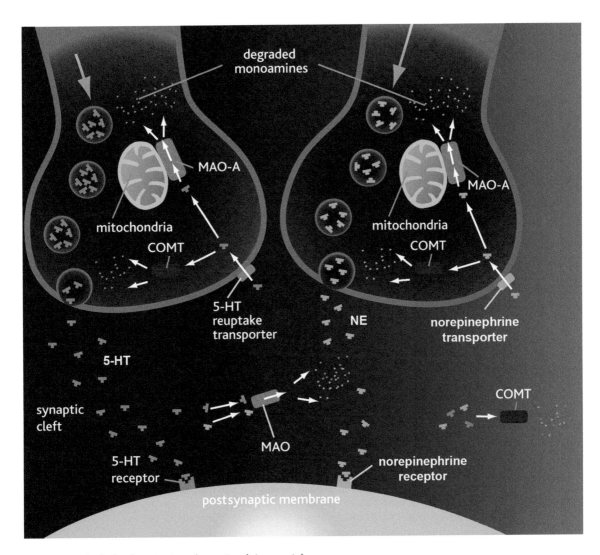

Figure 5.16 Blockade of serotonin and norepinephrine reuptake

This figure illustrates how levels of neurotransmitters, such as serotonin (5-HT) and norepinephrine (NE), are regulated, and how they can be modified by therapeutic drugs. For example, depression is associated with reduced levels of serotonin and norepinephrine in the brain. Serotonin and norepinephrine reuptake inhibitors such as Prozac are thought to restore the levels of these neurotransmitters in the synaptic cleft by blocking their reuptake and subsequent degradation. This reuptake blockade theoretically leads to the accumulation of catecholamines in the synaptic cleft and the concentration returns to within the normal range. The figure also suggests that blockade of monoamine oxidase (MAO)- or catechol o-methyl transferase (COMT)-induced degradation would also increase synaptic levels of serotonin and norepinephrine. Copyright CNSforum, Lundbeck Institute: www.cnsforum.com.

5.8.2 Drugs that act at postsynaptic receptors

Another way in which synaptic transmission can be altered is by using drugs that bind to postsynaptic receptors and act as neurotransmitter *agonists* or *antagonists*. *Agonist* drugs can mimic or even magnify the effect of the neurotransmitter. At an excitatory synapse, an agonist will increase the excitation of the postsynaptic cell. If the synapse is inhibitory, an

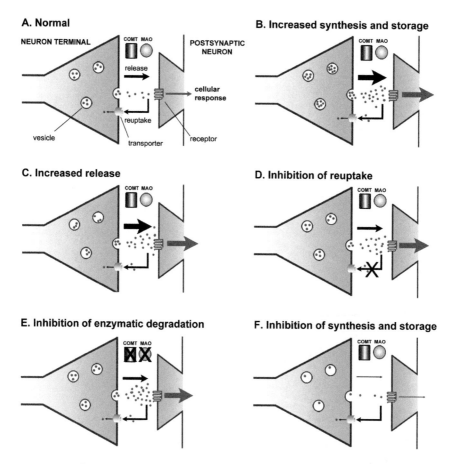

Figure 5.17 The mechanisms through which drugs alter the amount of synaptic catecholamine neurotransmitter available in the synapse and the resulting postsynaptic cellular response
A. This represents the normal situation, where storage, release, reuptake and enzymatic metabolism of norepinephrine (by MAO and COMT enzymes) induce a postsynaptic cellular response (red arrow) via a postsynaptic G-protein-coupled receptor; B. increased synthesis and storage, for example, following tyrosine or L-Dopa treatment, may increase release and postsynaptic cellular response (large arrows) to norepinephrine or dopamine; C. increased secretion of norepinephrine, for example, following increased nerve stimulation or treatment with amphetamine, induces an enhanced postsynaptic cellular response; D. inhibition of reuptake, for example, by antidepressants such as imipramine, increases synaptic levels of norepinephrine, which in turn induces an elevated postsynaptic cellular response; E. inhibitors of MAO- or COMT-induced metabolism of catecholamines such as clorgyline also increase synaptic levels of neurotransmitter and the postsynaptic cellular response; F. inhibition of the storage of norepinephrine (by reserpine), or inhibition of synthesis, reduces synaptic levels of the neurotransmitter and therefore inhibits the postsynaptic cellular response.
   Abbreviations: COMT, catechol o-methyl transferase; MAO, monoamine oxidase.

agonist will cause increased inhibition. A neurotransmitter *antagonist*, on the other hand, blocks the receptors for that neurotransmitter. The antagonist has a chemical structure that allows it to bind to the receptor, but when it does so it does not stimulate the cell. The real neurotransmitter that is present in the synapse will not be able to bind to the receptor because it is blocked by the antagonist. Treatment with a neurotransmitter agonist has the

Table 5.4 **Drugs altering the amount of neurotransmitter available in the synapse and some neurological disorders they are used to treat**

Transmitter	Drug	Action
GABA	Valproate (epilepsy; bipolar disorder)	(+) enhances GABA function
	Tiagebine (epilepsy)	(+) blocks reuptake
	Vigabatrine (epilepsy)	(+) blocks GABA transaminase
	Gabapentin (epilepsy; pain)	(+) enhances GABA function
	Benzodiazepine (anxiety)	(+) enhances GABA function
	Tetanus toxin	(−) inhibits release
	Allylglycine	(−) inhibits synthesis
Acetylcholine	Choline	(+) stimulates synthesis and release
	Donepezil (Alzheimer's)	(+) acetylcholinesterase inhibitor
	Black Widow spider venom	(+) promotes release
	Botulinum toxin	(−) inhibits release
	Chantix (smoking cessation)	(+) nicotinic agonist/antagonist
	Pyridines	(−) block synthesis
	Scopolamine (motion sickness)	(−) blocks muscarinic receptor
Dopamine	Tyrosine and L-Dopa (Parkinson's)	(+) increase dopamine synthesis
	Clorgyline	(+) inhibits MAO_A
	Deprenyl (Parkinson's)	(+) inhibits MAO_B
	Entacapone (Parkinson's)	(+) inhibits COMT
	Amphetamines	(+) blocks reuptake; increases release
	Cocaine	(+) blocks reuptake
	Ritalin (ADHD)	(+) blocks reuptake
	Reserpine	(−) causes depletion by inhibiting vesicle uptake
	α-methyltyrosine	(−) inhibits dopamine synthesis
Norepinephrine	Imipramine (tricyclic antidepressant)	(+) blocks reuptake
	Cocaine	(+) blocks reuptake
	Amphetamines	(+) blocks reuptake
	Effexor (antidepressant)	(+) mixed NE/5-HT uptake inhibition
	Clorgyline; Deprenyl	(+) MAO inhibition
	Entacapone	(+) COMT inhibition
	α-methyltyrosine	(−) inhibits synthesis
	Reserpine	(−) depletes stores by inhibiting vesicle reuptake
Serotonin	Tryptophan	(+) increases synthesis
	Iproniazid	(+) inhibits MAO
	Imipramine (antidepressant)	(+) blocks reuptake
	Prozac (antidepressant)	(+) blocks reuptake
	Ecstasy (recreational drug)	(+) blocks reuptake and is a neurotoxin
	p-chlorophenylalanine	(−) inhibits synthesis
	Reserpine	(−) depletion by blocking reuptake into vesicles
	LSD	(−) inhibits release

Data obtained from Iversen *et al.* 2009; and Nestler *et al.* 2009.

same effect as the neurotransmitter itself: it binds to the postsynaptic cell and either excites or inhibits it. Some synthetic agonists may be much more powerful than the neurotransmitter. For example, LSD (lysergic acid diethylamide) is a serotonin agonist that exerts its psychedelic effects by binding to serotonin receptors of the 5-HT$_{2A}$ subtype (Iversen *et al.* 2009). Other psychedelic drugs, like PCP (phencyclidine) and ketamine, bind as antagonists to the NMDA (N-methyl-D-aspartate) subtype of glutamate receptors. Table 5.5 provides examples of drugs acting as agonists and antagonists at various neurotransmitter receptors.

Table 5.5 Examples of drugs acting as neurotransmitter agonists and antagonists

Receptor	Drug	Action
Excitatory amino acids Glutamate and aspartate	N-methyl-D-aspartate (NMDA)	(+) agonist
	Kainic acid	(+) agonist
	Quinolinic acid	(+) agonist
	Domoic acid	(+) agonist
GABA$_A$	Muscimol	(+) agonist
GABA$_B$	Baclofen	(+) agonist
GABA$_A$	Bicuculline	(−) antagonist
GABA$_B$	Phaclophen	(−) antagonist
Acetylcholine *Muscarinic*	Oxotremorine	(+) agonist
	Pilocarpine	(+) agonist
	Bethanechol	(+) agonist
	Various synthetic Componds	(+) agonist
	Scopolamine	(−) antagonist
	Pirenzepine	(−) antagonist
	Methoctramine	(−) antagonist
Nicotinic	Epibatidine	(+) agonist
	Anatoxin A	(+) agonist
	Mecamylamine	(−) antagonist
	α-bungarotoxin	(−) antagonist
Dopamine	Bromocryptine	(+) agonist
	Quinpirole	(+) agonist
	Haloperidol	(−) antagonist
	Olanzapine	(−) antagonist
	Pimozide	(−) antagonist
	Raclopride	(−) antagonist
Norepinephrine	Clonidine	(+) α$_2$ receptor agonist
	Isoproterenol	(+) β$_1$ receptor agonist
	Propranolol	(−) β receptor antagonist
	Prazosin	(−) α$_1$ receptor antagonist
	Yohimbine	(−) α$_2$ receptor antagonist

Receptor	Drug	Action
Serotonin	Sumatriptan	(+) 5-HT$_{1B}$ agonist
	Dimethyltryptamine (DMT)	(+) 5-HT$_2$ agonist
	Lysergic acid diethylamide (LSD)	(+) 5-HT$_{2A}$ agonist
	Ketanserin	(−) 5-HT$_2$ antagonist
	Isamoltane	(−) 5-HT$_{1B}$ antagonist
Histamine	Impromidine	(+) H$_2$ agonist
	Bromophenyl histamine	(+) H$_1$ agonist
	Mepyramine	(−) H$_1$ antagonist
	Ranitidine	(−) H$_2$ antagonist
*Opioid	Enkephalins	(+) δ-agonist
	β-endorphins	(+) μ-agonist
	Endomorphins	(+) μ-agonist
	Dynorphins	(+) κ-agonist
	Naloxone	(−) antagonist all subtypes
	Naltrexone	(−) antagonist all subtypes
**Endocannabinoids	Tetrahydrocannabinol	(+) agonist
	Anandamide	(+) CB1 agonist
	Rimonabant	(−) CB1 antagonist

Table 5.5 **(Cont.)**

Data summarized from Iversen *et al.* (2009); and Nestler *et al.* (2001).

* Julius (1997).

** van Gaal *et al.* (2005).

5.8.3 Drugs, the neuroendocrine system and behavior

Drugs that alter the release or action of neurotransmitters also influence hormone release (Chapter 6), and alter arousal levels, emotional states and motivated behavior. For example, depression is associated with low catecholamine levels and schizophrenia with high levels, so drugs that alter catecholaminergic activity also influence these clinical symptoms. Altering catecholamine levels also alters sexual and aggressive motivation, and the activity of the neuroendocrine system. In general, those drugs that alter neurotransmitter levels in the brain may also induce neuroendocrine side effects such as compromised appetite, sex drive and prolactin secretion (Papakostas *et al.* 2006).

5.9 Can nutrients modify neurotransmitter levels and behavior?

The biosynthesis of some neurotransmitters, such as norepinephrine and serotonin, depends on the availability of precursor amino acids from the diet. This research was initiated by Richard Wurtman and his colleagues (Wurtman 1982; Milner and Wurtman 1986) and has been continued by Fernstrom and co-workers (Fernstrom *et al.* 2007; Choi *et al.* 2009; Fernstrom 2012). Their hypothesis is that amino acids in the diet influence the rate of synthesis of the neurotransmitters norepinephrine (from tyrosine; see Figure 5.8) and serotonin (from tryptophan) with possible functional consequences. For example, *high carbohydrate* meals increase blood insulin levels that in turn increase brain serotonin levels. This occurs because insulin decreases the levels of other amino acids entering the brain (by increasing their uptake into muscle, for example), thus allowing tryptophan to enter the CNS unimpeded. Thus, a high carbohydrate intake may be followed by tiredness and lethargy, and serotonin promotes sleep. In contrast, a *high protein* meal elevates blood levels of all amino acids and, therefore, lowers the ratio of tryptophan to other amino acids entering the brain. A similar argument suggests that foods high in tyrosine stimulate the synthesis of dopamine and other catecholamines. Such diet therapy may be of use in treating disorders such as depression, hyperactivity and hypertension. In particular, dietary *reductions* of tyrosine or tryptophan may reduce neurotransmitter levels in humans. For example, reduced tyrosine intake in healthy volunteers induced a fall in brain dopamine levels and a concomitant impairment in working memory (Mehta *et al.* 2005). For tryptophan, an *acute* reduced intake induced a significant lowering of mood in humans and this is consistent with the relationship between serotonin and depression (Riedel *et al.* 2002; Mendelsohn *et al.* 2009; Crockett *et al.* 2012). These results raise interesting questions about whether starvation – due to famine, extreme dieting or anorexia nervosa – might lead to abnormal brain serotonin levels. Reductions in brain serotonin or dopamine could have effects on the neuroendocrine system. For example, patients with anorexia nervosa have no menstrual cycles (see section 4.3.3; Figure 4.5) and have abnormal GH and adrenal output, which suggests abnormal hypothalamic function (Miller 2011). There is also evidence that increases in brain tryptophan induce prolactin secretion and reduce cortisol release, particularly in stress-sensitive humans (Markus *et al.* 2000).

Dietary proteins and amino acids also affect hormone secretion from the gastrointestinal (GI) tract and these GI hormones may regulate catecholamine and serotonin synthesis in the brain (Jahan-Mihan *et al.* 2011; Choi *et al.* 2009; Fernstrom and Fernstrom 2007). A specific diet (the *ketogenic diet*, low in protein and high in fat), used for many years to treat childhood epilepsy, is also effective in modifying brain neurotransmitters. This diet reduces the levels of serotonin and dopamine metabolites, but not norepinephrine, in cerebrospinal fluid of children given the diet for three months (Dahlin *et al.* 2012). The results suggest that serotonin and dopamine neurotransmission in the brain are reduced by the ketogenic diet. Other studies indicate that the ketogenic diet also alters the metabolic state by increasing leptin and decreasing insulin levels (Thio 2012). It seems likely that other hormones, especially from the GI tract, implicated in body weight control, will be modified by the ketogenic diet. The involvement of *neuropeptides* in GI tract feedback to the brain, and the regulation of food intake and body weight, is discussed in Chapter 12.

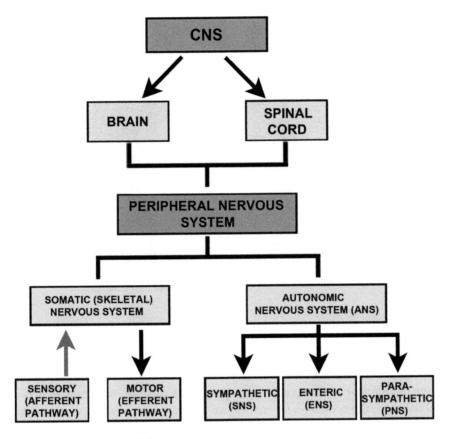

Figure 5.18 The divisions of the nervous system

5.10 Divisions of the nervous system

The nervous system can be divided into a number of divisions and sub-divisions, as shown in Figure 5.18. The *central nervous system* (CNS) consists of neurons and neural pathways of the brain and spinal cord described in section 5.7 and the *peripheral nervous system* consists of the peripheral sensory and motor nerves that extend throughout the body from the central nervous system. The peripheral nervous system has a somatic (skeletal) division and an autonomic division.

5.10.1 The somatic nervous system

The somatic division of the peripheral nervous system receives sensory (afferent) input from the sense organs and controls motor (efferent) output to the skeletal muscles through the cranial nerves and the spinal nerves. The somatic sensory nerves receive tactile input from the skin and muscles, while the somatic motor nerves form neuro-muscular synapses that use acetylcholine as a neurotransmitter. The somatic nervous system is concerned with conscious (voluntary) actions and reflexes and sends motor

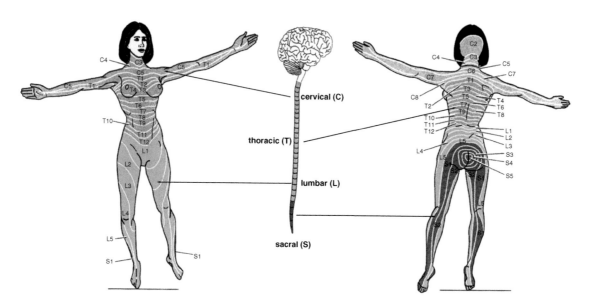

Figure 5.19 Location and distribution of spinal nerves Reproduced with permission (Boron and Boulpaep 2005).

nerves to those muscles that are under voluntary control. The cranial sensory and motor nerves include the *olfactory* nerve (smell), the *optic* nerve (vision), the *acoustic* nerve (hearing and balance), the *oculomotor, trochlear* and *abducens* nerves (eye movement), the *trigeminal* nerve (receives facial sensations and controls jaw muscles), the *facial* nerve (receives taste stimuli and controls the facial muscles), the *glossopharyngeal* nerve (receives taste stimuli and controls the muscles of the larynx used in speech), the *vagus* nerve (supplies heart, lungs, stomach, kidneys and intestines), the *accessory* nerve (controls head and neck movement) and the *hypoglossal* nerve (controls tongue movements).

The somatic spinal sensory and motor nerves extend from the spinal cord in four groups as shown in Figure 5.19. These are the cervical (nerves which innervate the throat, chest, arms and hands), thoracic (which innervate the trunk of the body), lumbar (which innervate the front of the legs and the feet) and sacral regions (which innervate the soles of the feet and back of the legs and the coccygeal nerve). The nerves of the somatic nervous system are myelinated and monosynaptic, extending directly from neurons in the spinal cord to their target tissues.

5.10.2 The autonomic nervous system

The ANS, as shown in Figure 5.20, receives sensory input from the viscera and sends motor nerves to the visceral (smooth) muscles of the heart, arteries and gastro-intestinal system, etc., to the glandular tissues of the body, such as the salivary glands, lachrymal (tear) glands, sweat glands, and to the peripheral endocrine glands (thymus, adrenal, thyroid and gonads). The sensory nerves of the ANS are non-myelinated nerves that project from the heart, lungs and gastrointestinal tract to

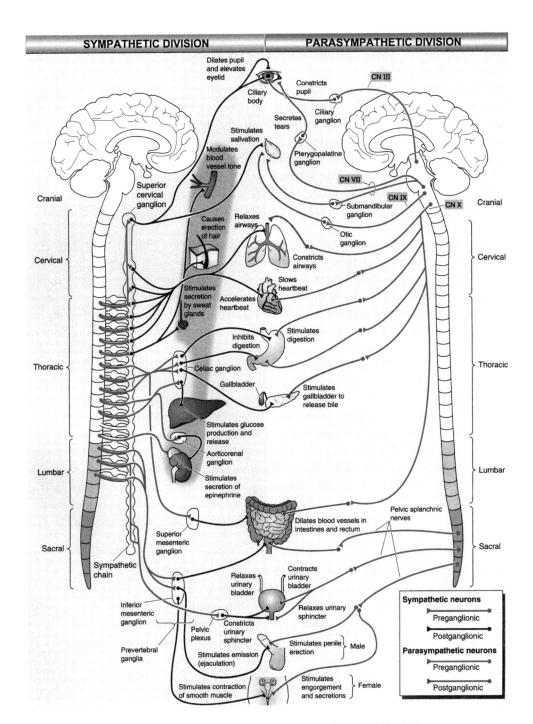

Figure 5.20 Organization of the sympathetic and parasympathetic divisions of the ANS
The *left panel* shows the sympathetic division. The cell bodies of sympathetic preganglionic neurons (*red*) are in the intermediolateral column of the thoracic and lumbar spinal cord (T1-L3). Their axons project to paravertebral ganglia (the sympathetic chain) and prevertebral ganglia. Postganglionic neurons (*blue*) therefore have long projections to their targets. The *right panel* shows the parasympathetic division. The cell bodies of parasympathetic preganglionic neurons (*orange*) are either in the brain (midbrain, pons, medulla) or in the sacral spinal cord (S2-S4). Their axons project to ganglia very near (or even inside) the end organs. Postganglionic neurons (*green*) therefore have short projections to their targets. Reproduced with permission (Boron and Boulpaep 2005).

the autonomic sensory ganglia. Sensory afferents are carried to the spinal cord and brain by the vagus, pelvic, splanchnic and other autonomic nerves. Neural transmission in the visceral afferent nerves involves a wide range of neuropeptides, including substance P, the endogenous opioid peptides, cholecystokinin, neuropeptide Y and others (see Chapters 11 and 12). The efferent branch of the ANS, that is, the nerves conveying information from the brain, has three divisions: the sympathetic (SNS), parasympathetic (PNS) and enteric (ENS) nervous systems, which regulate those activities over which there is no conscious control, the involuntary muscles and glands. The visceral muscles and glands receive paired inputs, one from the sympathetic and one from the parasympathetic branches of the ANS. The enteric nervous system is that branch of the ANS that innervates the gastrointestinal tract.

The nerves of the sympathetic ANS synapse at ganglia outside of the spinal cord and projections are sent from these autonomic ganglia to the target organs. The autonomic ganglia include the cervical ganglia, the celiac ganglia and the mesenteric ganglia (see Figure 5.20). The sympathetic nerves extend from the thoracic and lumbar regions of the spinal cord to the sympathetic ganglia and then to their target organs. The nerves of the parasympathetic ANS extend from the cranial and sacral regions of the spinal cord to the autonomic ganglia and then to their target organs. The SNS activates the visceral muscles and glands during stress and thus the SNS deals with emergency responses by increasing blood pressure, causing sweating and stimulating adrenaline release. The PNS acts to maintain the body's systems at a steady state and opposes the excitatory actions of the SNS. The PNS is involved in growth promoting and energy conserving functions, while the SNS increases energy expenditure. The SNS causes pupil dilation, relaxes the bronchi in the lungs, increases heart rate, stimulates adrenaline secretion, inhibits gastrointestinal activity and relaxes the bladder. The PNS constricts the pupils, stimulates salivation, constricts the bronchi in the lungs, slows heart rate, stimulates gastrointestinal activity and constricts the bladder. The SNS nerves are adrenergic, while the PNS nerves are cholinergic (Boron and Boulpaep 2005).

5.10.3 The enteric nervous system

The third branch of the ANS, the enteric nervous system (ENS), controls intestinal motility and secretion. It consists of over 100 million neurons in the form of two nerve plexuses located within the wall of the GI tract itself (Figure 5.21; Benarroch 2007). Although this local nerve network is innervated by neurons of the CNS, it can operate independently and has been referred to as a *minibrain*. Nevertheless, the entire system of SNS, PNS and ENS exerts fine control over the GI tract, with neural information traveling both ways, to and from the CNS (Figure 5.22) (Goyal and Hirano 1996). Thus, sensory nerves of the enteric nervous system send information from the intestinal walls and monitor the chemical environment of the gastrointestinal tract, while the motor nerves control the muscles, vasculature and glandular secretions of the gastrointestinal tract. More than 20 neurotransmitters and neuropeptides have been localized as components of the ENS, including catecholamines, GABA, ATP, nitric oxide, opioid peptides, peptide

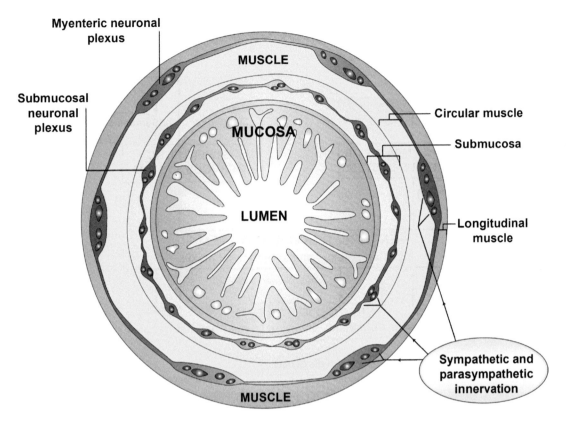

Figure 5.21 The enteric nervous system
Schematic representation of a transverse section through the small intestine. Enteric neurons are organized in ganglia (*green*) found within two main plexi. An outer myenteric plexus occupies a position between the longitudinal and circular muscle layers. An inner submucosal plexus resides within the submucosa. The enteric nervous system is connected with the CNS via neurons of the ANS. Reproduced with permission (Heanue and Pachnis 2007).

YY and galanin. In addition to these signaling molecules, endocrine cells of the GI tract and stomach release many additional hormones that act as paracrine mediators (see section 2.2.7).

5.10.4 Central regulation of the autonomic nervous system

All three branches of the ANS are regulated by the CNS as shown in Figure 5.20. Nuclei of the hypothalamus and limbic system regulate the efferent pathways of the SNS, PNS and ENS through the nucleus of the solitary tract that sends fibers to the brainstem nuclei and spinal autonomic ganglia (see Figure 5.15). Visceral afferents from the periphery project to the autonomic ganglia, which send fibers to the spinal cord and brain stem nuclei, and on to hypothalamic and limbic nuclei through pathways shown in Figure 5.15. The hypothalamus thus integrates the activity of the ANS as well as the neuroendocrine system as discussed in section 4.1 (Boron and Boulpaep 2005).

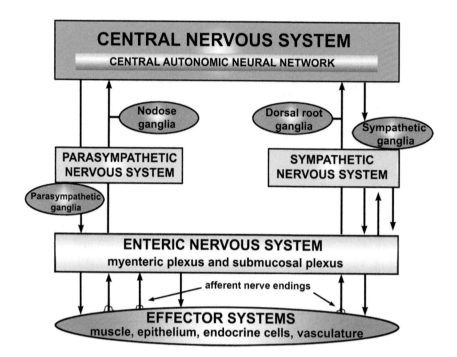

Figure 5.22 Diagram of innervation of the GI tract
As noted in Figure 5.21, the nerve plexuses in the gut are an independently functioning network. The enteric nervous system (ENS) can regulate the effector systems in the gut (e.g. muscle) directly or may do so indirectly through its action on intermediate cells, which include endocrine cells. Information passes both ways, from CNS to ENS and vice versa. Cell bodies of the vagal and splanchnic afferent neurons are located in the nodose ganglia and the dorsal-root ganglia, respectively. Redrawn with permission of Professor R. K. Goyal (Goyal and Hirano 1996).

5.11 Summary

This chapter examined the nature of communication between nerve cells through synaptic activity of neurotransmitters. It also defined the criteria necessary to establish whether chemicals found in the brain act as neurotransmitters. Details of the synthesis, storage, release and deactivation of amino acid, amine and peptide neurotransmitters are described, as well as the mechanism for neurotransmitter release and the actions of neurotransmitters at their post-synaptic receptors. The different types of receptors for neurotransmitters are described and the mechanisms that deactivate the neurotransmitters are summarized. Neurotransmitters are found in specific nerve pathways in the brain and the location of some of these pathways is outlined. Finally, the effects of drugs on the synthesis, storage, release, reuptake and deactivation of neurotransmitters was discussed. Other drugs act on neurotransmitter receptors as agonists or antagonists and the mechanism of action of these drugs was examined. Neurotransmitters and the drugs that influence them regulate the release of hypothalamic hormones and thus affect hormone-dependent behaviors. Because drugs are used both clinically and experimentally to alter neuroendocrine activity, they are important tools in the study of neuroendocrinology. Since dietary amino acids may influence neurotransmitter synthesis, diet may also influence neuroendocrine secretion. Finally, the somatic and autonomic divisions of the nervous system were described and the differences between the sympathetic, parasympathetic and enteric branches of the ANS were defined. The hypothalamus regulates both the autonomic and endocrine systems through a number of different neurotransmitters and neuropeptides.

FURTHER READING

Besser, G. M. and Thorner, M. O. (2002). *Comprehensive Clinical Endocrinology*, 3rd edn. (St. Louis, MO: Mosby).

Brady, S. T., Siegel, G. J., Albers, R. W. and Price, D. L. (2012). *Basic Neurochemistry*, 8th edn. (New York: Elsevier).

Burgoyne, R. D. and Morgan, A. (2003). "Secretory granule exocytosis," *Physiol Rev* 83, 581–632.

Iversen, L. L., Iversen, S. D., Bloom, D. E. and Roth, R. H. (2009). *Introduction to Neuropsychopharmacology* (New York: Oxford University Press).

Scheller, R. H. (2013). "In search of the molecular mechanism of intracellular membrane fusion and neurotransmitter release," *Nat Med* 19, 1232–1235.

Vincent, S. R. (2010). "Nitric oxide neurons and neurotransmission," *Prog Neurobiol* 90, 246–255.

Ward, S. J. and Raffa, R. B. (2011). "Rimonabant redux and strategies to improve the future outlook of CB1 receptor neutral-antagonist/inverse-agonist therapies," *Obesity* 19, 1325–1334.

REVIEW QUESTIONS

5.1 Name three catecholamine neurotransmitters.

5.2 Name three ways in which drugs can increase the amount of neurotransmitter available at a synapse.

5.3 Is GABA an excitatory or inhibitory neurotransmitter?

5.4 What is the cholinergic neurotransmitter?

5.5 Which neurotransmitter is synthesized from dopamine by the enzyme dopamine beta-hydroxylase (DBH)?

5.6 What is the difference between a neurotransmitter agonist and an antagonist?

5.7 Which two enzymes deactivate norepinephrine?

5.8 Which neurotransmitter is synthesized from the amino acid tryptophan?

5.9 Would an antagonist for dopamine stimulate production of the second messenger cyclic AMP at a D1 receptor in a postsynaptic cell? (yes or no)

5.10 Where are neurotransmitters stored before release and what triggers their release?

5.11 Name two neurotransmitters that are not stored in vesicles.

5.12 How does a transmitter cause a postsynaptic cell to "fire"?

5.13   Which neurotransmitter acts at nicotinic receptors?

5.14   Which neurotransmitter has its primary cell bodies in the locus ceruleus?

5.15   What two general ways can neurotransmitters be deactivated?

ESSAY QUESTIONS

5.1   Are the eight criteria that define a neurotransmitter necessary and sufficient? Could they be reduced or expanded?

5.2   Discuss the mechanisms involved in the release of a neurotransmitter into the synapse.

5.3   Discuss the functions of synaptic vesicles.

5.4   How do the alpha and beta adrenergic receptors differ in their functions?

5.5   Discuss the five major acetylcholine pathways in the brain with reference to the neuroendocrine system.

5.6   MAO inhibitors and tricyclic antidepressants are both used to treat depression, yet act through different mechanisms. Discuss these mechanisms and explain how these two types of drugs both reduce depression.

5.7   Discuss the problems in determining how a particular type of drug, such as a beta blocker, actually affects the person treated with them. That is, what receptors are influenced and what are the physiological results? Why do two drugs with the same action, such as two beta blockers, have different physiological effects?

5.8   Do dietary nutrients alter neurotransmitter levels enough to influence memory or other neural processes in humans?

REFERENCES

Baranano, D. E., Ferris, C. D. and Snyder, S. H. (2001). "Atypical neural messengers," *Trends Neurosci* 24, 99–106.

Belelli, D. and Lambert, J. J. (2005). "Neurosteroids: endogenous regulators of the $GABA_A$ receptor," *Nature Rev Neurosci* 6, 565–575.

Benarroch, E. E. (2007). "Enteric nervous system: functional organization and neurologic implications," *Neurol* 69, 1953–1957.

Brann, D. W. (1995). "Glutamate: A major excitatory transmitter in neuroendocrine regulation," *Neuroendocrinology* 61, 213–225.

Brann, D. W., Bhat, G. K., Lamar, C. A. and Mahesh, V. B. (1997). "Gaseous transmitters and neuroendocrine regulation," *Neuroendocrinology* 65, 385–395.

Breivogel, C. S. and Sim-Selley, L. J. (2009). "Basic neuroanatomy and neuropharmacology of cannabinoids," *Int Rev Psych* 21, 113–121.

Boron, W. F. and Boulpaep, E. L. (2005). *Medical Physiology,* updated edn. (Philadelphia, PA: Elsevier Saunders).

Burgoyne, R. D. and Morgan, A. (2003). "Secretory granule exocytosis," *Physiol Rev* 83, 581–632.

Choi, S., Disilvio, B., Fernstrom, M. H. and Fernstrom, J. D. (2009). "Meal ingestion, amino acids and brain neurotransmitters: effects of dietary protein source on serotonin and catecholamine synthesis rates," *Physiol Behav* 98, 156–162.

Christensen, R., Kristensen, P. K., Bartels, E. M., Bliddal, H. and Astrup, A. (2007). "Efficacy and safety of the weight-loss drug rimonabant: a meta-analysis of randomised trials," *Lancet* 370, 1706–1713.

Christie, M. J. and Vaughn, C. W. (2001). "Cannabinoids act backwards," *Nature* 410, 527–530.

Crockett, M. J., Clark, L., Roiser, J. P., Robinson, O. J., Cools, R., Chase, H. W. *et al.* (2012). "Converging evidence for central 5-HT effects in acute tryptophan depletion," *Mol Psych* 17, 121–123.

Dahlin, M., Mansson, J.-E. and Amark, P. (2012). "CSF levels of dopamine and serotonin, but not norepinephrine, metabolites are influenced by the ketogenic diet in children with epilepsy," *Epilepsy Res* 99, 132–138.

de Kloet, A. D. and Woods, S. C. (2009). "Minireview: Endocannabinoids and their receptors as targets for obesity," *Endocrinology* 150, 2531–2536.

Deng, J., Lei, C., Chen, Y., Fang, Z., Yang, Q., Zhang, H. *et al.* (2014). "Neuroprotective gases – fantasy or reality for clinical use?" *Prog Neurobiol* 115, 210–245.

Eyigor, O., Centers, A. and Jennes, L. (2001). "Distribution of ionotropic glutamate receptor subunit mRNAs in the rat hypothalamus," *J Comp Neurol* 434, 101–124.

Fernstrom, J. D. (2012). "Large neutral amino acids: dietary effects on brain neurochemistry and function," *Amino Acids* 45, 419–430.

Fernstrom, J. D. and Fernstrom, M. H. (2007). "Tyrosine, phenylalanine, and catecholamine synthesis and function in the brain," *J Nutr* 137, 1539S–1547S; discussion 1548S.

Goyal, R. K. and Hirano, I. (1996). "The enteric nervous system," *N Engl J Med* 334, 1106–1115.

Guzmán, M. (2003). "Cannabinoids: potential anticancer agents," *Nat Rev Cancer* 3, 745–755.

Haas, H. L., Sergeeva, O. A. and Selbach, O. (2008). "Histamine in the nervous system," *Physiol Rev* 88, 1183–1241.

Heanue, T. A. and Pachnis, V. (2007). "Enteric nervous system development and Hirschsprung's disease: advances in genetic and stem cell studies," *Nat Rev Neurosci* 8, 466–479.

Hill, M. N. and Tasker, J. G. (2012). "Endocannabinoid signaling, glucocorticoid-mediated feedback and regulation of the hypothalamic-pituitary-adrenal axis," *Neuroscience* 204, 5–16.

Iremonger, K. J., Constantin, S., Liu, X. and Herbison, A. E. (2010). "Glutamate control of GnRH neuron excitability," *Brain Res* 1364, 35–43.

Iversen, L. L., Iversen, S. D., Bloom, D. E. and Roth, R. H. (2009). *Introduction to Neuropsychopharmacology* (New York: Oxford University Press).

Jahan-Mihan, A., Luhovyy, B. L., El Khoury, D. and Anderson, G. H. (2011). "Dietary proteins as determinants of metabolic and physiologic functions of the gastrointestinal tract," *Nutrients* 3, 574–603.

Jorgensen, H. S. (2007). "Studies on the neuroendocrine role of serotonin," *Dan Med Bull* 54, 266–288.

Julius, D. (1997). "Another opiate for the masses?" *Nature* 386, 442.

Krsmanovic, L. Z., Mores, N., Navarro, C. E., Saeed, S. A., Arora, K. K. and Catt, K. J. (1998). "Muscarinic regulation of intracellular signaling and neurosecretion in gonadotrophin-releasing hormone neurons," *Endocrinology* 139, 4037–4043.

Kuhar, M., De Souza, E. B. and Unnerstall, J. R. (1986). "Neurotransmitter receptor mapping by autoradiography and other methods," *Annu Rev Neurosci* 9, 27–59.

Levy, B. H. and Tasker, J. G. (2012). "Synaptic regulation of the hypothalamic-pituitary-adrenal axis and its modulation by glucocorticoids and stress," *Front Cell Neurosci* 6, 24–32.

Maccarrone, M. and Wenger, T. (2005). "Effects of cannabinoids on hypothalamic and reproductive function," *Hand Exp Pharmacol* 168, 555–571.

Mackie, K. (2005). "Distribution of cannabinoid receptors in the central and peripheral nervous system," *Handb Exp Pharmacol* 168, 299–325.

Mansour, A., Fox, C. A., Akil, H. and Watson, S. J. (1995). "Opioid-receptor mRNA expression in the rat CNS: anatomical and functional implications," *Trends Neurosci* 18, 22–29.

Markus, C. R., Olivier, B., Panhuysen, E. M., Van der Gugten, J., Alles, M. S., Tuiten, A. *et al.* (2000). "The bovine protein α-lactalbumin increases the plasma ratio of tryptophan to the other large neutral amino acids, and in vulnerable subjects raises brain serotonin activity, reduces cortisol concentration, and improves mood under stress," *Am J Clin Nutr* 71, 1536–1544.

Mehta, M. A., Gumaste, D., Montgomery, A. J., McTavish, S. F. and Grasby, P. M. (2005). "The effects of acute tyrosine and phenylalanine depletion on spatial working memory and planning in healthy volunteers are predicted by changes in striatal dopamine levels," *Psychopharmacology (Berl)* 180, 654–663.

Mendelsohn, D., Riedel, W. J. and Sambeth, A. (2009). "Effects of acute tryptophan depletion on memory, attention and executive functions: a systematic review," *Neurosci Biobehav Rev* 33, 926–952.

Miller, K. K. (2011). "Endocrine dysregulation in anorexia nervosa update," *J Clin Endocr Metab* 96, 2939–2949.

Milner, J. D. and Wurtman, R. J. (1986). "Catecholamine synthesis: physiological coupling to precursor supply," *Biochem Pharmacol* 35, 875–881.

Nestler, E. J., Hyman, S. E. and Malenka, R. C. (2009). *Molecular Neuropharmacology: A Foundation for Clinical Neuroscience*, 2nd edn. (New York: McGraw-Hill).

Nicholls, J. G., Martin, A. R., Wallace, B. G. and Fuchs, P. A. (2001). *From Neuron to Brain*, 4th edn. (Sunderland: Sinauer Associates).

Olson, K. R. (2011). "The therapeutic potential of hydrogen sulfide: separating hype from hope," *Am J Physiol Regul Integr Comp Physiol* 301, R297–312.

Otto, M. W. (2011). "Expanding findings on D-cycloserine augmentation of therapeutic learning: a role for social learning relative to autism spectrum disorders?" *Biol Psych* 70, 210–211.

Pagotto, U., Marsicano, G., Cota, D., Lutz, B. and Pasquali, R. (2006). "The emerging role of the endocannabinoid system in endocrine regulation and energy balance," *Endocr Rev* 27, 73–100.

Papakostas, G. I., Miller, K. K., Petersen, T., Sklarsky, K. G., Hilliker, S. E., Klibanski, A. *et al.* (2006). "Serum prolactin levels among outpatients with major depressive disorder during the acute phase of treatment with fluoxetine," *J Clin Psychiatry* 67, 952–957.

Perry, E., Walker, M., Grace, J. and Perry, R. (1999). "Acetylcholine in mind: a neurotransmitter correlate of consciousness?" *Trends Neurosci* 22, 273–280.

Prevot, V., Bouret, S., Stefano, G. B. and Beauvillain, J. (2000). "Median eminence nitric oxide signalling," *Brain Res Rev* 34, 27–41.

Purves, D., Augustine, G. J., Fitzpatrick, D., Hall, W. C., LaMantia, A.-S., McNamara, J. O. *et al.* (2008). *Neuroscience*, 4th edn. (Sunderland: Sinauer Associates).

Rettori, V., De Laurentiis, A. and Fernandez-Solari, J. (2010). "Alcohol and endocannabinoids: neuroendocrine interactions in the reproductive axis," *Exp Neurol* 224, 15–22.

Richard, D. M., Dawes, M. A., Mathias, C. W., Acheson, A., Hill-Kapturczak, N. and Dougherty, D. M. (2009). "L-tryptophan: basic metabolic functions, behavioral research and therapeutic indications," *Int J Tryptophan Res* 2, 45–60.

Riedel, W. J., Klaassen, T. and Schmitt, J. A. (2002). "Tryptophan, mood, and cognitive function," *Brain Behav Immun* 16, 581–589.

Rodrigo, J., Springall, D. R., Uttenthal, O., Bentura, M. L., Abadia-Molina, F., Riveros-Moreno, V. *et al.* (1994). "Localization of nitric oxide synthase in the adult rat brain," *Phil Trans Royal Soc Lond B* 345, 175–221.

Scheller, R. H. (2013). "In search of the molecular mechanism of intracellular membrane fusion and neurotransmitter release," *Nat Med* 19, 1232–1235.

Squire, L. R., Berg, D. E., Bloom, F. E., du Lac, S., Ghosh, A. and Spitzer, N. C. (2008). *Fundamental Neuroscience*, 3rd edn. (London: Academic Press).

Sullivan, J. (2003). "No going back," *Nat Neurosci* 6, 905–906.

Thio, L. L. (2012). "Hypothalamic hormones and metabolism," *Epilepsy Res* 100, 245–251.

Van den Pol, A. N., Wuarin, J. P. and Dudek, F. E. (1990). "Glutamate, the dominant excitatory transmitter in neuroendocrine regulation," *Science* 250, 1276–1278.

Van den Pol, A. N., Hermans-Borgmeyer, I., Hofer, M., Ghosh, P. and Heinemann, S. (1994). "Ionotropic glutamate-receptor gene expression in hypothalamus: localization of AMPA, kainate, and NMDA receptor RNA with in situ hybridization," *J Comp Neurol* 343, 428–444.

Van Gaal, L. F., Rissanen, A. M., Scheen, A. J., Ziegler, O. and Rössner, S. (2005). "Effects of cannabinoid-1 receptor blocker rimonabant on weight reduction and cardiovascular risk factors in overweight patients: 1-year experience from the RIO-Europe study," *Lancet* 365, 1389–1397.

Vincent, S. R. (2010). "Nitric oxide neurons and neurotransmission," *Prog Neurobiol* 90, 246–255.

Watanabe, M., Inoue, Y., Sakimura, K. and Mishina, M. (1993). "Distinct distribution of five N-methyl-D aspartate receptor channel subunit mRNAs in the forebrain," *J Comp Neurol* 338, 377–390.

Watkins, J. C. and Jane, D. E. (2006). "The glutamate story," *Brit J Pharmacol* 147, S100–S108.

Wittman, G., Lechan, R. M., Liposits, Z. and Fekete, C. (2005). "Glutamatergic innervation of corticotropin-releasing hormone- and thyrotropin-releasing hormone-synthesizing neurons in the hypothalamic paraventricular nucleus of the rat," *Brain Res* 1039, 53–62.

Wurtman, R. J. (1982). "Nutrients that modify brain function," *Sci Am* 246, 50–59.

Ward, S. J. and Raffa, R. B. (2011). "Rimonabant redux and strategies to improve the future outlook of CB1 receptor neutral-antagonist/inverse-agonist therapies," *Obesity* 19, 1325–1334.

Neurotransmitter and neuropeptide control of hypothalamic, pituitary and other hormones

Previous chapters have discussed the endocrine glands and their hormones (Chapter 2), the hormones of the pituitary gland (Chapter 3), the hypothalamic hormones (Chapter 4) and neurotransmitters (Chapter 5). This chapter will describe how neurotransmitters influence the release of hypothalamic and pituitary hormones and the hormones of the adrenal medulla, pancreas, thymus and gastrointestinal tract. It will also examine the electrophysiological properties of neurosecretory cells and the effects of drugs on the release of neurohormones.

6.1 The cascade of chemical messengers

As shown in Figure 6.1, and using the adrenal gland as an example, there is a cascade of chemical messengers that regulate target tissue function from the brain to the endocrine glands. For example, neurons release a neurotransmitter that regulates the secretion of neurohormones (such as CRH) from hypothalamic neurosecretory cells. These hypothalamic hormones stimulate the cells of the adenohypophysis (anterior pituitary) to synthesize and release their hormones. Many pituitary hormones, such as ACTH, act on endocrine target cells, such as the adrenal cortex, causing them to synthesize and release their own hormones (e.g. cortisol) which then stimulate biochemical changes in target cells elsewhere in the body, including the brain. In each step of this pathway, the individual neurotransmitters and peptide hormones bind to membrane receptors that activate a second messenger, such as cAMP, within the target cell (see Chapter 10). Steroid hormones (such as cortisol) act on receptors located inside the target cells (see Chapter 9). This chapter describes the effects of neurotransmitters on hypothalamic neurosecretory cells.

6.2 Neural control of hypothalamic neurosecretory cells

6.2.1 Neural input to the endocrine hypothalamus

Figure 4.1 illustrates the different nuclei of the hypothalamus, many of which have neurons, or nerve terminals, which release a variety of neurotransmitters including GABA, glutamate, kisspeptin, opioids, dopamine, norepinephrine and serotonin. These neurotransmitters all bind to receptors on hypothalamic neurosecretory cells. This chapter describes the *magnocellular* hypothalamic neurosecretory cells of the paraventricular and

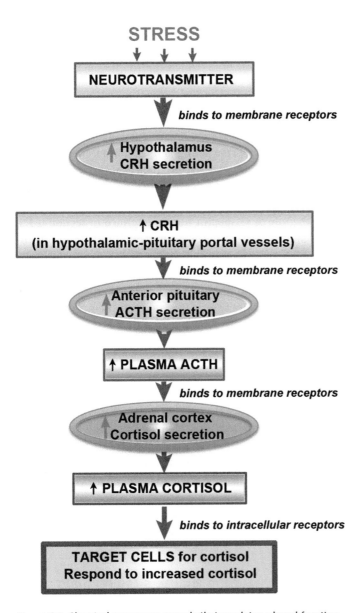

Figure 6.1 Chemical messenger cascade that regulates adrenal function
Neurons release neurotransmitters that regulate the secretion of neurohormones (such as CRH) from
hypothalamic neurosecretory cells. These hypothalamic hormones stimulate the cells of the anterior pituitary
to synthesize and release their hormones. Many of these pituitary hormones, such as ACTH, act on endocrine
target cells, such as the adrenal cortex, causing them to synthesize and release their own hormones (e.g. cortisol)
which then stimulate biochemical changes in target cells elsewhere in the body, including the brain.
Abbreviations: ACTH, adrenocorticotropic hormone; CRH, corticotropin-releasing hormone.

supraoptic nuclei (PVN and SON), whose axons terminate in the posterior pituitary, and the
parvicellular hypothalamic neurosecretory cells, whose axons terminate in the median
eminence (see Figure 3.1). Neural input to the endocrine hypothalamus comes from sensory
receptors (olfaction, taste, hearing, vision and touch); exteroreceptors which detect pain,

temperature changes, and suckling stimulation; interoreceptors such as chemoreceptors, blood volume receptors (baroreceptors), osmoreceptors and glucoreceptors; sensory receptors in the uterus and cervix; other nuclei in the brain which control time-dependent and sleep-wake rhythms; and from psychological arousal as occurs during emotional states such as stress. All of these stimuli are mediated to some degree by neurotransmitters. For example, as noted in Chapter 3, oxytocin is released from the posterior pituitary by the stimulus from the infant suckling at the breast. This stimulus is exerted in the PVN/SON via several neurotransmitters, including norepinephrine acting at α- and β-adrenergic receptors (see Figure 6.9) (Bealer *et al.* 2010). Similarly, the parvicellular neurons are regulated by many neurotransmitters and neuropeptides (e.g. see Figure 4.3). Figure 6.2 is a summary of how sensory input and different neurotransmitters (e.g. GABA, glutamate) and neuropeptides (e.g. kisspeptin, opioids – the role of kisspeptin will be described in section 6.3.3) from neuron terminals might regulate the secretion of hypothalamic GnRH in the reproductive system of males and females. Neurosecretory cells occur in different neurotransmitter pathways and the neurotransmitters in these pathways may stimulate or inhibit hypothalamic hormone release. One problem in studying the neurotransmitters that regulate hypothalamic and pituitary hormone release is that the same neurotransmitters may be involved in different modes of hormone secretion. Thus, different neurotransmitters may regulate the baseline secretion of a hormone, the circadian rhythms of hormone secretion, hormone responses to stress, pain or cold, and feedback regulation by other hormones such as gonadal hormones and glucocorticoids.

Because hypothalamic releasing and inhibiting hormones are released into the pituitary portal system in very small quantities (approximately 10^{-10} M, or 10^{-10} gm/ml) and have very short half-lives in the bloodstream (minutes), it is simpler to measure peripheral, circulating levels of pituitary hormones after the manipulation of neurotransmitter levels; for example, CRH released from the hypothalamus into the portal blood induces the anterior pituitary to secrete ACTH into the bloodstream, where it can be measured. Thus, the following discussion will focus on pituitary hormone responses to changes in neurotransmitter input, bearing in mind that the plasma levels of pituitary hormones reflect changes in hypothalamic hormone release via the cascade of chemical messengers illustrated in Figure 6.1.

6.2.2 Mechanisms through which neurotransmitters can regulate hypothalamic and pituitary hormone release

As illustrated in Figure 6.3, and using GnRH neurons as an example, there are four mechanisms through which neurotransmitters can act to regulate hormone release from the hypothalamus and pituitary gland: (A) The neurotransmitter can be released at a synapse in a neural pathway involving more than one neuron, carrying input from sensory receptors, such as the olfactory epithelium, from other brain regions, such as the limbic system, or from the spinal cord, to the neurosecretory cell (indirect stimulation); (B) The neurotransmitter (or neuropeptide) can be released at a synapse between axons from other hypothalamic neurons and the dendrites or cell body of the neurosecretory cell (direct stimulation); (C) The neurotransmitter/neuropeptide can be released from nerve cells having synapses on the axons or nerve terminals of the neurosecretory cell; and (D) neurotransmitters, such as dopamine, can be released into the pituitary portal vessels to

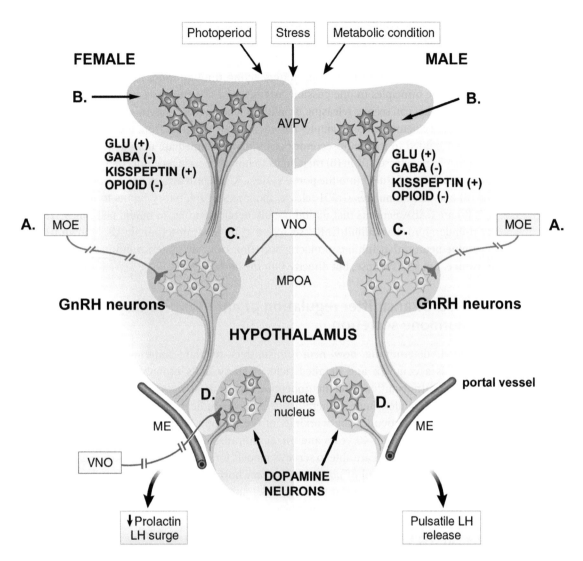

Figure 6.2 Sites of chemosensory and neurochemical modulation in the male and female rodent hypothalamus

The anteroventral periventricular nucleus (B. AVPV) is a pivotal relay for signals that regulate gonadotropin-releasing hormone (GnRH) neurons. GABA, glutamate, kisspeptin and opioid inputs from the AVPV, together with estradiol, regulate GnRH neurons in the medial preoptic area (C. MPOA). Chemosensory input from the vomeronasal organ (VNO) as well as other sensory inputs (e.g. photoperiod) regulate endocrine responses in the hypothalamus. GnRH neurons also receive polysynaptic input that originates in the main olfactory epithelium (A. MOE). GnRH neurons project to the median eminence (ME) blood capillaries, and in the male, synchronized release of GnRH produces pulsatile luteinizing hormone (LH) release from the pituitary. In the female, dopamine neurons receive a polysynaptic input originating from the vomeronasal organ (D). These neurons also project to the ME, where dopamine release inhibits prolactin, a prelude to inducing female cyclicity and an ovulatory LH surge. Reproduced with permission (Keverne 2005).

act directly on the pituitary gland. For example, as we saw in Chapter 4, prolactin secretion from the anterior pituitary is regulated primarily by dopamine that is released from neurons directly into the pituitary portal vessels. The secretion of both anterior and posterior pituitary hormones is regulated by neurotransmitters acting through these mechanisms. Figure 6.3 outlines these principles in one possible arrangement of neurons: a hypothalamic neurose-cretory cell, that can secrete releasing or inhibiting hormones, is regulated by both excitatory and inhibitory neurons. For example, a noradrenergic stimulus from elsewhere in the brain (A) is transmitted to the neurosecretory cell via an intermediate noradrenergic neuron (B). Norepinephrine released from (B) then binds to noradrenergic receptors on GnRH neurons to stimulate release of GnRH into the portal system. Kisspeptin acts as an excitatory neuropep-tide (B) and also stimulates GnRH release. Examples of inhibitory inputs to neurosecretory cells (C) are GABA neurons that inhibit noradrenergic neurons, or opioid neurons (E; releas-ing β-endorphin) that inhibit GnRH secretion at the neuron terminal. Also shown (D) is a dopamine neuron that inhibits prolactin secretion from the anterior pituitary by releasing dopamine, not at a synapse, but directly into the portal vessels.

6.3 Neurotransmitter regulation of anterior pituitary hormone secretion

In general, determining how neurotransmitters regulate anterior pituitary hormone secretion is a complex and detailed field of study. It is probably true to say that all known neurotransmitters and neuropeptides have at some time been implicated. The information provided here is therefore merely an indication of some of the available information. In particular, the workings of the reproductive system have received close attention for more than 40 years and the neurotransmitter systems that regulate pituitary secretion of LH and FSH are able to serve as models for other pituitary hormones (Smith and Jennes 2001). The control of *posterior pituitary* hormone secretion is described by Sladek and Kapoor (2001) (see section 6.4). Table 6.1 summarizes some of the complexity asso-ciated with the neural control of pituitary secretion.

6.3.1 ACTH and β-endorphin

Figure 6.4 shows some of the neural influences that regulate the stimulation of ACTH secretion (see also Figure 1.2). Two major control systems are stress and a circadian (daily) rhythm controlled by the light-dark cycle. As with GnRH/LH secretion, release of CRH and ACTH also occurs in bursts and the rate of these bursts shows an increase in the few hours before waking (6 am to 9 am in humans; Figure 6.5). β-endorphin is released simulta-neously with ACTH because they are biosynthesized together, from the prohormone POMC, in corticotroph cells (see Figure 3.4). As shown in Table 6.1 and Figure 6.4, ACTH and cortisol secretion is controlled by a variety of neurotransmitters (further details can be found in Carrasco and Van de Kar 2003; Van de Kar and Blair 1999). Figure 6.4 also illustrates that ADH, released from the posterior pituitary, acts synergistically with CRH to increase the release of ACTH from the anterior pituitary. A good overall description of the regulation of ACTH/cortisol secretion is provided by Khoo and Grossman (2014). For example, opiates and opioid peptides, and endocannabinoids, inhibit ACTH secretion, whereas norepinephrine stimulates CRH release through α1 adrenergic receptors.

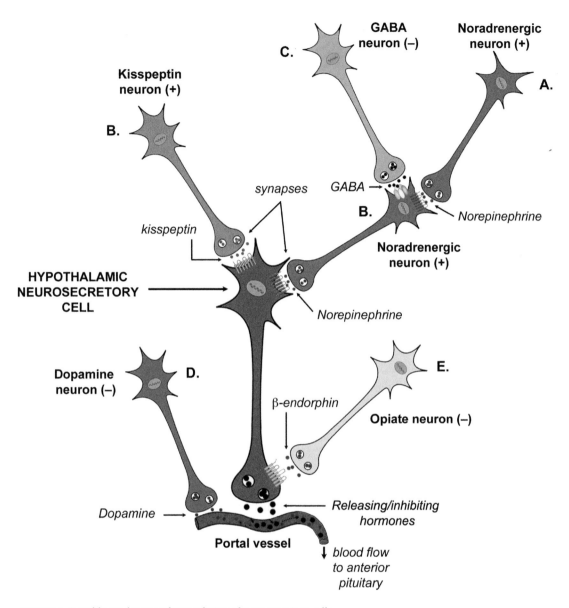

Figure 6.3 Possible mechanisms for regulation of neurosecretory cells
This figure outlines one possible arrangement by which neurotransmitters might regulate secretion from a hypothalamic neurosecretory cell that secretes releasing (e.g. GnRH) or inhibiting (e.g. somatostatin) hormones. The neurosecretory cell is regulated by both excitatory and inhibitory neurons. For example, norepinephrine (from B) binds to noradrenergic receptors on GnRH neurons to stimulate release of GnRH into the portal system. Additional pathways, such as A, relay information to the neurosecretory cell via an intermediate noradrenergic neuron B. Kisspeptin acts as an excitatory neuropeptide (B) and also stimulates GnRH release. Examples of inhibitory inputs (C) to neurosecretory cells are GABA neurons that inhibit noradrenergic neurons, or opioid neurons (E; releasing β-endorphin) that inhibit GnRH secretion via an effect at the neuron terminal. Also shown (D) is a dopamine neuron that inhibits prolactin secretion from the anterior pituitary by releasing dopamine, not at a synapse, but directly into the portal vessels. In these examples, the neurotransmitters/neuropeptide exert their effects by binding to G-protein-coupled receptors in the cell membrane except for GABA that binds to an ion channel.

Table 6.1 **Some examples of neurotransmitter control of pituitary gland hormone secretion**								
NEUROTRANSMITTER	ACTH	TSH	LH/FSH	PRL	GH	α-MSH	OXYTOCIN	ADH
Norepinephrine	+/–	+	+	+/–?	+	+/–	+/–	+/–
Dopamine	+/–	–	+/–	–	+	–	+/–	+/–
Serotonin	+	+/–	+/–	+	+/–	+?	?	?
Acetylcholine	+	+	–?	–?	+	+?	+	+
GABA	–	–	–	–	+/–	–	–	–
Glutamate	+	+	+	+	+	+	+	+
Nitric oxide	–	+	+	+	+	?	+/–	+/–

+ = increase; – = decrease; ? = effect uncertain

However, drugs such as sumatriptan, acting at the serotonin receptor subtype $5\text{-HT}1_D$, can inhibit ACTH secretion (Rainero *et al.* 2001).

6.3.2 TSH

As shown in Figure 6.6, TSH release from thyrotroph cells of the anterior pituitary is stimulated by TRH released from neurosecretory cells localized in the hypothalamic paraventricular nuclei (see Chapter 4). TRH release, and therefore TSH, is controlled by a variety of neurotransmitters (Table 6.1). In addition, neuropeptides such as ADH, CCK, opioids, NPY and α-MSH have all been implicated in the control of TRH secretion (Mariotti 2011). Somatostatin, which negatively regulates GH secretion (see section 6.3.5), is also an inhibitory factor for TSH release. Thyroid function can be regulated by changes in body temperature, by stress and by nutrition. For example, the body responds to cold through a norepinephrine-induced release of TRH that leads to T3/T4 secretion and an increase in thermogenesis (Figure 6.6). Dopamine reduces baseline TSH levels and inhibits the cold-induced release of TSH in two ways: by inhibiting the release of TRH from the hypothalamus; and by inhibition of thyrotroph cells in the pituitary. Stress reduces TSH secretion via an inhibitory effect of cortisol directly on the anterior pituitary. The hormone leptin (see Chapter 2), released from adipose tissue following food intake, binds to leptin receptors on hypothalamic TRH neurons and stimulates the release of TRH (Nillni 2010; see also Fekete and Lechan 2013).

6.3.3 LH and FSH

As shown in Table 6.1, the reproductive system is regulated by a variety of neurotransmitters that control the secretion of LH and FSH through their effect on GnRH neurons. The pathways involved are complex and the effect of each neurotransmitter on GnRH neurons is influenced by species and sex differences, by the reproductive cycle in females, by seasonal variations and by steroid feedback. As noted in Chapter 4, GnRH neurons are localized to the basal hypothalamus in humans and primates, but are found in the preoptic area in rodents. Smith and Jennes (2001) provide a summary of the neural pathways and signals that regulate GnRH cells. For example, they express receptors for a variety of neurotransmitters/neuromodulators, including catecholamines, GABA, glutamate, NPY, VIP and β-endorphin. Serotonin is also known to stimulate GnRH secretion, but norepinephrine is the key neurotransmitter that

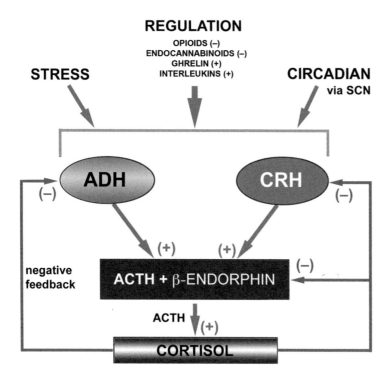

Figure 6.4 Regulation of ACTH and cortisol secretion
Two major control systems are stress and a circadian (daily) input via the light-dark cycle and the suprachias-matic nuclei (SCN). Secretion of corticotropin-releasing hormone (CRH) and adrenocorticotropic hormone (ACTH) occurs in bursts and the rate of these bursts shows an increase in the few hours before we awake in the morning (6 am to 9 am; see Figure 6.5). β-endorphin is released simultaneously with ACTH because they are biosynthesized together in corticotroph cells.

CRH and ACTH secretion, and therefore cortisol, is controlled by a variety of neurotransmitters. This figure also illustrates that vasopressin (ADH), released from the posterior pituitary, acts synergistically with CRH to increase the release of ACTH from the anterior pituitary. Reproduced with permission from Dr. L. J. De Groot, Endotext.org.

"primes" the GnRH neuron to respond to incoming stimulation by other neurotransmitters, especially glutamate (Herbison 1997). In all species so far examined the primary neural control systems for GnRH release consist of an excitatory drive via glutamate and an inhibitory input via GABA neurons (Figure 6.7). Other influences such as nutrition, exercise, stress, etc. are also mediated through these systems. As well as these classical neurotransmitters, two neuropeptides regulate GnRH secretion. First, there is an excitatory stimulus from neurons that express the *kiss1* gene that produces kisspeptin (Roa *et al.* 2008; Oakley *et al.* 2009) (Figure 6.7). This discovery of kisspeptin offers clinical possibilities for the treatment of infertility (Jayasena and Dhillo 2009). Second, a neuropeptide that effectively counters the effects of kisspeptin has been called *gonadotropin inhibitory hormone (GnIH)* (Clarke and Parkington 2013). This control system for LH/FSH is similar to that for GH secretion (section 6.3.5) where two neuropeptides, GHRH (stimulatory) and somatostatin (inhibitory), are responsible for pituitary secretion of GH (see Figure 6.9). The electrophysiological effects of kisspeptin and GnIH on GnRH neurons are discussed in Chapter 12 (Figure 12.6).

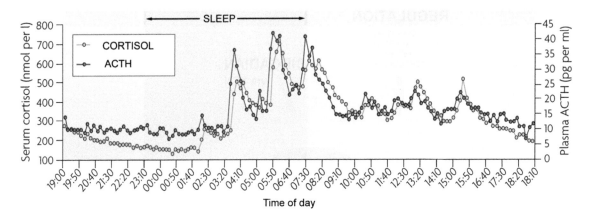

Figure 6.5 Pulsatile secretion of ACTH and cortisol in humans
The temporal relationship between adrenocorticotropic hormone (ACTH) and cortisol pulses. This 24-hour automated blood sampling profile in humans reveals the positive, delayed relationship between ACTH and cortisol. There is a diurnal rhythm in hormone secretion such that secretion begins to increase in the late stages of sleep. Note the delay between each pulse of ACTH and the subsequent response of the adrenal in releasing a pulse of cortisol. Also note the rapid negative feedback effect of each cortisol pulse on ACTH secretion, which rapidly decreases. Reproduced with permission (Lightman and Conway-Campbell 2010).

A critical property of GnRH secretion, and one that is indispensable for the onset of puberty and for adult fertility, is its episodic mode of release that generates a corresponding pulsatile secretion of LH/FSH from the pituitary gland. As shown in Figures 4.4 and 4.5, data from studies *in vivo* on humans and experimental animals indicate that the GnRH pulse rate is about one pulse every 1 to 2 hours. What kind of neurotransmitter input could produce such a pattern of secretion? The cellular origin of this pulsatility is not completely understood, but GnRH pulsatility – the frequency of neuronal firing – is an intrinsic property of GnRH neurons, and perhaps of other releasing hormone neurons as well (Gan and Quinton 2010). The basic pulse frequency of GnRH neurons could be modified by neurotransmitter/neuropeptide input. For those students who might appreciate a more biochemical explanation for the cellular and molecular origin of GnRH pulses, a recent review provides this information (Krsmanovic *et al.* 2009).

6.3.4 Prolactin

Prolactin is not just a pituitary hormone that regulates milk secretion during suckling. It is a multifunctional hormone involved in a vast array of activities in the body. Indeed, some authors have named this hormone "versatilin" (Freeman *et al.* 2000). Table 6.1 indicates that prolactin secretion is controlled by an array of excitatory and inhibitory neurotransmitters and a detailed description of this regulation can be found in Freeman *et al.* (2000). As discussed in Chapter 4, PRL secretion from anterior pituitary lactotroph cells is controlled *primarily* by dopamine released from hypothalamic neurons into the portal vessels (see Figures 6.2 and 6.3). Dopamine acts as an inhibitory factor, so that when dopamine levels are reduced, PRL release is increased. Freeman *et al.* (2000) summarize the influence of other neurotransmitters on PRL secretion as the ability of each of them to regulate dopamine release; that is, norepinephrine and serotonin *increase* PRL secretion by

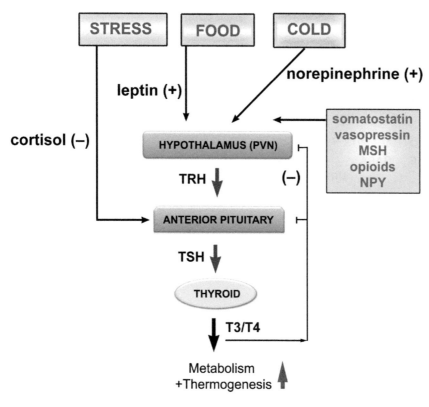

Figure 6.6 Regulation of thyroid hormone secretion
TSH release from thyrotroph cells of the anterior pituitary is stimulated by TRH released from neurosecretory cells localized in the hypothalamic paraventricular nuclei. TRH release, and therefore TSH, is controlled by a variety of neurotransmitters/neuropeptides such as somatostatin, ADH, MSH, opioids and NPY. Thyroid function can be regulated by changes in body temperature, by stress and by nutrition. For example, the body responds to cold through a norepinephrine-induced release of TRH that leads to T3/T4 secretion and an increase in thermogenesis. Stress reduces TSH secretion via an inhibitory effect of cortisol directly on the anterior pituitary. Leptin feedback from fat cells stimulates the release of TRH from hypothalamic neurons. Thus, food intake and energy expenditure are linked via leptin secretion from fat cells.

inhibiting dopamine neurons. Other neurotransmitters/neuromodulators, such as ACh, NPY and glutamate, *stimulate* dopamine neurons and therefore *inhibit* PRL release. The discovery of a prolactin-releasing hormone (PRH) has proved to be elusive, although a *prolactin-releasing peptide* (PrRP) has attracted attention because of its apparent role in energy homeostasis and stress responses (Onaka *et al.* 2010). Figure 6.8 illustrates many other factors that regulate PRL secretion, including neuropeptides. Stimulatory inputs include TRH, estradiol, mating, lactation/suckling and oxytocin. Inhibitory factors include opioids, somatostatin, CCK and stress (see Egli *et al.* 2010). In addition to these other factors, PRL can modify (inhibit) its own secretion by a short-loop negative feedback on the tuberoinfundibular hypothalamic dopamine neurons (Figure 6.8). In addition, a peptide called ghrelin (see section 6.3.5), which regulates growth hormone secretion as well as appetite, is a potent regulator of PRL secretion (Lim *et al.* 2010). This effect is probably a direct influence on the anterior pituitary.

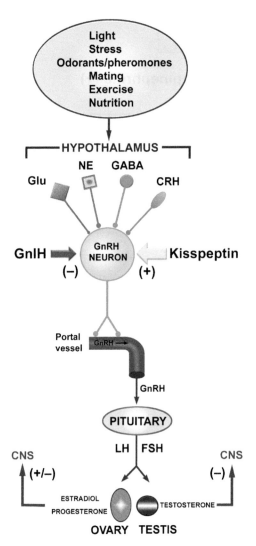

Figure 6.7 Schematic diagram highlighting the control of GnRH neurons and GnRH secretion
Note that GnRH neurons are subjected to multiple neurochemical inputs, both stimulatory and inhibitory, as well as being sensitive to light, stress, mating, exercise, etc. The major stimulatory input is glutamate and the primary inhibitory pathway is via GABA, although many other neurotransmitters are implicated, including dopamine, 5-HT, NO and β-endorphin. Also included in this diagram are the inhibitor GnIH and the stimulatory influence of kisspeptin.

Abbreviations: CNS, central nervous system; CRH, corticotropin-releasing hormone; FSH, follicle-stimulating hormone; GABA, γ-aminobutyric acid; Glu, glutamate; GnIH, gonadotropin inhibitory hormone; GnRH, gonadotropin-releasing hormone; LH, luteinizing hormone; NE, norepinephrine.

6.3.5 Growth hormone

The primary regulatory factors for the control of GH secretion are two hypothalamic hormones: growth hormone releasing hormone (GHRH) and somatostatin. GHRH is the stimulatory releasing hormone, whereas somatostatin exerts an inhibitory effect. GH also regulates its own secretion through a short-loop negative feedback effect (Figure 6.9). The peptide *ghrelin* is also a powerful stimulator of GH secretion. The role of ghrelin as a gut-derived peptide with effects on appetite was discussed in Chapter 2. Ghrelin is now regarded as a key, circulating regulator of GH and PRL secretion (see section 6.3.4; Lim *et al.* 2010). An overview of the control of GH secretion is shown in Figure 6.9, which illustrates the complex interplay and feedback systems between GH and three hypothalamic neuropeptides (GHRH, ghrelin and somatostatin) which regulate its release. As shown in Table 6.1 and Figure 6.9, there are additional neural inputs to the GHRH/somatostatin neurons

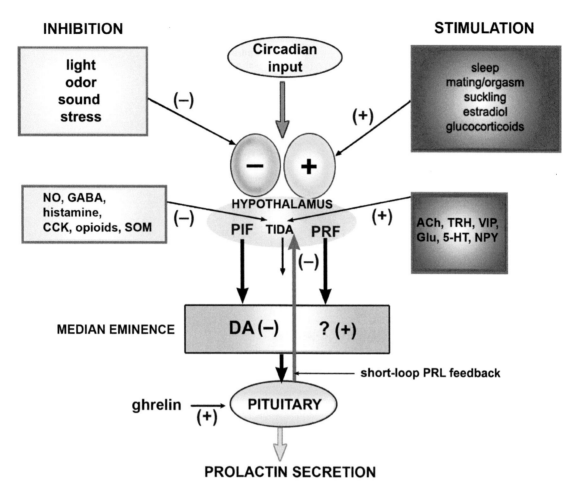

Figure 6.8 Regulation of prolactin secretion
Prolactin (PRL) secretion is regulated by a light-entrained circadian rhythm, which is modified by environmental input (e.g. light; sound), with the internal milieu (e.g. sleep; mating) and reproductive stimuli (estradiol) affecting stimulatory elements of the hypothalamic regulatory circuit. The final common pathways of the central inhibitory control of prolactin secretion are neuroendocrine neurons producing the prolactin releasing inhibiting factor (PIF; dopamine (DA)), which is released from tuberoinfundibular DA (TIDA) neurons, and various factors such as somatostatin (SOM) and γ-aminobutyric acid (GABA). PRL also regulates its own secretion through a short-loop negative feedback. The stimulatory pathways may include a so-far unknown prolactin releasing factor (PRF), but include also, for example, thyrotropin-releasing hormone (TRH) and glutamate (Glu).

Abbreviations: 5-HT, 5-Hydroxytryptamine (serotonin); ACh, acetylcholine; CCK, cholecystokinin; GABA, gamma-aminobutyric acid; Glu, glutamate; NO, nitric oxide; PIF, prolactin releasing inhibiting factor; PRF, prolactin releasing factor; SOM, somatostatin; TIDA, tuberoinfundibular DA; TRH, thyroid hormone releasing hormone; VIP, vasoactive intestinal polypeptide.

that may impose further levels of regulation (see Khoo and Grossman 2014). As noted already for LH secretion, glutamate (Glu; excitatory) and GABA (inhibitory) are important neurotransmitters for hypothalamic regulation of releasing/inhibiting hormone secretion. However, both Glu *and* GABA induce secretion of GH (Aguilar *et al.* 2005; Orio *et al.* 2001).

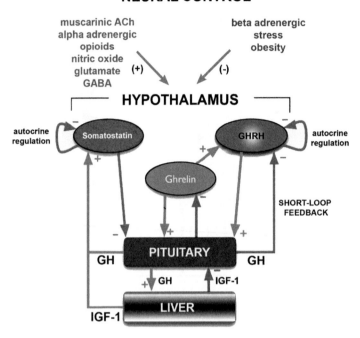

Figure 6.9 Regulation of growth hormone secretion

Primary regulation of GH secretion is imposed by a stimulatory input from GHRH and a negative control from somatostatin. Additional regulation of GH secretion is accomplished by *positive* and *negative* short-loop feedback. High level GH secretion imposes a negative, short-loop feedback on GHRH neurons, limiting further secretion of GHRH and GH. In addition, GH *stimulates* secretion of somatostatin through positive feedback that further inhibits GH release from the anterior pituitary. The GH regulatory system also includes negative autocrine feedback loops to regulate GHRH output. A variety of neurotransmitters/neuropeptides regulate GHRH and somatostatin release and GH secretion is inhibited by stress and obesity. Another peptide, *ghrelin*, is also a powerful stimulator of GH secretion, acting at the pituitary and at GHRH neurons. IGF-1 also provides negative feedback on GH secretion by stimulating release of somatostatin and by a short negative feedback loop to the pituitary. Reproduced with permission from Dr. L. J. De Groot, Endotext.org.

This paradox is explained by assuming that GABA acts as an inhibitory neurotransmitter on somatostatin neurons, thus inducing GH secretion. Other stimulatory inputs include norepinephrine (acting at α2 adrenergic receptors), acetylcholine (acting at muscarinic receptors) and opioid peptides such as enkephalin and β-endorphin (Khoo and Grossman 2014). Serotonin receptors are also stimulatory to GH secretion (Valverde *et al.* 2000). Norepinephrine may also be inhibitory by stimulating somatosatin neurons through β-adrenergic receptors.

6.3.6 α-MSH

As discussed in Chapter 4, α-MSH and β-endorphin are both derived from the POMC peptide and are released from melanotroph cells in the intermediate lobe of the pituitary. It was assumed that their secretion is controlled by melanocyte-stimulating hormone – releasing hormone (MSH-RH) and by melanocyte-stimulating hormone – release-inhibiting hormone (MSH-RIH). However, neither MSH-RH nor MSH-RIH has been identified. The secretion of α-MSH from the intermediate lobe of the pituitary is under both excitatory and inhibitory control (Table 6.1) and it seems likely that nerve fibers releasing dopamine and GABA exert inhibitory control of α-MSH secretion, whereas CRH, serotonin and epineprine may stimulate α-MSH release (Saland 2001). In humans, norepinephrine stimulates α-MSH by binding to β-adrenergic receptors, whereas binding to $α_2$-adrenergic receptors is inhibitory (Catania *et al.* 2000). The neurotrophic factors brain-derived neurotrophic factor (BDNF) and glial-derived neurotrophic factor (GDNF) have also been identified in most cells of the intermediate lobe and may be involved in α-MSH and β-endorphin secretion. In the absence of clearly defined releasing/inhibiting factors such as MSH-RH, it is unclear whether the neurochemical regulation outlined above is exerted directly on the pituitary or on hypothalamic neurons that control pituitary intermediate lobe secretion.

6.3.7 Other neuroregulators

A variety of other factors are implicated in the hypothalamic regulation of anterior pituitary hormone release. Using TSH as an example, Mariotti (2011) lists 21 agents that act as inhibitors or stimulators of TSH secretion. This list includes prostaglandins, leptin, steroids, glucagon-like peptide (GLP-1), galanin, CCK and opioids. Some of the hypothalamic effects of neuropeptides are discussed in Chapter 12. Effects of prostaglandins on hypothalamic GnRH neurosecretory cells have been studied in detail by Ojeda and co-workers (Ojeda and Negro-Vilar 1985; Clasadonte *et al.* 2011). Prostaglandins are released from hypothalamic astrocytes to trigger responses in neighboring neurosecretory cells. In this way, prostaglandins act as *gliotransmitters*; this topic will be discussed in section 6.9.

6.4 Neurotransmitter regulation of neurohypophyseal hormone secretion

As discussed in Chapter 3, oxytocin and vasopressin (antidiuretic hormone; ADH) are biosynthesized in magnocellular neurons of the hypothalamic SON and PVN. These two hormones reach the posterior pituitary via axonal transport via the infundibulum (Figure 6.10). Note that there are specific oxytocin-producing neurons as well as vasopressin-producing neurons, although each nucleus (SON and PVN) contains both kinds of neurons. The SON and PVN receive information from the periphery to maintain water balance (ADH) and to regulate some aspects of reproduction (oxytocin: childbirth and lactation). These feedback links are mediated through classical neurotransmission of various types (Table 6.1; Sladek and Kapoor 2001; Sladek and Song 2012), although glutamate appears to be the primary excitatory neurotransmitter (Van den Pol *et al.* 1990). However, release of both hormones is stimulated by ACh, CCK, norepinephrine, dopamine and nitric oxide. In particular, norepinephrine is important in the release of oxytocin during childbirth (see Chapter 12; Figure 12.8). Inhibition of secretion of the two hormones is exerted by GABA, glycine, dopamine, norepinephrine, dynorphin, somatostatin and

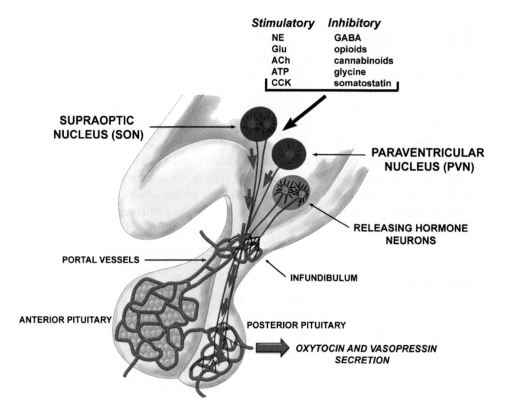

Stimulatory	Inhibitory
NE	GABA
Glu	opioids
ACh	cannabinoids
ATP	glycine
CCK	somatostatin

Figure 6.10 Regulation of oxytocin and vasopressin secretion
Oxytocin and vasopressin (ADH) are biosynthesized in magnocellular neurons of the hypothalamic SON and PVN. The two hormones reach the posterior pituitary via axonal transport (red arrows). Note that there are specific oxytocin-producing neurons as well as vasopressin-producing neurons, although each nucleus (SON and PVN) contains *both* kinds of neurons. The SON and PVN receive information from the periphery to maintain water balance (ADH) and to regulate some aspects of reproduction (oxytocin; childbirth; lactation). These feedback links are mediated through classical neurotransmission of various types (*stimulatory*: e.g. NE, Glu; *inhibitory*: e.g. GABA, opioids, cannabinoids), as well as neuropeptides such as CCK and somatostatin. Copyright Alex Pincock.

nitric oxide. Some of these factors both stimulate and inhibit secretion of oxytocin and ADH depending on: (a) which receptor is used (e.g. norepinephrine can act through α- and β-adrenergic receptors, and their subtypes); and (b) which neural pathway is used. Oxytocin and vasopressin neurons are known to express a large variety of neuro-transmitter/neuropeptide receptors, consistent with the list given above (Sladek and Kapoor 2001; Sladek and Song 2012).

Endocannabinoids (see Chapter 5) are also implicated in the control of oxytocin secretion. It has long been known that cannabis intake interferes with normal lactation by inhibiting oxytocin secretion. Cannabinoid receptors on SON/PVN neurons respond to both cannabis and to endogenous endocannabinoids by inhibiting oxytocin secretion (Rettori *et al.* 2010). Oxytocin itself, released into the vicinity of oxytocin neurons, also provides feedback (see Figure 12.9; autocrine/paracrine control), which modulates the secretion of oxytocin release from the posterior pituitary (Bealer *et al.* 2010).

6.5 Electrophysiology of neurosecretory cells

As discussed in Chapter 5, the response of a postsynaptic cell to neurotransmitter stimulation at excitatory synapses is a change of electrical potential, causing the cell to "fire" an action potential (see Figure 5.10). When a neuron generates an action potential, it releases neurotransmitters from its axon terminals into the synapse. However, when a hypothalamic neurosecretory cell fires, it releases *neurohormones* such as GnRH and oxytocin. GnRH is secreted into the hypophyseal portal system, whereas oxytocin enters the peripheral circulation directly from the posterior pituitary. Neurosecretory cells show different baseline patterns of firing (generation of action potentials), so by taking recordings from these cells and measuring changes in hormone levels, one can show how the release of neurohormones is related to the electrical activity of the neurosecretory cell. For example, bursts of electrical activity have been recorded from neurons in the arcuate nucleus/median eminence of the hypothalamus in rhesus monkeys, sheep, rats and goats (e.g. Okamura *et al.* 2010; Nishihara *et al.* 1999). Figure 6.11 illustrates a typical experiment. Electrical recordings of multiple unit activity (MUA; i.e. neuronal firing) from electrodes placed in the arcuate/median eminence are correlated with blood levels of LH measured in the circulation (Figure 6.11A). Figure 6.11B shows real-time measurements of electrical activity (blue) *preceding* the appearance of LH pulses. It is therefore concluded that electrical activity in the GnRH neurons, monitored via the recording electrodes, induces pulsatile release of GnRH from hypothalamic neurons and this stimulates the pulsatile release of LH.

This functional relationship between nerve cell activity and neurohormone release has also been studied *in vitro*; that is, individual neurons, or groups of neurons, can be electrically stimulated in tissue culture and subsequent hormone release measured. If the neuron stimulated is involved in the control of the hormone being measured, increased (or decreased) hormone release should be correlated with the electrical activity of the nerve cell.

In terms of posterior pituitary hormone secretion, classical studies were performed on suckling-induced release of oxytocin in lactating female rats. The effect of pups suckling at the nipples is to stimulate the ascending mammary nerves in the spinal cord that in turn elicit a burst of firing in the magnocellular neurosecretory cells of the SON and PVN (see Figure 8.3). Figure 6.12 shows that these electrical bursts are followed about 18 seconds later by a rise in intramammary pressure (milk ejection) that also coincides with a marked suckling response from the pups (Wakerley and Lincoln 1973). The period of 18 seconds represents the time taken for the neurosecretory neurons to release oxytocin in sufficient quantities into the bloodstream to stimulate the mammary glands to induce milk ejection.

6.6 Neurotransmitter regulation of other endocrine glands

Neurotransmitters also regulate hormone release from the adrenal medulla, the pineal gland, thymus and thyroid glands, the gastrointestinal tract, pancreas and the ovary. The release of epinephrine and norepinephrine from the adrenal medulla is stimulated by acetylcholine released from sympathetic nerves that innervate this gland (Berne and

A.

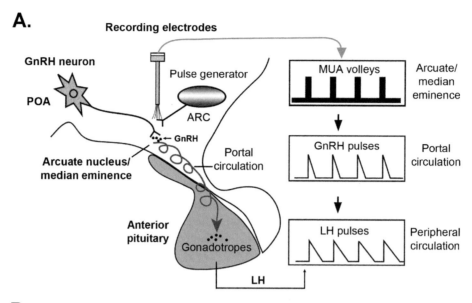

B.

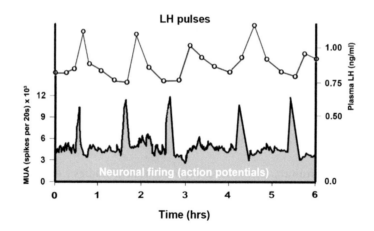

Figure 6.11 Electrical activity of GnRH neurons

A. Schematic representation of the association of neuronal multiple unit activity (MUA), representing neuronal firing, originating from the so-called "pulse generator," with gonadotropin-releasing hormone (GnRH) and luteinizing hormone (LH) pulses. An array of recording electrodes is stereotaxically implanted into the arcuate nucleus/median eminence region, where GnRH axons terminate. In freely moving, non-anaesthetized animals, recordings can be obtained at the same time as blood collections are made. B. MUA recorded in a conscious goat is electrically processed and displayed on a personal computer in real time. Because increases in MUA are always associated with LH pulses, they are considered to be the electrophysiological manifestation of GnRH pulses secreted from GnRH neurosecretory neurons. Note that each MUA volley is followed by a LH pulse.

Abbreviations: ARC, arcuate nucleus; POA, preoptic area. Part A of the figure reproduced with permission (Okamura et al. 2010).

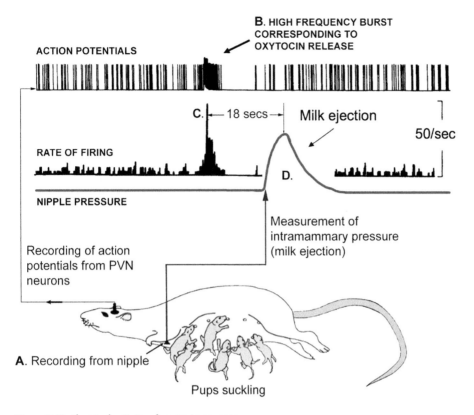

ACTION POTENTIALS

B. HIGH FREQUENCY BURST CORRESPONDING TO OXYTOCIN RELEASE

C. ⟵ 18 secs ⟶ | Milk ejection

50/sec

RATE OF FIRING

NIPPLE PRESSURE

D.

Measurement of intramammary pressure (milk ejection)

Recording of action potentials from PVN neurons

A. Recording from nipple

Pups suckling

Figure 6.12 Electrical activity of oxytocin neurons
This figure illustrates data from a recording of the electrical activity of an oxytocin neurosecretory cell in the hypothalamic paraventricular nucleus (PVN) of a lactating female rat, showing the rapid increase in firing rate at the time of oxytocin release. The milk ejection response at the nipple is measured by a pressure transducer (A). The top line of the figure shows the electrical activity of the PVN neuron illustrating the rapid burst of activity associated with oxytocin release (B) and the period of inhibition of electrical activity following this burst. The second line shows that the peak firing rate of this cell during the firing burst is about 50 pulses/second (C). The third line shows that there is a latency of about 18 seconds between the burst of firing in the PVN cell and the onset of the milk ejection response (D); that is, the increase in intramammary pressure corresponds with milk ejection and a marked suckling response of the pups. The period of 18 seconds represents the time taken for the neurosecretory neurons to release oxytocin from the posterior pituitary into the bloodstream in sufficient quantities to reach the mammary glands to induce milk ejection. Reproduced with kind permission from Dr. Jonathan Wakerley (Wakerley and Lincoln 1973).

Levy 2000). Likewise, the synthesis and release of melatonin from the pineal gland is stimulated primarily by norepinephrine released from superior cervical postganglionic neurons (Simonneaux and Ribelayga 2003; see Figure 8.5). Catecholamines are also important in the neural control of the thymus gland (Nance and Sanders 2007). The thyroid gland is innervated by adrenergic, cholinergic and peptide releasing neurons. Norepinephrine stimulates thyroid hormone secretion, at least in animals, and VIP facilitates TSH-induced thyroid hormone release, while acetylcholine inhibits thyroid hormone secretion (Ahren 1986). The pancreas is also innervated by cholinergic and adrenergic nerves from the autonomic nervous system (ANS) (Kiba 2004). Acetylcholine stimulates insulin release, while norepinephrine inhibits insulin release and stimulates glucagon release. The insulin-secreting β-cells of the pancreas also synthesize dopamine which acts as an autocrine inhibitor of insulin secretion (Simpson *et al.* 2012; Ustione *et al.* 2013). As shown in Figures 5.20 and 5.21, the GI tract is innervated through the sympathetic, parasympathetic and enteric nervous systems. The ovary also receives complex innervation and possesses a network of so-called intrinsic, or local, neurons (D'Albora *et al.* 2002; Gerendai *et al.* 2005). Thus, "true" endocrine glands are regulated by neural activity and the release of neurotransmitters and neuropeptides, as well as by hypothalamic hormones and other chemicals. It seems likely that the innervation of endocrine glands through the ANS, which is controlled by the hypothalamus, means that hormone release from these glands can be stimulated by psychological factors, such as stress or depression, and conditioned to external stimuli. In the same way that Pavlov showed that salivation could be conditioned to the expectation of food, so hormone release may be conditioned to a variety of mental states (Stockhorst 2005).

6.7 Complications in the study of neurotransmitter control of hypothalamic hormone release

As illustrated in Table 6.1, and in the descriptions provided above, many of the effects of neurotransmitters on hypothalamic and pituitary hormone secretion are complex and often appear to be contradictory. Why should this be so, when we know so much about neurotransmitters and their pathways in the brain? Nine possible reasons for these complications are given here.

1. *Multiple neurotransmitter interactions.* It is naive to assume that a single neurotransmitter controls the release of a specific hypothalamic hormone. It has been estimated that a single human neuron can receive up to 100,000 synaptic connections, many of which might be operated through different neurotransmitters (Purves *et al.* 2008). As far as hypothalamic neurons are concerned, it should be obvious from this chapter that multiple interactions between neurotransmitters are involved in stimulating the synthesis and release of hypothalamic hormones. Thirty years ago, one of the founders of neuroendocrinology, S. M. McCann, proposed the following hypothesis for the neural control of the pulsatile secretion of LH release:

   "We would postulate that pulsatile LH release involves at least several steps: (1) discharge of cholinergic neurons in the arcuate region which would synapse with the tuberoinfundibular dopaminergic neurons in the arcuate nucleus and trigger the release of dopamine; (2) dopamine released from terminals in axo-axonal contact with LH-RH terminals in the median eminence would then depolarize these terminals resulting in the release of LH-RH which would evoke the pulsatile release of LH; (3) ... noradrenergic tone would be necessary to maintain the central excitatory state of the LH-RH neurons."   (McCann 1980).

   Such a view is now obsolete because: (a) many newly discovered peptides have been shown to regulate LH secretion (Figure 6.7); and (b) GnRH neurons possess an inherent pulsatility (see section 6.3.3). What does the future hold? Veldhuis *et al.*

(2010) have proposed a series of important next steps: "the need for a more comprehensive ensemble of mathematical constructs that meld macroscopic (blood-borne), microscopic (cellular) and molecular (genomic, proteomic) components of a neuroendocrine system." This is not to say that McCann was wrong, only that in 1980 he did not have all the answers.

2. *Interactions with other hormones*. The response of hypothalamic neurosecretory cells to particular neurotransmitters and neuropeptides depends on the presence or absence of other hormones. For example, the effects of norepinephrine and dopamine on the release of LH depend on the presence of gonadal steroids. We know this because without ovaries, and therefore estradiol, the female reproductive system will not work; that is, the menstrual cycle will be absent. This principle applies to all hormone feedback systems, including the hypothalamic-pituitary-thyroid and hypothalamic-pituitary-adrenal systems.

3. *Non-specific drug effects*. Most studies that seek to investigate neuroendocrine systems use various drugs to manipulate neurotransmitter levels, and these drugs may not be specific in their action. This is one of the difficult issues faced by the pharmaceutical industry in their attempts to manipulate neurotransmitter systems for therapeutic purposes. There are many adverse side effects of all clinically useful drugs. It is not possible to specifically target a given receptor or neurotransmitter pathway without affecting other pathways or receptors. This principle also applies to research to study, for example, the involvement of dopamine in the reproductive system. Molecular biology offers far more precise techniques for establishing whether a specific receptor, or neurotransmitter, is involved in a particular neuroendocrine system. For example, and sticking with the reproductive system, one of the newest additions to our understanding of how the brain controls reproduction is the *kiss1/kisspeptin* system (see Figures 6.7 and 12.11). It was relatively simple to construct "knockout" mice in which the *kiss1* gene or the gene for the kisspeptin receptor are disabled (i.e. knocked out). The resulting mice are infertile, but can be restored to fertility by treatment with kisspeptin (Oakley *et al.* 2009). These experiments nicely complement the knowledge that humans with a defective *kiss1* gene are also infertile and can theoretically be treated with kisspeptin (Chan *et al.* 2009). However, these studies raised additional problems that add to the complexity of neurotransmitter control. An important principle to remember is that some neural systems are so important for the survival of the species – for example, the regulation of fertility or the control of body weight – that redundancies are often built into the regulatory pathways. In other words, if one pathway, or neurotransmitter, is somehow rendered inoperative, then another one will take over its function to maintain homeostasis. This is another reason, or complication, that makes the study of the control of hypothalamic hormone release so difficult.

4. *Anaesthetic use*. Many neuroendocrine studies in experimental animals require that the animals be anaesthetized. Several anaesthetics influence the secretion of anterior and posterior pituitary hormones, primarily through their actions on the nervous system. For example, Figure 5.2 shows that some common anaesthetics (e.g. isoflurane; thiopental) act via the GABA receptor ion channel, and Table 6.1 reveals that GABA is a powerful inhibitory neurotransmitter in several hormone pathways. This means that the act of

anaesthetizing an experimental animal could change hormone levels and compromise the influence of the drug under investigation. To overcome this, it is sometimes possible to train monkeys to sit quietly, without restraint or anaesthesia, while drugs are injected intravenously and blood samples collected (e.g. see Ramaswamy *et al.* 2010).

5. In vivo *versus* in vitro *studies. In vivo* studies, that is, experiments performed in living animals, may give different results from those performed in tissue culture (i.e. *in vitro*). For example, it is possible to do experiments on hypothalamic neurons in culture dishes (Mayer *et al.* 2009). However, although this technical triumph has greatly simplified the difficult study of hypothalamic control mechanisms, such as energy homeostasis and reproduction, these hypothalamic neurons are separated from all incoming neuronal influences, both excitatory and inhibitory.

6. *Species differences.* There are species differences in the responses of the neuroendocrine system to neural input. In general terms, again using reproduction as an example, regulation of the reproductive cycle in sheep (seasonal) is quite different from that in humans. In terms of hormones, stress elevates GH levels in primates, but inhibits GH release in rodents. The point here is that it is not always easy to extrapolate experimental findings from animals to humans.

7. *Sex differences.* It should be obvious that there are sex differences in neuroendocrine secretion patterns. The most obvious example is in the reproductive system, where women, and many female animals, possess a neural mechanism that induces cyclic release (e.g. monthly) of pituitary gonadotropins to cause ovulation. It is logical to conclude that males and females may show different neuroendocrine responses to changes in the same neurotransmitter. It is probably true to say that the majority of experiments in animals use males because of their perceived simpler hormonal systems. Unfortunately, this spills over into clinical trials where women are relatively neglected, giving rise to unexpected, different responses to therapeutic drugs (Fisher and Ronald 2010; Manson 2010).

8. *Psychological factors.* Psychological factors play an important role in the endocrine response to neurotransmitters. The emotional state of the animal (e.g. fear or stress); the time of the day-night cycle when the test is done; and the relationship with feeding patterns, all influence hormone responses to changes in neurotransmitter levels. Likewise, the test environment and the stimuli presented may influence the responses.

9. *Multiple hypothalamic-pituitary interactions.* As shown in later chapters, one must also remember that some "hypothalamic hormones" such as GnRH, TRH and soma-tostatin are released, not just into the pituitary portal system to regulate pituitary secretion, but as neuropeptides throughout the brain, where they act as neuromodu-lators to regulate neurotransmitter release. A good example of this is the finding that GnRH is intimately involved in the estrogen-facilitated activation of sexual behavior (Sakuma 2002).

Understanding the effects of neurotransmitters on specific pituitary hormones is further complicated by the fact that hypothalamic hormones are not as specific in their pituitary stimulation as was once thought. For example, TRH releases PRL as well as TSH; and somatostatin inhibits TSH release as well as GH (see Figure 4.9). Thus, it may be difficult to determine the exact relationship between changes in neurotransmitter release and changes in hypothalamic-pituitary hormone levels for the following reasons: (1) more than one neurotransmitter affects a single neurosecretory cell type; (2) one neurotransmitter may affect several types of hypothalamic neurosecretory cells; (3) more than one hypothalamic-releasing hormone may affect a single pituitary hormone; and (4) other hormones and neuropeptides may also alter neurohormone release.

6.8 Neuroendocrine correlates of psychiatric disorders and psychotropic drug treatment of these disorders

6.8.1 Psychotropic drugs alter neurohormone release

As shown in Table 5.4, psychoactive drugs used as antidepressants (e.g. MAO inhibitors, tricyclic antidepressants and reuptake inhibitors) such as imipramine and prozac, and antipsychotics (e.g. haloperidol and olanzapine) alter the levels, or action, of dopamine, norepinephrine and serotonin in the central nervous system. Because these neurotransmitters are implicated in the regulation of the neuroendocrine system, it is not surprising that such drugs may also have adverse effects on hormone release. In addition to their effects on these neurotransmitter systems, most of these drugs are also non-specific and will also affect other pathways, including acetylcholine and histamine signaling (Iversen *et al.* 2009). For example, the antipsychotic drug olanzapine not only binds to dopamine D2 receptors, but also to D1, α1, H1, 5-HT_2 and muscarinic receptors (Boyda *et al.* 2010). Thus, psychotropic drugs will predictably have side effects in the neuroendocrine system. A serious clinical concern for patients being treated with antipsychotic drugs such as olanzapine is a rapid weight gain and increased risk of heart disease and type II diabetes (Reynolds and Kirk 2010). This effect of antipsychotics is almost certainly exerted via the hypothalamus, not just through changes in catecholamine signaling, but also through alterations in neuropeptide systems. The role of hypothalamic neuropeptides, such as NPY and α-MSH, in controlling body weight and energy balance is discussed in Chapter 12.

A second significant side effect of psychotropic drugs is the induction of high levels of prolactin secretion (*hyperprolactinaemia*). As discussed in section 6.3.4, the primary inhibitory input to prolactin secretion is dopamine. This means that the dopamine receptor-blocking properties of antipsychotics have a direct *stimulatory* influence on prolactin release from lactotrophs in the anterior pituitary (O'Keane 2008). The consequences of antipsychotic-induced hyperprolactinaemia are not trivial. There is a suppression of the hypothalamic-pituitary-gonadal axis and associated menstrual cycle abormalities and sexual dysfunction in women. Young Caucasian women are particularly vulnerable to developing both hyperprolactinaemia and the associated hypogonadism and bone pathology. However, men are also affected, as they show breast enlargement if prolactin levels are increased.

We can speculate that in future, as new antidepressant and antipsychotic drugs become available, they will have additional and unexpected influences on the neuroendocrine system. For example, psychedelic drugs, such as LSD and ketamine, have antidepressant efficacy (Vollenweider and Kometer 2010). The neuroendocrine effects of LSD are little studied (Horowski and Graf 1979), but ketamine anaesthesia is known to increase secretion of cortisol and prolactin (Krystal *et al.* 1999), probably via blocking glutamate receptors.

6.8.2 Neuroendocrine correlates of psychiatric disorders

The previous section addressed some of the neuroendocrine problems associated with the use of antipsychotic drugs. This section considers the opposite question: that is, given that mood disorders (e.g. depression; bipolar disorder) are thought to be a result of abnormal neurotransmitter signaling (Taylor *et al.* 2005), are there neuroendocrine problems that result from such psychological disorders? Attention has been focused on women with bipolar disorder who have increased rates of menstrual and metabolic dysfunction. Since such patients are treated with antipsychotic drugs, it becomes difficult to determine whether the neuroendocrine disturbances are related to the disease or its treatment (Kenna *et al.* 2009). Another neuroendocrine system that is dysfunctional in depressed patients is the hypothalamic-pituitary-adrenal (H-P-A) axis. Many such patients have chronic high levels of cortisol that do not respond to the negative feedback effects of the synthetic glucocorticoid dexamethasone (De Kloet *et al.* 2005). In other words, in depressed patients there appears to be no negative feedback control of CRH/vasopressin-induced release of ACTH (see Figure 6.4). Antidepressant treatment is able to correct this deficiency (Schule 2007). Although H-P-A axis dysfunction in depression is the most studied neuroendocrine aspect of the disease, abnormalities in thyroid and growth hormone secretion also occur in depressed patients (Jackson 1998; Tichomirowa *et al.* 2005).

6.9 Glial cells and the regulation of hormone release

6.9.1 Glial cells

Previous chapters focused specifically on the role of neurons in neuroendocrine regulation. However, the brain contains a large number of non-neuronal cells called *glia*, or *neuroglia* (Nicholls *et al.* 2011; Stern 2010). These cells, which do not possess axons or dendrites and do not generate action potentials as neurons do, were originally thought of as "brain glue" that provided an inert structural scaffold for neurons within the brain (Allen and Barres 2009). In fact, glial cells possess ion channels and neurotransmitter receptors and they play a critical role in the maintenance of neuronal function in key systems such as sleep (Halassa and Haydon 2010) and reproduction. The latter involves astrocytic regulation of neurosecretory cells of the hypothalamic neuroendocrine system (see section 6.9.3).

There are three main types of glia: (1) *oligodendrocytes* are responsible for the myelination of axons. Myelination provides the insulation (the myelin sheath) necessary for the rapid transmission of action potentials along the axon (Squire *et al.* 2008); (2) *microglia* regulate the brain's immune system. These cells survey the brain for damage, infection and inflammation (Aguzzi *et al.* 2013) and act by rapid transformation into macrophages (a detailed description of the immune system, including macrophages, is found in Chapter 13). In addition, microglia play a physiological role in shaping neuronal activity, synaptic function and plasticity (Graeber 2010; Wake *et al.* 2011; Wake *et al.* 2012). (3) *Astrocytes* are the most abundant cell type in the brain and represent the most plentiful glial cell in any brain region (20 to 50 percent by volume). In the hypothalamus, astrocytes are important in regulating neurosecretory cell activity, such as secretion of oxytocin, and this aspect of their function will be described in more detail in section 6.9.3.

6.9.2 The role of glial cells at the synapse

Subsequent sections will outline the important role of astrocytes in neuroendocrine function, but this section will provide some background and focus specifically on the physical and neurochemical relationship between astrocytes and neurons. Astrocytes possess several important properties that facilitate their interaction with neurons: (a) astrocytes exist in close proximity to neurons and blood vessels (Figure 6.13). Since astrocytes have many processes, it is possible for a single astrocyte to make contact with as many as 30,000 other cells, including neurons and other astrocytes (Smith 2010). Also, the number of astrocytes per neuron increases dramatically with brain complexity and Figure 6.14 illustrates how the human brain compares with those from lower species (Nedergaard *et al.* 2003); (b) astrocytes provide neurons with energy; (c) astrocytes are closely associated with blood vessels and enable oxygen and glucose to be delivered to neurons; (d) astrocytes remove excess amounts of neurotransmitters from the synapse via specific reuptake mechanisms such as the glutamate reuptake transporter; (e) astrocytes have neurotransmitter

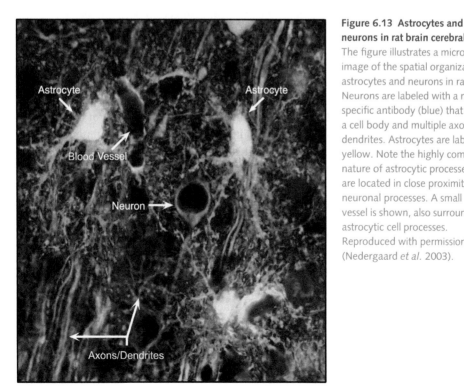

Figure 6.13 Astrocytes and neurons in rat brain cerebral cortex
The figure illustrates a microscopic image of the spatial organization of astrocytes and neurons in rat brain. Neurons are labeled with a neuron-specific antibody (blue) that reveals a cell body and multiple axons and dendrites. Astrocytes are labeled in yellow. Note the highly complex nature of astrocytic processes that are located in close proximity to neuronal processes. A small blood vessel is shown, also surrounded by astrocytic cell processes. Reproduced with permission (Nedergaard *et al.* 2003).

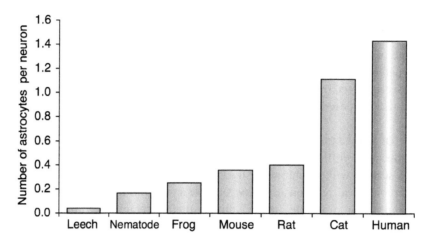

Figure 6.14 Increased number of astrocytes correlates with increased brain complexity
This figure reveals the ratio of astrocytes to neurons for different species, showing an evolutionary increase in the number of astrocytes per neuron with increasing brain complexity and size. It seems likely that the greater abundance of astrocytes indicates the presence of increasingly complex synaptic networks. Reproduced with permission (Nedergaard *et al.* 2003).

receptors that respond to stimulation by neurotransmitters released from neurons; and (f) astrocytes also secrete neurotransmitters and neuroactive substances that influence neuronal activity.

Properties (e) and (f) are illustrated in the physical connection between astrocytes and neurons, called *the tripartite synapse* (Figure 6.15; Perea *et al.* 2009; Haydon 2001). In brief, astrocytes are activated by neurotransmitters released from neurons. These neurotransmitters bind to receptors located on the astrocyte that then induce the release of intracellular calcium ions. This change in intracellular calcium ions induces secretion of neuroactive molecules from astrocytes, called *gliotransmitters*, which then bind to neuronal receptors (Figure 6.15). Not shown in Figure 6.15 is the ability of astrocytes to communicate with each other; that is, a rise in internal calcium ions is communicated, not just to neurons, but also to other astrocytes, joined together by structures called *gap junctions*. Gap junctions are contact points between cells that allow rapid transfer of ions from cell to cell. Thus, calcium ions in one astrocyte pass freely through gap junctions to simultaneously activate many other astrocytes in the form of a calcium wave (Haydon 2001).

Figure 6.15 illustrates that glutamate is released from a neuronal presynaptic terminal and binds postsynaptically to neuronal ionotropic glutamate receptors (see section 5.5 and Figure 5.3), as well as to G-protein-coupled metabotropic glutamate receptors located on the astrocyte. The astrocyte responds to glutamate by releasing D-serine, which is a co-agonist at glutamate (NMDA; N-methyl-D-aspartate) receptors (see also Figure 5.6), but also releases calcium ions to activate other astrocytes. In this way, the astrocyte can send a regulatory signal to amplify the effect of glutamate neurotransmission. There are several other examples of gliotransmitters, such as GABA, ATP, prostaglandins, cytokines such as TNFα (tumor necrosis factor α) and neuropeptides

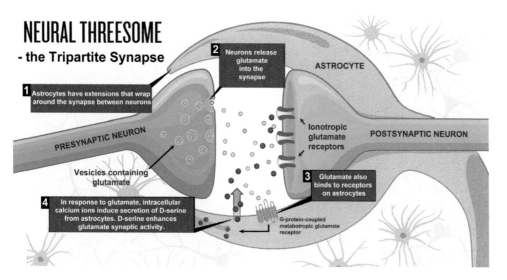

Figure 6.15 A neuron-astrocyte tripartite synapse
The figure illustrates how astrocyte processes enclose a glutamate synapse between two neurons (1). An activated presynaptic neuron (2) secretes the excitatory neurotransmitter glutamate into the synapse where it binds to postsynaptic ionotropic receptors that are located on postsynaptic neuron membranes, and to G-protein-coupled receptors on astrocytes (3). Following the binding of glutamate to its receptors the astrocyte releases calcium ions from intracellular sites (4). This mobilization of calcium ions induces secretion of D-serine which then enters the synaptic space to act as a co-agonist at ionotropic glutamate receptors of the NMDA (N-methyl-D-aspartate) subtype. D-serine enhances the postsynaptic action of glutamate. *Not shown* are glutamate reuptake transporters located on the astrocyte cell membrane. These reduce the accumulation of excess glutamate in the synapse by internalizing glutamate, thereby avoiding overstimulation of postsynaptic glutamate receptors. Reproduced with permission (Smith 2010).

(Martin 1992; Volterra and Meldolesi 2005; Perea *et al.* 2009). To summarize, astrocytes engage in bidirectional communication with neurons; that is, they respond to neuronal synaptic activity with specific feedback to modulate synaptic transmission and activate the astrocytic network.

6.9.3 Astrocytes and the regulation of neurohypophyseal hormone release

As discussed in section 6.4, oxytocin and vasopressin (ADH) are biosynthesized in magnocellular neurons of the hypothalamic SON and PVN that project to the posterior pituitary (Figure 6.10). The hormones are then secreted from nerve endings in the posterior pituitary to control birth, lactation and body fluid balance. This section will describe how astrocytes play a critical role in regulating the secretory activity of magnocellular SON and PVN neurons, with a specific focus on oxytocin neurons. Astrocytes do this in two ways: (1) they modify neurotransmission as part of a tripartite synapse; and (2) they modulate neurotransmission by rapidly changing the morphology of the synaptic complex. As noted in section 6.4, glutamate is an excitatory neurotransmitter that regulates neuronal secretion of oxytocin and ADH (Van den Pol *et al.* 1990;

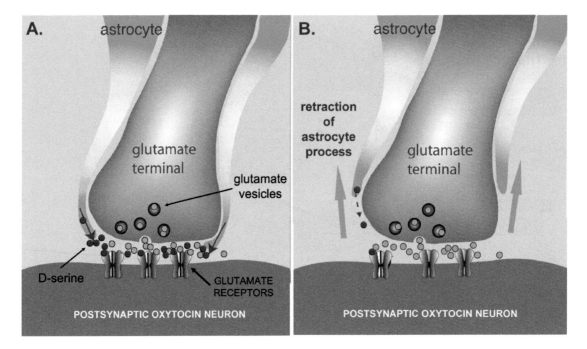

Figure 6.16 Glutamate-induced morphological changes in a tripartite synapse
This example, taken from the SON of a mother rat, shows a glutamatergic-astrocyte-oxytocinergic tripartite synapse. Increased firing of the presynaptic glutamate neuron, for example as a result of pups suckling at the mother's nipples, releases glutamate that induces the secretion of D-serine from the astrocyte (A). Continued excitation of the oxytocin neuron, via D-serine gliotransmission and glutamate (NMDA) receptors, is accompanied by retraction of astrocyte processes, thereby exposing the postsynaptic membrane (B). This morphological plasticity (retraction of the astrocyte processes) results in a reduction of D-serine concentration in the synapse and it also prevents reuptake of synaptic glutamate into the astrocyte. The latter effect increases glutamate levels in the synapse. The reversible alterations in astrocytic coverage of the tripartite synapse thus induce variable neurotransmission at the oxytocin neuron. Reproduced with permission (Oliet *et al.* 2008).

Majdoubi *et al.* 1996; Busnardo *et al.* 2012). Glutamate neurons form tripartite synapses with astrocytes and magnocellular neurons and D-serine acts as a gliotransmitter, as illustrated in Figure 6.15. Such synapses also exhibit *morphological plasticity*; that is, in certain physiological states when the neurons are activated (e.g. during lactation; Figure 6.12), the astrocytic processes are retracted from the tripartite synapse, effectively unwrapping and exposing the glutamate nerve terminals, as well as the postsynaptic oxytocin neuron (Figure 6.16) (Theodosis *et al.* 2008). Figure 6.16A illustrates how glutamate activation of oxytocin neurons, via postsynaptic glutamate receptors, is enhanced by the gliotransmitter D-serine released from astrocytes. This strong stimulation of oxytocin neuron firing is responsible for the milk-ejection reflex shown in Figure 6.12. Over a period of several hours, during multiple bursts of neuron firing and milk ejections, the astrocytic processes wrapped around the tripartite synapse are retracted to expose the neuron-neuron synapse (Figure 6.16B). When pups are weaned (i.e. when suckling ceases), the process is completely reversed. Similar changes occur when oxytocin is released during the birth process (parturition). However, no such alterations in morphology take place at synapses that control ADH secretion (Chapman *et al.* 1986).

What are the neurochemical consequences of astrocytic process retraction? First, D-serine concentration in the synapse is *reduced* (Figure 6.16B). Second, glutamate levels in the synapse *increase* because there is no reuptake by the now absent astrocytic processes (see Figure 6.15). Third, the resulting elevated synaptic glutamate levels exert inhibitory effects on the presynaptic terminal causing a reduction in glutamate secretion (presynaptic inhibition). All three of these effects contribute to alterations in neuronal excitability and electrical activity of oxytocin neurons (Panatier 2009). Students interested in further details are referred to the work of Theodosis and co-workers (Theodosis 2002; Theodosis *et al.* 2008).

6.9.4 Astrocytes and the regulation of hypothalamic hypophyseal hormone release

As outlined earlier in this chapter, hormone secretion from the adenohypophysis (anterior pituitary) is controlled by releasing and inhibiting hormones that are released by neurosecretory neurons in the hypothalamus (e.g. see Figure 6.7). In particular, the secretory activity of GnRH neurons is regulated by a variety of neurotransmitters and neuromodulators, including kisspeptin, opioids, glutamate and GABA. However, this section will introduce the role of glial cells in modulating the secretion of GnRH from nerve terminals located in the median eminence. It is possible that secretion of other releasing hormones, such as TRH and CRH, could also have a glial component, but this has not yet been studied (Ojeda *et al.* 2008). *Astrocytic* involvement in the mechanism for the release of GnRH parallels that outlined already for oxytocin, namely the involvement of a gliotransmitter (in this case PGE2; see below). The median eminence contains additional glial cells, called *tanycytes*. Tanycytes line the floor of the third ventricle and extend their processes down to the capillaries of the portal system (Figure 6.17; Prevot *et al.* 2010; Ojeda *et al.* 2008). As described previously, GnRH nerve terminals were assumed to be located in close apposition to capillaries of the portal vasculature to permit secretion of GnRH and other releasing hormones directly into the portal blood (see Figures 3.3 and 4.3). A more accurate schematic arrangement of astrocytes, tanycytes and GnRH terminals is shown in Figure 6.17. Note that the tanycyte processes are strategically placed to prevent GnRH nerve terminals from close contact with the capillaries. Astrocytes are also in intimate contact with GnRH axons and terminals, and signaling from astrocytes induces morphological changes (retraction) in the tanycyte processes. This allows the GnRH nerve terminals to release GnRH into the capillaries, a process analogous to that already described for oxytocin neurons (Figure 6.16). The mechanism underlying changes in morphology is outlined in Figure 6.18 as follows: GnRH release is initiated by neuronal glutamate that stimulates astrocytes to release TGFα (1; Figure 6.18A). (2) TGFα is responsible for the release of the gliotransmitter PGE2 by acting as an *autocrine* signal on astrocytes, and as a *paracrine* signal on tanycytes, to secrete PGE2 (3). PGE2 then causes secretion of GnRH by a stimulatory effect on

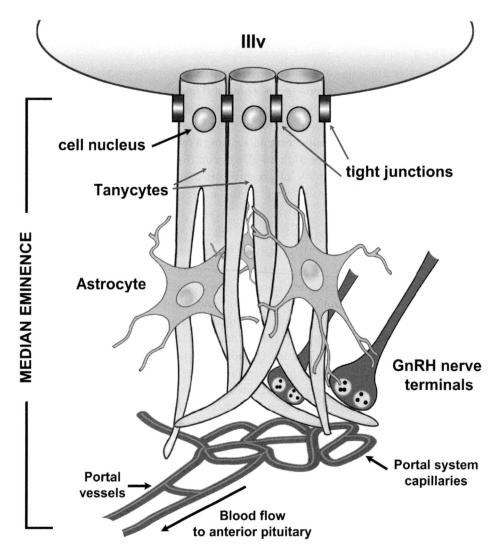

Figure 6.17 Outline of astrocyte-tanycyte-GnRH nerve terminal arrangement in median eminence of the hypothalamus

This figure is a schematic representation of the cell types (astrocytes and tanycytes) and nerve terminals (GnRH) that coexist in the median eminence of the hypothalamus. As shown previously (see Figure 4.1), the median eminence is the brain structure where releasing/inhibiting hormones, such as GnRH, are released from nerve endings in proximity to the capillaries of the pituitary portal system. A major component of the median eminence, in addition to astrocytes, is another glial cell called the tanycyte. Tanycytes form the floor of the third ventricle and communicate with each other via tight junctions. Tanycyte processes are wrapped around GnRH neuroendocrine terminals to create a diffusion barrier that hampers GnRH entry into the pituitary portal circulation.

    Abbreviation: IIIv, third ventricle. Reproduced with permission, and redrawn (Prevot *et al.* 2010).

the GnRH nerve terminals (4). PGE2 also has an autocrine effect on tanycytes (5), inducing them to synthesize and release TGFβ1 (6). TGFβ1 is the signal for tanycyte processes to retract and expose the portal capillaries to GnRH nerve endings (7). This latter step is illustrated in Figure 6.18B.

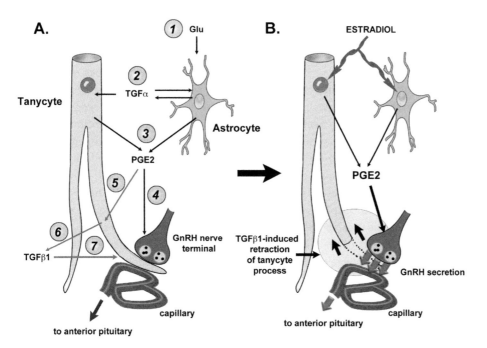

Figure 6.18 Gliotransmission regulates GnRH secretion from the median eminence
PGE2 acts as a gliotransmitter to induce secretion of GnRH (A), but also indirectly causes a retraction of tanycyte processes (B) to allow GnRH to enter the portal capillary system. The release of GnRH is initiated by neuronal glutamate that stimulates astrocytes to release TGFα (1). (2) TGFα then acts as an autocrine signal on astrocytes, and as a paracrine signal on tanycytes, to release PGE2 as a gliotransmitter (3). PGE2 causes secretion of GnRH by a stimulatory effect on the GnRH nerve terminals (4). PGE2 also has an autocrine effect on tanycytes (5), inducing them to synthesize and release TGFβ1 (6). TGFβ1 is the signal for tanycyte processes to retract and expose the portal capillaries to GnRH nerve endings (7). This latter step is illustrated in B. Both astrocytes and tanycytes have receptors for estradiol, and estradiol amplifies the production of PGE2 to induce a large secretion of GnRH. *Note that receptors for glutamate, PGE2, TGFα, TGFβ1 and estradiol have been omitted for the sake of clarity.*
   Abbreviations: Glu, glutamate; GnRH, gonadotropin-releasing hormone; PGE2, prostaglandin E2; TGFα, transforming growth factor α; TGFβ1, transforming growth factor β1.

The process outlined in Figure 6.18 is designed to generate a surge of GnRH in the portal blood leading to a large secretion of LH and FSH from the anterior pituitary. As described in section 4.3.3, this surge of LH/FSH is responsible for inducing ovulation (Figure 4.6). An important component of the LH/FSH surge is estradiol that acts by exerting a positive feedback on GnRH release (Figure 4.6). As seen in Figure 6.18B, one of the effects of estradiol is to increase the production of PGE2. Both astrocytes and tanycytes have receptors for estradiol, and estradiol amplifies the production of PGE2 to induce a large secretion of GnRH. To summarize, the neuronal drive to generate an ovulatory surge of LH and FSH is facilitated by astrocytes and tanycytes responding to glutamate through production of TGFα and the gliotransmitter PGE2. PGE2 subsequently causes tanycytes to retract and uncover GnRH neuron terminals so that they can secrete a surge of GnRH in response to PGE2, directly into the portal capillaries.

6.10 Summary

This chapter examined the ways in which neurotransmitters and neuropeptides regulate the release of hypothalamic and pituitary hormones. Neural input to the endocrine hypothalamus arises from neurotransmitter pathways from many parts of the brain. The endocrine hypothalamus is affected by changes in acetylcholine, dopamine, noradrenaline, serotonin, GABA, enkephalin and other neurotransmitters and neuropeptides. Newly discovered neuropeptides such as kisspeptin have provided further insight into the control of the reproductive system. Neuroendocrine cells can be influenced indirectly, through neural pathways passing through the hypothalamus; directly, through axo-dendritic or axo-axonal synapses; or by release of neurotransmitters and neuropeptides into the hypophyseal portal vessels to act directly on the pituitary gland. The use of electrophysiological stimulation and recording methods demonstrates some aspects of the neural control of pituitary hormone release. The effects of neurotransmitters on hormone release are, however, complicated by multiple neurotransmitter interactions, the non-specific effects of the drugs given to alter neurotransmitter levels, the species and sex of subject used, interactions with other hormones and psychological factors. Further progress in attempts to clarify how the neuroendocrine hypothalamus is regulated will undoubtedly depend on techniques of molecular biology and genetics. The assay of neurotransmitter effects is further complicated by the multiple interactions between hypothalamic and pituitary hormones. Neuroendocrine abnormalities may be associated with psychiatric disorders and the drugs used to treat them. Finally, neurons are not the only brain cells involved in regulating the secretion of hypothalamic and pituitary hormones. Two examples of the role of glial cells are described. Thus, astrocytes and tanycytes are intimately involved in the control of oxytocin and GnRH secretion.

FURTHER READING

Araque, A., Carmignoto, G., Haydon, P. G., Oliet, S. H. R., Robitaille, R. and Volterra, A. (2014). "Gliotransmitters travel in time and space," *Neuron* 81, 728–739.

Barres, B. A. (2008). "The mystery and magic of glia: a perspective on their roles in health and disease," *Neuron* 60, 430–440.

Egli, M., Leeners, B. and Kruger, T. H. (2010). "Prolactin secretion patterns: basic mechanisms and clinical implications for reproduction," *Reprod* 140, 643–654.

Emery, B. (2010). "Regulation of oligodendrocyte differentiation and myelination," *Science* 330, 779–782.

Fekete, C. and Lechan, R. M. (2013). "Central regulation of the hypothalamic-pituitary-thyroid axis under physiological and pathophysiological conditions," *Endocr Revs* 35, 159–194.

Khoo, B. and Grossman, A. B. (2014). "Normal and abnormal physiology of the hypothalamus and anterior pituitary" in *Neuroendocrinology, Hypothalamus, and Pituitary*, www.endo text.org/chapter/normal-physiology-of-acth-and-gh-release-in-the-hypothalamus-and-anterior/?singlepage=true.

Rodriguez, E. M., Blazquez, J. L. and Guerra, M. (2010). "The design of barriers in the hypothalamus allows the median eminence and the arcuate nucleus to enjoy private

milieus: the former opens to the portal blood and the latter to the cerebrospinal fluid,"
Peptides 31, 757–776.

Spergel, D. J., Kruth, U., Shimshek, D. R., Sprengel, R. and Seeburg, P. H. (2001). "Using
reporter genes to label selected neuronal populations in transgenic mice for gene promoter,
anatomical, and physiological studies," *Prog Neurobiol* 63, 673–686.

Theodosis, D. T. (2002). "Oxytocin-secreting neurons: a physiological model of
morphological neuronal and glial plasticity in the adult hypothalamus," *Front Neuroendocr*
23, 101–135.

REVIEW QUESTIONS

6.1   What is a neuroendocrine transducer?

6.2   Give two reasons why pituitary hormones, rather than hypothalamic hormones, are used
to assess the effects of changes in hypothalamic neurotransmitter levels.

6.3   What are the four ways by which neurotransmitters can regulate the release of
hypothalamic and pituitary hormones?

6.4   What are the two hypothalamic peptides that control secretion of ACTH?

6.5   What effect does norepinephrine have on TSH release?

6.6   How does GABA influence GnRH release?

6.7   Name two stimulatory peptides that regulate GnRH secretion.

6.8   Norepinephrine can inhibit and stimulate growth hormone secretion. What is the
mechanism for this?

6.9   Name two neurotransmitters that stimulate oxytocin release in response to infant
suckling?

6.10   Which neurotransmitter is released through the Sympathetic Nervous System to
stimulate hormone release from the adrenal medulla and the pineal gland?

6.11   What effect would an injection of morphine, an opiate agonist, have on the electrical
activity of the magnocellular neurosecretory cells of a lactating female rat during a bout of
suckling?

6.12   The release of prolactin and thyroid-stimulating hormone is inhibited by which
neurotransmitter?

6.13   Do opioid peptides stimulate or inhibit the release of (a) prolactin and (b) LH?

6.14   Neurotransmitters can be released into the hypophyseal portal veins to stimulate or
inhibit the pituitary gland. True or False?

6.15   Glutamate inhibits the release of LH and FSH. True or False?

6.16   Which hypothalamic neurosecretory cells respond to kisspeptin stimulation?

6.17   How do psychiatric drugs such as chlorpromazine affect prolactin levels?

6.18   What is the name given to the physical association between astrocytes and neuronal
synapses?

6.19   Which gliotransmitter is responsible for the release of GnRH from hypothalamic neurons?

ESSAY QUESTIONS

6.1   I want to do a study on the effects of prolactin on parental behavior in rats and decide to use three groups of subjects: a control group; a group with elevated prolactin levels; and a group with prolactin release inhibited. What procedures should I use to alter prolactin levels in these three groups; what side effects should I worry about; and what experimental and control procedures should I consider with respect to the use of drugs to alter prolactin release?

6.2   You are under a great deal of stress (from cold, exercise or psychological pressure). Compare the effects of this stress on your ACTH and LH levels and explain the neural pathways and neurotransmitters involved in these neuroendocrine changes.

6.3   Compare the effects of glutamate and GABA on the release of hormones from the anterior pituitary gland.

6.4   Discuss the neurotransmitter control of the release of TSH, explaining how cold activates transmitter pathways controlling TRH release and how the thyroid hormones modulate the transmitter control of TSH release by negative feedback.

6.5   Describe the neuroendocrine reflex involved in suckling-induced milk ejection, outlining the nerve pathways, neurons and neurotransmitters involved.

6.6   Describe the neural control of the pineal and thyroid glands.

6.7   Present a learning theory model for the conditioned release of hormones, such as oxytocin.

6.8   Outline the costs and benefits of using *in vivo* versus *in vitro* methods for the study of the neurotransmitter control of neurohormone release.

6.9   Discuss the advantages and disadvantages of using electrophysiological methods in the study of the neural control of neurohormone secretion.

6.10   Discuss the hormonal changes associated with depression.

6.11   Outline the connection between astrocytes and the milk-ejection reflex.

6.12   Compare the role of glial cells in the regulation of (1) oxytocin and (2) GnRH secretion from the hypothalamus.

REFERENCES

Aguzzi, A., Barres, B. A. and Bennett, M. L. (2013). "Microglia: scapegoat, saboteur, or something else?" *Science* 339, 156–161.

Ahren, B. (1986). "Thyroid neuroendocrinology: neural regulation of thyroid hormone secretion," *Endocr Rev* 7, 149–155.

Aguilar, E., Tena-Sempere, M. and Pinilla, L. (2005). "Role of excitatory amino acids in the control of growth hormone secretion," *Endocrinology* 28, 295–302.

Allen, N. J. and Barres, B. A. (2009). "Glia – more than just brain glue," *Nature* 457, 675–677.

Bealer, S. L., Armstrong, W. E. and Crowley, W. R. (2010). "Oxytocin release in magnocellular nuclei: neurochemical mediators and functional significance during gestation," *Am J Physiol Regul Integr Comp Physiol* 299, R452–458.

Berne, R. M. and Levy, M. N. (2000). *Principles of Physiology*, 3rd edn. (St. Louis, MO: Mosby).

Boyda, H. N., Tse, L., Procyshyn, R. M., Honer, W. G. and Barr, A. M. (2010). "Preclinical models of antipsychotic drug-induced metabolic side effects," *Trends Pharmacol Sci* 31, 484–497.

Busnardo, C., Crestani, C. C., Resstel, L. B., Tavares, R. F., Antunes-Rodrigues, J. and Correa, F. M. (2012). "Ionotropic glutamate receptors in hypothalamic paraventricular and supraoptic nuclei mediate vasopressin and oxytocin release in unanesthetized rats," *Endocrinology* 153, 2323–2331.

Carrasco, G. A. and Van de Kar, L. D. (2003). "Neuroendocrine pharmacology of stress," *Eur J Pharmacol* 463, 235–272.

Catania, A., Airaghi, L., Colombo, G. and Lipton, J. M. (2000). "Alpha-melanocyte-stimulating hormone in normal human physiology and disease states," *Trends Endocrinol Metab* 11, 304–308.

Chan, Y. M., Broder-Fingert, S. and Seminara, S. B. (2009). "Reproductive functions of kisspeptin and GPR54 across the life cycle of mice and men," *Peptides* 30, 42–48.

Chapman, D. B., Theodosis, D. T., Montagnese, C., Poulain, D. A. and Morris, J. F. (1986). "Osmotic stimulation causes structural plasticity of neurone-glia relationships of the oxytocin but not vasopressin secreting neurones in the hypothalamic supraoptic nucleus," *Neuroscience* 17, 679–686.

Clarke, I. J. and Parkington, H. C. (2013). "Gonadotropin inhibitory hormone (GnIH) as a regulator of gonadotropes," *Mol Cell Endocr* 385, 36–44.

Clasadonte, J., Poulain, P., Hanchate, N. K., Corfas, G., Ojeda, S. R. and Prevot, V. (2011). "Prostaglandin E2 release from astrocytes triggers gonadotropin releasing hormone (GnRH) neuron firing via EP2 receptor activation," *Proc Natl Acad Sci USA* 108, 16104–16109.

D'Albora, H., Anesetti, G., Lombide, P., Dees, W. L. and Ojeda, S. R. (2002). "Intrinsic neurons in the mammalian ovary," *Microsc Res Tech* 59, 484–489.

Egli, M., Leeners, B. and Kruger, T. H. (2010). "Prolactin secretion patterns: basic mechanisms and clinical implications for reproduction," *Reprod* 140, 643–654.

Fekete, C. and Lechan, R. M. (2013). "Central regulation of the hypothalamic-pituitary-thyroid axis under physiological and pathophysiological conditions," *Endocr Revs* 35, 159–194.

Fisher, J. A. and Ronald, L. M. (2010). "Sex, gender, and pharmaceutical politics: from drug development to marketing," *Gend Med* 7, 357–370.

Freeman, M. E., Kanyicska, B., Lerant, A. and Nagy, G. (2000). "Prolactin: structure, function, and regulation of secretion," *Physiol Rev* 80, 1523–1631.

Gan, E. H. and Quinton, R. (2010). "Physiological significance of the rhythmic secretion of hypothalamic and pituitary hormones," *Prog Brain Res* 181, 111–126.

Gerendai, I., Banczerowski, P. and Halasz, B. (2005). "Functional significance of the innervation of the gonads," *Endocrine* 28, 309–318.

Graeber, M. B. (2010). "Changing face of microglia," *Science* 330, 783–788.

Halassa, M. M. and Haydon, P. G. (2010). "Integrated brain circuits: astrocytic networks modulate neuronal activity and behavior," *Ann Rev Physiol* 72, 335–355.

Haydon, P. G. (2001). "Glia: listening and talking to the synapse," *Nat Rev Neurosci* 2, 185–193.

Herbison, A. E. (1997). "Noradrenergic regulation of cyclic GnRH secretion," *Revs Reprod* 2, 1–6.

Horowski, R. and Graf, K. J. (1979). "Neuroendocrine effects of neuropsychotropic drugs and their possible influence on toxic reactions in animals and man – the role of the dopamine-prolactin system," *Arch Toxicol* (Suppl. 2), 93–104.

Iversen, L. L., Iversen, S. D., Bloom, F. E. and Roth, R. H. (2009). *Introduction to Neuropsychopharmacology* (New York: Oxford University Press).

Jackson, I. M. (1998). "The thyroid axis and depression," *Thyroid* 8, 951–956.

Jayasena, C. N. and Dhillo, W. S. (2009). "Kisspeptin offers a novel therapeutic target in repro-
    duction," *Curr Opin Investig Drugs* 10, 311–318.
Kenna, H. A., Jiang, B. and Rasgon, N. L. (2009). "Reproductive and metabolic abnormalities
    associated with bipolar disorder and its treatment," *Harv Rev Psychiatry* 17, 138–146.
Keverne, E. B. (2005). "Odor here, odor there: chemosensation and reproductive function,"
    Nat Neurosci 8, 1637–1638.
Khoo, B. and Grossman, A. B. (2007). "Normal and abnormal physiology of the hypothalamus and
    anterior pituitary" in *Neuroendocrinology, Hypothalamus, and Pituitary*, www.endotext.org/
    neuroendo/index.htm.
Kiba, T. (2004). "Relationships between the autonomic nervous system and the pancreas including
    regulation of regeneration and apoptosis: recent developments," *Pancreas* 29, e51–58.
de Kloet, E. R., Joels, M. and Holsboer, F. (2005). "Stress and the brain: from adaptation to
    disease," *Nat Rev Neurosci* 6, 463–475.
Krsmanovic, L. Z., Hu, L., Leung, P. K., Feng, H. and Catt, K. J. (2009). "The hypothalamic GnRH
    pulse generator: multiple regulatory mechanisms," *Trends Endocrinol Metab* 20, 402–408.
Krystal, J. H., D'Souza, D. C., Karper, L. P., Bennett, A., Abi-Dargham, A., Abi-Saab, D. *et al.*
    (1999). "Interactive effects of subanesthetic ketamine and haloperidol in healthy humans,"
    Psychopharmacology (Berl) 145, 193–204.
Lightman, S. L. and Conway-Campbell, B. L. (2010). "The crucial role of pulsatile activity of the
    HPA axis for continuous dynamic equilibration," *Nat Rev Neurosci* 11, 710–718.
Lim, C. T., Kola, B., Korbonits, M. and Grossman, A. B. (2010). "Ghrelin's role as a major regulator
    of appetite and its other functions in neuroendocrinology," *Prog Brain Res* 182, 189–205.
Majdoubi, M. E., Poulain, D. A. and Theodosis, D. T. (1996). "The glutamatergic innervation of
    oxytocin- and vasopressin-secreting neurons in the rat supraoptic nucleus and its contribution
    to lactation-induced synaptic plasticity," *Europ J Neurosci* 8, 1377–1389.
Manson, J. E. (2010). "Pain: sex differences and implications for treatment," *Metabolism* 59
    (Suppl. 1), S16–S20.
Mariotti, S. (2011). "Physiology of the hypothalamic-pituitary thyroidal system" in A. B. Grossman
    (ed.), *Neuroendocrinology, Hypothalamus, and Pituitary*, Endotext.org, www.thyroidmana
    ger.org/chapter/physiology-of-the-hypothalmic-pituitary-thyroidal-system/.
Martin, D. L. (1992). "Synthesis and release of neuroactive substances by glial cells," *Glia* 5,
    81–94.
Mayer, C. M., Fick, L. J., Gingerich, S. and Belsham, D. D. (2009). "Hypothalamic cell lines to
    investigate neuroendocrine control mechanisms," *Front Neuroendocr* 30, 405–423.
McCann, S. M. (1980). "Fifth Geoffrey Harris memorial lecture. Control of anterior pituitary
    hormone release by brain peptides," *Neuroendocrinology* 31, 355–363.
Nance, D. M. and Sanders, V. M. (2007). "Autonomic innervation and regulation of the
    immune system (1987–2007)," *Brain Behav Immun* 21, 736–745.
Nedergaard, M., Ransom, B. and Goldman, S. A. (2003). "New roles for astrocytes: redefining
    the functional architecture of the brain," *Trends Neurosci* 26, 523–530.
Nicholls, J. G., Martin, A. R., Fuchs, P. A., Brown, D. A., Diamond, M. E. and Weisblat, D. (2011).
    From Neuron to Brain, 5th edn. (Sunderland, MA: Sinauer).
Nillni, E. A. (2010). "Regulation of the hypothalamic thyrotropin releasing hormone (TRH) neuron
    by neuronal and peripheral inputs," *Front Neuroendocr* 31, 134–156.
Nishihara, M., Takeuchi, Y., Tanaka, T. and Mori, Y. (1999). "Electrophysiological correlates of
    pulsatile and surge gonadotrophin secretion," *Rev Reprod* 4, 110–116.
Oakley, A. E., Clifton, D. K. and Steiner, R. A. (2009). "Kisspeptin signaling in the brain," *Endocr
    Rev* 30, 713–743.

Ojeda, S. R. and Negro-Vilar, A. (1985). "Prostaglandin E2-induced luteinizing hormone releasing hormone release involves mobilization of intracellular Ca^{2+}," *Endocrinology* 116, 1763–1770.

Ojeda, S. R., Lomniczi, A. and Sandau, U. S. (2008). "Glial-gonadotrophin hormone (GnRH) neurone interactions in the median eminence and the control of GnRH secretion," *J Neuroendocr* 20, 732–742.

Okamura, H., Murata, K., Sakamoto, K., Wakabayashi, Y., Ohkura, S., Takeuchi, Y. *et al.* (2010). "Male effect pheromone tickles the gonadotrophin-releasing hormone pulse generator," *J Neuroendocr* 22, 825–832.

O'Keane, V. (2008). "Antipsychotic-induced hyperprolactinaemia, hypogonadism and osteoporosis in the treatment of schizophrenia," *J Psychopharmacol* 22, 70–75.

Oliet, S. H. R., Panatier, A., Piet, R., Mothet, J.-P., Poulain, D. A. and Theodosis, D. T. (2008). "Neuron-glia interactions in the rat supraoptic nucleus," *Prog Brain Res* 170, 109–117.

Onaka, T., Takayanagi, Y. and Leng, G. (2010). "Metabolic and stress-related roles of prolactin-releasing peptide," *Trends Endocrinol Metab* 21, 287–293.

Orio, F., Palomba, S., Colao, A., Tenuta, M., Dentico, C., Pettreta, M. *et al.* (2001). "Growth hormone secretion after baclofen administration in different phases of the menstrual cycle in healthy women," *Horm Res* 55, 131–136.

Panatier, A. (2009). "Glial cells: indispensable partners of hypothalamic magnocellular neurones," *J Neuroendocr* 21, 665–672.

Perea, G., Navarette, M. and Araque, A. (2009). "Tripartite synapses: astrocytes process and control synaptic information," *Trends Neurosci* 32, 421–431.

Prevot, V., Bellefontaine, N., Baroncini, M., Sharif, A., Hanchate, N. K., Parkash, J. *et al.* (2010). "Gonadotrophin-releasing hormone nerve terminals, tanycytes and neurohaemal junction remodelling in the adult median eminence: functional consequences for reproduction and dynamic role of vascular endothelial cells," *J Neuroendocr* 22, 639–649.

Purves, D., Augustine, G. J., Fitzpatrick, D., Hall, W. C., LaMantia, A.-S., McNamara, J. O. *et al.* (2008). *Neuroscience*, 4th edn. (Sunderland: Sinauer Associates).

Rainero, I., Valfrè, W., Savi, L., Gentile, S., Pinessi, L., Gianotti, L. *et al.* (2001). "Neuroendocrine effects of subcutaneous sumatriptan in patients with migraine," *J Endocrinol Invest* 24, 310–314.

Ramaswamy, S., Seminara, S. B., Ali, B., Ciofi, P., Amin, N. A. and Plant, T. M. (2010). "Neurokinin B stimulates GnRH release in the male monkey (*Macaca mulatta*) and is colocalized with kisspeptin in the arcuate nucleus," *Endocr* 151, 4494–4503.

Rettori, V., De Laurentiis, A. and Fernandez-Solari, J. (2010). "Alcohol and endocannabinoids: neuroendocrine interactions in the reproductive axis," *Exp Neurol* 224, 15–22.

Reynolds, G. P. and Kirk, S. L. (2010). "Metabolic side effects of antipsychotic drug treatment–pharmacological mechanisms," *Pharmacol Ther* 125, 169–179.

Roa, J., Aguilar, E., Dieguez, C., Pinilla, L. and Tena-Sempere, M. (2008). "New frontiers in kisspeptin/GPR54 physiology as fundamental gatekeepers of reproductive function," *Front Neuroendocr* 29, 48–69.

Saland, L. C. (2001). "The mammalian pituitary intermediate lobe: an update on innervation and regulation," *Brain Res Bull* 54, 587–593.

Sakuma, Y. (2002). "GnRH in the regulation of female rat sexual behavior," *Prog Brain Res* 141, 293–301.

Schule, C. (2007). "Neuroendocrinological mechanisms of actions of antidepressant drugs," *J Neuroendocrinol* 19, 213–226.

Simonneaux, V. and Ribelayga, C. (2003). "Generation of the melatonin endocrine message in mammals: a review of the complex regulation of melatonin synthesis by norepinephrine, peptides, and other pineal transmitters," *Pharmacol Rev* 55, 325–395.

Simpson, N., Maffei, A., Freeby, M., Borroughs, S., Freyberg, Z., Javitch, J. *et al.* (2012). "Dopamine-mediated autocrine inhibitory circuit regulating human insulin secretion *in vitro*," *Mol Endocr* 26, 1757–1772.

Sladek, C. D. and Kapoor, J. R. (2001). "Neurotransmitter/neuropeptide interactions in the regulation of neurohypophyseal hormone release," *Exp Neurol* 171, 200–209.

Sladek, C. D. and Song, Z. (2012). "Diverse roles of G-protein coupled receptors in the regulation of neurohypophyseal hormone secretion," *J Neuroendocr* 24, 554–565.

Smith, K. (2010). "Settling the great glia debate," *Nature* 468, 160–162.

Smith, M. J. and Jennes, L. (2001). "Neural signals that regulate GnRH neurones directly during the oestrous cycle," *Reprod* 122, 1–10.

Squire, L. R., Berg, D. E., Bloom, F. E., du Lac, S., Ghosh, A. and Spitzer, N. C. (2008). *Fundamental Neuroscience*, 3rd edn. (London: Academic Press).

Stern, P. (2010). "Glee for glia," *Science* 330, 773.

Stockhorst, U. (2005). "Classical conditioning of endocrine effects," *Curr Opin Psych* 18, 181–187.

Taylor, C., Fricker, A. D., Devi, L. A. and Gomes, I. (2005). "Mechanisms of action of antidepressants: from neurotransmitter systems to signaling pathways," *Cell Signal* 17, 549–557.

Theodosis, D. T. (2002). "Oxytocin-secreting neurons: a physiological model of morphological neuronal and glial plasticity in the adult hypothalamus," *Front Neuroendocr* 23, 101–135.

Theodosis, D. T., Poulain, D. A. and Oliet, S. H. R. (2008). "Activity-dependent structural and functional plasticity of astrocyte-neuron interactions," *Physiol Rev* 88, 983–1008.

Tichomirowa, M. A., Keck, M. E., Schneider, H. J., Paez-Pereda, M., Renner, U., Holsboer, F. *et al.* (2005). "Endocrine disturbances in depression," *J Endocrinol Invest* 28, 89–99.

Ustione, A., Piston, D. W. and Harris, P. E. (2013). "Minireview: dopaminergic regulation of insulin secretion from the pancreatic islet," *Mol Endocr* 27, 1198–1207.

Valverde, I., Penalva, A. and Dieguez, C. (2000). "Influence of different serotonin receptor subtypes on growth hormone secretion," *Neuroendocrinology* 71, 145–153.

Van de Kar, L. D. and Blair, M. L. (1999). "Forebrain pathways mediating stress-induced hormone secretion," *Front Neuroendocr* 20, 1–48.

Van den Pol, A. N., Wuarin, J. P. and Dudek, F. E. (1990). "Glutamate, the dominant excitatory transmitter in neuroendocrine regulation," *Science* 250, 1276–1278.

Veldhuis, J. D., Keenan, D. M. and Pincus, S. M. (2010). "Regulation of complex pulsatile and rhythmic neuroendocrine systems: the male gonadal axis as a prototype," *Prog Brain Res* 181, 79–110.

Vollenweider, F. X. and Kometer, M. (2010). "The neurobiology of psychedelic drugs: implications for the treatment of mood disorders," *Nat Rev Neurosci* 11, 642–651.

Volterra, A. and Meldolesi, J. (2005). "Astrocytes, from brain glue to communication elements: the revolution continues," *Nat Rev Neurosci* 6, 626–640.

Wake, H., Moorhouse, A. J. and Nabekura, J. (2011). "Functions of microglia in the central nervous system – beyond the immune response," *Neuron Glia Biol* 7, 47–53.

Wake, H., Moorhouse, A. J., Miyamoto, A. and Nabekura, J. (2012). "Microglia: actively surveying and shaping neuronal circuit structure and function," *Trends Neurosci* 36, 209–217.

Wakerley, J. B. and Lincoln, D. W. (1973). "The milk-ejection reflex of the rat: a 20- to 40-fold acceleration in the firing of paraventricular neurones during oxytocin release," *J Endocrinol* 57, 477–493.

Regulation of hormone synthesis, storage, release, transport and deactivation

7

The endocrine system is driven by hormones released into the bloodstream in order to regulate other, distant organs (see Figure 1.3). These hormones are chemical messages that are decoded by specific recognition sites, or receptors, located in the target cells. Hormones are synthesized and stored in endocrine cells and, when required, they are released into the circulatory system. A number of hormones are transported in the bloodstream by carrier proteins. For example, *sex hormone binding globulin* is specifically responsible for transport of estradiol and testosterone. Other proteins such as albumin are less specific and serve to transport a variety of hormones. Hormone synthesis, storage, release, transport and deactivation occur through a variety of different mechanisms, depending on the chemical structure of the hormone. For example, peptides such as oxytocin are different in almost every respect from steroid hormones like estradiol. For this reason, the first section of this chapter will examine the chemical structure of hormones.

7.1 The chemical structure of hormones

In terms of chemical structure, hormones can be divided into three major groups: (1) steroid hormones; (2) amines; and (3) peptide hormones (Table 7.1). Steroid hormones, like estradiol, are different from the other two groups because of their very low solubility in blood. This is because they are essentially hydrocarbons and therefore need a binding protein to carry them in the bloodstream. A further distinction is the mechanism of their synthesis; steroids and amines are produced from precursors (such as cholesterol and tyrosine, respectively) via specific enzymes, whereas peptides are encoded by specific genes (DNA) that transcribe messenger RNA (mRNA) that is then translated into precursor peptides.

7.1.1 Steroid hormones
The steroid hormones are biosynthesized from cholesterol in the adrenal cortex and gonads. Adrenal steroids include cortisol and aldosterone and the gonadal steroids are progesterone, testosterone and estradiol (see Table 7.1). Note that some steroids are also made in the brain, fat tissue and placenta.

7.1.2 Amine hormones
These hormones are modified amino acids and include the catecholamines, indoleamines and thyroid hormones. The catecholamines norepinephrine and epinephrine are

Table 7.1 **Chemical structure of hormones**

a. STEROID HORMONES: synthesized from cholesterol

Gonadal steroids

Progestins (progesterone)

Androgens (testosterone)

Estrogens (estradiol)

Adrenal steroids

Glucocorticoids (corticosterone, cortisol)

Mineralocorticoids (aldosterone)

b. AMINES: hormones synthesized from single amino acids

Catecholamines – epinephrine, norepinephrine and DA are synthesized from tyrosine as described in section 5.3.1

Indoleamines – melatonin and serotonin are synthesized from tryptophan

Thyroid hormones – thyroxine (T4) and triiodothyronine (T3) are synthesized from iodinated thyroglobulin

c. PEPTIDES: hormones formed from chains of amino acids

Small peptides

Thyrotropin-releasing hormone (TRH) (3 amino acid residues)

Angiotensin II (8 amino acid residues)

Vasopressin (9 amino acid residues)

Oxytocin (9 amino acid residues)

GnRH (10 amino acid residues)

Somatostatin (14 amino acid residues)

Large peptides

Gastrins (17 and 34 amino acid residues)

Secretin (27 amino acid residues)

Glucagon (29 amino acid residues)

Calcitonin (32 amino acid residues)

Andrenocorticotropic hormone (ACTH) (39 amino acid residues)

Kisspeptin (145 amino acids)

Proteins (polypeptides)

Insulin (51 amino acid residues)

Parathyroid hormone (84 amino acid residues)

Human growth hormone (GH) (191 amino acid residues)

Leptin (167 amino acids)

Glycoproteins: pituitary hormones synthesized from proteins and carbohydrates

Follicle-stimulating hormone (FSH)

Luteinizing hormone (LH)

Thyroid-stimulating hormone (TSH)

Human chorionic gonadotropin (HCG) (placenta)

synthesized from the amino acid tyrosine in the adrenal medulla, from which they are secreted as hormones. As we saw in Chapter 5 (Figures 5.8 and 5.9), they are also made in the same way in the central and peripheral nervous system, where they act as neurotransmitters. The indoleamines include serotonin and melatonin. As we discussed in Chapter 5, serotonin (5-HT) acts as a neurotransmitter and is synthesized from the amino acid tryptophan in the central nervous system. However, 95 percent of mammalian serotonin is made in the enterochromaffin cells of the gut where it is probably best regarded as a hormone (Sanger 2008). Melatonin is a hormone made in the pineal gland from serotonin. The thyroid hormones, thyroxine (T4) and triiodothyronine (T3), are synthesized in the thyroid gland from the glycoprotein precursor, thyroglobulin. Thyroid hormones are formed by iodination of the amino acid tyrosine, which is an integral part of the thyroglobulin molecule (Widmaier *et al.* 2010).

7.1.3 Peptide hormones

Peptides are synthesized from amino acids in many different organs, including the peripheral endocrine glands, the pituitary gland, the hypothalamus, neurons, fat tissue, the lungs, placenta and in the gut (see Tables 2.1 and 2.2). They function as peptide hormones, neurohormones or neuropeptides, as described in Chapter 1. Peptides consist of amino acid chains that can vary in length from three to more than 200 amino acids and are categorized according to their size into small peptides, large peptides, and proteins or polypeptides (Table 7.1). Glycoproteins are special forms of the large polypeptides; that is, they are linked to carbohydrate side-chains. For example, luteinizing hormone (LH) and thyroid-stimulating hormone (TSH) are both glycoprotein hormones.

7.2 Hormone synthesis

7.2.1 Biosynthesis of peptide hormones

Most peptide hormones are synthesized from large pre-propeptides that are long amino acid chains (polypeptides). The exact sequence of the amino acids that make up these pre-propeptides is determined by the mRNA that leaves the cell nucleus to then bind with ribosomes to make the pre-propeptide (Figure 7.1). The pre-propeptide has a signal sequence that is removed as the peptide moves through the endoplasmic reticulum. The remaining propeptide exits from the Golgi contained in secretory vesicles. Then proteolytic enzymes, called *proprotein convertases*, in the secretory vesicles cleave the propeptides at specific sites to convert them into shorter amino acid chains that become the active peptides and mature hormones (Figure 7.1; see also Figure 4.2) (Bicknell 2008; Nillni 2007). We saw an example of this in Chapter 3 for the processing of the POMC gene product (see Figures 3.4 and 7.2). Other examples are shown in Figure 7.2.

  Cleavage of the propeptides by the convertases takes place at specific recognition sites along the polypeptide chain. These are shown in Figure 7.2 labeled K and R; that is, K is the symbol for the amino acid lysine and R is the symbol for arginine. The figure shows the location of these two amino acids and allows us to predict where the cleavage will take place; that is, cleavage occurs at pairs of amino acids such as lys-arg, lys-lys and arg-lys. Thus, in the case of proTRH, we can see why several copies of TRH are produced, whereas only one copy of CRH is cleaved. Note that while small peptides have the same chemical structure in almost all species of vertebrates, the larger peptides are more species specific. For example, human insulin, synthesized from pro-insulin (Figure 7.2), is structurally different from rat insulin.

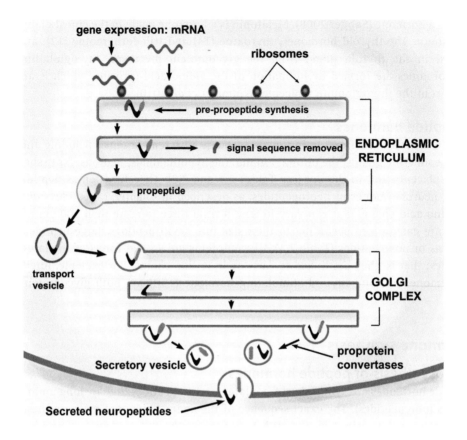

Figure 7.1 Features of peptide hormone synthesis, storage and secretion
Peptide hormone biosynthesis begins with the translation of the mRNA to a pre-propeptide on the ribosomes of the rough endoplasmic reticulum (ER). Still within the ER the signal peptide is cleaved from the pre-propeptide to form the propeptide which is then transported, via a transport vesicle, to the Golgi complex. Propeptides (prohormones) are packaged into secretory vesicles that bud off from the Golgi bodies. Peptide hormone synthesis is then completed within the secretory vesicles under the influence of proprotein convertases. Peptide hormones are released into the circulatory system from the secretory vesicles by exocytosis.

One might ask why peptide hormones should be synthesized in such a roundabout way: first, a very large molecule (pre-propeptide) is made, then a piece of this is cut off to make a propeptide, then this is cut up to make a hormone plus residual peptides. It turns out that, in terms of ease of handling during synthesis, propeptides have a number of advantages over active hormones. (1) Propeptides enable the three-dimensional structure of the peptide molecule to be stabilized during synthesis (like carpenters bracing a wall during building). (2) Propeptides are easier to transport and package into granules than active hormones. (3) Propeptides act as storage reserves. (4) Propeptides are more resistant to degradation than active hormones and have longer half-lives. (6) Perhaps the most important reason is that the propeptides can be processed to a large variety of active hormones depending on the tissue. This principle is most easily seen in the case of the POMC gene (Figures 3.4 and 7.2). In Figure 7.2, we see that POMC is cleaved into

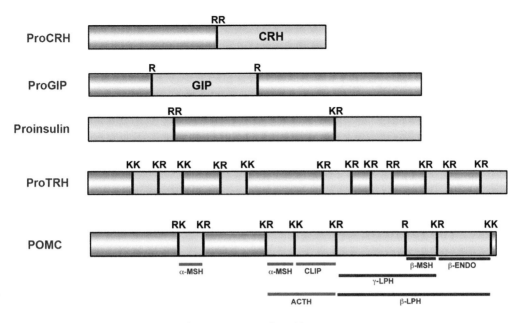

Figure 7.2 Schematic representation of some neuropeptide and hormone precursors
Proteolytic processing of propeptides occurs at various basic sites. The location of Arginine (R) and
Lysine (K) residues allows us to predict where the cleavage will take place; that is, cleavage occurs at
pairs of amino acids such as lys-arg, lys-lys and arg-lys. The precursor proteins may contain one copy of the
active neuropeptide (proCRH or proGIP), multiple copies of the active neuropeptide (proTRH), or distinct
peptide hormones (POMC). The bioactive molecules and the Arginine (R) and Lysine (K) residues
are indicated by yellow color and black bars, respectively. Reproduced with permission (Rholam and
Fahy 2009).

many active peptides, including ACTH, α-MSH and β-endorphin. Processing of this type
takes place in the pituitary gland and ACTH and β-endorphin are secreted into the
bloodstream. However, this same gene is also expressed in the hypothalamus, but this
time the propeptide is cleaved into α-MSH and β-endorphin, but not into ACTH. In this
case, α-MSH and β-endorphin are neurotransmitters secreted from neurons. Thus, we can
appreciate that propeptides provide a critical degree of flexibility depending on the tissue
of origin and the availability of different convertases.

7.2.2 Synthesis of steroid hormones

Steroid hormones are synthesized from cholesterol in the smooth endoplasmic reticulum of
tissues such as the ovaries and adrenal cortex. All steroids have a similar four-ring carbon
structure as shown in Figure 7.3. We can see that those steroids that we have mentioned
already – such as cortisol, estradiol and testosterone – are closely related in structure to the
parent compound cholesterol. Cholesterol is converted to these hormones via a series of
enzymes, depending on the tissue. Steroid-secreting cells take up cholesterol from the
blood, or synthesize it from acetate, and convert it to biologically active steroids via the
mitochondria. These conversion steps are mediated by a variety of so-called P450 enzymes
that are responsible for processing the cholesterol side-chains into the final hormone

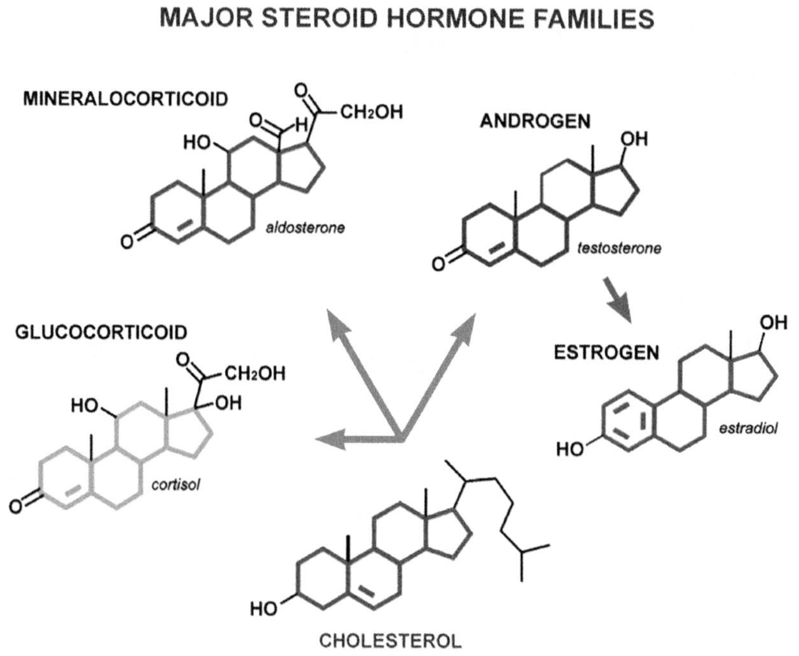

MAJOR STEROID HORMONE FAMILIES

Figure 7.3 Chemical structures of the major steroid hormones
Four primary families of steroid hormones: androgens (e.g. testosterone), estrogens (e.g. estradiol), glucocorticoids (e.g. cortisol) and mineralocorticoids (e.g. aldosterone) are derived from cholesterol through various enzymatic steps. An important family not shown here are the progestogens (e.g. progesterone). Note that the female sex hormone estradiol is formed from the male hormone testosterone. Copyright Alex Pincock.

structures as shown in Figure 7.3. Note that the female sex hormone estradiol is formed from the male hormone testosterone. Those students who are particularly interested in the organic chemistry of these steps are referred to a clear description elsewhere (Nussey and Whitehead 2001).

7.3 Storage and intracellular transport of hormones

As we have noted already, propeptides/hormones are packaged in secretory vesicles in the Golgi apparatus as shown in Figure 7.1. The vesicles serve a number of functions: (1) they provide intracellular storage and transport for the peptides. (2) In many cases, peptide synthesis is completed within the vesicle by convertases that act on propeptides (e.g. conversion of POMC to ACTH). Other enzymes such as aminopeptidases and carboxypeptidases may also be present. (3) The vesicle protects the propeptide from deactivation by non-specific peptidases, and (4) the packaging of propeptides in vesicles may help to regulate their rate of synthesis through negative feedback effects on the convertases. (5) The secretion of the peptide/hormone into the bloodstream is achieved by the secretory vesicle fusing with the cell membrane (Figure 7.1). Note that this process is identical in many ways to that described for neurotransmitter secretion (see Figure 5.11). However, an important difference between neurotransmitter secretion (e.g. norepinephrine) and the release of peptide hormones or neuropeptide neurotransmitters is that there is no reuptake mechanism to replenish the empty vesicles (see Figures 5.9 and 5.14). In other words, peptide hormone vesicles have to be refilled with newly synthesized molecules rather than those recycled from the extracellular fluid.

In contrast to the picture for peptide hormones, steroid hormones are not packaged and stored in secretory vesicles for release, but in general are simply made on demand. For example, there is only enough testosterone stored in the testes for perhaps 2 hours of secretion. The thyroid is the only gland that stores sufficient hormone for a two-week period. Following biosynthesis, steroids simply diffuse out of the secreting cells and into the circulation.

7.4 Hormone release

Peptide hormone release occurs when the endocrine cell is activated, usually by hormonal stimulation. For example, corticotrophs of the anterior pituitary secrete ACTH following stimulation by CRH (e.g. Figure 6.4). Stimulation of peptide hormone-secreting cells causes a change in cell membrane permeability so that the hormone is released into the circulatory system by exocytosis when the secretory granule fuses with the cell membrane (Figure 7.1). Depolarization-induced entry of sodium and calcium ions into the cell is essential for peptide secretion, as is the case for neurotransmitter secretion that we have covered already in Chapter 5. Calcium ion (Ca^{2+}) binding is necessary to initiate the fusion of the secretory granule to the cell membrane during exocytosis. Some peptidergic cells release their hormones in bursts or pulses, rather than as a continuous secretion. The level of circulating hormones thus rises dramatically and then gradually declines as the hormone is deactivated. There are many examples of this phenomenon, but it is clearly seen in Figure 6.5, where bursts of ACTH are released from the anterior pituitary to stimulate the adrenal to release corresponding peaks of cortisol. The bursts of ACTH are induced by pulsatile secretion of CRH from CRH neurons.

Another example is the secretion of luteinizing hormone from the pituitary gland as shown in Figures 6.11 and 4.4. As discussed in detail in Chapter 4, this pattern of LH secretion reflects the pulsatile release of GnRH from hypothalamic neurosecretory cells. Pulsatile release of GnRH/LH appears to be an optimal mechanism for hormonal communication. For example, we saw in Figure 4.5 that the presence of LH pulses is necessary for normal menstrual/ovulatory cycles; that is, when LH pulses are inhibited by weight loss, as in patients with anorexia nervosa, the menstrual cycle is abolished. An interesting corollary of this is that when pulses are abolished by stimulation with *continuous* GnRH infusion, the levels of LH and FSH fall rapidly because pituitary GnRH receptors become desensitized to continuous stimulation with GnRH (Figure 7.4). However, as shown, the situation is readily reversed through reinstatement of pulsatile stimulation with GnRH.

By way of comparison with peptide secretion, and as we noted already, steroid hormone release is controlled by the rate of synthesis. For example, estradiol release from the ovary occurs in response to stimulation with LH and FSH. Steroid

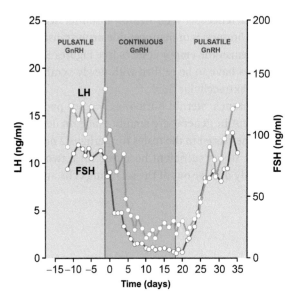

Figure 7.4 Non-pulsatile, continuous stimulation with GnRH suppresses luteinizing hormone (LH) and follicle-stimulating hormone (FSH) secretion from the anterior pituitary
Continuous infusion of GnRH abolishes release of LH and FSH from the pituitary. This occurs because the GnRH receptors become over-stimulated causing desensitization to further treatment with GnRH. However, restoration of pulsatile stimulation gradually restores a normal secretion pattern.
   Abbreviation: GnRH, gonadotropin-releasing hormone. Reproduced with permission (Besser and Thorner 2002).

hormones are not stored in the secretory cells, since steroids are not packaged in vesicles. Steroids are, however, secreted in pulses due to the response of the steroid-secreting cells to pituitary hormones such as ACTH released as pulses (see Figure 6.5).

7.5 Hormone transport

Carrier proteins facilitate transportation of some hormones in the blood and may serve to prolong their active life by protecting them from deactivation. The steroid and thyroid hormones have particularly long half-lives (Table 7.2). However, the primary role of such carrier proteins is to enable the steroid to be solubilized in the blood. Hormones such as estradiol and cortisol are insoluble in aqueous media. Plasma albumin acts as a general, rather low-affinity, carrier protein, but most steroid hormones also have specific, high-affinity carrier proteins. For example, the thyroid hormones T3 (triiodothyronine) and T4 (thyroxine) are transported via the thyroid binding globulin. The sex steroids estradiol and testosterone are reversibly bound to both albumin and to sex hormone binding globulin. These transport proteins are glycoproteins produced in the liver. A large percentage of the hormone is inactive since it is bound to the carrier protein. The very low percentage of "free," non-bound hormone

Table 7.2 **Transport of steroids in plasma**

Steroid	Total conc (nmols/L)	% unbound	% bound CBG	% bound albumin	% bound SHBG	$t_{1/2}$ (mins)
Cortisol	400	4	90	6	0.1	100
Aldosterone	0.4	40	20	40	0.1	10
Progesterone	0.6	2.4	17	80	0.6	5
Testosterone	20	2.0	3	40	55	10
Estradiol	0.1	2.0	0	68	30	20

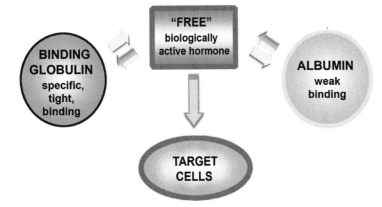

Figure 7.5 Equilibrium binding of hormones to circulating binding globulins Hormones, including water-insoluble hormones such as sex steroids, circulate in the blood bound to specific binding globulins (e.g. sex hormone binding globulin, SHBG and corticosteroid binding globulin, CBG). Hormone binding is in equilibrium so that a very small amount of "free" hormone becomes biologically active and available for binding to specific receptors.

constitutes the biologically active hormone; that is, only free hormone can interact with the target cell to induce a biological response (Figure 7.5). The high levels of protein-bound hormones might then act as readily available storage reserves. Table 7.2 shows how some steroids are bound to different binding proteins (Nussey and Whitehead 2001).

It is generally thought that peptide hormones are not bound to carrier proteins and thus have short half-lives in the bloodstream. For example, detailed studies on the kinetics of ACTH and cortisol secretion in humans showed that the half-life of ACTH is about 20 minutes, whereas that for cortisol is 49 minutes (Keenan *et al.* 2004). Nevertheless, there is good evidence for some peptides being bound to carrier proteins, including GH and IGF-1 (Fisker 2006; Banerjee and Clayton 2006). Oxytocin and vasopressin may be special cases of peptide hormones bound to specific binding proteins. As we noted in Chapter 3, oxytocin and vasopressin are secreted from the posterior pituitary together with two proteins called neurophysins. Neurophysin I is specific for oxytocin and vasopressin is secreted with neurophysin II. The neurophysins are components of the prohormone structures and their presence is thought to be a protective device to ensure stability of oxytocin and vasopressin within the secretory vesicles (Legros and Geenen 1996).

Table 7.3 **Hormone metabolites in blood and urine**	
Hormone	Metabolite
Serotonin	5-hydroxyindoleacetic acid (5-HIAA)
Melatonin	6-sulphatoxy melatonin
Dopamine	homovanillic acid (HVA)
Norepinephrine	3-methoxy-4-hydroxyphenylglycol (MHPG)
Testosterone	various conjugated sulphates and glucuronides
Estradiol	various conjugated sulphates and glucuronides
Growth hormone	peptide fragments
Prostaglandins	fatty acids

7.6 Deactivation of hormones

Most circulating steroid, thyroid and peptide hormones are deactivated by enzymes in the liver, kidney or blood and their metabolites are then excreted in the urine or feces. Some proteins, however, may be taken up into target cells by endocytosis and deactivated by lysosomes within the target cells. Since hormone metabolites excreted in the urine reflect the levels of these hormones released from their tissue of origin, it is often clinically important to measure circulating hormone levels in humans and animals, as well as measuring the hormone metabolites in the urine (see Table 7.3). Hormone metabolites in the urine and feces also provide the basis for some of the pheromones excreted by mammals (see section 1.4.5).

7.7 Methodology for neuroendocrine research

A wide variety of analytical techniques have been developed and adapted for use in the study of neuroendocrine systems. There is no adequate single source for this type of information, but we have listed an array of methods available to neuroendocrinologists (Table 7.4). All of these approaches are currently exploited in neuroendocrine studies. Several of them represent classical techniques that have been vastly improved by modern technology; for example, cellular localization of immunofluorescence has been greatly facilitated by the availability of highly specific monoclonal antibodies. However, despite the enormous amount of information generated by such classical approaches, perhaps the most incisive breakthroughs arise from the application of molecular techniques. These include the polymerase chain reaction (PCR; Mullis 1990) and *in situ* hybridization for quantifying and localizing mRNA, and the breeding of experimental animals with targeted changes in specific genes. For example, hypothalamic control of the anterior pituitary can now be studied in living animals by selectively manipulating the genes for releasing hormones such as CRH and TRH (Wells and Murphy 2003).

Table 7.4 Methods used to investigate cellular sites of hormone biosynthesis, storage and secretion

Method	Information gained
1. Identification of neuroendocrine cells	
Gross anatomy	anatomical location, size, vascularization and innervation of glands
Surgical removal of glands	physiological effects
Light microscopy and histology	identification of neural and endocrine cells
Confocal microscopy	identification of subcellular structure in neuroendocrine cells
Scanning electron microscope	three-dimensional image of cell structure
2. Localization of hormones, receptors and their genes within cells	
Histological staining	identify cell bodies (nucleus and cytoplasm) and axons
Fluorescence immunocytochemistry	localize hormones in histological sections by specific fluorescent-labeled antibodies
Autoradiography	localize radioactively labeled hormones and receptors
In situ hybridization	detection and cellular localization of mRNA
3. Molecular biology of neuroendocrine systems	
Polymerase chain reaction (PCR)	identification and quantification of specific gene expression
Silencing of genes	RNA interference; antisense methods
Gene "knockouts"	experimental animals with inactivated genes
4. Examining biosynthetic pathways	
Antibody injection	to detect and isolate molecules
Drug injection: agonists or antagonists	to study physiological activity of cells
Radioisotope injection	to measure enzyme activity, trace chemical pathways during synthesis, and determine biochemical responses in a cell
5. Determining the chemical structure of hormones	
Chromatography	separates proteins into different fractions
Electrophoresis	determines the size and structure of proteins
Amino acid sequencers of peptides	determine amino acid sequences of peptides
6. Quantification (assay) of hormones	
Mass spectrometry	quantification of peptides, hormones, proteins
Automated immunometric and chemiluminescent assays	clinical measure of hormone levels

7.8 Summary

This chapter outlines the factors regulating the synthesis, storage, release, transport and deactivation of hormones. Hormones can be divided into two general types: steroids and non-steroids. The non-steroid hormones include modified amino acids, peptides and glyco-proteins. Peptide and protein hormones are biosynthesized as preprohormones from amino acid precursors in the endoplasmic reticulum of the cell and packaged into secretory vesicles in the Golgi bodies. Often, the prohormones are stored in vesicles together with proteolytic enzymes called convertases and the final synthesis of the hormone occurs inside the secretory vesicles. As well as the completion of hormone synthesis, vesicles store hormones, regulate their rate of synthesis and control the release of hormones, serving many of the same functions as secretory vesicles do for neurotransmitters. Steroid hormones are synthesized from cholesterol in the adrenal cortex, gonads, adipose tissue and placenta and are released directly into the circulation without being packaged in secretory vesicles. The thyroid, adrenal and gonadal hormones are transported through the blood by specific carrier proteins (binding globulins), and this may also be true for peptides such as GH. Oxytocin and vasopressin rely on carrier proteins (neurophysins) for their stability within the vesicles. It is important to remember that binding to the carrier proteins is in equilibrium, so that a small amount of "free," biologically active hormone is available to bind to target cells. Some hormones such as LH, FSH and ACTH are released from the pituitary in bursts, or pulses. Hormones are deactivated by metabolic enzymes in the kidney, liver and blood and their metabolites are excreted in the urine and feces.

REVIEW QUESTIONS

7.1   Name *four* major classes of steroid hormones.

7.2   What is the primary precursor molecule giving rise to steroid hormones?

7.3   Peptide and protein hormones are made up of which essential chemical unit?

7.4   Where in the cell are peptides biosynthesized?

7.5   Peptide hormone precursors – *pre-propeptides* – are processed where in the cell?

7.6   The male sex hormone testosterone is made in which two *female* endocrine glands?

7.7   Are both steroid hormones and peptides stored in secretory granules?

7.8   The posterior pituitary hormone, oxytocin, is secreted bound to which protein?

7.9   Pituitary hormones such as vasopressin are released from secretory vesicles. What is this process called?

7.10   The prohormone POMC is the precursor for which three hormones?

7.11   Hormones are transported in the blood bound to carrier molecules. Are these hormones biologically active?

ESSAY QUESTIONS

7.1   Discuss the significance of pre-propeptides in the biosynthesis of peptide hormones such as ACTH.

7.2   Discuss the physiological importance of pulsatile secretion of peptide hormones.

7.3   Discuss the importance of carrier proteins in the transport of hormones in the bloodstream.

7.4   Describe the biosynthesis and secretion of oxytocin.

7.5   Describe two techniques used to localize which neurons produce specific peptide hormones.

REFERENCES

Banerjee, I. and Clayton, P. E. (2006). "Clinical utility of insulin-like growth factor-I (IGF-I) and IGF binding protein-3 measurements in paediatric practice," *Pediatr Endocrinol Rev* 3, 393–402.

Besser, G. M. and Thorner, M. O. (2002). *Comprehensive Clinical Endocrinology*, 3rd edn. (St. Louis, MO: Mosby).

Bicknell, A. B. (2008). "The tissue-specific processing of pro-opiomelanocortin," *J Neuroendocrinol* 20, 692–699.

Fisker, S. (2006). "Physiology and pathophysiology of growth hormone-binding protein: methodological and clinical aspects," *Growth Horm IGF Res* 16, 1–28.

Keenan, D. M., Roelfsema, F. and Veldhuis, J. D. (2004). "Endogenous ACTH concentration-dependent drive of pulsatile cortisol secretion in the human," *Am J Physiol Endocrinol Metab* 287, E652–661.

Legros, J. J. and Geenen, V. (1996). "Neurophysins in central diabetes insipidus," *Horm Res* 45, 182–186.

Mullis, K. (1990). "The unusual origin of the polymerase chain reaction," *Sci Amer* 262, 56–61.

Nillni, E. A. (2007). "Regulation of prohormone convertases in hypothalamic neurons: implications for prothyrotropin-releasing hormone and proopiomelanocortin," *Endocr* 148, 4191–4200.

Nussey, S. S. and Whitehead, S. A. (2001). *Endocrinology: An Integrated Approach* (Oxford: Bios Scientific Publishers), www.ncbi.nlm.nih.gov/books/NBK20.

Rholam, M. and Fahy, C. (2009). "Processing of peptide and hormone precursors at the dibasic cleavage sites," *Cell Mol Life Sci* 66, 2075–2091.

Sanger, G. J. (2008). "5-hydroxytryptamine and the gastrointestinal tract: where next?" *Trends Pharmacol Sci* 29, 465–471.

Wells, S. and Murphy, D. (2003). "Transgenic studies on the regulation of the anterior pituitary gland function by the hypothalamus," *Front Neuroendocr* 24, 11–26.

Widmaier, E. P., Raff, H. and Strang, K. T. (2010). *Vander's Human Physiology: The Mechanisms of Body Function*, 12th edn. (New York: McGraw-Hill).

Regulation of hormone levels in the bloodstream

As illustrated in Figures 6.5 and 6.11, pituitary (LH, FSH and ACTH) and steroid (cortisol) hormone levels in the bloodstream fluctuate dramatically over short periods of time (minutes to hours). In addition, hormones such as GH, ACTH and melatonin show marked circadian variations in their secretion patterns (Figure 6.5). These patterns are physiologically important; for example, we saw in the case of LH secretion, a continuous release, rather than a pulsatile secretion, will not stimulate the ovaries or testes correctly (Figure 7.4). In other words, fertility is dependent on an appropriate pulsatile LH signal reaching the gonads. This principle might be generally applicable to all pituitary hormone secretions. The measurement, or assay, of hormone levels is therefore an important clinical goal, as well as a crucial aid in understanding how hormone levels in blood are regulated and how the neuroendocrine system functions in health and disease. This chapter thus begins with an examination of the methods for measuring hormone levels in the circulation.

8.1 Analysis of hormone levels

The level of a circulating hormone can be measured directly in blood samples or estimated by measuring hormone levels in the *saliva, urine* or *feces*, measuring urinary metabolites, or by using bioassays. The determination of glucocorticoids levels in *hair*, for example, is a way to detect long-term exposure to stress.

8.1.1 Direct measurement of circulating hormones

In the past 20 years, there have been striking changes in the analytical techniques used to estimate hormone levels. Until recently, the benchmark in determination of hormone levels was the radioimmunoassay. However, this method, employing antibodies specific to each hormone, and radioactively labeled hormones, is slow, labor-intensive and raised safety problems in the use and disposal of radioactive materials. It is now routine to analyze hormone levels using rapid and automated chemiluminescent or immunometric assays that produce data in a matter of hours, rather than days.

A widely used assay is the *Enzyme-linked Immunosorbent Assay (ELISA)*. One form of this technique (there are several other variations) is illustrated in Figure 8.1 for the quantification of ACTH levels in blood. To perform this assay, a specific ACTH antibody

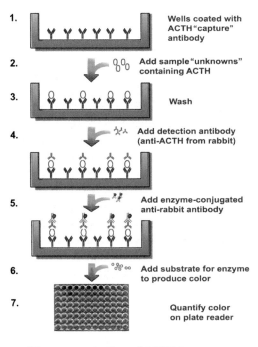

1. Wells coated with ACTH "capture" antibody

2. Add sample "unknowns" containing ACTH

3. Wash

4. Add detection antibody (anti-ACTH from rabbit)

5. Add enzyme-conjugated anti-rabbit antibody

6. Add substrate for enzyme to produce color

7. Quantify color on plate reader

The concentration of ACTH is proportional to color density

Figure 8.1 Measurement of hormone levels
Schematic representation of a widely used method for automated assay of hormone levels: the *Enzyme-linked Immunosorbent Assay (ELISA)*. This example illustrates the quantification of adrenocorticotropin (ACTH) levels in blood. A specific ACTH antibody (the "capture" antibody) is irreversibly attached to the bottom of a plastic plate well (Step 1). The "unknown" ACTH samples (i.e. obtained from blood samples) are then added and the ACTH allowed to complex with the bound antibody (Step 2). Everything except the ACTH can then be removed with a wash step, leaving ACTH bound to the antibodies (Step 3). A labeled second antibody (the "detection" antibody that also recognizes ACTH) is then added and this binds to the ACTH (Step 4). This detection antibody is itself now labeled with a further antibody that contains an enzyme capable of generating a color response to a particular substrate (Step 5). ACTH is quantified by measuring the amount of color produced at Step 6 and compared to a set of color values using known amounts of ACTH. This is done automatically in an electronic plate reader (Step 7) and serves to quantify the levels of ACTH.

(the "capture" antibody) is fixed to the bottom of a plastic plate well (Step 1). The "unknown" ACTH samples are then added and allowed to complex with the bound antibody (Step 2). Unbound products are then removed with a wash step, leaving ACTH bound to the antibodies (Step 3). A labeled second antibody (the "detection" antibody) is allowed to bind to the ACTH (Step 4). This detection antibody is itself now labeled with an antibody that contains an enzyme capable of generating a color response to a particular substrate (Step 5). The assay is then quantified by measuring the amount of color produced at Step 6. This is done automatically in an electronic plate reader (Step 7). Major advantages of this technique are that the samples do not need to be purified prior to use, and these assays are very specific and can be automated to handle large numbers of samples. The data illustrated in Figure 6.5 were obtained using this type of assay (Henley *et al*. 2009). Another more practical form of an ELISA is the over-the-counter pregnancy test that uses an immobilized antibody for the human chorionic gonadotropin (HCG) (see section 2.2.12) and a simple color reaction (http://webphysics.iupui.edu/webscience/bio_archive/goodfor3.html).

Mass spectrometry is a second powerful assay technique used to quantify hormones in biological fluids such as blood. The biggest advantage it possesses over the ELISA is that it does not require antibodies in order to detect the hormone of interest. This is not to say that the ELISA is not useful; there are many hormones for which highly specific antibodies are available and therefore can be measured with ELISAs. However, even the most specific antibodies can still cross-react with other hormones and therefore give abnormal results. The mass spectrometer, on the other hand, can determine absolute and specific values of many substances in blood based on their differences in molecular weight. What used to be a slow and cumbersome process can now be automated and, when coupled to high perfor-mance liquid chromatography (HPLC), is an efficient and rapid way to assay hormones in a clinical setting (Soldin and Soldin 2009). In fact, mass spectrometry may be the only way to determine very low levels of hormones in blood; e.g. estradiol levels in prepubertal children (Stanczyk and Clarke 2014). Studies are available that compare ELISA with mass spectrom-etry and show that the two methods yield different results, especially when the blood levels in sick patients are abnormal to begin with (Zerikly *et al.* 2010; Tractenberg *et al.* 2010; Mitchell 2012). Another example of the power of such a technique is its ability to measure several hormones in the same sample; that is, it may not be diagnostically useful to measure ACTH in isolation, without knowing what the levels of cortisol might be (e.g. see Figure 6.5). In other words, ACTH levels should be low in the presence of high cortisol, or vice versa (negative feedback; see Chapter 4). If ACTH is not low, this might suggest an abnormal hypothalamic-pituitary-adrenal system in which normal feedback is not working.

8.1.2 Analysis of hormones in saliva, urine, feces and hair

The collection of biological samples such as saliva, urine, feces and hair for the analysis of circulating hormone levels is easy, non-invasive and stress-free, compared to taking blood samples, and can be used to measure hormone levels in free-ranging animals, newborn children and adults outside of the laboratory or hospital setting. These methods are particularly useful for obtaining baseline measures of stress-related hormones, such as cortisol. Also, since cortisol is deposited in the growing hair shaft, an estimate of cortisol concentrations in hair serves as a measure of chronic exposure over weeks and months (Meyer and Novak 2012). Further, with the aid of HPLC-mass spectrometry, several hormones can be measured simultaneously not only in the same sample, but in urine, saliva and blood all from the same patient (McWhinney *et al.* 2010). However, note that urine, saliva, fecal and hair analysis can only provide a measure of the *average* hormone secretion over a number of hours, or weeks in the case of hair, and would not permit the minute-by-minute analysis of hormone levels that can be determined from blood samples. Nevertheless, urine samples obtained several times each day, over several days, can still provide useful measures of hormone release. Conversely, blood samples taken every few minutes is the only way to demonstrate that pituitary secretion is significantly variable over a time-period of minutes/hours (Figure 6.5), especially those with a circadian period. For example, if urine samples were only obtained once each day, say after lunch, then the rhythm of hormone secretion shown in Figure 6.5 would be completely missed. In contrast, cortisol measurement in hair provides a biomarker for prolonged activity of the hypothalamic-pituitary-adrenal axis (e.g. during stress) over periods of weeks. This approach has been applied to human patients, wild animals, laboratory animals and archival specimens that may be centuries old (Meyer and Novak 2012).

Gonadal steroid hormones, as well as FSH, LH, melatonin and human chorionic gona-dotropin, are excreted in the urine and detection of these hormones in the urine can provide important information. Urinary assays of LH, for example, are often used to detect the onset of the preovulatory LH surge in clinical settings. In this way, artificial insemination, or the collection of oocytes for *in vitro* fertilization (Figure 8.2A), can be carefully timed. This technique is also invaluable in attempts to breed threatened species such as giant pandas (Figure 8.2B). In this latter case, urinary hormone levels are followed until the surges in estradiol and LH are detected. This should coincide with the optimal time for mating.

One problem with urine analysis is that many hormones, including oxytocin, vasopressin, TSH, GH, prolactin, insulin and progesterone, are modified or degraded before being excreted in the urine (Table 7.3). Thus, urine analysis is not capable of measuring circulating levels of metabolites, but, as noted earlier, HPLC-Mass spectrometry can readily be used.

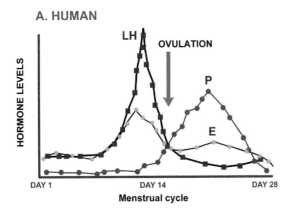

A. HUMAN

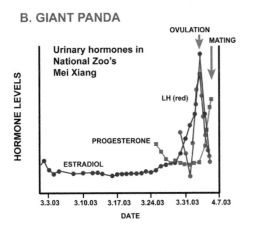

B. GIANT PANDA

Figure 8.2 Measurement of hormone levels in a clinical/breeding situation

In A we see the typical changes in hormone values in women through a menstrual cycle. Routine measurement of the rise in estradiol levels, and the subsequent peak of LH secretion, allows clinicians to accurately predict the occurrence of ovulation. The peak of progesterone confirms that ovulation occurred. This is important for those women, for example, who are having difficulty conceiving. Ova released from ovarian follicles can be collected and used for *in vitro* fertilization. In B, a similar procedure is used to predict if, and when, the female Giant Panda Mei Xiang may come into heat and be ready for mating. Data were obtained at the Smithsonian National Zoo in Washington, DC. Collection of urine samples is simple and non-invasive and the assay is powerful enough to perhaps ensure the survival of the species. In this figure, there is a clear rise in estradiol, LH and progesterone that implies successful ovulation.

Abbreviations: E, estradiol; LH, luteinizing hormone; P, progesterone. Reproduced with permission (Smithsonian National Zoological Park).

8.1.3 Bioassays

Simple, but relatively quantitative, estimates of hormone levels can be obtained by deter-mining the changes in endocrine gland weights following hormone injections. This is called a *bioassay*. Such an approach is obviously far removed from the sophisticated techniques mentioned already, such as the ELISA. Nevertheless, the bioassay does have the advantage of a biological endpoint; that is, injection of a female rat with estradiol, for example, will increase the size of the uterus. In the same way, TSH stimulates growth of the thyroid gland, so thyroid weight could be used as a bioassay for TSH levels. Likewise, gonadal weights can be used as a bioassay for gonadotropin levels, and adrenal gland weights as bioassays for ACTH levels. These can only be used in animal experiments, but in human terms Figure 1.4 reveals that body weight and fat mass are potential endpoints for the effects of leptin. These traditional *in vivo* bioassays, however, are time consuming, require large numbers of experimental animals, are relatively insensitive and lack specificity (www.ncbi.nlm. nih.gov/bookshelf/br.fcgi?book=endocrin&part=A972#A1095). Another form of bioassay, and one which is much more versatile, involves the use of cells grown in culture (*in vitro*). For example, the activity of hypothalamic hormones such as CRH can be measured by the amount of pituitary hormone (ACTH) released from pituitary tissue or pituitary cells in culture, while the action of pituitary hormones, such as LH, can be assayed by their effect on the release of their target hormones, such as testosterone, from isolated endocrine (e.g. testicular) tissue. Other bioassays use cultures of specific cells whose growth rate is sensitive to a particular hormone. For example, the NB2 lymphoma cell (a special line of cells) is sensitive to prolactin and the level of prolactin in a blood sample can be determined by the rate of cell division in a culture of these cells (Murphy *et al.* 1989).

Given that bioassay techniques may not be as simple, or practical, to perform compared to immunoassays, is there a case to be made for the continued use of bioassays? By way of example there are many clinical studies that have compared *bioactive* LH and FSH levels with those obtained from immunoassays and there appear to be clear differences between the results of the two types of assay. One reason for this is that glycoprotein hormones such as LH and FSH may be present in serum in different isoforms; that is, the structures may have slightly different degrees of glycosylation so that the antibodies used in the assays may not detect them all. There is also evidence that the ratio of bioactive to immunoreactive hormones can change with the onset of puberty (Dunger *et al.* 1991). Whether this is an important phenomenon or just a methodological issue remains to be determined.

8.2 Mechanisms regulating hormone levels

Blood levels of hormones are determined not just by how much hormone is secreted into the bloodstream from any given gland, but also on the rate of clearance from the blood. Thus, absolute concentrations of hormones represent a balance between how much is being secreted into the blood, and how much is being removed, either by uptake into target tissues or by enzymatic inactivation mechanisms in the liver and kidney, for example. The following section concentrates on how the amount of a hormone entering the blood can be regulated by at least four different mechanisms: (1) the autonomic nervous system (ANS); (2) non-hormonal chemicals in the blood or gastrointestinal tract; (3) hormonal feedback; and (4) neurotransmitters and neuropeptides in the brain and central nervous system. We have already covered a large part of #4 in Chapter 6.

8.2.1 Autonomic nervous system (ANS)

This topic was essentially covered in section 6.6. Briefly, a good example of hormone regulation by the ANS is the adrenal medulla, which is stimulated to release epinephrine by the splanchnic nerves of the sympathetic branch of the ANS. The vagus nerve of the ANS also regulates hormone release from endocrine glands, including the release of insulin and glucagon from the pancreas and the release of some of the gastro-intestinal hormones, such as gastrin. The synthesis of melatonin in the pineal gland is also regulated by noradrenergic neurons in the sympathetic branch of the ANS. The thyroid gland, thymus and other endocrine glands are also innervated by nerves of the ANS.

Neuroendocrine reflexes

A neuroendocrine reflex occurs when a stimulus from the central nervous system results in the release of a hormone. Such a stimulus may or may not be mediated by the ANS. Thus, a fear-inducing stimulus causing neurotransmitter release from the sympathetic nerves of the ANS results in the reflexive secretion of epinephrine from the adrenal medulla (section 2.2.9). In this respect, the adrenal medulla is sometimes called a neuroendocrine transducer, in that it converts a neural input to a hormonal output. An excellent example of a neuroendocrine reflex occurs in the magnocellular neurosecretory cells of the hypothalamus. As we noted in section 4.2, oxytocin and vasopressin are biosynthesized in these neurosecretory cells (SON and PVN) and then released from the posterior pituitary (Figure 6.10). In the case of oxytocin, one neuroendocrine reflex originates in the mother's nipples to generate a surge of oxytocin that then releases milk for the baby by causing myoepithelial cells in the breasts to contract (Figure 8.3). This critical reflex performs a double duty, since it also induces the secretion of PRL, via the anterior pituitary, which ensures the production of more milk for the next feeding. This reflex arc inhibits dopamine release that allows PRL to increase.

Also shown here is the remarkable effect of the baby crying, and inducing oxytocin secretion. Figure 8.4 illustrates this reflex; some mothers respond with oxytocin secretion without actual contact with the baby. So, when the baby cries, or even becomes agitated, higher brain centers are recruited to release oxytocin, but not prolactin (McNeilly *et al.* 1983).

A final example of hormone secretion induced by neural input is the release of melatonin from the pineal gland (see section 2.2.2). In humans, this hormone is secreted at night in a distinct circadian rhythm controlled via the suprachiasmatic nuclei located close to the hypothalamus. As shown in Figure 8.5, light impinging on the retina inhibits melatonin synthesis in the pineal gland. This phenomenon is well described in humans (Gooley *et al.* 2010). A detailed description of this system can be found in an authoritative review by Arendt (Arendt 2006; and for further reading see Reiter *et al.* 2007; and Hardeland *et al.* 2011).

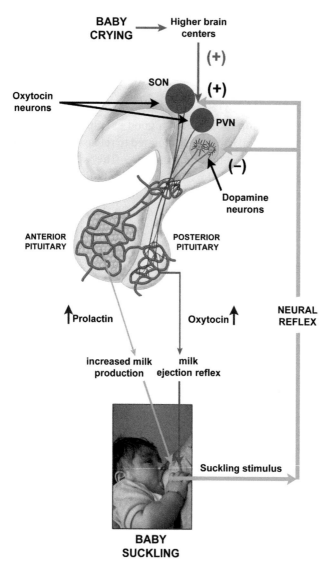

Figure 8.3 Suckling-induced reflex release of two different pituitary hormones

For oxytocin, secreted from the posterior pituitary, a neuroendocrine reflex originates in the mother's nipples to stimulate SON/PVN neurons. This stimulus generates pulses of oxytocin that then release milk for the baby by causing myoepithelial cells in the breasts to contract and expel milk. This is called the *milk ejection reflex* and it performs a double duty, since it also induces the secretion of prolactin (PRL), via the anterior pituitary, which ensures the production of more milk for the next feeding. This latter reflex arc *inhibits* dopamine release from the hypothalamus that allows increased PRL secretion from the anterior pituitary. This figure also shows an additional reflex that stimulates SON/PVN neurons following the sound of the baby crying (see Figure 8.4 for further details).

Abbreviations: PVN, paraventricular nucleus; SON, supraoptic nucleus.

8.2.2 Non-hormonal stimuli

The secretion of some hormones is regulated by stimuli other than hormonal or neural input. For example, sunlight is responsible for the production of vitamin D by the skin (see section 2.2.10). Another example is the release of gastrin from the stomach, caused by distension as a result of food or liquid intake. The feeling of hunger causes release of ghrelin from the stomach (see Chapter 12). Similarly, ingestion of proteins and fat in food stimulates cholecystokinin release from the duodenum (for further reading see sections 2.2.7 and 2.2.8; and Chaudhri *et al.* 2008). Several other cases in which chemical, and non-chemical, circulatory stimuli regulate hormone release are given in Table 8.1. Hormones function to maintain the levels of glucose, calcium and sodium in the circulation at constant (homeostatic) levels and the levels of the regulatory hormones fluctuate in

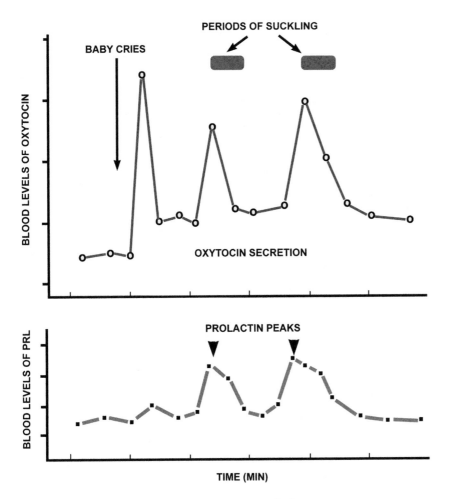

Figure 8.4 Oxytocin and prolactin release during breast feeding
Schematic representation of the timing of PRL and oxytocin secretion during breast feeding. When the baby actively suckles, there is a corresponding secretory burst of oxytocin in the mother's blood. This is accompanied by release of PRL, as referred to in Figure 8.3. However, under circumstances where the mother hears the child's cry, there is also a release of oxytocin. In this case, however, the reflex does not influence PRL secretion and is probably mediated through higher cortical centers. This figure is based on data from McNeilly *et al.* (1983).
  Abbreviation: PRL, prolactin.

response to the levels of these chemicals in the blood. Thus, these chemicals also regulate the release of the hormones that regulate them. The endocrine gland releases its hormone in response to an increase of some chemical in the blood (e.g. calcium) and reduces the level of that chemical. When the level of that chemical decreases, stimulation of hormone release declines. Thus, circulating hormone levels are constantly fluctuating by small amounts, increasing or decreasing in response to chemicals in the blood. For example, a decline in blood calcium level stimulates the release of parathyroid hormone (PTH) that will cause bone calcium to be released into the blood (see Figure 2.2). The resulting rise in blood calcium levels reduces the stimulation for the release of PTH, and thus less of it is secreted.

Table 8.1 **Non-hormonal stimuli regulating hormone levels**

Stimulus	Hormone	Gland
Sunlight	Vitamin D	Skin
Light	Melatonin decreased	Pineal gland
Food/liquids: distension	Gastrin	Stomach
Protein and fats	Cholecystokinin	Duodenum
High protein diet	PYY	Small and large intestine
Hunger	Ghrelin	Stomach
Blood glucose:		
high	Insulin	β cells of pancreas
low	Glucagon	α cells of pancreas
Blood calcium:		
high	Calcitonin	Thyroid gland
low	Parathyroid hormone	Parathyroid gland
Blood sodium:		
high	Vasopressin	Posterior pituitary
low	Aldosterone	Adrenal cortex

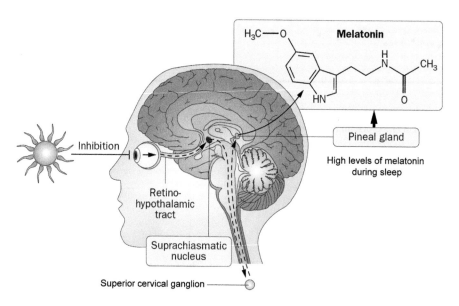

Figure 8.5 Light inhibits melatonin production from pineal gland
Melatonin secretion from the human pineal gland normally occurs each night and is regulated by signals from the suprachiasmatic nuclei (SCN) in the brain that reach pinealocytes via adrenergic nerves and adrenergic receptors. However, when light reaches the retina, it is able to interfere with the SCN signal, stopping the production of melatonin. Reproduced with permission (Koch *et al.* 2009).

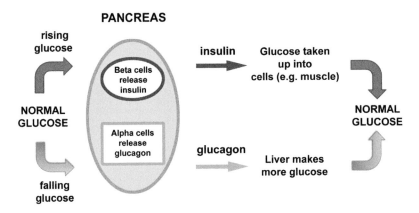

Figure 8.6 Maintenance of blood glucose levels
The pancreas responds to fluctuating glucose levels in blood by releasing two hormones that homeostatically maintain circulating glucose within healthy limits. When blood glucose levels rise, after a meal for example, this stimulates insulin release from the beta cells of the pancreas. These high glucose levels in blood are reduced by insulin promoting glucose uptake into various tissues such as muscle and fat. In contrast, low blood glucose levels signal the release of glucagon from the alpha cells of the pancreas and this hormone instructs the liver to increase glucose production.

Conversely, high levels of blood calcium decrease PTH secretion and, at the same time, increase calcitonin release from the thyroid, which also acts to reduce blood calcium levels. Likewise, blood glucose levels are maintained within healthy limits by the hormones insulin and glucagon. An increase in blood glucose levels stimulates insulin release from the beta cells of the pancreas to ensure glucose uptake by various tissues (Figure 8.6). In contrast, low blood glucose levels stimulate the release of glucagon from the alpha cells of the pancreas which then instructs the liver to increase glucose production. As a final example, and as we covered in section 3.2, high blood sodium levels increase blood osmolality and stimulate vasopressin release from the posterior pituitary. This in turn promotes water reabsorption in the kidney. Conversely, low blood sodium stimulates aldosterone release from the adrenal cortex, which in turn promotes sodium retention by the kidney.

8.2.3 Hormonal feedback systems

In order to maintain an optimal internal environment (*homeostasis*), the neuroendocrine system has built-in regulatory devices that function to ensure that hormone secretion is switched on and off as needed. As we have seen already, hormonal output can be adjusted in several ways, including neural and non-neural input. However, the most common mechanism is called hormone feedback, which tends to be *negative*. We have already touched on several examples of such control: for example, Figures 6.4 (adrenal hormones), 6.6 (thyroid hormones) and 6.9 (GH) all illustrate how the target tissues release hormones that exert negative feedback on the hypothalamic-pituitary system to prevent excess hormones being released. Figure 8.7 illustrates the principle of *negative feedback*.

Further description of negative feedback will be restricted to those hormone systems controlled by the anterior, rather than the posterior, pituitary. The reason for this is that

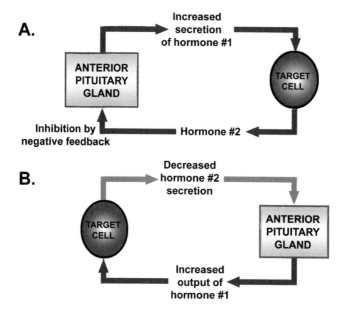

Figure 8.7 Principle of hormone negative feedback

This figure uses the anterior pituitary response to a circulating hormone as an example of negative feedback control of pituitary secretion. In A, anterior pituitary cells release hormone #1 (such as ACTH) that stimulates a target tissue (such as adrenal cortex) which in turn secretes hormone #2 (such as cortisol). Hormone #2 then exerts negative feedback on anterior pituitary cells to reduce the output of hormone #1. In this way, an excessive production of hormone #2 is avoided. In contrast, B, if there is a deficit in production of hormone #2, the negative feedback on the pituitary is lifted and the gland responds by making more hormone #1, which in turn restores the correct level of hormone #2.

oxytocin and vasopressin release from the *posterior pituitary* is regulated by mechanisms such as breast stimulation (oxytocin) and high sodium levels (vasopressin). Strictly speaking, there is no hormonal feedback to limit further secretion. In other words, oxytocin release ceases when the baby stops suckling, and vasopressin secretion is reduced when sodium balance is achieved. In contrast, hormone release from the *anterior pituitary* is strictly regulated by a combination of neural and hormonal feedback.

There are three examples of negative feedback that conform to that shown in Figure 8.7. These are the adrenal, thyroid and reproductive systems. Figure 4.10 illustrates how the hypothalamic-pituitary system stimulates the adrenal glands to release steroid hormones, such as cortisol, which exerts feedback to inhibit both the hypothalamus and anterior pituitary; that is, cortisol acts on the hypothalamus to reduce CRH secretion and on pituitary ACTH cells to reduce sensitivity to CRH. The net result is that ACTH secretion is inhibited. A similar mechanism is at work in controlling thyroid hormone release. Figure 6.6 shows that thyroid hormones such as T4 (thyroxine) are released from the thyroid gland and feed back into the hypothalamus and pituitary to restrain further stimulation of thyroid output. The mechanism of the negative feedback is analogous to that shown for cortisol; that is, T4 inhibits TRH release and also prevents TRH from stimulating thyrotrophs in the anterior pituitary. What would happen if negative feedback were not present? Our bodies would be flooded with high levels of either cortisol (as in

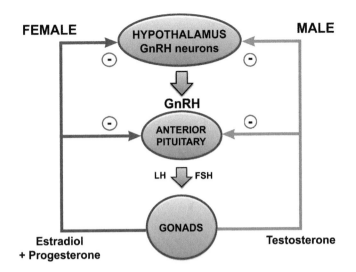

Figure 8.8 Control of the reproductive system by negative feedback
This figure illustrates the principle for both male and female reproduction: estradiol and progesterone are released from the ovary and exert negative feedback in the female, whereas in the male the inhibition is mediated by testosterone secreted from the testes. In both cases, negative feedback (a) reduces GnRH secretion from the hypothalamus and (b) inhibits the response of the pituitary to stimulation by GnRH. Such a two-pronged feedback effectively prevents overstimulation of the gonads.
    Abbreviations: FSH, follicle-stimulating hormone; GnRH, gonadotropin-releasing hormone; LH, luteinizing hormone.

Cushing's Disease) or thyroxine (*hyperthyroidism* – high body temperature; elevated heart rate), both of which are incompatible with good health. So negative feedback is crucial to maintain homeostasis and survival.

The reproductive system is also regulated by a classic negative feedback mechanism. Figure 8.8 illustrates this principle for both male and female reproduction; estradiol and progesterone exert negative feedback in the female, whereas in the male the inhibition is mediated by testosterone. In both cases, feedback reduces GnRH secretion and inhibits the response of the pituitary to stimulation by GnRH. By way of example, Figure 8.9 represents the temporal response of LH secretion when negative feedback is removed (e.g. by castration) and reinstated by an injection of testosterone. This principle also applies to estradiol. There are not many examples of hormonal *positive feedback*, but in section 4.3.3, we saw that estradiol could generate a surge of GnRH secretion from the hypothalamus that, in turn, caused a release of LH from the pituitary. This is the positive feedback effect of estradiol that is responsible for female fertility and the induction of ovulation. A good description of the hormonal sequence through the menstrual cycle, leading up to estradiol's effect on ovulation, can be found elsewhere (Widmaier *et al.* 2010).

Feedback loops: the long, the short and the ultra-short

As described above, negative feedback control of pituitary hormone secretion is termed *long loop feedback* (Figure 8.8): for example, testosterone negatively regulates LH/FSH secretion from the pituitary and the secretion of GnRH from the hypothalamus. However,

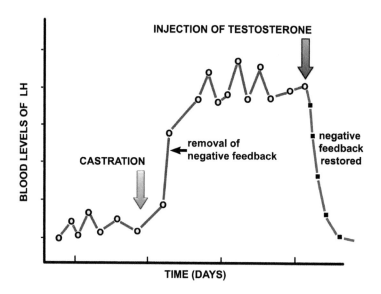

Figure 8.9 Illustration of temporal sequence of testosterone negative feedback
When negative feedback of testosterone is removed (e.g. by castration), the normal low level of luteinizing hormone (LH) secretion from the pituitary increases dramatically. Negative feedback properties of testosterone are shown when this sex hormone is injected. LH release from the pituitary is sharply decreased (see also Figure 8.10). This principle also applies to estradiol in the female.

there are some pituitary hormones that exert feedback effects directly on hypothalamic neurons. One example is growth hormone (GH), as illustrated in Figure 6.9. Here, we can see that GH secretion is normally inhibited, via somatostatin, through a long loop negative feedback imposed via IGF-I released from the liver following GH stimulation. In addition, GH exerts a *short loop* negative feedback effect on GHRH as well as stimulating the inhibitory hormone somatostatin. Another example of short loop feedback can be seen in the regulation of PRL secretion (Figure 6.8). PRL is secreted from the pituitary and exerts a short loop feedback effect on the hypothalamus. In this case, PRL induces the release of dopamine (or possibly a PRL release-inhibitory factor) that reduces PRL secretion. Indeed, there is evidence that all anterior pituitary hormones (e.g. TSH, ACTH and LH/FSH) may also have short loop inhibitory effects. Bear in mind that experiments designed to examine such influences are difficult to perform in animal studies and may be impossible in humans. Nevertheless, although the physiological significance of short loop feedback remains obscure, this type of control could be important in the minute-to-minute regulation of pulsatile secretion of some pituitary hormones. To complicate the issue even further, there is additional evidence for a so-called *ultra-short feedback loop*. Such an example is also shown in Figure 6.9, where GHRH and somatostatin exert negative feedback on their own secretion. From the description in Chapter 1, we can immediately recognize this as an example of *autocrine regulation* (Figure 1.3; example 2). As noted above, experimental verification of ultra-short feedback loops is difficult to obtain, although such experiments have now been performed for GnRH (Xu *et al.* 2008). Again, the physiological relevance of such control remains to be determined, although for GnRH it is likely to be a component of the autoregulatory mechanism that drives pulsatile secretion of GnRH.

Effects of altering one hormone in a multihormonal feedback system

If the level of one hormone is altered in a feedback system, what happens to the level of the other hormones? We can see from Figure 8.9 that the removal of the testes will cause testosterone levels to decline to zero. This, in turn, removes negative feedback, so that GnRH and therefore LH levels will no longer be inhibited and will rise. On the other hand, if LH is removed (e.g. by surgical removal of the pituitary), testosterone levels will decline because of the absence of LH stimulation of the testes. Without testosterone, hypothalamic GnRH levels will also increase. Thus, by definition, altering the level of one hormone in a feedback loop influences the level of the other hormones within the system.

8.2.4 Neurotransmitter and neuropeptide control of hormone levels

Central neurotransmitter regulation of hypothalamic and pituitary hormone secretion was discussed in detail in Chapter 6. In particular, variations in the release of neurotransmitters and neuropeptides are instrumental in changing hormone release by modifying neurosecretory activity and the secretion of releasing/inhibiting hormones. For example, we can see in Figure 6.4 that increased activity of opioid neurons will decrease secretion of CRH. This effectively closes down the entire hypothalamic-pituitary-adrenal system, inhibiting not only the release of CRH, but also the release of pituitary ACTH and cortisol from the adrenal cortex.

Hypothalamic integration of stimuli influencing neuroendocrine release

The hypothalamus is responsible for a large array of endocrine and non-endocrine homeostatic functions and receives input from every major site of the CNS. This integrative capacity includes regulation of blood flow, regulation of energy balance, control of reproduction and the response to stress (Squire *et al.* 2008). As we have noted already (see section 8.2.2), the release of hypothalamic hormones is under neurochemical control as well as by input from chemicals in the blood (e.g. glucose, sodium ions, hormones). We can add to these a variety of sense organs (light), and information from a variety of intero- (i.e. within the body) and extero-receptors. The hypothalamus therefore performs a profound integrative role in processing all of these signals in the homeostatic control of the neuroendocrine system (Table 8.2).

8.3 Hormonal modulation of neurotransmitter release

We have discussed the critical importance of neurotransmitters in the regulation of hypothalamic-pituitary control of the neuroendocrine system (Chapter 6). In the present chapter, we introduced the concept of hormonal feedback as an additional regulatory influence on pituitary secretion. For example, Figures 8.8 and 8.9 used testosterone and estradiol as examples of how the reproductive system deals with variability in blood hormone levels. This poses an important question: do circulating hormones such as testosterone exert their feedback effects via changes in hypothalamic neurotransmission? The answer to this question is yes, and a huge literature exists describing innumerable studies that have sought to explain how steroid hormone feedback modifies pituitary hormone secretion through a variety of neurochemical pathways. It is beyond the scope

Table 8.2 **Integration functions of the endocrine hypothalamus**
Neural inputs
Sense organs: vision, smell, taste, hearing, touch.
Exteroceptors: suckling, coitus, pain, temperature.
Interoceptors: uterus, cervix, blood volume, chemoreceptors.
Stress: physical, emotional, overcrowding.
Circadian rhythms and sleep-wake cycles.
Exercise
Sleep
Cerebrospinal fluid inputs
Hormones
Ions
Other substances, such as peptides
Circulatory system inputs
Hormones: gonadal and adrenal steroids, thyroid hormones, pituitary and hypothalamic peptides, melatonin.
Neuropeptides: Substance P, bombesin, etc.
Other substances: glucose, amino acids, blood gases, nutrition.
Other properties of blood: for example, temperature, osmotic pressure.

of this book to adequately deal with this field. However, the reproductive system stands as an excellent model for the interaction of sex steroid hormones and central neurotransmission in regulating fertility. Figure 6.7 illustrates that estradiol and testosterone feedback is mediated through a variety of CNS neurotransmitters, including excitatory (glutamate; kisspeptin) and inhibitory (GABA) neurons (for further reading, see Herbison 2006). As a final point here, remember that hormonal feedback into the brain is not solely a means to modify pituitary secretion; steroid hormones also have profound effects on emotion and behavior, and these will be discussed later.

8.4 The cascade of chemical messengers revisited

Figure 6.1 illustrated the stress response as an example of the cascade of chemical and hormonal signals, from the initial release of a neurotransmitter through to the stimulation of hormones from the hypothalamus, pituitary and peripheral endocrine glands. Whereas the nervous system responds very rapidly (milliseconds) to external stimulation, the endocrine system is relatively slow to react. Thus, in response to a stressful stimulus, epinephrine is released almost instantaneously by sympathetic nervous system stimulation of the adrenal medulla, whereas cortisol is released much more slowly from the adrenal cortex and reaches peak values in the blood at about 30 minutes. The time it takes for cortisol to be released from the adrenal cortex is dependent on several steps. As can be seen in Figure 6.1: (1) CRH is secreted from hypothalamic neurons; (2) CRH is carried to the anterior pituitary gland by the portal blood; (3) CRH binds to membrane receptors on corticotrophs and induces the secretion of ACTH; (4) ACTH enters the peripheral blood and eventually stimulates the adrenal cortex to biosynthesize and release cortisol. We can also estimate the time it takes for ACTH to induce cortisol release by inspection of the data in Figure 6.5; that is, the time between one of the peaks of ACTH and the subsequent surge in cortisol is about 10 minutes. Once a hormone is released, it may remain, in free and in bound form, for some time in the blood. Thus, the active life of a hormone is much longer than that of a neurotransmitter, especially for hormones bound to carrier proteins. The result is that, after a stimulus occurs, a hormone such as cortisol may take about 2 hours to be restored to basal levels, whereas a neurotransmitter will take only milliseconds to be released into a synapse. Note also that inspection of Figure 6.5 reveals that as cortisol pulse levels increase, ACTH levels fall rapidly through a combination of cortisol negative feedback, as well as by metabolic clearance of ACTH. All of the hypothalamic-pituitary systems that we have covered, such as thyroid and reproductive systems, operate under similar temporal control. Figure 8.10 shows, in schematic form, the effects of negative feedback not only on LH levels in blood, but also on the *frequency* of the LH pulses. In Figure 8.10A, we see high LH secretion in multiple pulses resulting from castration; that is, testosterone is not present to provide negative feedback (see also Figure 8.9). However, four days after implantation of testosterone, the LH pulse frequency is greatly reduced and the average level of LH in the blood drops (B). Eight days after the implantation of testosterone (C), the pulses of LH are almost completely suppressed. The negative feedback action of testosterone on LH pulse frequency occurs through inhibition of the firing rate of hypothalamic GnRH pulse generator neurons (Plant 1986; for further reading, see Maeda *et al.* 2010).

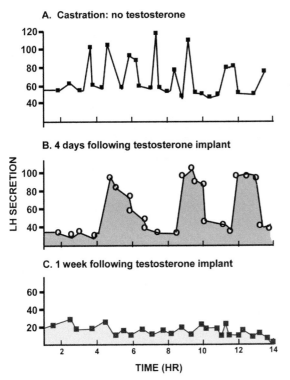

A. Castration: no testosterone

B. 4 days following testosterone implant

LH SECRETION

C. 1 week following testosterone implant

TIME (HR)

Figure 8.10 Negative feedback on pulsatile secretion of luteinizing hormone (LH)

This figure is a schematic illustration of the pronounced negative feedback effects of testosterone on LH secretion in the castrated monkey. However, in addition to LH secretion being inhibited, we can see that negative feedback on GnRH neurons in the hypothalamus actually slows the rate of pulsatile secretion (A vs B). The frequency of pulses in A is about one pulse per 2 hours, whereas in B it has slowed to approximately one pulse every 5 hours. The slowing of pulsatile secretion represents a negative feedback effect directly on the firing of GnRH neurons.

The figure was drawn based on data published by Plant (1986).

8.5 When hormone regulatory mechanisms fail

Hormone secretion may be reduced (hyposecretion) or elevated (hypersecretion) due to disease or damage to an endocrine gland. In such cases, the feedback mechanisms that normally regulate hormone levels in the blood may fail to operate properly. Such abnormal hormone secretion may result in the under- or over-stimulation of target cells, causing a wide range of endocrine-related physiological, behavioral, psychological and psychiatric disorders. Endocrine disorders may occur as a result of abnormal hypothalamic, pituitary or peripheral endocrine gland secretion. For example, pituitary tumors can produce excess GH secretion to cause *acromegaly* (abnormal growth of bones and soft tissue), whereas an *absence* of GH causes severely reduced growth in children. Clinical aspects of neuroendocrine disorders can be found in the textbook by Nussey and Whitehead (Nussey and Whitehead 2001), but most standard library texts are suitable. By way of example, an interesting neuroendocrine disorder that illustrates the feedback principles that we have covered is Cushing's Syndrome. All Cushing's patients exhibit excessive cortisol secretion that induces a range of characteristic physical symptoms (Table 8.3). Most often, this disorder is due to a *pituitary tumor* that hypersecretes ACTH, which in turn causes the adrenal cortex to secrete chronic, elevated levels of cortisol. This is called *Cushing's Disease*. In this case, therefore, we are confronted with high levels of ACTH *and* cortisol at the same time. Our knowledge of negative feedback tells us that the cortisol should inhibit the release of ACTH. Since it does not, we have to conclude that pituitary tumors do not respond to negative feedback. An alternative cause of Cushing's Syndrome is an *adrenal tumor* that continuously hypersecretes cortisol. In these patients, however, there is very little ACTH in the blood because of the powerful negative feedback of cortisol on the normal pituitary gland. In a third variation of the disease, patients may have a *lung tumor* that secretes ACTH. In this case, the hormonal profile looks exactly like the first example; that is, high ACTH levels stimulate high cortisol secretion, but there is no negative feedback on the source of the ACTH, in the lung tumor. Students interested in the clinical tests that can distinguish between the first and third examples (i.e. high ACTH and cortisol) are referred elsewhere (Nussey and Whitehead 2001).

Tumors of the endocrine glands can often be treated by surgical removal, although taking out an adrenal tumor is simpler than removing a pituitary tumor. Alternatively, excess cortisol secretion can be treated by drugs that inhibit cortisol synthesis. Similarly, a

Table 8.3 Clinical features of Cushing's Syndrome

- Weight gain, central obesity (100% of patients)
- Hypertension (80%)
- Mental changes (80%)
- Purple skin striations (60%)
- Acne (60%)
- "Moon" face (100%)
- Osteoporosis (50%)
- Hypogonadism (e.g. no menstruation; 70%)

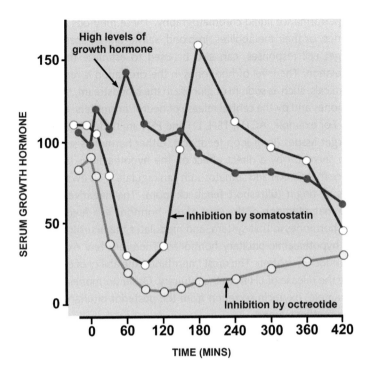

Figure 8.11 Inhibition of growth hormone (GH) secretion in a patient with high levels of GH in the blood
GH secretion from the anterior pituitary can be inhibited using the natural inhibiting hormone, somatostatin. However, we can see that in this patient, suffering from acromegaly, the inhibition is very short-lived because the injected somatostatin is rapidly metabolized.
In contrast, a synthetic long-acting analog of somatostatin, called Octreotide, is very effective in suppressing GH secretion over many hours.
Reproduced with permission (Besser and Thorner 2002).

pituitary tumor that is secreting high levels of GH can now be treated with somatostatin-like drugs. Remember that in section 4.3.4 (see Figure 4.11) we described the inhibitory influence of hypothalamic somatostatin on pituitary secretion of GH. Synthetic, long-acting analogs of somatostatin are now available with which to treat patients with GH-secreting pituitary tumors (Melmed 2009). Figure 8.11 illustrates the effect of somatostain, and one of its analogs (Octreotide), on GH secretion in an acromegalic patient with high circulating levels of GH (Besser and Thorner 2002). Somatostatin is rapidly metabolized and exerts only a brief inhibitory effect (2 to 3 hours), whereas Octreotide – a long-acting analog of somatostatin – is highly effective in inhibiting GH secretion over many hours.

There are several other examples of clinically abnormal hormonal feedback mechanisms that succeed in disrupting endocrine systems. We will not cover them here, but they include the thyroid (*hyper- and hypothyroidism*), the adrenal medulla (*pheochromocytoma*), the adrenal cortex (*Addison's disease*) and the reproductive system (*precocious puberty; pituitary prolactinoma*). Descriptions of these can be found in many clinical endocrinology textbooks, but a concise coverage is available elsewhere (Nussey and Whitehead 2001).

8.6 Summary

This chapter examines the methods used for measuring hormone levels in the circulation and the mechanisms through which these hormone levels are regulated. Hormone levels can be measured directly from blood samples using automated, highly specific and sensitive immunoassays such as the ELISA. Equally powerful, and perhaps more versatile, is the technique of mass spectrometry coupled to high performance liquid chromatography. These methods are readily applied to the assay of hormones, or their metabolites, in blood, saliva, urine, feces and hair. Bioassays, which quantify target cell responses, can also be used to estimate the level of bioactive hormones in the circulation. The level of hormones in the circulation is regulated by the ANS, by non-hormone chemicals, such as sodium or glucose in the bloodstream, by light, by feedback from circulating hormones and by the central release of neurotransmitters. In general, hormone feedback is negative. For example, ACTH, TSH, LH and FSH are under tight negative feedback control from their target tissues (long-loop feedback). Other hormones such as PRL and GH can control their own secretion by a direct effect on the hypothalamus (short-loop feedback). There is also evidence that GHRH and somatostatin can regulate their own secretion by an effect on their neurons of origin (ultrashort feedback loop). The negative feedback mechanism is sensitive enough that alteration of the level of one hormone in a feedback loop changes the levels of the other hormones in that system, and modulates the neurotransmitters controlling the release of the hypothalamic-pituitary hormones. There are few examples of positive feedback in the neuroendocrine system. The most important is the ability of estradiol to induce ovulation by stimulating the release of LH from the pituitary. Other hormones are under so-called reflex control. For example, oxytocin secretion from the posterior pituitary is stimu-lated by babies suckling at their mother's breast.

Hypothalamic neurons controlling the neuroendocrine system are able to integrate stimuli from a wide variety of sources in order to regulate the release of hypothalamic hormones. These inputs include neurotransmitters, hormonal feedback and non-hormone chemicals in the circu-lation, which are detected by osmoreceptors, glucoreceptors and other chemoreceptors. Hormonal feedback can alter the rate of pulsatile release of hypothalamic hormones and the responsiveness of the endocrine cells in the pituitary to hypothalamic hormones. Diseases of the hypothalamus, pituitary or peripheral endocrine glands result in hypo- or hyper-secretion of hormones that can be detected by routine blood or urine analysis. Abnormal hormone levels result in a number of physiological and psychological disorders that can be reversed by surgical intervention or drug treatment. Once the abnormal hormone secretion is treated, the symptoms usually disappear.

FURTHER READING

Hardeland, R., Cardinali, D. P., Srinivasan, V., Spence, D. W., Brown, G. M. and Pandi-Perumal, S. R. (2011). "Melatonin – a pleiotropic, orchestrating regulator molecule," *Progress in Neurobiology* 93, 350–384.

Herbison, A. E. (2006). "Physiology of the GnRH neuronal network" in Jimmy D. Neill (ed), *Knobil and Neill's Physiology of Reproduction*, 3rd edn. (San Diego, CA: Academic Press), pp. 1415–1482.

Maeda, K., Ohkura, S., Uenoyama, Y., Wakabayashi, Y., Oka, Y., Tsukamura, H. *et al.* (2010). "Neurobiological mechanisms underlying GnRH pulse generation by the hypothalamus," *Brain Res* 1364, 103–115.

Reiter, R. J., Tan, D. X., Korkmaz, A., Erren, T. C., Piekarski, C., Tamura, H. *et al.* (2007). "Light at night, chronodisruption, melatonin suppression, and cancer risk: a review," *Crit Rev Oncog* 13, 303–328.

REVIEW QUESTIONS

8.1   Name two sensitive techniques for the assay/quantification of hormones in the blood.

8.2   What is the difference between a bioassay and a radioimmunoassay?

8.3   Give two advantages for measuring hormone levels in saliva or urine, rather than in blood.

8.4   What is the effect of removing testosterone (e.g. by castration) on the secretion of LH?

8.5   What is the name of the phenomenon that governs the release of LH in question 8.4?

8.6   Which two pancreatic hormones regulate blood glucose levels?

8.7   What effect do the changes in Column A, below, have on the hormones listed in Column B?

A.	B.
(a) High blood glucose	Insulin
(b) Increased blood thyroxine	TSH
(c) Increased blood estradiol	LH
(d) Increased sunlight	Vitamin D
(e) Hunger	Ghrelin
(f) Stimulation of nipple by baby suckling	Oxytocin
(g) Stimulation of nipple by baby suckling	Prolactin

8.8   Give an example of a short loop negative feedback system.

8.9   Give an example of a positive feedback hormonal system.

8.10 Name two tissue types that you would predict should contain estradiol receptors.

ESSAY QUESTIONS

8.1   Compare the advantages and disadvantages of radioimmunoassays versus bioassays in the measurement of, for example, cortisol in blood.

8.2   Compare and contrast the use of mass spectrometry versus immunoradiometric assay of blood hormones such as estradiol.

8.3   Describe and illustrate the hormonal feedback loops, both negative and positive, that regulate the menstrual cycle in female primates such as the Rhesus monkey.

8.4   Discuss the importance of pulsatile secretion of GnRH from the hypothalamus. What is the likely mechanism for the inhibition of LH secretion from the pituitary by chronic stimulation with GnRH?

8.5   Discuss the role of light in the regulation of melatonin secretion from the pineal gland. Can melatonin be used to combat "jet lag"?

8.6   Describe the importance of long, short and ultra-short feedback systems in the regulation of GH secretion.

REFERENCES

Arendt, J. (2006). "Melatonin and human rhythms," *Chronobiol Int* 23, 21–37.

Besser, G. M. and Thorner, M. O. (2002). *Comprehensive Clinical Endocrinology*, 3rd edn. (St. Louis, MO: Mosby).

Chaudhri, O. B., Salem, V., Murphy, K. G. and Bloom, S. R. (2008). "Gastrointestinal satiety signals," *Annu Rev Physiol* 70, 239–255.

Dunger, D. B., Villa, A. K., Matthews, D. R., Edge, J. A., Jones, J., Rothwell, C. *et al*. (1991). "Pattern of secretion of bioactive and immunoreactive gonadotrophins in normal pubertal children," *Clin Endocrinol (Oxf)* 35, 267–275.

Gooley, J. J., Chamberlain, K., Smith, K. A., Khalsa, S. B., Rajaratnam, S. M., Van Reen, E. *et al*. (2010). "Exposure to room light before bedtime suppresses melatonin onset and shortens melatonin duration in humans," *J Clin Endocrinol Metab* 96, 463–472.

Henley, D. E., Leendertz, J. A., Russell, G. M., Wood, S. A., Taheri, S., Woltersdorf, W. W. *et al*. (2009). "Development of an automated blood sampling system for use in humans," *J Med Eng Technol* 33, 199–208.

Koch, B. C. P., Nagtegaal, J. E., Kerkhof, G. A. and ter Wee, P. M. (2009). "Circadian sleep-wake rhythm disturbances in end-stage renal disease," *Nat Rev Nephrol* 5, 407–416.

McNeilly, A. S., Robinson, I. C., Houston, M. J. and Howie, P. W. (1983). "Release of oxytocin and prolactin in response to suckling," *Br Med J (Clin Res Ed)* 286, 257–259.

McWhinney, B. C., Briscoe, S. E., Ungerer, J. P. and Pretorius, C. J. (2010). "Measurement of cortisol, cortisone, prednisolone, dexamethasone and 11-deoxycortisol with ultra high performance liquid chromatography-tandem mass spectrometry: application for plasma, plasma ultrafiltrate, urine and saliva in a routine laboratory," *J Chromatogr B Analyt Technol Biomed Life Sci* 878, 2863–2869.

Melmed, S. (2009). "Acromegaly pathogenesis and treatment," *J Clin Invest* 119, 3189–3202.

Meyer, J. S. and Novak, M. A. (2012). "Minireview: hair cortisol: a novel biomarker of hypothalamic-pituitary-adrenocortical activity," *Endocr* 153, 4120–4127.

Mitchell, F. (2012). "Mass spectrometry 'gold standard' for measuring steroid sex hormones?" *Nat Rev Endocr* 8, 320.

Murphy, P. R., Friesen, H. G., Brown, R. E. and Moger, W. H. (1989). "Verification of NB2 lymphoma cell bioassay for the measurement of plasma and pituitary prolactin in the Mongolian gerbil (*Meriones unguiculatus*)," *Life Sci* 45, 303–310.

Nussey, S. S. and Whitehead, S. A. (2001). *Endocrinology: An Integrated Approach* (Oxford: Bios Scientific Publishers), www.ncbi.nlm.nih.gov/books/NBK20.

Plant, T. M. (1986). "Gonadal regulation of hypothalamic gonadotropin-releasing hormone release in primates," *Endocr Rev* 7, 75–88.

Soldin, S. J. and Soldin, O. P. (2009). "Steroid hormone analysis by tandem mass spectrometry," *Clin Chem* 55, 1061–1066.

Squire, L. R., Berg, D. E., Bloom, F. E., du Lac, S., Ghosh, A. and Spitzer, N. C. (2008). *Fundamental Neuroscience*, 3rd edn. (London: Academic Press).

Stanczyk, F. Z. and Clarke, N. J. (2014). "Measurement of estradiol – challenges ahead," *J Clin Endocr Metab* 99, 56–58.

Tractenberg, R. E., Jonklaas, J. and Soldin, S. J. (2010). "Agreement of immunoassay and tandem mass spectrometry in the analysis of cortisol and free t4: interpretation and implications for clinicians," *Int J Anal Chem* 2010, article no. 234808.

Widmaier, E. P., Raff, H. and Strang, K. T. (2010). *Vander's Human Physiology: The Mechanisms of Body Function*, 12th edn. (New York: McGraw-Hill).

Xu, C., Roepke, T. A., Zhang, C., Ronnekleiv, O. K. and Kelly, M. J. (2008). "Gonadotropin-releasing hormone (GnRH) activates the m-current in GnRH neurons: an autoregulatory negative feedback mechanism?" *Endocr* 149, 2459–2466.

Zerikly, R. K., Amiri, L., Faiman, C., Gupta, M., Singh, R. J., Nutter, B. *et al.* (2010). "Diagnostic characteristics of late-night salivary cortisol using liquid chromatography-tandem mass spectrometry," *J Clin Endocrinol Metab* 95, 4555–4559.

9 Steroid and thyroid hormone receptors

In previous chapters, we focused on the neuroendocrine system in terms of a variety of hypothalamic neurotransmitter and hormonal messengers. In Chapter 1, these messengers were seen to act via endocrine, paracrine, autocrine, intracrine and neuroendocrine mechanisms. So far, however, we have not discussed in detail how target cells detect and respond to such messages. Chapter 5 introduced this story by illustrating how neurotransmitters, neurohormones and peptide hormones affect their target cells through *receptors* localized to the cell membrane. Examples of these receptors are illustrated in Figures 5.2 (ion channel; GABA receptor) and 5.13 (G-protein-coupled receptor) and this type of receptor will be covered in more detail in Chapter 10. The location of receptors on the outside of cells, that is, in the cell membrane, is important for at least two reasons: (1) because peptide hormones are large, water soluble (hydrophilic) molecules which cannot easily pass through the cell membrane; and (2) because cells such as neurons must respond very quickly (seconds) to neurotransmitters like GABA or glutamate that do not have to enter the cell. In marked contrast, steroid hormones (testosterone, estradiol, progesterone, glucocorticoids and mineralocorticoids; Figure 7.3), and thyroid hormones, are small lipophilic (fat soluble) molecules that can readily diffuse through the cell membranes into any cell in the body. As we shall see in this chapter, target cells for steroid and thyroid hormones have receptors that are located *inside* the cell. These cells therefore respond relatively slowly (minutes to hours) to hormonal stimulation (see Figure 9.1). In brief, the steroid hormone is transported in the blood and released from a binding globulin before freely moving through the cell membrane. Unoccupied steroid hormone receptors (R) are coupled to a molecular chaperone (HSP90; heat shock protein 90) that stabilizes R in the correct shape. When the hormone binds to the receptor-HSP complex, the HSP dissociates and the remaining steroid hormone-receptor complex dimerizes before it enters the cell nucleus. The steroid-R dimer complex then binds to responsive genes via specific hormone response elements (HRE). Various factors such as general transcription factors (GTF) and RNA polymerase II (POL II) assist in inducing gene transcription and the export of messenger RNA (mRNA) into the cell cytosol where it is translated into protein. We will see later that steroid hormones can also act at the cell membrane.

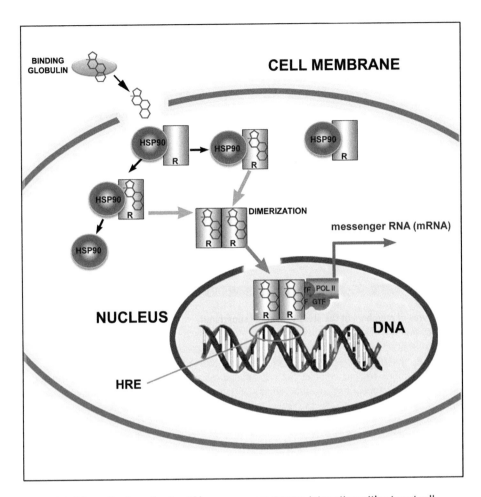

Figure 9.1 Schematic view of a steroid hormone, progesterone, interacting with a target cell
The steroid is transported in the blood and then released from a binding globulin before freely moving through the cell membrane. The unoccupied steroid receptor (R) is coupled to a molecular chaperone (HSP90; heat shock protein 90) that stabilizes R in the correct shape. When the steroid binds to the receptor-HSP complex, the HSP dissociates and the remaining steroid-receptor complex dimerizes before it enters the cell nucleus. The steroid-R dimer complex then binds to responsive genes via a specific hormone response element (HRE). Various factors such as general transcription factors (GTF) and RNA polymerase II (POL II) assist in inducing gene transcription and the export of messenger RNA (mRNA) into the cell cytosol where it is translated into protein.

9.1 The intracellular receptor superfamily

Although steroid and thyroid hormones have different chemical structures and perform very different biological functions, their intracellular receptors have common structural elements; that is, these receptors belong to a *superfamily* of related receptor proteins. Some of the receptors relevant to this chapter are illustrated in Figure 9.2 (Weigel and Moore 2007). This superfamily now numbers in excess of 40, and they are involved in a remarkable variety of regulatory processes in many tissues, including reproduction, metabolism

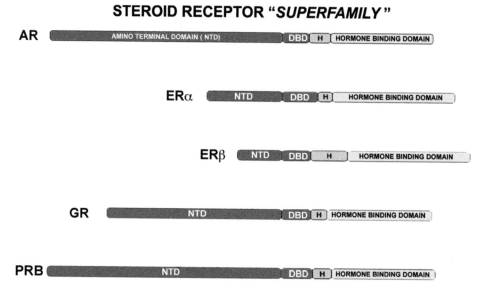

Figure 9.2 Structures of members of the steroid receptor superfamily
The receptors consist of a single polypeptide chain and all of the family members contain a *hormone binding domain* (HBD) that is specific for each hormone. For example, GR (glucocorticoid receptor) binds cortisol in the HBD, whereas the progesterone receptor (PR) will only recognize progesterone. The *DNA binding domain* (DBD) enables the hormone-receptor complex to bind to the hormone response element (HRE) on a target gene. The hinge region (H) is important, together with the DBD, for nuclear localization of the receptor, and the amino terminal domain (NTD) is crucial for the activation of gene transcription once the hormone-receptor complex reaches the nucleus and binds with the HRE.
Abbreviations: AR, androgen receptor; ER, estrogen receptor; GR, glucocorticoid receptor; PRB, progesterone receptor type B.

and energy homeostasis (Figure 9.3; for further reading, see Bookout *et al.* 2006). Figure 9.2 shows that they all contain a *hormone binding domain* (HBD) that is specific for each hormone. Thus, for example, the glucocorticoid receptor (GR) binds cortisol in the HBD, whereas the progesterone receptor (PR) will only recognize progesterone. *Note that in general the receptors themselves have no biological activity until they bind to a hormone;* that is, this hormone-receptor complex interacts with specific sites on target DNA called *hormone response elements* (HRE; shown in Figure 9.1). The hormone-receptor complex binds to the HRE via a highly conserved region called the *DNA binding domain* (DBD). The hinge region (H) appears to be important, together with the DBD, for nuclear localization of the receptor, and the amino terminal domain (NTD) is crucial for the activation of gene transcription once the hormone-receptor complex reaches the nucleus and binds with the HRE (Brinkman 2009). Other hormones that have receptors as part of this family include thyroid and mineralocorticoid hormones, and vitamin D (Mangelsdorf *et al.* 1995).

As noted already, steroid and thyroid hormones are lipid soluble and are bound to transport proteins (binding globulins; see Figure 7.5) in the blood to allow them to be carried to their target tissues. Using progesterone and its receptor, PR, as an example, Figure 9.1 illustrates the sequence of events that occur when progesterone stimulates a target tissue (such as the uterus, or brain). *Note that a similar sequence of events occurs for*

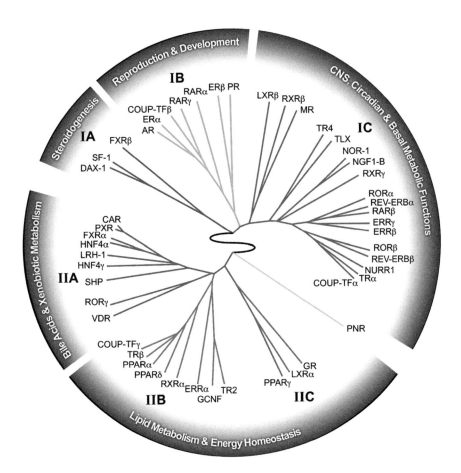

Figure 9.3 The nuclear receptor ring of physiology
The figure shows the remarkable links between large families of nuclear receptors and critical characteristics of animal physiology. This type of analysis reveals an extensive network tying nuclear receptor function to reproduction, development, central and basal metabolic functions, dietary-lipid metabolism and energy homeostasis. Reproduced with permission (Bookout *et al.* 2006).

estradiol and testosterone at their receptors. Progesterone is released from the binding globulin and moves through the cell membrane, after which it binds to a progesterone receptor (PR) in the cell cytosol. The unoccupied PR is coupled to a so-called molecular chaperone (HSP90; heat shock protein 90) that ensures the receptor is stabilized in the correct shape, enabling progesterone to bind tightly to the HBD. Once P is bound to the receptor-HSP complex, the HSP dissociates and the remaining progesterone-receptor complex dimerizes. This dimer then enters the cell nucleus, where it attaches itself to a progesterone-responsive gene via a HRE, and with the assistance of various factors such as general transcription factors (GTF) and RNA polymerase II (POL II), induces gene transcription and the export of mRNA into the cell cytosol where it is translated into protein. A variety of co-activators and co-repressors are also required at this stage, but a detailed consideration of these mechanisms is beyond the scope of this book (for further reading, see Stanisic *et al.* 2010; Picard 2006).

9.1.1 Estradiol receptors

Women are exposed to high levels of estradiol for much of their lives. This occurs through the normal menstrual cycle, via oral contraceptives, during pregnancy and also as hormonal replacement therapy in the post-menopausal period. As we saw in Figure 2.7, estradiol has widespread biological effects, not just in the brain, but throughout the body (Gruber *et al.* 2002). Insofar as the brain is concerned, estradiol is a critical component of the reproductive system (see, e.g., Figure 8.2), but in addition has important influences on learning and memory, on aging and in neuroprotection (see section 9.9.2, below; for further reading, see Behl 2002; and Casadesus 2010). Two important discoveries with regard to estradiol receptors (ER) have changed the accepted view of estradiol action. First, an additional intracellular receptor, ERβ, is now known to exist together with the original receptor, now designated ERα (Figure 9.2). These receptors are the products of two distinct genes located on two different chromosomes. Second, a new type of ER, structurally different from those in Figure 9.2, appears to function like a neurotransmitter receptor within the cell membrane (Arbogast 2008; see Figure 9.9). The two types of receptor mechanisms are referred to as: (1) *genomic* (for ERα and ERβ) because their site of action is in the cell nucleus; and (2) *non-genomic* because of the membrane site of stimulation. In general terms, the genomic action of estradiol, and other steroid hormones, is slow (minutes to hours), whereas the non-genomic action is fast (seconds). Further discussion of genomic and non-genomic mechanisms can be found in subsequent sections (see sections 9.8 and 9.9).

9.2 How are steroid hormone target cells identified?

Steroid hormones exert biological effects throughout the body. This is readily seen in the neuroendocrine system where, for example, estradiol control in the brain, pituitary, uterus, ovary and placenta is indispensable for the reproductive process. However, the influence of estradiol extends to many other organs, including liver, pancreas, skeletal muscle, fat, bone and blood vessels (Gruber *et al.* 2002; Mauvais-Jarvis 2011). By way of illustration, Figure 9.4 shows some tissue targets of estradiol's actions in the regulation of metabolism and body weight control in females. Note that in *males* testosterone is converted to estradiol, which exerts similar effects on metabolism.

Our understanding of how sex steroid hormones affect specific tissue function is greatly enhanced by techniques that allow us to determine the precise cellular location of hormone receptors. Pioneering studies with radioactively labeled probes have now given way to safer, and faster, non-radioactive methods that use specific antibodies raised against the receptor proteins. This technique allows us to localize hormone receptor-positive cells in tissue sections (*immunohistochemistry*). This versatile method has been used to provide brain maps for many different types of hormone receptors, including glucocorticoid, mineralocorticoid and androgen receptors.

A second important technique for localizing cells that possess hormone receptors is called *in situ hybridization*. This method identifies the messenger ribonucleic acid (mRNA) that enables the cell to manufacture the receptor protein. Such an approach depends on knowing the nucleotide sequence of the gene responsible for the protein of interest. The

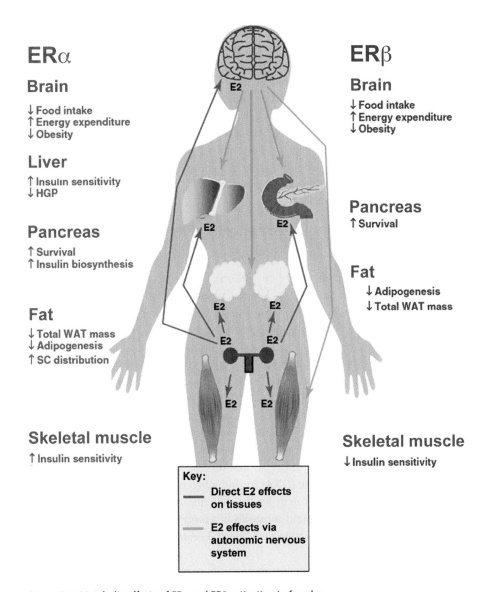

ERα

Brain

↓ Food intake
↑ Energy expenditure
↓ Obesity

Liver

↑ Insulin sensitivity
↓ HGP

Pancreas

↑ Survival
↑ Insulin biosynthesis

Fat

↓ Total WAT mass
↓ Adipogenesis
↑ SC distribution

Skeletal muscle

↑ Insulin sensitivity

ERβ

Brain

↓ Food intake
↑ Energy expenditure
↓ Obesity

Pancreas

↑ Survival

Fat

↓ Adipogenesis
↓ Total WAT mass

Skeletal muscle

↓ Insulin sensitivity

Key:

— Direct E2 effects on tissues

— E2 effects via autonomic nervous system

Figure 9.4 Metabolic effects of ERα and ERβ activation in females

Many tissues, including liver, fat and pancreas, contain estradiol receptors. Activation of estrogen receptor alpha (ERα) in the brain suppresses food intake, increases energy expenditure and decreases body weight. Estradiol also acts at ERα to improve peripheral energy and glucose homeostasis in several ways, such as: (i) suppressing liver (hepatic) glucose production (HGP) and improving liver insulin sensitivity; (ii) enhancing skeletal muscle insulin sensitivity; (iii) enhancing subcutaneous (SC) white adipose fat (WAT) distribution while decreasing overall WAT mass by decreasing adipocyte formation (adipogenesis); and (iv) favoring pancreatic β cell survival, and increasing insulin biosynthesis. Activation of brain ERβ also suppresses food intake, increases energy expenditure and prevents obesity in animals on a high fat diet (HFD). In addition, activation of ERβ affects peripheral energy and glucose homeostasis by (i) favoring pancreatic β cell survival and (ii) preventing obesity and decreasing WAT mass. The metabolic actions of ERα and ERβ on peripheral tissues result from direct activation of ERs in these tissues, or from a central ER action affecting peripheral tissues via the autonomic nervous system. Reproduced with permission (Mauvais-Jarvis 2011).

successful sequencing of the human genome allows the detection of any receptor gene of interest. Using ERα as an example, once the sequence of the gene was known it was then possible to design labeled probes (these can be radioactive, but non-radioactive labels are now commonly used) to detect ERα mRNA in thin sections of brain tissue mounted on microscope slides. The process is outlined in cartoon form in Figure 9.5. This also contains a photomicrograph of ERα mRNA localized to the rat mediobasal hypothalamus, a site already known to contain ERα protein (Shughrue *et al.* 1997) (for further reading, see http://en.wikipedia.org/wiki/Fluorescent_in_situ_hybridization). Most, if not all, steroid receptor genes have now been mapped in this way in a variety of tissues. Examples of brain distribution of other steroid receptors are described below.

9.3 How are steroid hormone target cells differentiated from non-target cells?

As noted, steroid and thyroid hormones are fat-soluble compounds that can readily diffuse from the blood into cells, and just as readily move out again. They can therefore be found in both target and non-target cells. However, the hormone will only exert a biological effect in a cell if that cell expresses the hormone receptor gene; that is, these cells will make hormone receptors, and the hormone will bind to them, forming hormone-receptor complexes. The hormone-receptor complex then regulates the biochemical activity of the target cell by affecting other genes (Figure 9.1). As in section 9.2, the action of estradiol at its target cells will be used as an example of a "typical" steroid hormone-receptor interaction. Target cells for estradiol can readily be discriminated from non-target cells. First, target cells for estradiol have ER proteins in the cytosol and the nucleus of the cell, while non-target cells do not. Second, target cells will accumulate estradiol, while non-target cells will not. Third, the nuclear hormone-receptor dimer stimulates the replication (transcription) of certain information on the target gene (DNA) by inducing mRNA synthesis (Figure 9.1). This mRNA is then translated into protein by the cell ribosomes (see Figure 7.1). Thus, estradiol will stimulate mRNA and protein synthesis in a target cell, but not in a non-target cell. The type of protein produced will obviously depend on the target cell, some of which are seen in Figure 9.4.

Another property of estradiol target cells is their ability to respond rapidly (in seconds) to estradiol; that is, in the discussion so far, we have described only those intracellular receptors that respond slowly (minutes to hours). The rapidly responsive receptors represent a new class of steroid receptor that is located in the cell membrane. These will be covered in the next section.

9.4 Genomic and non-genomic actions of steroid hormones

As described above, steroid hormones modulate mRNA and protein synthesis through their actions on nuclear receptors in neurons. These effects are called *genomic* because the steroid-receptor complex acts as a *transcription factor* to change target gene activity (see Figure 9.1). Some genes are activated, whereas others are repressed. Genomic effects could result in structural or enzymatic changes, for example, but can also be seen as alterations in

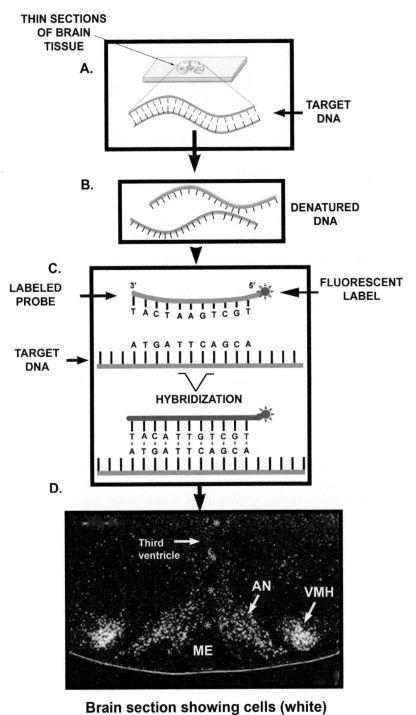

THIN SECTIONS OF BRAIN TISSUE

A.

TARGET DNA

B.

DENATURED DNA

C.

LABELED PROBE

3'   5'

FLUORESCENT LABEL

T A C T A A G T C G T

A T G A T T C A G C A

TARGET DNA

HYBRIDIZATION

T A C A T T G T C G T
A T G A T T C A G C A

D.

Third ventricle

AN   VMH

ME

Brain section showing cells (white) positive for ERα mRNA

Figure 9.5 Schematic representation of the technique for visualization of estrogen receptor alpha (ERα) mRNA by *in situ* hybridization

neurotransmitter receptors or as neuronal levels of neurotransmitter, neuropeptide or neurohormone synthesized and stored in the cell. However, not all of the effects of steroid hormones on neurons can be accounted for by changes in genomic activity (Pfaff and Levine 2008). *Non-genomic*, or fast, direct effects of steroid hormones on the nerve cell membrane, resulting in changes in electrical potential and firing rate, also occur. These rapid effects of steroids on neurons are usually of brief duration, very much like the effect of a neurotransmitter. For example, Figure 9.6 illustrates one of the first demonstrations of very fast electrical effects of estradiol on hypothalamic neurons. We can see that the firing rate of action potentials changes almost instantaneously following application of estradiol; one of these neurons is inhibited (A), whereas the other is excited (C) (Kelly *et al.* 1977; and see Micevych and Kelly 2012). Note that neuronal activity quickly returns to normal when the estradiol is removed.

An illustration of how an estradiol-induced increase in neuronal activity is transduced into a critical cellular response is shown in Figure 9.7, where a low, physiological dose of estradiol (10^{-9} M) rapidly stimulates the secretion of GnRH from hypothalamic neurons *in vitro* (Noel *et al.* 2009).

These results provide two reasons for thinking that some neurons respond to estradiol at the level of the membrane, and not the nucleus. The *first* is the speed of response, which suggests that estradiol does not enter the cell, bind to intracellular receptors and then travel to the nucleus. This would take far too long. The *second* reason is that the estradiol used in the experiments in Figure 9.6 is in the form of a water-soluble succinate derivative that is unlikely to penetrate the neuron cell membrane. There is now evidence that many, if not all, steroid hormones have a membrane site of action in addition to the classical nuclear receptors (e.g. see Ferris and Stolberg 2010; Rahman and Christian 2007) (for further reading, see de Kloet *et al.* 2008; Brinton *et al.* 2008). A striking demonstration of a rapid effect of the glucocorticoid, corticosterone, in the rat brain is shown in Figure 9.8 (Ferris and Stolberg 2010). In this experiment, the authors used functional magnetic resonance imaging (fMRI) to investigate the effects of stress levels of corticosterone on brain activity. Figure 9.8 illustrates a widespread, rapid activation within 60 seconds of injection.

Further examples of the membrane effects of glucocorticoids can be seen elsewhere (de Kloet *et al.* 2008; Groeneweg *et al.* 2011).

Figure 9.5 (cont.)

Briefly, thin frozen sections (5–20 microns) of brain tissue are placed on microscope slides (A). They are then processed in such a way as to break open (*denature*) the DNA strands (B). The denatured DNA is treated with a strand of synthetic DNA that is specific (complementary) to the target DNA and which is also labeled with a fluorescent tag that will be visible in a microscope (C). This complementary DNA will stick (*hybridize*) to the target DNA, but not to any other DNA. The final step is to examine the tissue sections in a fluorescence microscope to view the localization of the label. The visible fluorescence (D) is the site of the ERα mRNA and therefore the neurons that possess ERα receptors.

Abbreviations: AN, arcuate nucleus; ME, median eminence; VMH, ventromedial nucleus of hypothalamus; letters in C represent nucleotides (A = adenine; T = thymine, etc.). Image in D. reproduced with permission (Shughrue *et al.* 1997).

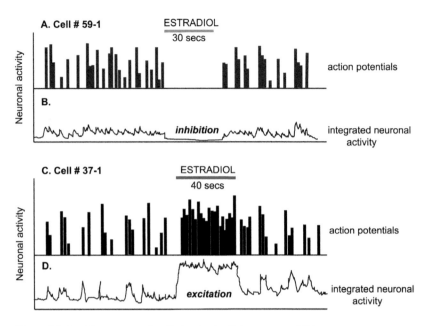

Figure 9.6 Rapid, non-genomic effects of estradiol on neuronal firing
Electrical recordings of action potentials from estradiol-treated neurons from the preoptic area of the hypothalamus. Estradiol was applied for approximately 30 to 40 seconds. The upper trace (A; cell #59-1) shows an estradiol-induced deceleration of the firing rate, clearly seen in the lower trace of integrated activity (B). The action potentials are totally abolished. A second cell (C; #37-1) is excited by a similar pulse of estradiol. In this case, the action potentials are greatly accelerated. Note that the onset of estradiol's effects is very fast (seconds) and is quickly reversible. Reproduced with permission, and redrawn (Kelly et al. 1977).

9.4.1 What is the evidence for steroid hormone membrane receptors?

Rapid effects of steroid hormones on neuronal membranes are now well described and it is logical to assume that some type of membrane receptor must exist. However, the nature of the membrane receptor(s) is far from clear. For example, some evidence, at least for estradiol, indicates that the membrane receptor is the same as the nuclear receptor. In other words, cells from animals that lack ERα and ERβ genes (*ERα/ERβ knockouts*) do not possess the corresponding receptor proteins and do not respond to estradiol either through nuclear signaling or via the membrane (Levin 2009). Nevertheless, this view is controversial and there may be other types of estrogen receptor molecules localized to the cell membrane (for further reading, see Kelly and Ronnekleiv 2008; Kelly and Qiu 2010). If membrane receptors for estradiol do exist, what is the mechanism of the fast cellular response? As described in Chapter 10, membrane receptors for neurotransmitters/neuropeptides are often coupled with G proteins (see Figure 5.13). Membrane receptors for estradiol (ER$_M$; Figure 9.9) are also physically associated with G protein complexes, and the binding of estradiol to this complex elicits fast intracellular signaling cascades. This type of signaling is implicated in the physiological control of neural, reproductive, endocrine and immune systems, as

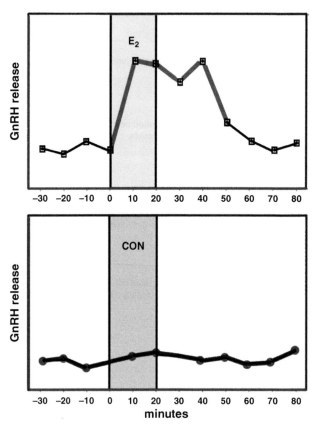

Figure 9.7 Rapid effect of estradiol on GnRH secretion from GnRH neurons

A low, physiological concentration of estradiol (E2; 10^{-9} M) stimulated GnRH secretion from monkey GnRH neurons in tissue culture. Notice how secretion is increased within minutes of applying the E2 stimulus (blue bar; 20 minutes). In contrast, perfusion with the vehicle (CON; grey bar) has no effect on GnRH release.

Abbreviation: GnRH, gonadotropin-releasing hormone. The figure is drawn based on data from Noel *et al.* 2009.

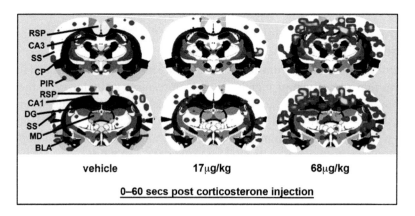

Figure 9.8 Fast, global effect of injected corticosterone on brain cells

Functional magnetic resonance imaging (fMRI) activation maps showing dose-dependent changes in neuronal activity 60 seconds following intravenous corticosterone injection. This figure illustrates only two of the many brain levels studied, but clearly demonstrates the widespread and rapid effect of the higher dose of corticosterone, especially in cerebral cortex and hippocampus.

Abbreviations: BLA, basolateral amygdala; CA1 and CA3, hippocampus; CP, caudate/putamen; DG, dentate gyrus; MD, dorsomedial thalamus; PIR, piriform cortex; RSP, retrosplenial cortex; SS, somatosensory cortex. Reproduced with permission (Ferris and Stolberg 2010).

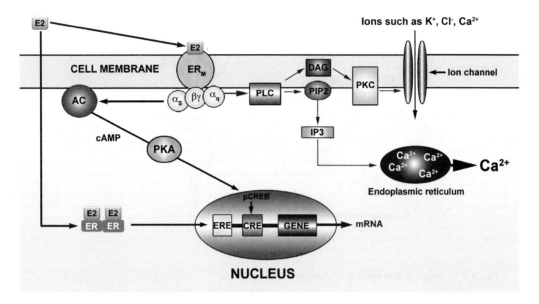

Figure 9.9 Estradiol acts at both extracellular (membrane) and intracellular receptors
The classical form of estradiol action is shown on the left: estradiol (E2) crosses the cell membrane to bind with internal receptors (ER) that dimerize and enter the cell nucleus (see also Figure 9.1). Here, the receptor-E2 complex binds to an estrogen response element (ERE) to regulate gene expression and the production of messenger RNA (mRNA). E2 also binds to G-protein-coupled membrane receptors (ER$_M$; yellow) which elicit fast cellular responses via the α_q subunit and changes in intracellular ions such as calcium (Ca^{2+}) and potassium (K$^+$) ions. This form of membrane signaling also influences the nucleus via stimulation of cAMP formation, protein kinase A (PKA) and the transcription factor pCREB. Such signaling also causes changes in gene expression.
   Abbreviations: AC, adenyl cyclase; CREB, cAMP responsive element binding protein; DAG, diacylglycerol; IP3, inositol triphosphate; PIP2, phosphatidylinositol diphosphate; PLC, phospholipase C; α_s, α_q and $\beta\gamma$ are G protein subunits.

well as in disease states such as depression, obesity, heart disease and breast cancer (Prossnitz and Barton 2011). A typical outline of such signaling is shown in Figure 9.9. On the left side of the figure, estradiol enters the cell to bind to intracellular receptors (ER). These dimerize and eventually enter the nucleus and bind to estrogen response elements (ERE), as previously described. Also shown is an estrogen receptor located in the cell membrane and coupled to G protein subunits α_s, α_q and $\beta\gamma$. This type of receptor is described in Chapter 10. Briefly, the α-subunits are responsible for activating various signals that in turn switch on enzyme activity such as phospholipase C (PLC) and adenyl cyclase (AC). Each of these regulates other pathways. For example, the cAMP pathway activates protein kinase A (PKA), which subsequently activates (phosphorylates) a transcription factor called CREB (cAMP binding protein). Phosphorylated CREB (pCREB) modifies target gene expression through its own response element, CRE. An alternate route of cell activation through the ER$_M$ G protein complex is the stimulation of the enzyme phospholipase C (Figure 9.9). In this case, PLC produces a second

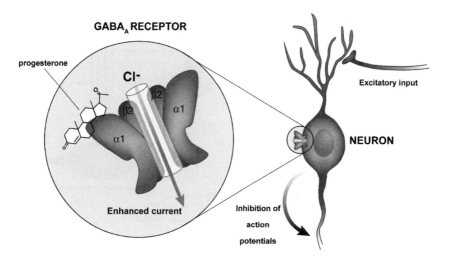

Figure 9.10 Progesterone binds to membrane GABA receptors
In contrast to the previous example, where estradiol binds to its own receptor in cell membranes, progesterone binding sites are a component of GABA receptors. In this example, progesterone binds to the α-subunit of the GABA$_A$ receptor. This interaction enhances the GABA-mediated inward passage of chloride ions into the neuron. This hyperpolarizes the cell and prevents the excitatory drive coming from the dendrites.
Abbreviation: GABA, γ-aminobutyric acid. Reproduced with permission (McCarthy 2007).

messenger, inositol triphosphate (IP3), which then controls intracellular calcium ion levels through release from the endoplasmic reticulum. PLC also produces another messenger, diacylglycerol (DAG), and this in turn activates protein kinase C (PKC), which modifies ion movement through ion channels.

There is also evidence that *progesterone* acts at the cell membrane, in addition to its intracellular receptor sites (see section 9.6.2, below). These membrane receptors (PR), in common with those for estradiol, possibly operate through rapid intracellular signals (Mani 2008) and affect behavioral, endocrine and sensory systems through membrane receptors that are found in specific brain regions (Zuloaga *et al.* 2012). In addition, progesterone and some of its metabolites bind to a site that is not just a progesterone receptor, but is part of the GABA receptor (see also Figure 5.2). Progesterone therefore modifies the membrane potential of neurons (Figure 9.10). In this example, progesterone binds to the α-subunit of the GABA receptor and enhances the inward passage of chloride ions into the neuron. This hyperpolarizes the cell and reduces the excitatory drive coming from the dendrites. This field of study is large and controversial, although there is no doubt that progesterone, and related molecules such as allopregnanolone and pregnenolone, are active in the nervous system by modifying several membrane receptors. These include glutamate, L-type calcium channels, GABA and dopamine receptors (for further reading, see Melcangi and Panzica 2006; and Zheng 2009). These steroids have profound effects on the release of many neurotransmitters and have been implicated in the mechanism of action of anxiolytic (anti-anxiety), analgesic (pain) and anticonvulsant (epilepsy) drugs (Belelli *et al.* 2006; Melcangi and Panzica 2014). An important issue that will be discussed later is the evidence that such steroids are also synthesized in the brain;

these are called *neurosteroids* (see section 9.9.6). A final point to emphasize is that nuclear PR can be activated by other stimuli in the absence of progesterone. For example, vaginal stimulation can stimulate PR activity when progesterone is not present (Auger *et al.* 1997). This is likely to occur through a neurotransmitter/neuropeptide effect at the cell membrane.

9.5 Measurement and regulation of hormone receptor numbers

Pioneering studies in the 1960s, with the then newly available radioactively labeled hormones such as $[^3H]$- or $[^{125}I]$-estradiol, allowed the brain and pituitary to be mapped for areas sensitive to steroid feedback. The same technique was also used to assay receptor numbers in attempts to correlate them with cell responsiveness to hormone stimulation. Just as importantly, the ability to localize steroid and thyroid hormone receptors in specific subcellular sites, such as the cell nucleus, led directly to our current understanding of how these hormones modify gene expression (Figure 9.1). In a wider context, the labeling techniques were adapted to the study of, for example, breast tumors. It quickly became possible to determine which tumors were sensitive to estradiol and to design treatments accordingly. In other words, patients whose tumors were estradiol-dependent (i.e. they would grow in response to estradiol) could be treated with anti-estrogens. More recently, hormones such as estradiol have been labeled with positron-emitting isotopes such as ^{18}F, so that hormone localization (binding) is now possible in living patients using positron emission tomography (PET) (Dunphy and Lewis 2009).

This field underwent yet another revolution with the successful isolation and structural determination of the receptor proteins (Figure 9.2). This was a critical step that facilitated the preparation of specific antibodies for each receptor. As noted (section 9.2), these antibodies are routinely used to specifically localize receptors to cells in various brain regions (see also section 9.6, below), and to allow quantification of, for example, sex differences in ERα and ERβ in the human brain (Kruijver *et al.* 2003). This versatile method is obviously well suited for the study of many receptors in several species.

The number of receptors in a steroid hormone responsive cell is not fixed, but can be altered by age, by the level of hormones in the circulation and by sex. For example, the synthesis of hypothalamic progesterone receptors is increased by changes in estradiol levels, and this effect requires both ERα and ERβ (Kudwa *et al.* 2004; Simerly *et al.* 1996). Hormone receptor numbers undergo developmental changes and exhibit sexually dimorphic patterns as well (Ikeda *et al.* 2003). Changes in the amount of hormone in the circulation can also regulate gene expression of receptors for that hormone. For example both ERα and ERβ mRNA are modified by estradiol treatment and it is possible that such regulation takes place normally, for example, through the menstrual cycle when estradiol and progesterone levels vary significantly (Scott *et al.* 2000; Patisaul *et al.* 1999). Finally, some neurons appear to express more than one receptor type. For example, using double-label immunohistofluorescence, with specific antibodies for ERα and ERβ, it is possible to see that some neurons in the ventromedial nucleus of the hypothalamus are ERα-specific or ERβ-specific (Ikeda *et al.* 2003), and that some cells have both ERα and ERβ. The authors suggest that estradiol regulates these cells via so-called *receptor heterodimers*. In other words, instead of estradiol binding to ERα–ERα dimers (see Figures 9.1 and 9.9), in some cells these dimers are ERα–ERβ. It is likely that these heterodimers exert different effects on target gene expression compared to the homodimers (Pettersson *et al.* 1997; Katzenellenbogen and Korach 1997). In a slightly different way, although not in the same cells, many steroid receptors coexist in the same brain area (see below) and therefore probably interact to regulate physiological pathways.

The complexity of this type of interaction can now be probed by generating mouse mutants that lack individual receptor types (for further reading, see Hewitt *et al.* 2005). For example, mice in which the ERα gene was knocked out (ERα KO) were infertile and the positive feedback of estradiol on GnRH secretion was absent. On the other hand, ERβ KO appeared to have a normal estradiol positive feedback (Wintermantel *et al.* 2006), although litter sizes were smaller (Couse and Korach 2001). This powerful technology can be applied to any receptor type.

9.6 Gonadal steroid hormone target cells in the brain

In section 9.2, we outlined some approaches to detecting and localizing steroid receptors. We also emphasized how such detection systems were crucial in probing the signaling pathways by which these hormones influenced target cells in tissues throughout the body. The present section describes the specific neural sites that possess estradiol, androgen and progesterone receptors. Section 9.7 will examine adrenal steroid hormone receptors in the brain. Bear in mind that the original studies on receptor localization focused on the receptor proteins, either by binding with radioactively labeled hormones or via immunohistochemistry with specific receptor antibodies. At the present time, molecular techniques, such as *in situ* hybridization, have added a new, complementary, approach to identifying specific cells that express the genes for each receptor. Other molecular biological tools, such as Northern analysis and quantitative RT-PCR (reverse transcriptase – polymerase chain reaction) are routinely used to quantify receptor mRNA levels in many tissues, including brain (for further reading, see Kopchick *et al.* 2007). Thus, it is possible to localize both the receptor gene and the protein product of that gene in the same cell.

9.6.1 Estradiol receptors

Figure 9.11 illustrates the distribution of ERα and ERβ mRNA in the rat brain as determined by *in situ* hybridization (Shughrue *et al.* 1997). This figure only shows a brain section that contains the hypothalamus, but the study mapped the entire rat CNS. ERα and ERβ mRNA were detected throughout the brain and spinal cord, the pineal gland and the retina. As is clear from Figure 9.11, the distribution of the two receptors is significantly different. Some areas have only one of the receptors, whereas others express both of them. For example, ERβ mRNA was plentiful in areas such as PVN, SON, pineal gland, hippocampus and cerebellum, where ERα mRNA was sparse. In contrast, ERα mRNA was enriched in VMH and amygdala. As noted in section 9.5, in those areas where both genes are expressed there may be cells that express both receptors. The human brain has also been mapped for these receptors, revealing many similarities with the rat brain (Gonzalez *et al.* 2007; Kruijver *et al.* 2003).

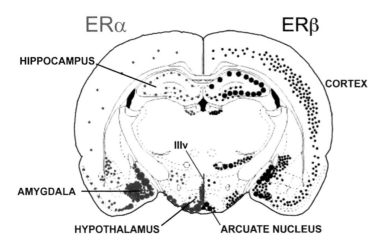

Figure 9.11 Distribution of ERα and ERβ mRNA in female rat brain
The location of cells expressing the mRNA for ERα and ERβ was determined by *in situ* hybridization on thin sections of rat brain (see Figure 9.5). Coronal (frontal) sections are schematically represented, with cells positive for ERα mRNA (left side; red dots) and ERβ mRNA (right side; black dots). These sections are from the level of the hypothalamus, but the entire rat brain was mapped in this study. The information shown here is simplified so that the distribution of labeled cells is depicted by dots that represent both the number and distribution of labeled cells; that is, small dots – 1–5 labeled cells; medium dots – 6–10 labeled cells; large dots – approximately 50 labeled cells.

Abbreviation: IIIv, third ventricle. Reproduced with permission (Shughrue *et al.* 1997).

What do these studies tell us about the role of ER in the nervous system? It should be obvious that estradiol and its receptors operate on a wide field and are not restricted to reproductive control. In the remainder of this section, we will touch on just three of the many systems that may be regulated by ER. First, the beneficial effect of estradiol on the heart is well described; the incidence of high blood pressure and heart disease is much lower in premenopausal women compared to men. However, after the menopause, when estradiol levels fall in women, the incidence of cardiovascular disease increases (Mendelsohn and Karas 1999). ERα and ERβ have been localized to brain nuclei associated with autonomic control, such as the PVN, nucleus tractus solitarius and spinal cord. And these nuclei respond to estradiol treatment by modifying blood pressure and sympathetic nerve activity (Spary *et al.* 2009). Second, in terms of *behavior*, much information is now available to delineate the individual contributions of ERα and ERβ (Tetel and Pfaff 2010). A powerful technique by which we can deduce some of the functional attributes of these receptors is to breed mutant mice that lack them. In their review of these studies, Tetel and Pfaff (2010) emphasize the highly complex nature of the interactions between ERα and ERβ in modifying behavior. In their words: "The extreme scenarios are that ERα and ERβ act in an obligatory synergistic fashion ... or that they exert equal and opposite function." By way of example, female sexual behavior (lordosis) is dependent upon ERα, but not ERβ. However, the presence of ERβ seems to reduce sexual receptivity. Similarly, male aggression is stimulated by ERα, but is suppressed by ERβ. A final example concerns the ability of estradiol to increase the number of progesterone receptors (PR) in the brain (see section 9.5). In females, this process requires both ERα and ERβ, but in males estradiol is only effective in the absence of ERβ. Third, estradiol appears to have significant neuroprotective properties and may offer therapeutic options for the aging brain. Much experimental data has led to the conclusion that estradiol, and other estrogens, regulate the transcription of genes that code for proteins necessary for neuronal survival, including those for neurotrophic factors (e.g. NGF, BDNF and IGF-1), proteins which suppress apoptosis and those affecting neuronal structure (Behl 2002). A possible role for estradiol in the pathogenesis of Alzheimer's Disease has inevitably attracted considerable attention. Although there are some inconsistencies between the results and conclusions of various clinical studies, it is now accepted that an age-related decline in estradiol levels could be an important component of cognitive impairment, perhaps in men as well as in women (Janicki and Schupf 2010; Luine and Frankfurt 2012). The timing, dosage and method of administration of estradiol are critical variables that are currently part of clinical trials. Equally important are suggestions that estrogens offer some potential for the treatment of depression, sleep problems, anxiety, schizophrenia and some aspects of pain (Hughes *et al.* 2009; Joffe *et al.* 2011). These authors also believe that specifically targeting the ERα and ERβ subtypes may offer additional therapeutic efficacy (for further reading, see Gosselin and Rivest 2011).

9.6.2 Progesterone receptors

Compared to the attention paid to estradiol receptors, those for progesterone have been somewhat neglected (Blaustein 2008a; 2008b). Nevertheless, it is now known that the PR, like their ER counterparts, exist in several forms. Progesterone can act at the cell membrane via the GABA receptor (see section 9.4.1), but there are also two nuclear receptors, PRA and PRB, which are members of the steroid receptor superfamily (Figure 9.2). The mammalian brain has been mapped for progesterone receptors using techniques for detection of the receptor protein (e.g. immunohistochemistry; Western blotting), as well as mRNA (*in situ* hybridization; RT-PCR) (Brinton *et al.* 2008). Primary locations for PR are the hypothalamus and the median eminence, but significant populations are also found in cortex, hippocampus and amygdala (Figure 9.12). This figure shows that there is a significant overlap between the areas positive for ER and PR (compare Figures 9.11 and 9.12). This is consistent with a point made earlier (see section 9.5), that expression of PR in arcuate nucleus, VMH and POA is under the control of ERα and ERβ. This means that some neurons will express both ER and PR. Although hypothalamic PR are particularly important for the control of reproduction and sexual behavior (Pfaff 1997), Figure 9.12 illustrates that PR are also found in brain areas not usually associated with reproduction, including the hippocampus and cerebral cortex. In contrast to the hypothalamus, the expression of PR in cortical neurons, and those in the cerebellum, is not under estrogenic control (Mani 2008).

What is known about the roles of the individual nuclear progesterone receptors, PRA and PRB? First, PRA and PRB proteins can dimerize and bind to the DNA hormone response elements (see Figure 9.1) as three distinct species: A:A and B:B homodimers and A:B heterodimers. This suggests that the dimers may have distinct effects on gene transcription (Mani 2008). Functionally, in the adult female rat, PRA and PRB mRNA are both induced by estradiol and down-regulated by progesterone in the basal hypothalamus. However, only

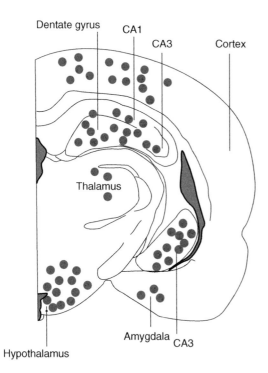

Dentate gyrus CA1
CA3 Cortex
Thalamus
Amygdala CA3
Hypothalamus

Figure 9.12 Distribution of brain progesterone receptors
This schematic summarizes several studies, using various techniques, which localized progesterone receptors (PR) or PR mRNA in regions of rat brain.
   Abbreviations: CA1 and CA3 are regions of the hippocampus. Reproduced with permission (Brinton *et al.* 2008).

PRB mRNA expression was hormonally modulated in the preoptic area. In the hippocampus, however, estradiol induced only PRA expression (Camacho-Arroyo *et al.* 1998). A detailed description of the physiological roles of brain PRA and PRB awaits the production of specific gene knockout mice. Nevertheless, it is becoming clearer that progesterone has multiple influences in the brain in regulating cognition, mood, inflammation, neurogenesis and regeneration, myelination and recovery from traumatic brain injury (for further reading, see Brinton *et al.* 2008; and Junpeng *et al.* 2011).

9.6.3 Androgen receptors
Figure 9.13 shows the distribution of androgen receptor (AR) mRNA in the rat brain, as determined by *in situ* hybridization (Simerly *et al.* 1990). This image is at the level of the hypothalamus, although the study mapped the entire brain. As we have seen for ER and PR, there is labeling of the hypothalamus, amygdala, hippocampus and cortex, indicating that many cells express the AR gene. Using a similar technique, male and female brains had a similar distribution of AR except for a significant increase in the male, compared to female, POA (McAbee and Doncarlos 1998). The increase appears to be consistent with this area being important for male sexual behavior. The overlap of areas that contain ER, PR and AR is striking (Figures 9.11, 9.12 and 9.13), especially within the hypothalamus, as would be expected because of their critical roles in the control of reproduction. The two most important androgens are testosterone and 5α-dihydrotestosterone (DHT; see section 9.6.4), both of which bind to the androgen receptor. Different target tissues respond to one or the other of these hormones depending upon the presence of the 5α-reductase enzyme; that is, in some tissues, such as the prostate and penis, 5α-reductase converts testosterone to DHT

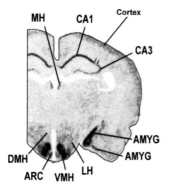

Figure 9.13 Distribution of brain androgen receptor (AR) mRNA
This figure is an autoradiogram (an X-ray image) which shows the location of AR mRNA in female rat brain following *in situ* hybridization with a radioactively labeled probe specific for AR mRNA.

Abbreviations: AMYG, amygdala; ARC, arcuate nucleus; CA1 and CA3, hippocampus; DMH, dorsomedial hypothalamus; LH, lateral hypothalamus; MH, medial habenula. Reproduced with permission (Simerly *et al.* 1990).

which then binds to the androgen receptor. In other tissues, testosterone is the active hormone (e.g. vas deferens) (Brinkman 2009). The brain and hypothalamus are also targets for both hormones and recent studies indicate that they influence not only the reproductive system, but may be involved in energy homeostasis as well, perhaps offering insights into the treatment of obesity (Mauvais-Jarvis 2011).

9.6.4 Testosterone is a prohormone

As noted in section 9.6.3, certain androgen target cells such as those in the brain, pituitary gland, gonads and secondary sex organs (including the prostate gland, seminal vesicles and penis) use the enzyme 5α-reductase to convert testosterone to 5α-dihydrotestosterone. The active hormone in these tissues is therefore DHT and testosterone can be considered to be a *prohormone*. In other tissues such as brain and ovary, testosterone is also converted to *estradiol* by the enzyme aromatase so that many of the known biological effects of testosterone are actually mediated by the female sex hormone acting through estradiol receptors (Figure 9.14; Morris *et al.* 2004). In addition to the brain, aromatization of testosterone to estradiol occurs in the gonads, placenta, adrenal cortex, breast tissue, fat cells, kidney, liver and other tissues in both mammalian and non-mammalian vertebrates (Boon *et al.* 2010). As far as the brain is concerned, the conversion of testosterone to estradiol by aromatase is not restricted to the regulation of sexual differentiation, gonado-tropin secretion and copulatory behavior alone (Garcia-Segura 2008). In fact, the biosynthesis of estradiol from testosterone in various brain areas is a regulated process that is dependent on developmental stage, sex and circulating hormone levels (Roselli *et al.* 2009). For example, male animals sustain far more neural damage following brain trauma than do females and it remains possible that local aromatase, as well as circulating estradiol, might exert a neuroprotective effect. Other studies suggest that rapid changes in brain aromatase can modulate synaptic activity and plasticity, with concomitant modification of mood and behavior (Garcia-Segura 2008).

9.7 Adrenal steroid target cells in the brain

Like the gonadal steroid hormones, adrenal steroids such as cortisol bind to intracel-lular receptors to form a hormone-receptor dimer which then binds to a nuclear

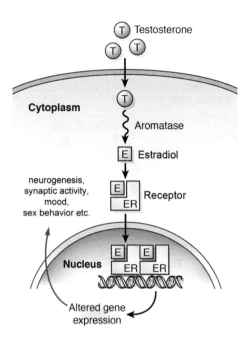

Figure 9.14 Conversion of testosterone to estradiol in the brain

Testosterone is transported into the brain and some of it binds to androgen receptors. However, some testosterone is also converted to the female hormone estradiol by an enzyme called *aromatase*. The resulting estradiol then binds to estradiol receptors that dimerize and modify brain function through changes in gene expression. Testosterone thus affects many systems such as mood, sex behavior and synaptic activity. Reproduced with permission (Morris *et al.* 2004).

glucocorticoid response element in target genes (de Kloet *et al.* 2005). There are, however, two major steroid hormones secreted from the adrenal glands: the *glucocorticoid* cortisol, and the *mineralocorticoid* aldosterone (Figure 2.5). They bind to two related receptor molecules, the glucocorticoid receptor (GR) and the mineralocorticoid receptor (MR), respectively. In some tissues, such as kidney, aldosterone appears to be the primary hormone that binds exclusively to MR. However, both GR and MR are present in the brain and both receptors bind cortisol *and* aldosterone (see below). Since MR and GR are encoded by two distinct genes, this has facilitated the localization of specific GR mRNA and MR mRNA in brain tissue by *in situ* hybridization. Detailed brain maps can be found elsewhere (Seckl *et al.* 1991; Morimoto *et al.* 1996; Sanchez *et al.* 2000), but a summary is shown in Figure 9.15. GR gene expression is distributed widely in the mammalian brain compared to MR mRNA; but note the high concentration of both receptors in the hippocampus and hypothalamus. In the hypothalamus, the highest density is found in the PVN, where cortisol exerts a powerful negative feedback on ACTH secretion by reducing the secretion of CRH from PVN neurons (see Figure 6.4).

9.7.1 Mineralocorticoid receptors (MR)

MR have their highest density in the hippocampus, lateral septum, the dentate gyrus of the hippocampus, brain stem and in some areas of the cerebral cortex, such as the entorhinal cortex (see Figure 9.15). MR are largely absent from the pituitary gland, which has a high concentration of GR (Ozawa *et al.* 1999). Note that brain MR bind equally well to aldosterone and to corticosteroids such as cortisol. This is quite different

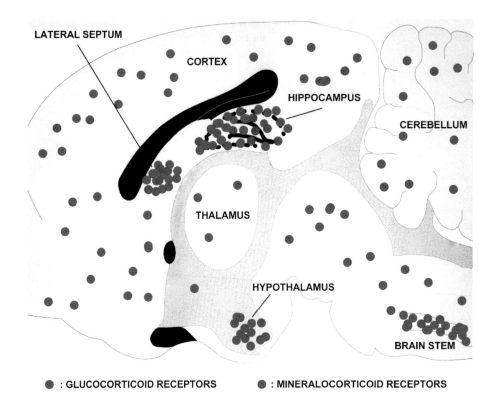

Figure 9.15 Distribution of glucocorticoid and mineralocorticoid receptors in rat brain
Schematic representation of expression profiles of glucocorticoid (GR) and mineralocorticoid (MR)
receptors in the rat brain. Note that whereas GR are ubiquitously expressed, MR expression is confined to
distinct brain areas, especially in the limbic brain. This figure was redrawn from an original generously provided
by Dr. O. F. X. Almeida.

from the situation in the kidney tubules where MR respond only to aldosterone, promoting reuptake of sodium ions. The reason for this seems to be that the kidney is able to
rapidly metabolize cortisol, but not aldosterone, thus preventing cortisol from binding to
MR (for further reading, see Funder 2010). In the brain, cortisol binds to MR with high
affinity compared to GR; in other words, low cortisol levels will activate MR, but not GR
(see section 9.7.3). This type of feedback regulates CRH and ACTH secretion, vasopressin
secretion, locomotor activity, neurogenesis and mental performance under normal baseline (non-stress) fluctuations in glucocorticoid levels (de Kloet 2008; for further reading,
see de Kloet *et al.* 2005).

9.7.2 Glucocorticoid receptors

GR are widely distributed throughout the brain (Figure 9.15), but are particularly concentrated in the hippocampus, septum, amygdala, PVN, certain areas of the cerebral
cortex and in the brain stem, as well as many cells of the anterior pituitary. GR have a
lower affinity for glucocorticoids than the MR and are only activated by the high level of

glucocorticoids secreted during stress. They provide negative feedback to terminate stress-activated neural and endocrine (hypothalamic-pituitary-adrenal) activation and regulate behavioral responses to stress, and thus function to protect the organism from its own stress response (de Kloet *et al.* 2005). GR may also modulate the rhythm of circadian behaviors that are influenced by the hippocampus, such as sleep-wake cycles, appetite and mood (Conway-Campbell *et al.* 2010; Nader *et al.* 2010). Glucocorticoids also function as negative feedback signals to inhibit the immune system (Figures 1.1 and 2.6; see also section 13.6.2).

9.7.3 Why are there two different receptors for glucocorticoids in the brain?

When equilibrium is threatened by various stressors, the immediate autonomic response is an increase in alertness and vigilance that is probably mediated through the hippocampus, amygdala and prefrontal cortex. At the same time, the adrenal medulla rapidly secretes epinephrine and the adrenal cortex begins to release glucocorticoids such as cortisol. Peak cortisol concentrations are achieved in 15 to 30 minutes, and resting, pre-stress levels are reinstated by about 90 minutes. There is evidence that the initial neural response to increasing cortisol levels is mediated by MR, whereas GR mediates the recovery phase (de Kloet *et al.* 2005; de Kloet 2008; Lightman 2008). Under normal conditions, as we have seen (Figure 6.5), pulsatile secretion of ACTH from the pituitary is closely mirrored by cortisol secretion; that is, cortisol exerts a negative feedback influence on ACTH release following each peak. This is very likely exerted through MR. However, when cortisol levels begin to increase, following stress, GR are then recruited to damp down the stress response (Lightman 2008). This represents the most straightforward view of the need for two types of glucocorticoid receptor, and additional information can be found elsewhere (de Kloet *et al.* 2005).

A third signaling device, although not necessarily an additional receptor, should also be considered. As noted (see section 9.4), several steroid hormones exert fast electrical effects on neurons via receptors located in the cell membrane. Such a mechanism may operate in the rapid feedback effect of cortisol on MR in the initial stress response (see also Figure 9.9), probably through membrane ion channels and/or MR localized to the membrane (de Kloet *et al.* 2005).

It is appropriate to consider here what happens when the second phase of the stress response is not controlled by cortisol acting on GR; that is, during a period of chronic stress. The adverse effects of unremitting stress are well described (for further reading, see de Kloet *et al.* 1998; Sousa *et al.* 2008; Chrousos 2009), including a marked down-regulation of GR and MR, atrophy of hippocampal dendrites, impaired neurogenesis, compromised learning ability and increased risk of depression. Thus, disruption of the GR/MR regulation of the hypothalamic-pituitary-adrenal axis has serious consequences. In human terms, the fetus may be particularly vulnerable. For example, the routine treatment of pregnant mothers with synthetic glucocorticoids, to enhance fetal lung development, exposes the fetal brain to possibly neurotoxic levels of glucocorticoid with adverse consequences in terms of hippocampal development (Yu *et al.* 2010).

9.8 Steroid hormone-induced changes in neurotransmitter release

When the distribution of steroid hormone receptors in the brain (illustrated in Figures 9.11 to 9.13 and 9.15) is compared with the distribution of neurotransmitter pathways (illustrated in Figure 5.15), an obvious overlap is observed between the pattern of steroid hormone receptors and the nerve pathways (see also human brain patterns: Dudas and Merchenthaler 2006). For example, the co-distribution of estradiol and glucocorticoid receptors in norepinephrine, dopamine and serotonin nerve pathways has been known for some time (Grant and Stumpf 1975; for further reading, see Woolley 1999). It is now routine practice to combine *in situ* hybridization and immunohistochemistry, or to use double-label immunohistochemistry, to co-localize steroid receptors and various neurotransmitters/neuropeptides in the investigation of steroid effects in the brain. For example, the negative feedback of progesterone on GnRH secretion is mediated through hypothalamic dopamine neurons (Dufourny *et al.* 2005). In other words, some hypothalamic PR-positive cells use dopamine as a neurotransmitter, and progesterone feedback causes release of dopamine. Estradiol also modifies dopamine levels, and perhaps dopamine neurotransmission, via an effect on dopamine transporters in astrocytes (Karakaya *et al.* 2007). Another example is the repression of noradrenergic neurotransmission by glucocorticoids (Yang *et al.* 2007). A wide variety of steroid/neurotransmitter interactions in the brain have been described (Zheng 2009). What is the mechanism by which steroid hormones control neurotransmitter release? Many examples are presented in the review by Zheng (2009). As we saw earlier, a hormone such as estradiol may exert multiple effects on target neurons; some effects are fast and mediated via the cell membrane, whereas others are genomic (Figure 9.9). Both of these inputs may affect membrane potential and/or ion channels, thereby modulating electrical activity and neurotransmitter secretion. Figure 9.7 illustrates that estradiol has just such a rapid stimulatory effect on GnRH secretion. A similar mechanism is also proposed for glucocorticoids (de Kloet *et al.* 2005 – see Box 2). In contrast, progesterone can inhibit neurotransmitter secretion by its action at GABA receptors (Figure 9.10) and estradiol can directly affect neurotransmission by binding to transmembrane ion channels such as $GABA_A$, NMDA (N-methyl-D-aspartate) and $5-HT_3$ (5-hydroxytryptamine type 3) receptors (Behl 2002). Androgens, such as testosterone, may also modify neurotransmission via effects on ion channels (for further reading, see Foradori *et al.* 2008).

9.8.1 Steroid hormone induced changes in protein synthesis in nerve cells (genomic effects)

As outlined in Figure 9.1, steroid hormones such as progesterone form hormone-receptor dimers that bind to specific hormone response elements in target genes. This step induces (but sometimes represses) mRNA and protein synthesis in the target cell. These proteins can be any of a variety of neurotransmitters, hormones, neuropeptides, enzymes, neurohormones, growth factors or the receptors for these chemical messengers. For example, as described earlier (section 9.6.2), estradiol stimulates the synthesis of progesterone receptors in target cells of the hypothalamus and in the uterus. Estradiol also regulates the synthesis of opioid peptides in the hippocampus (Williams *et al.* 2011). Of significant clinical interest is the possibility that estradiol is a neuroprotective agent, regulating genes that code for proteins that enhance neurotrophic support and affect neuronal structure (e.g. see Figure 9.17). In contrast to the influence of steroids in the mature brain, steroid hormone activation of protein synthesis is particularly important during development, when the effect on protein synthesis results in permanent changes in neural growth and connections during sexual differentiation of the brain (see section 9.9.5). As noted previously, the effects of steroid hormones on protein synthesis (genomic effects) are not immediate and may take hours or days to occur. For example, the activation of sexual behavior in ovariectomized female rats following estradiol injection takes 24 to 48 hours. This enhancement of sexual behavior can be blocked by drugs such as anisomycin which inhibits protein synthesis. For example, the sexual behavior of female rats treated with estradiol and progesterone can be blocked by injections of anisomycin into the ventromedial hypothalamus (VMH), indicating that the stimulation of sexual behavior by gonadal steroids requires protein synthesis in the VMH (Rainbow *et al.* 1982). Blocking protein synthesis also inhibits estradiol stimulation of the preovulatory LH surge (positive feedback on LH), the negative feedback effects of corticosteroids on ACTH and the effects of progesterone on female sexual behavior (McEwen *et al.* 1979).

9.9 Functions of steroid hormone modulation of nerve cells

Steroid hormone modulation of nerve cells can take many forms, from the well-described feedback regulation of pituitary hormone secretion, the regulation of food intake and body weight (Clegg 2012), to the possibility that estradiol might interfere with the progression of Alzheimer's Disease (Behl 2002). In this section, we will summarize five different functions: (1) the feedback regulation of hypothalamic and pituitary hormone secretion; (2) neuropathological and neuroprotective effects of steroids; (3) the modulation of emotional, motivational, sensory and behavioral changes; (4) the regulation of adaptive behaviors for coping with stress; and (5) the organization of neural pathways in the brain during perinatal development.

9.9.1 Feedback regulation of hypothalamic and pituitary hormone secretion

As discussed in detail in Chapter 8 (section 8.2.3), steroid hormones exert feedback effects on hypothalamic and pituitary hormone release. Steroid feedback is usually negative (see Figure 8.9), with the exception of a positive feedback effect of estradiol at the midpoint of the menstrual cycle. Steroid hormone feedback is linked to neurohormone/hormone release through modulation of neurotransmitter release in the hypothalamus and other brain regions. Steroid hormone feedback controls neurohormone secretion by regulating the rate of synthesis, release and degradation of hypothalamic hormones, by altering the firing rate of neurons which control neurohormone release from the hypothalamus, and by regulating the number of receptor proteins in the cells of the pituitary gland, thus altering the responsiveness of these cells to hypothalamic hormones. For example, estradiol increases the number of pituitary GnRH receptors, so that the pituitary is maximally responsive to GnRH during the midcycle LH surge (Strauss and Barbieri 2004).

The feedback effects of gonadal steroids on hypothalamic and pituitary hormones are illustrated in several figures in Chapter 6. In addition, and using the reproductive axis as an example, there are also sex differences in the pattern of GnRH and LH secretion and in the effects of gonadal steroid feedback on the release of these hormones. *Adult males* exhibit pulsatile secretion of GnRH, LH and FSH (Figure 4.4) and castration (i.e. removal of testosterone negative feedback) is followed by an increase in the pulsatile secretion of these hormones. Testosterone treatment (adding negative feedback) inhibits pulsatile secretion and the circulating levels of gonadotropic hormones as shown in Figure 8.10. The most likely neural sites for the negative feedback action of testosterone are in the hypothalamus (see section 9.6.3). In contrast to males, *adult females* have a *cyclic* release of GnRH, LH and FSH and this is responsible for the reproductive cycle. In female rodents, and in women, estradiol feedback can either increase or decrease the secretion of these hormones, depending on the dose and treatment schedule. Thus, negative feedback of estradiol and progesterone is as shown in Figure 8.9; this is the type of feedback obtained during use of oral contraceptives. On the other hand, positive feedback of estradiol generates a surge of LH secretion, but only under certain hormonal conditions. Neurons of the medial basal hypothalamus (called the pulse generator) control the baseline pulsatile secretion of GnRH and neurons in the preoptic area (of the rat) or medial basal hypothalamus (in primates) provide the stimulus for the ovulatory surge of GnRH. In primates, the site of estradiol positive feedback is at the anterior pituitary *and* the medial basal hypothalamus (Plant 2008; for further reading, see Plant 1986). Of course, other factors also influence the release of GnRH. During stress, for example, the increased release of glucocorticoids from the adrenal glands inhibits firing of the hypothalamic neurons responsible for the GnRH rhythm, thus reducing the amount of LH released (see Figure 6.7). This figure also reveals that *kisspeptin*, a new hypothalamic factor, positively controls GnRH secretion and has become the focus of much clinical interest (for further reading, see Pineda *et al.* 2010).

This outline of feedback effects of sex steroids on the reproductive system serves as an example of the many effects of other steroids, such as glucocorticoids, which we have covered already (see Chapter 6 for further details).

9.9.2 Steroids and neuropathology/neuroprotection

In section 9.7, we described the importance of adrenal steroid hormone receptors (GR and MR) in the homeostatic control of the stress response. This system can be regarded as protective under acute conditions and such short-term adaptation has been termed *allostasis* (McEwen and Gianaros 2011). In contrast, under chronically stressful conditions (such as following a sexual assault, or caring for a spouse with dementia), the body and brain undergo maladaptive changes that increase vulnerability to brain pathophysiology, including mental health. This is called *allostatic load* (for further reading, see Cohen *et al.* 2007; McEwen and Gianaros 2010). Animal experiments have provided valuable information on the effects of glucocorticoids on brain function. For example, different behavioral effects occur depending on whether the applied stress is acute or chronic. Acute release of glucocorticoids has a positive influence on some behaviors, whereas sustained stress induces deficits (e.g. in spatial working memory; McEwen and Gianaros 2011). Chronic stress also has deleterious structural effects in the brain and high levels of glucocorticoids render hippocampal neurons vulnerable to excitotoxic damage from the neurotransmitter glutamate (Lu *et al.* 2003). Hippocampal function also seems to be compromised through glucocorticoid-induced dendritic atrophy and loss of synapses (McEwen and Gianaros 2011). In terms of the receptors involved, specific activation of GR has serious effects on hippocampal cell loss and atrophy of dendrites, whereas concomitant GR and MR activation is much less damaging. Note that other brain areas possessing GR and MR, such as amygdala, are similarly affected (Sousa *et al.* 2008). McEwen and Gianaros (2010) have suggested that the known involvement of these brain regions should enable us to target them – for example, through exercise – in efforts to deal with some physical and mental health issues. Similarly, de Kloet and co-workers (de Kloet *et al.* 2005) review some clinical studies that indicate that blockade of GR with antagonists ameliorates some forms of depression.

In marked contrast to the effects of glucocorticoids, steroids such as *estradiol and progesterone* may offer a degree of neuroprotection. For example, in animal studies, progesterone is effective in promoting and enhancing repair mechanisms that follow traumatic brain injury and stroke (Stein 2008; Liu *et al.* 2012). Estradiol has been studied in more detail (Simpkins *et al.* 2012) and much is known about the stimulation of neurogenesis not only in the hippocampus, but in brain areas such as hypothalamus and amygdala that also contain estradiol receptors (Fowler *et al.* 2008). Perhaps the most important clinical studies have focused on the possibility that estradiol may be beneficial in the treatment of Alzheimer's Disease. Both women and men are at risk for Alzheimer's Disease because of the age-related loss of sex steroid hormones and there is evidence that estradiol or testosterone replacement may be a valuable hormone therapy (Pike *et al.* 2009; Amtul *et al.* 2010). (For further reading, see Behl 2002; Garcia-Segura and Balthazart 2009; and Azcoitia *et al.* 2011.) Janicki and Schupf (2010) concluded that: "Although individual study results are inconsistent, overall data from epidemiologic studies, observational studies and clinical trials of HRT (hormone replacement therapy), studies of endogenous hormones, and evaluations of genetic variants involved in estrogen biosynthesis and receptor activity indicate that estrogen plays an important role in the pathogenesis of cognitive decline and risk for AD in both men and women."

Synthetic chemical pollutants that mimic estradiol and testosterone are of great clinical concern (see Chapter 15). These chemicals, sometimes called "Neuroendocrine Disruptors," seem unavoidable and they are present in soy supplements consumed by humans (phytoestrogens), they leach from plastic water bottles (bisphenol A), are present in tap water, house dust, toys, fabrics, cookware and food containers. Yet others have deleterious effects on thyroid function (Gore and Patisaul 2010; for further reading, see Wuttke *et al.* 2010). The possibility that human endocrine systems might be disregulated by such chemicals, especially during development, is a serious issue that is attracting increasing attention (Nadal 2012; Andersson *et al.* 2014).

9.9.3 Modulation of emotional, motivational, sensory and behavioral changes

In adult animals, sex hormones activate the neural pathways which coordinate sexual and parental behaviors as well as a wide range of non-sexual behaviors, including

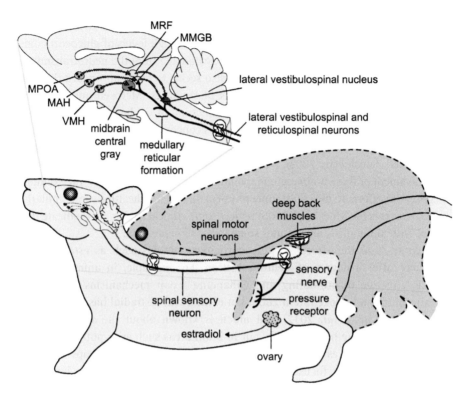

Figure 9.16 Schematic illustration depicting male-induced sensory and motor signals that regulate female sexual response (lordosis) in estradiol-primed female rats

Hypothalamic neurons in the female are stimulated both by sensory inputs and by circulating estradiol. When the male stimulates the skin of the female flanks and rump, pressure receptors provide sensory input to the spinal cord via the dorsal root ganglia. This information is then relayed via the spinal cord to the reticular formation in the medulla and the midbrain central gray area. In the presence of estradiol, neurons of the ventromedial hypothalamus (VMH) are activated and those of the medial preoptic area (MPOA) are inhibited. Axons from the VMH descend to the midbrain central gray area, the midbrain reticular formation (MRF) and the medial (magnocellular) region of the medial geniculate body (MMGB). The motor neurons for lordosis (the lateral vestibulospinal and the reticulospinal neurons) are then activated. These spinal motor neurons stimulate the lateral longissimus dorsi and the transverso spinalis (multifidus) muscles of the lumbar region that causes dorsiflexion of the rump and tailbase, resulting in the lordosis response.

locomotor activity, play, aggression, feeding, scent-marking, sensory perception, learning and memory (Beatty 1979). Male sexual behavior and other male-specific behaviors such as scent-marking are regulated by androgen and estrogen receptors in the medial preoptic area and anterior hypothalamus (McEwen *et al.* 1979; Hull and Dominguez 2007). Female sexual behavior is activated by the actions of estrogen and progesterone in the ventromedial hypothalamus, medial preoptic area, anterior hypothalamus and midbrain central gray (McEwen 1981; Blaustein 2008a; 2008b). Figure 9.16 shows a model for the neural and endocrine interactions in the control of sexual behavior in the female rat that can be briefly summarized as follows. The sensory input from the male, via the cutaneous stimulation of the somatosensory (touch) receptors on the female's flanks, is carried to the spinal cord by the pudendal nerve. The spinal sensory nerves then stimulate the reticular formation in the medulla and the midbrain central gray area of the brain. Estradiol-activated neurons in the ventromedial hypothalamus send axons to the midbrain central gray area, the reticular formation and the medial geniculate body. When activated by the incoming tactile stimulation, these estrogen-primed neurons activate spinal motor neurons in the reticular formation. These spinal motor neurons contract the lateral longissimus dorsi and multifidus muscles, resulting in the *lordosis reflex*, the sexual posture of the female rat (Pfaff 1989; Pfaff *et al.* 2008). If a male approaches a non-estrous female rat (i.e. a female with a low estradiol level), she is aggressive and does not show sexual behavior. However, if she is primed with estradiol and progesterone injections, she responds to his touch by arching her back in the "lordosis posture" and allowing him to mount.

Female parental behavior is elicited by estradiol and progesterone acting at the medial preoptic area in the rat (Numan 2007; Pfaff *et al.* 2011). Cyclic fluctuations of estradiol and progesterone that occur during the female reproductive cycle also induce changes in a wide range of behaviors, including aggression, locomotion, food preferences, emotional state, olfactory and visual perception (Messent 1976; Rubinow and Schmidt 2006; Asarian and Geary 2006; Bos *et al.* 2012).

9.9.4 Regulation of adaptive behaviors for coping with stress

Glucocorticoids are released in stressful situations, resulting in very high levels of these steroids in the circulation (see section 9.7). Stress-induced increases in glucocorticoid levels provide negative feedback to the brain and immune system and dampen the over-activity of the hypothalamic-pituitary-adrenal system, thus protecting the body from its own endocrine and immune responses. The role of glucocorticoids in regulating immune responses to stress is described in more detail in Chapter 13 (section 13.6.2).

Neural responses to glucocorticoids provide behavioral as well as neuroendocrine defenses against excessive stress responses (see sections 9.7.2 and 9.7.3). Glucocorticoids influence learning and memory in aversive situations, such as taste aversion and passive avoidance conditioning, and this may be mediated through modulation of norepinephrine and serotonin release in cells with type II glucocorticoid receptors (McEwen *et al.* 1986; de Kloet *et al.* 2005; 2008). Corticosteroids also facilitate exploration of a novel environment, possibly through their action on hippocampal type I glucocorticoid receptors.

Corticosteroids influence mood, sleep, sensory perception and stress-related behaviors by modulating the release of neurotransmitters such as catecholamines (McEwen 1981; de Kloet *et al.* 2005). Prolonged stress-induced hypersecretion of glucocorticoids, however, suppresses neural activity and may cause neural degeneration (cell loss) due to the inhibition of glucose uptake and protein synthesis and the depletion of energy reserves in glucocorticoid target cells (see sections 9.7.2 and 9.7.3). Thus, prolonged glucocorticoid secretion may also lead to stress-related diseases such as depression and other affective disorders by altering dopamine levels at glucocorticoid sensitive cells in the hippocampus, thus inhibiting their neural functioning (McEwen *et al.* 1986). High levels of glucocorticoids may also accelerate brain aging and inhibit mental performance (Yau and Seckl 2012). It is, therefore, essential for normal emotional and cognitive functioning to have an efficient mechanism for regulating the adrenocortical stress response. Chronic stress may interfere with this adaptive response (McEwen *et al.* 1986).

9.9.5 Steroid-induced organization of neural pathways during pre- and perinatal development

During prenatal and early postnatal mammalian development, sex hormones influence the growth and differentiation of neurons, the formation of synapses and the synthesis of hormone and neurotransmitter receptors, all resulting in sex differences in the structure of the adult brain (McCarthy 2010). Major structural differences in the human brain are thought to account for sex differences in human behavior (Ingalhalikar *et al.* 2014; Cahill 2014). For example, men have better motor and spatial abilities, whereas females possess enhanced memory and social cognition skills. Structural variations also account for sex differences in gonadotropin secretion, such as in the female menstrual cycle, but may also determine sexual orientation and the prevalence of adult neurological and psychiatric diseases. For example, anorexia nervosa (93 percent) and bulimia (75 percent) are predominantly seen in women, whereas Tourette's Syndrome (90 percent) and autism (80 percent) largely occur in men (Swaab *et al.* 2003; Schaafsma and Pfaff 2014). What causes these differences to occur? Testosterone secreted from the testes causes masculinization of the brain during a sensitive period around the time of birth in rodents, although this is likely to occur during fetal life in children (Swaab *et al.* 2003). In contrast, the *absence* of testosterone in females permits the development of a female brain, behavior and reproductive system. However, as noted in section 9.6.4, testosterone is really a prohormone and is converted to estradiol in the brain. It is now accepted that estradiol, the female sex hormone, is responsible for the effects of testosterone on sexual differentiation of the brain (McCarthy *et al.* 2009). Indeed, treatment of neonatal female rat pups with estradiol, or testosterone, can masculinize the female brain. Whether this is true in the human brain is unknown, but in the rhesus monkey brain the conversion of testosterone to estradiol is not important. In other words, masculinization can be achieved by androgens such as dihydrotestosterone, which cannot be metabolized to estradiol (Thornton *et al.* 2009).

What are the mechanisms by which sex hormones modify brain function? Profound sex differences in the size of various hypothalamic nuclei have been reported. The most obvious of these, first observed in birds, is the sexually dimorphic nucleus (SDN) (McCarthy *et al.* 2009). The SDN is several-fold larger in males than in females, and the female SDN can be enlarged following estradiol treatment at birth. Many other brain sites have now been found to be different in males and females (Simerly 2002) and some evidence suggests similar differences in the human brain (LeVay 1991; Swaab *et al.* 2003). Sex differences in neurochemistry are also readily observed, and a good example is the distribution of ERβ in the anteroventral area of the hypothalamus. There are far more ERβ-positive neurons in the female brain compared to males (Orikasa *et al.* 2002). Simerly (2002) concludes that estradiol not only affects the number of neurons in various nuclei, but also regulates many developmental changes, including neurogenesis, neural connectivity, synaptogenesis, cell death and neurotransmitter plasticity. An example of the effect of estradiol on neural tissue is shown in Figure 9.17. The figure reveals a pronounced stimulation of neurite outgrowth from neurons of the mouse hypothalamus in tissue culture (Toran-Allerand 1980). Investigations into the cell signaling events that regulate these processes are beginning to yield fascinating mechanistic insights. For

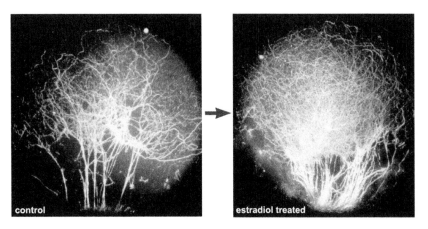

Figure 9.17 Estradiol stimulates neurite outgrowth in hypothalamic neurons
Hypothalamic tissue fragments in tissue culture were treated with low concentrations of estradiol.
This hormone accelerated and enhanced neuritic proliferation. Reproduced with permission (Toran-Allerand 1980).

example, McCarthy and co-workers have so far implicated growth factors, prostaglandins and epigenetic influences (for further reading, see McCarthy 2010; and McCarthy *et al.* 2009).

Glucocorticoids may also be capable of modifying the immature brain. As we discussed in section 9.9.2, chronic stress has deleterious structural effects in the *adult* brain, including glucocorticoid-induced dendritic atrophy and loss of synapses. It seems logical that neurons in the fetal and neonatal brain may also be vulnerable (Matthews 2001). In fact, there is growing evidence that exposure of the embryonic brain to glucocorticoids has lifelong effects, not only on neuroendocrine function, but on the development of obesity and diabetes (Reynolds 2010). This is especially relevant given that synthetic glucocorticoids, such as dexamethasone, are routinely administered during human pregnancy to reduce morbidity and mortality in preterm infants by accelerating fetal lung maturation (Tegethoff *et al.* 2009). Experiments in animals have revealed that prenatal glucocorticoid treatment reduces hippocampal GR and MR mRNA and induces neuronal hippocampal degeneration, and human data indicate that hippocampal volume may be reduced (Matthews 2001). Of great clinical significance is the fact that these effects of synthetic glucocorticoids can also be seen in offspring of animals subjected to certain types of stress during pregnancy or in the neonatal period; that is, when the fetus/newborn is exposed to *endogenous* glucocorticoids, for example as a result of under-nutrition *in utero*, or maternal stress. In humans, there seems no doubt that smaller size at birth, and perhaps during maturation, results in elevated rates of coronary heart disease, stroke, Type 2 diabetes, adiposity and osteoporosis in adult life (Gluckman *et al.* 2008; for further reading, see Seckl and Meaney 2004).

Much less is known about the possible effects of progesterone on the pre- and neonatal brain. Progesterone receptors (PR) are present in the fetal brain, particularly in the cerebral cortex, and marked sex differences have been reported. These data suggest that PR may play an important role in neural development (Wagner 2008). The human fetus and neonate are often exposed to progesterone or progesterone-like molecules (progestins). For example, progesterone is frequently used in attempts to prevent preterm birth (Meis and Connors 2004) and there is evidence that the children of these mothers have higher IQs and develop earlier than untreated babies (Wagner 2008). The routine exposure of the human fetus, and neonate, to progestins should guide future research into the precise effects of these steroids on brain maturation.

9.9.6 Are steroid hormones ("neurosteroids") synthesized in the brain?

This chapter has emphasized the crucial importance of steroid hormones, and their receptors, in the regulation of many aspects of the neuroendocrine system. Steroid hormones, released from tissues such as ovary and adrenal, act as circulating hormones and target many brain regions, but especially the hypothalamic-pituitary system. However, an additional site of of steroid hormone synthesis is the brain itself (Baulieu *et al.* 2001; Taves *et al.* 2011). Note that steroid hormones released from neurons and glial cells (called *neurosteroids*) cannot be classed as hormones since they probably act in an autocrine or paracrine fashion within the brain and in many ways behave like neurotransmitters or neuromodulators (for an interesting development of this idea – *synaptocrine signaling* – and for further reading, see Saldanha *et al.* 2011). There is no doubt that brain cells possess the appropriate steroidogenic enzymes necessary for biosynthesizing steroids from precursors such as cholesterol, although this is not the sole route of production (Melcangi *et al.* 2008; Melcangi and Panzica 2014). For example, we already discussed the central synthesis of estradiol from testosterone acting as a prohormone (see section 9.6.4), and brain progesterone ("neuroprogesterone") synthesis also appears to be regulated by estradiol (Micevych and Sinchak 2008). But given the evidence that neurons and glia are capable of biosynthesizing steroid hormones such as estradiol, what would be the purpose of two sources of the hormone? In other words, since estradiol of ovarian origin readily enters the brain, the amount of brain estradiol must represent a mixture of gonadal and brain-derived estradiol. Many early experiments revealed the presence of brain-derived steroid hormones in surgically prepared animals lacking gonads and adrenals; that is, the only possible source of hormone was the brain. This might suggest that estradiol of brain origin could be important in conditions where peripheral estradiol levels were low or absent; for example, following the menopause when estradiol production by the ovaries is almost non-existent. Yet, other experiments have shown that when estradiol synthesis is blocked in cultured hippocampal neurons, there is a pronounced decrease in the number of dendritic spines and spine synapses (Rune and Frotscher 2005). This result reveals the importance of locally synthesized estradiol, but since the experiments were performed on hippocampal cells *in vitro*, in the absence of exogenous estradiol, the results cannot distinguish the relevance of brain versus peripheral estradiol. Overall, there is sound evidence that steroid hormones are made in the brain, but the true functional purpose of neurosteroids is far from clear.

9.10 Thyroid hormone receptors in the brain

In section 2.2.3 and Figure 6.6, we outlined the regulation of thyroid hormone secretion. The two major hormones released by the thyroid gland are T3 and T4. The receptor for these hormones is intracellular and is a member of the steroid receptor superfamily described in section 9.1. Thyroid hormones target most tissues in the body (e.g. liver, skeletal and cardiac muscle and fat cells), but they also bind to receptors in the brain and pituitary gland. T3 and T4 do not have a steroidal structure, but biochemically they behave in a similar way. T4 acts as a prohormone, and is normally converted to the biologically active hormone T3 by deiodinase enzymes in target cells (Figure 9.18; Dayan and Panicker 2009). The thyroid hormone receptor (THR) is therefore specific for T3. The T3/THR receptor/ hormone complex acts as a transcription factor. However, unlike estradiol, it does not dimerize with itself, but forms a dimer with a different molecule, the retinoid X receptor (RXR). This dimer, together with co-regulators, binds to hormone response elements on target genes, just as we saw for estradiol and progesterone. The chain of events that takes T3 and T4 from the bloodstream to the nuclear receptor is shown in Figure 9.18. In the bloodstream, T3 and T4 are reversibly bound to thyroid hormone binding globulin (TBG) and only the very small amounts of free, non-bound hormones can enter their target cells via specific transporters. When T4 enters the target cell, it is converted to T3 by two deiodinase enzymes, D1 and D2. When T3 and T4 enter the brain, they do so via specific transporters at the blood-brain barrier. Thyroid hormone receptors, like estradiol receptors, exist in several subtypes. Thus, in the fetal brain, TRα1 is the predominant subtype, but after birth the major subtype is TRβ1, but with TRβ2 being the primary receptor in hypothalamus (PVN) and pituitary (Li and Boyages 1996). This means that the negative feedback effects of T3/T4 on TSH secretion are mediated through the TRβ2 receptor (Harvey and Williams 2002). The brain and pituitary of rats and humans have also been mapped immunocytochemically for the presence of THR protein (Lechan *et al.* 1994; Alkemade *et al.* 2005). Of interest was the demonstration that THR immunoreactivity was also localized to many other regions of the rat brain, including the olfactory bulb, hippocampus, amygdala, neocortex and cerebellum (Puymirat *et al.* 1991). THR was observable in neurons and glial cells. In section 9.4.1, we considered the evidence that neurons could respond rapidly to steroids via steroid receptors located in the cell membrane. There is now evidence that thyroid hormone receptors may also exist in neuronal cell membranes (for further reading, see Davis *et al.* 2010).

9.10.1 Action of thyroid hormones at target cells

Thyroid hormones have diverse effects on target cells throughout the body. They influence basal metabolic and respiration rates, oxygen consumption, carbohydrate and protein metabolism, thermogenesis and sodium pump activity, and provide negative feedback regulation of TRH and TSH secretion (Boron and Boulpaep 2005). Thyroid hormones also regulate many aspects of growth and differentiation and are essential for stimulating the maturation of the brain, skeleton, heart and lungs during prenatal and early postnatal development. Abnormal thyroid hormone secretion can have profound pathological effects in children and adults. For example, low levels of T3/T4 (*hypothyroidism*) in the fetus or newborn can cause mental retardation, short stature and delay in

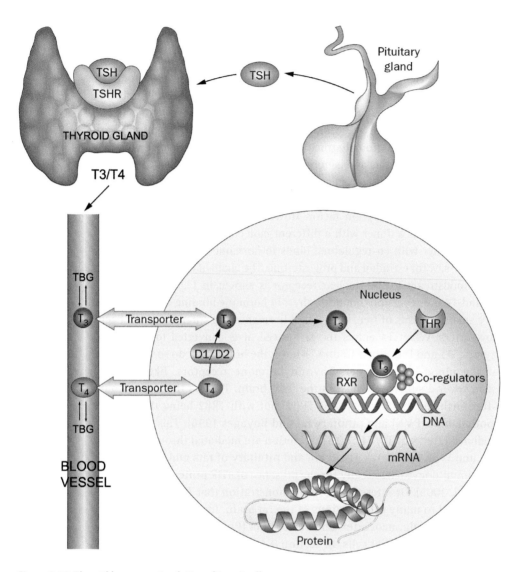

Figure 9.18 Thyroid hormone stimulation of target cells
Thyroid-stimulating hormone (TSH), released from the anterior pituitary gland, stimulates the thyroid gland via membrane TSH receptors (TSHR). The thyroid responds by secreting thyroxine (T4) and triiodothyronine (T3) that are transported in the blood bound to the thyroid hormone binding globulin (TBG). Free T3/T4 enter target cells via specific transporters. T4 is then converted to the active hormone T3 by deiodinases D1 and D2. T3, the bioactive hormone, binds to thyroid hormone receptors (THR) that dimerize and complex with target DNA. This final step activates gene transcription (mRNA) and the production of specific proteins. Reproduced with permission (Dayan and Panicker 2009).

motor development. Early intervention with T4 treatment, or treatment of the mother with T4 during pregnancy, can largely avoid these problems (Gruters and Krude 2012). In adults, hypothyroidism causes weight gain, lethargy, mental slowness and a decrease in metabolic rate. These symptoms are treatable with T4. In contrast, *high levels* of T3/T4 (*hyperthyroidism*) induce weight loss, tremors, sweating, muscle weakness and increased cardiac output. These patients are treated with drugs that reduce the synthesis of T3 and T4.

Important studies in mutant mice that lack various components of the thyroid feedback system have revealed many interesting insights into certain human pathologies. This work is in its infancy, but already there are clues as to how thyroid hormone receptors may be implicated in deafness and blindness. Other mutants suggest that THR are implicated in some forms of behavior, including sex and memory, and abnormal TRβ mutants appear to have a form of attention deficit hyperactivity disorder (for further reading, see Williams 2008).

9.11 Summary

Steroid and thyroid hormone receptors belong to a superfamily of nuclear receptor proteins. The target cells that possess these nuclear receptors can be identified by *in situ* hybridization (to detect receptor mRNA) and by immunocytochemistry (to detect the receptor protein). Steroid and thyroid hormones bind to specific receptors in the cytosol of their target cells and the hormone-receptor complex then binds to DNA and will activate mRNA and protein synthesis in these target cells, but not in non-target cells. Steroid receptors are found in many brain regions. Receptors also exist in the cell membrane to permit rapid responses to steroids. Estradiol, progesterone, testosterone, glucocorticoids, mineralocorticoids and thyroid hormones all have target cells in the brain and pituitary gland. Estradiol priming induces progesterone receptors and activates female sexual and parental behavior, while androgens (testosterone) stimulate male sexual behavior, aggression and other behaviors. Testosterone can act by direct action on androgen receptors, by reduction to dihydrotestosterone (DHT) and activation of androgen receptors, or by aromatization to estradiol and activation of estradiol receptors. Estradiol and progesterone receptors often occur in the same neural areas, as do estradiol and androgen receptors. There are two types of corticosteroid receptors: MR which bind to both mineralo-corticoids and glucocorticoids; and GR which bind only to glucocorticoids. The neural locations and functions of these two receptors differ, with GR more widely distributed in the brain than MR, and GR involved in more severe stress responses. Steroid hormones are able to alter electrophysiological activity of neural cells, regulate the release of neurotransmitters and influence protein synthesis. Neurons may also biosynthesize steroids and these "neurosteroids" may act on membrane receptors to modulate ion channels in the nerve cell membrane. As well as affecting the electrical activity and release of neurotransmitters from nerve cells, steroid hormones modulate the synthesis and release of hypothalamic and pituitary hormones and regulate the number of receptors for neurotransmitters and hormones. Steroid hormones may also be involved in neuroprotection (estradiol), but can also cause neuronal damage (glucocorticoids). Steroid and thyroid hormones also modulate emotional, motivational and behavioral changes and regulate adaptive responses to stress. Sex hormones (estradiol and testosterone) and thyroid hormones (thyroxine) play a major role in organizing neural development, especially in terms of sexual differentiation of the brain.

FURTHER READING

Azcoitia, I., Arevalo, M.-A., De Nicola, A. F. and Garcia-Segura, L. M. (2011). "Neuroprotective actions of estradiol revisited," *Trends Endocr Metab* 22, 467–473.

Behl, C. (2002). "Oestrogen as a neuroprotective hormone," *Nat Rev Neurosci* 3, 433–442.

Bookout, A. L., Jeong, Y., Downes, M., Yu, R. T., Evans, R. M. and Mangelsdorf, D. J. (2006). "Anatomical profiling of nuclear receptor expression reveals a hierarchical transcriptional network," *Cell* 126, 789–799.

Brinton, R. D., Thompson, R. F., Foy, M. R., Baudry, M., Wang, J., Finch, C. E. *et al.* (2008). "Progesterone receptors: form and function in brain," *Front Neuroendocr* 29, 313–339.

Casadesus, G. (2010). "Special issue on estrogen actions in the brain," *Biochim Biophys Acta* 1800, 1029.

Chrousos, G. P. (2009). "Stress and disorders of the stress system," *Nat Rev Endocrinol* 5, 374–381.

Cohen, S., Janicki-Deverts, D. and Miller, G. E. (2007). "Psychological stress and disease," *JAMA* 298, 1685–1687.

Davis, P. J., Zhou, M., Davis, F. B., Lansing, L., Mousa, S. A. and Lin, H. Y. (2010). "Mini-review: cell surface receptor for thyroid hormone and nongenomic regulation of ion fluxes in excitable cells," *Physiol Behav* 99, 237–239.

de Kloet, E. R., Joels, M. and Holsboer, F. (2005). "Stress and the brain: from adaptation to disease," *Nat Rev Neurosci* 6, 463–475.

de Kloet, E. R., Karst, H. and Joels, M. (2008). "Corticosteroid hormones in the central stress response: quick-and-slow," *Front Neuroendocr* 29, 268–272.

de Kloet, E. R., Vreugdenhil, E., Oitzl, M. S. and Joels, M. (1998). "Brain corticosteroid receptor balance in health and disease," *Endocr Rev* 19, 269–301.

Foradori, C. D., Weiser, M. J. and Handa, R. J. (2008). "Non-genomic actions of androgens," *Front Neuroendocr* 29, 169–181.

Funder, J. W. (2010). "Minireview: aldosterone and mineralocorticoid receptors: past, present, and future," *Endocr* 151, 5098–5102.

Garcia-Segura, L. M. and Balthazart, J. (2009). "Steroids and neuroprotection: new advances," *Front Neuroendocr* 30, v–ix.

Gluckman, P. D., Hanson, M. A., Cooper, C. and Thornburg, K. L. (2008). "Effect of in utero and early-life conditions on adult health and disease," *N Engl J Med* 359, 61–73.

Gosselin, D. and Rivest, S. (2011). "Estrogen receptor transrepresses brain inflammation," *Cell* 145, 495–497.

Hewitt, S. C., Harrell, J. C. and Korach, K. S. (2005). "Lessons in estrogen biology from knockout and transgenic animals," *Annu Rev Physiol* 67, 285–308.

Junpeng, M., Huang, S. and Qin, S. (2011). "Progesterone for acute traumatic brain injury," *Cochrane Database Syst Rev* Jan 19(1), CD008409.

Kelly, M. J. and Qiu, J. (2010). "Estrogen signaling in hypothalamic circuits controlling reproduction," *Brain Res* 1364, 44–52.

Kelly, M. J. and Ronnekleiv, O. K. (2008). "Membrane-initiated estrogen signaling in hypothalamic neurons," *Mol Cell Endocrinol* 290, 14–23.

Kopchick, J. J., Sackmann-Sala, L. and Ding, J. (2007). "Primer: molecular tools used for the understanding of endocrinology," *Nat Clin Pract Endocrinol Metab* 3, 355–368.

McCarthy, M. M., Auger, A. P., Bale, T. L., De Vries, G. J., Dunn, G. A., Forger, N. G. *et al.* (2009). "The epigenetics of sex differences in the brain," *J Neurosci* 29, 12815–12823.

McEwen, B. S. and Gianaros, P. J. (2010). "Central role of the brain in stress and adaptation: links to socioeconomic status, health, and disease," *Ann NY Acad Sci* 1186, 190–222.

Melcangi, R. C. and Panzica, G. C. (2006). "Neuroactive steroids: old players in a new game," *Neurosci* 138, 733–739.

Picard, D. (2006). "Chaperoning steroid hormone action," *Trends Endocrinol Metab* 17, 229–235.

Pineda, R., Aguilar, E., Pinilla, L. and Tena-Sempere, M. (2010). "Physiological roles of the kisspeptin/GPR54 system in the neuroendocrine control of reproduction," *Prog Brain Res* 181, 55–77.

Plant, T. M. (1986). "Gonadal regulation of hypothalamic gonadotropin-releasing hormone release in primates," *Endocr Rev* 7, 75–88.

Saldanha, C. J., Remage-Healey, L. and Schlinger, B. A. (2011). "Synaptocrine signaling: steroid synthesis and action at the synapse," *Endocr Revs* 32, 532–549.

Seckl, J. R. and Meaney, M. J. (2004). "Glucocorticoid programming," *Ann NY Acad Sci* 1032, 63–84.

Sousa, N., Cerqueira, J. J. and Almeida, O. F. (2008). "Corticosteroid receptors and neuroplasticity," *Brain Res Rev* 57, 561–570.

Stanisic, V., Lonard, D. M. and O'Malley, B. W. (2010). "Modulation of steroid hormone receptor activity," *Prog Brain Res* 181, 153–176.

Williams, G. R. (2008). "Neurodevelopmental and neurophysiological actions of thyroid hormone," *J Neuroendocrinol* 20, 784–794.

Woolley, C. S. (1999). "Electrophysiological and cellular effects of estrogen on neuronal function," *Crit Rev Neurobiol* 13, 1–20.

Wuttke, W., Jarry, H. and Seidlova-Wuttke, D. (2010). "Definition, classification and mechanism of action of endocrine disrupting chemicals," *Hormones (Athens)* 9, 9–15.

Zheng, P. (2009). "Neuroactive steroid regulation of neurotransmitter release in the CNS: action, mechanism and possible significance," *Prog Neurobiol* 89, 134–152, http://en. wikipedia.org/wiki/Fluorescent, *in situ hybridization*.

REVIEW QUESTIONS

9.1   What specific cellular characteristic differentiates a hormone target cell from a non-target cell?

9.2   With regard to the biological activity of testosterone, what is the aromatization hypothesis?

9.3   What are the three pathways by which testosterone regulates target cells?

9.4   Aldosterone binds to which of the two types of corticosteroid receptor?

9.5   "Steroid hormones are exclusively biosynthesized in endocrine glands, such as the adrenals." True or false?

9.6   What is the difference between genomic and non-genomic activity of steroid hormones?

9.7   Steroid hormone receptors can be localized to at least two distinct cellular sites. Where are they?

9.8   "Estradiol stimulates an increase in the number of progesterone receptors in the ventro-medial hypothalamus and preoptic area." True or false?

9.9   Which area of the rat brain (hypothalamus) regulates female sexual behavior when activated by estradiol?

9.10   *Briefly (two sentences)* describe the significance of the term: "intracellular receptor superfamily."

9.11   Name the *three* types of estradiol receptor.

9.12   Receptor proteins can be localized by immunocytochemistry. Which technique can be used to determine the cellular site of receptor gene expression?

9.13   "Intracellular steroid hormone receptors have no intrinsic biological activity." True or false?

9.14   Give three examples of changes in behavior induced by steroid hormones.

9.15   Steroid hormone feedback into the brain can regulate releasing hormone secretion. Give two mechanisms by which this is achieved.

9.16   Which thyroid hormone binds to nuclear receptors in thyroid hormone target cells?

9.17   In general, what effect does the prolonged exposure of target cells to a high concentration of hormone have on the hormone receptors?

ESSAY QUESTIONS

9.1   Describe the "superfamily" of nuclear receptors for steroid and thyroid hormones.

9.2   Discuss the importance of testosterone acting as a prohormone in the activation of male sexual behavior.

9.3   Discuss the role of steroid hormones in the organization of neural pathways during perinatal development.

9.4   Describe the distinction between type I and type II adrenal steroid hormone receptors in the brain with regard to their anatomical distribution and function.

9.5   How do steroid hormones influence the electrical activity of neurons?

9.6   Describe the origin and possible function of neurosteroids.

9.7   "Estradiol acts as a neuroprotectant that could be effective in preventing Alzheimer's Disease." Critically evaluate this hypothesis.

REFERENCES

Alkemade, A., Vuijst, C. L., Unmehopa, U. A., Bakker, O., Vennstrom, B., Wiersinga, W. M. *et al*. (2005). "Thyroid hormone receptor expression in the human hypothalamus and anterior pituitary," *J Clin Endocrinol Metab* 90, 904–912.

Amtul, Z., Wang, L., Westaway, D. and Rozmahel, R. F. (2010). "Neuroprotective mechanism conferred by 17beta-estradiol on the biochemical basis of Alzheimer's disease," *Neurosci* 169, 781–786.

Andersson, A.-M., Frederiksen, H., Grigor, K. M., Toppari, J. and Skakkebaek, N. E. (2014). "Special issue on the impact of endocrine disrupters on reproductive health," *Reprod* 147, E1.

Arbogast, L. A. (2008). "Estrogen genomic and membrane actions at an intersection," *Trends Endocrinol Metab* 19, 1–2.

Asarian, L. and Geary, N. (2006). "Modulation of appetite by gonadal steroid hormones," *Phil Trans R Soc B* 361, 1251–1263.

Auger, A. P., Moffatt, C. A. and Blaustein, J. D. (1997). "Progesterone-independent activation of rat brain progestin receptors by reproductive stimuli," *Endocr* 138, 511–514.

Baulieu, E. E., Robel, P. and Schumacher, M. (2001). "Neurosteroids: beginning of the story," *Int Rev Neurobiol* 46, 1–32.

Beatty, W. W. (1979). "Gonadal hormones and sex differences in nonreproductive behaviors in rodents: organizational activational influences," *Horm Behav* 12, 112–163.

Behl, C. (2002). "Oestrogen as a neuroprotective hormone," *Nat Rev Neurosci* 3, 433–442.

Belelli, D., Herd, M. B., Mitchell, E. A., Peden, D. R., Vardy, A. W., Gentet, L. *et al.* (2006). "Neuroactive steroids and inhibitory neurotransmission: mechanisms of action and physiological relevance," *Neurosci* 138, 821–829.

Blaustein, J. D. (2008a). "Progesterone and progestin receptors in the brain: the neglected ones," *Endocr* 149, 2737–2738.

Blaustein, J. D. (2008b). "Neuroendocrine regulation of feminine sexual behavior: lessons from rodent models and thoughts about humans," *Ann Rev Psychol* 59, 93–118.

Boon, W. C., Chow, J. D. and Simpson, E. R. (2010). "The multiple roles of estrogens and the enzyme aromatase," *Prog Brain Res* 181, 209–232.

Boron, W. F. and Boulpaep, E. L. (2005). *Medical Physiology*, updated edn. (Philadelphia, PA: Elsevier Saunders).

Bos, P. A., Panksepp, J., Bluthe, R. M. and van Honk, J. (2012). "Acute effects of steroid hormones and neuropeptides on human social-emotional behavior: a review of single administration studies," *Front Neuroendocr* 33, 17–35.

Brinkman, A. O. (2009). "Androgen physiology: receptor and metabolic disorders" in *Endocrinology of Male Reproduction* (South Dartmouth: Endotext.org).

Brinton, R. D., Thompson, R. F., Foy, M. R., Baudry, M., Wang, J., Finch, C. E. *et al.* (2008). "Progesterone receptors: form and function in brain," *Front Neuroendocr* 29, 313–339.

Cahill, L. (2014). "Fundamental sex differences in human brain architecture," *Proc Natl Acad Sci USA* 111, 577–578.

Camacho-Arroyo, I., Guerra-Araiza, C. and Cerbon, M. A. (1998). "Progesterone receptor isoforms are differentially regulated by sex steroids in the rat forebrain," *Neuroreport* 9, 3993–3996.

Clegg, D. J. (2012). "Minireview: the year in review of estrogen regulation of metabolism," *Mol Endocr* 26, 1957–1960.

Conway-Campbell, B. L., Sarabdjitsingh, R. A., McKenna, M. A., Pooley, J. R., Kershaw, Y. M., Meijer, O. C. *et al.* (2010). "Glucocorticoid ultradian rhythmicity directs cyclical gene pulsing of the clock gene period 1 in rat hippocampus," *J Neuroendocrinol* 22, 1093–1100.

Couse, J. F. and Korach, K. S. (2001). "Contrasting phenotypes in reproductive tissues of female estrogen receptor null mice," *Ann NY Acad Sci* 948, 1–8.

Dayan, C. M. and Panicker, V. (2009). "Novel insights into thyroid hormones from the study of common genetic variation," *Nat Rev Endocr* 5, 211–218.

de Kloet, E. R. (2008). "About stress hormones and resilience to psychopathology," *J Neuroendocrinol* 20, 885–892.

de Kloet, E. R., Joels, M. and Holsboer, F. (2005). "Stress and the brain: from adaptation to disease," *Nat Rev Neurosci* 6, 463–475.

de Kloet, E. R., Karst, H. and Joels, M. (2008). "Corticosteroid hormones in the central stress response: quick-and-slow," *Front Neuroendocr* 29, 268–272.

Dudas, B. and Merchenthaler, I. (2006). "Three-dimensional representation of the neurotransmitter systems of the human hypothalamus: inputs of the gonadotrophin hormone-releasing hormone neuronal system," *J Neuroendocrinol* 18, 79–95.

Dufourny, L., Caraty, A., Clarke, I. J., Robinson, J. E. and Skinner, D. C. (2005). "Progesterone-receptive dopaminergic and neuropeptide Y neurons project from the arcuate nucleus to gonadotropin-releasing hormone-rich regions of the ovine preoptic area," *Neuroendocrinology* 82, 21–31.

Dunphy, M. P. and Lewis, J. S. (2009). "Radiopharmaceuticals in preclinical and clinical development for monitoring of therapy with PET," *J Nucl Med* 50(Suppl. 1), 106S–121S.

Ferris, C. F. and Stolberg, T. (2010). "Imaging the immediate non-genomic effects of stress hormone on brain activity," *Psychoneuroendocr* 35, 5–14.

Fowler, C. D., Liu, Y. and Wang, Z. (2008). "Estrogen and adult neurogenesis in the amygdala and hypothalamus," *Brain Res Rev* 57, 342–351.

Garcia-Segura, L. M. (2008). "Aromatase in the brain: not just for reproduction anymore," *J Neuroendocrinol* 20, 705–712.

Gluckman, P. D., Hanson, M. A., Cooper, C. and Thornburg, K. L. (2008). "Effect of in utero and early-life conditions on adult health and disease," *N Engl J Med* 359, 61–73.

Gonzalez, M., Cabrera-Socorro, A., Perez-Garcia, C. G., Fraser, J. D., Lopez, F. J., Alonso, R. *et al.* (2007). "Distribution patterns of estrogen receptor alpha and beta in the human cortex and hippocampus during development and adulthood," *J Comp Neurol* 503, 790–802.

Gore, A. C. and Patisaul, H. B. (2010). "Neuroendocrine disruption: historical roots, current progress, questions for the future," *Front Neuroendocr* 31, 395–399.

Grant, L. D. and Stumpf, W. E. (1975). *Anatomical Neuroendocrinology* (Basel: Karger).

Groeneweg, F. L., Karst, H., de Kloet, E. R. and Joels, M. (2011). "Rapid non-genomic effects of corticosteroids and their role in the central stress response," *J Endocrinol* 209, 1–15.

Gruber, C. J., Tschugguel, W., Schneeberger, C. and Huber, J. C. (2002). "Production and actions of estrogens," *N Engl J Med* 346, 340–352.

Grüters, A. and Krude, H. (2012). "Detection and treatment of congenital hypothyroidism," *Nat Rev Endocrinol* 8, 104–113.

Harvey, C. B. and Williams, G. R. (2002). "Mechanism of thyroid hormone action," *Thyroid* 12, 441–446.

Hughes, Z. A., Liu, F., Marquis, K., Muniz, L., Pangalos, M. N., Ring, R. H. *et al.* (2009). "Estrogen receptor neurobiology and its potential for translation into broad-spectrum therapeutics for CNS disorders," *Curr Mol Pharmacol* 2, 215–236.

Hull, E. M. and Dominguez, J. M. (2007). "Sexual behavior in male rodents," *Horm Behav* 52, 45–55.

Ikeda, Y., Nagai, A., Ikeda, M. A. and Hayashi, S. (2003). "Sexually dimorphic and estrogen-dependent expression of estrogen receptor beta in the ventromedial hypothalamus during rat postnatal development," *Endocr* 144, 5098–5104.

Ingalhalikar, M., Smith, A., Parker, D., Satterthwaite, T. D., Elliot, M. A., Ruperal, K. *et al.* (2014). "Sex differences in the structural connectome of the human brain," *Proc Natl Acad Sci USA* 111, 823–828.

Janicki, S. C. and Schupf, N. (2010). "Hormonal influences on cognition and risk for Alzheimer's disease," *Curr Neurol Neurosci Rep* 10, 359–366.

Joffe, H., Petrillo, L. F., Koukopoulos, A., Viguera, A. C., Hirschberg, A., Nonacs, R. *et al.* (2011). "Increased estradiol and improved sleep, but not hot flashes, predict enhanced mood during the menopausal transition," *J Clin Endocr Metab* 96, E1044–E1054.

Karakaya, S., Kipp, M. and Beyer, C. (2007). "Oestrogen regulates the expression and function of dopamine transporters in astrocytes of the nigrostriatal system," *J Neuroendocrinol* 19, 682–690.

Katzenellenbogen, B. S. and Korach, K. S. (1997). "A new actor in the estrogen receptor drama – enter ER-beta," *Endocr* 138, 861–862.

Kelly, M. J., Moss, R. L. and Dudley, C. A. (1977). "The effects of microelectrophoretically applied estrogen, cortisol and acetylcholine on medial preoptic-septal unit activity throughout the estrous cycle of the female rat," *Exp Brain Res* 30, 53–64.

Kruijver, F. P., Balesar, R., Espila A. M., Unmehopa, U. A. and Swaab, D. F. (2003). "Estrogen-receptor-beta distribution in the human hypothalamus: similarities and differences with ER alpha distribution," *J Comp Neurol* 466, 251–277.

Kudwa, A. E., Gustafsson, J. A. and Rissman, E. F. (2004). "Estrogen receptor beta modulates estradiol induction of progestin receptor immunoreactivity in male, but not in female, mouse medial preoptic area," *Endocr* 145, 4500–4506.

Lechan, R. M., Qi, Y., Jackson, I. M. and Mahdavi, V. (1994). "Identification of thyroid hormone receptor isoforms in thyrotropin-releasing hormone neurons of the hypothalamic paraventricular nucleus," *Endocr* 135, 92–100.

LeVay, S. (1991). "A difference in hypothalamic structure between heterosexual and homosexual men," *Science* 253, 1034–1037.

Levin, E. R. (2009). "Plasma membrane estrogen receptors," *Trends Endocrinol Metab* 20, 477–482.

Li, M. and Boyages, S. C. (1996). "Detection of extended distribution of beta2-thyroid hormone receptor messenger ribonucleic acid (RNA) in adult rat brain using complementary RNA in situ hybridization histochemistry," *Endocr* 137, 1272–1275.

Lightman, S. L. (2008). "The neuroendocrinology of stress: a never-ending story," *J Neuroendocrinol* 20, 880–884.

Liu, A., Margaill, I., Zhang, S., Labombarda, F., Coqueran, B., Delespierre, B. *et al.* (2012). "Progesterone receptors: a key for neuroprotection in experimental stroke," *Endocr* 153, 3747–3757.

Lu, J., Goula, D., Sousa, N. and Almeida, O. F. (2003). "Ionotropic and metabotropic glutamate receptor mediation of glucocorticoid-induced apoptosis in hippocampal cells and the neuro-protective role of synaptic N-methyl-D-aspartate receptors," *Neurosci* 121, 123–131.

Luine, V. N. and Frankfurt, M. (2012). "Estrogens facilitate memory processes through membrane mediated mechanisms and alterations in spine density," *Front Neuroendocr* 33, 388–402.

Mangelsdorf, D. J., Thummel, C., Beato, M., Herrlich, P., Schutz, G., Umesono, K. *et al.* (1995). "The nuclear receptor superfamily: the second decade," *Cell* 83, 835–839.

Mani, S. (2008). "Progestin receptor subtypes in the brain: the known and the unknown," *Endocr* 149, 2750–2756.

Matthews, S. G. (2001). "Antenatal glucocorticoids and the developing brain: mechanisms of action," *Semin Neonatol* 6, 309–317.

Mauvais-Jarvis, F. (2011). "Estrogen and androgen receptors: regulators of fuel homeostasis and emerging targets for diabetes and obesity," *Trends Endocrinol Metab* 22, 24–33.

McAbee, M. D. and DonCarlos, L. L. (1998). "Ontogeny of region-specific sex differences in androgen receptor messenger ribonucleic acid expression in the rat forebrain," *Endocr* 139, 1738–1745.

McCarthy, M. M. (2007). "GABA receptors make teens resistant to input," *Nat Neurosci* 10, 397–399.

McCarthy, M. M. (2010). "How it's made: organisational effects of hormones on the developing brain," *J Neuroendocrinol* 22, 736–742.

McCarthy, M. M., Wright, C. L. and Schwarz, J. M. (2009). "New tricks by an old dogma: mechanisms of the Organizational/Activational Hypothesis of steroid-mediated sexual differentiation of brain and behavior," *Horm Behav* 55, 655–665.

McEwen, B. S. (1981). "Neural gonadal steroid action," *Science* 211, 1303–1311.

McEwen, B. S. and Gianaros, P. J. (2011). "Stress- and allostasis-induced brain plasticity," *Annu Rev Med* 62, 431–445.

McEwen, B. S., Davis, B. G., Parsons, B. and Pfaff, D. W. (1979). "The brain as a target for steroid hormone action," *Ann Rev Neurosci* 2, 65–112.

McEwen, B. S., de Kloet, E. R. and Rostene, W. (1986). "Adrenal steroid receptors and actions in the nervous system," *Physiol Rev* 66, 1121–1188.

Meis, P. J. and Connors, N. (2004). "Progesterone treatment to prevent preterm birth," *Clin Obstet Gynecol* 47, 784–795; discussion 881–782.

Melcangi, R. C., Garcia-Segura, L. M. and Mensah-Nyagan, A. G. (2008). "Neuroactive steroids: state of the art and new perspectives," *Cell Mol Life Sci* 65, 777–797.

Melcangi, R. C. and Panzica, G. C. (2014). "Allopregnanolone: state of the art," *Prog Neurobiol* 113, 1–5.

Mendelsohn, M. E. and Karas, R. H. (1999). "The protective effects of estrogen on the cardio-vascular system," *N Engl J Med* 340, 1801–1811.

Messent, P. R. (1976). "Female hormones and behavior" in B. Lloyd and J. Archer (eds.), *Exploring Sex Differences* (New York: Academic Press), pp. 185–212.

Micevych, P. and Sinchak, K. (2008). "Estradiol regulation of progesterone synthesis in the brain," *Mol Cell Endocrinol* 290, 44–50.

Micevych, P. E. and Kelly, M. J. (2012). "Membrane estrogen receptor regulation of hypothalamic function," *Neuroendocr* 96, 103–110.

Morimoto, M., Morita, N., Ozawa, H., Yokoyama, K. and Kawata, M. (1996). "Distribution of glucocorticoid receptor immunoreactivity and mRNA in the rat brain: an immunohistochemical and in situ hybridization study," *Neurosci Res* 26, 235–269.

Morris, J. A., Jordan, C. L. and Breedlove, S. M. (2004). "Sexual differentiation of the vertebrate nervous system," *Nat Neurosci* 7, 1034–1039.

Nadal, A. (2012). "Fat from plastics? Linking bisphenol A exposure and obesity," *Nat Rev Endocrinol* 9, 9–10.

Nader, N., Chrousos, G. P. and Kino, T. (2010). "Interactions of the circadian CLOCK system and the HPA axis," *Trends Endocrinol Metab* 21, 277–286.

Noel, S. D., Keen, K. L., Baumann, D. I., Filardo, E. J. and Terasawa, E. (2009). "Involvement of G protein-coupled receptor 30 (GPR30) in rapid action of estrogen in primate LHRH neurons," *Mol Endocrinol* 23, 349–359.

Numan, M. (2007). "Motivational systems and the neural circuitry of maternal behavior in the rat," *Dev Psychobiol* 49, 12–21.

Orikasa, C., Kondo, Y., Hayashi, S., McEwen, B. S. and Sakuma, Y. (2002). "Sexually dimorphic expression of estrogen receptor β in the anteroventral periventricular nucleus of the rat preoptic area: implication in luteinizing hormone surge," *Proc Natl Acad Sci USA* 99, 3306–3311.

Ozawa, H., Ito, T., Ochiai, I. and Kawata, M. (1999). "Cellular localization and distribution of glucocorticoid receptor immunoreactivity and the expression of glucocorticoid receptor messenger RNA in rat pituitary gland. A combined double immunohistochemistry study and in situ hybridization histochemical analysis," *Cell Tissue Res* 295, 207–214.

Patisaul, H. B., Whitten, P. L. and Young, L. J. (1999). "Regulation of estrogen receptor beta mRNA in the brain: opposite effects of 17β-estradiol and the phytoestrogen, coumestrol," *Brain Res Mol Brain Res* 67, 165–171.

Pettersson, K., Grandien, K., Kuiper, G. G. and Gustafsson, J. A. (1997). "Mouse estrogen receptor beta forms estrogen response element-binding heterodimers with estrogen receptor alpha," *Mol Endocrinol* 11, 1486–1496.

Pfaff, D. W. (1989). "Features of a hormone-driven defined neural circuit for a mammalian behavior," *Ann NY Acad Sci* 563, 131–147.

Pfaff, D. W. (1997). "Hormones, genes, and behavior," *Proc Natl Acad Sci USA* 94, 14213–14216.

Pfaff, D. W. and Levine, J. E. (2008). "Reconciling molecular neuroendocrine signals and the scientists who study them," *Front Neuroendocr* 29, 167–168.

Pfaff, D. W., Kow, L. M., Loose, M. D. and Flanagan-Cato, L. M. (2008). "Reverse engineering the lordosis behavior circuit," *Horm Behav* 54, 347–354.

Pfaff, D. W., Waters, E., Khan, Q., Zhang, X. and Numan, M. (2011). "Minireview: estrogen receptor-initiated mechanisms causal to mammalian reproductive behaviors," *Endocr* 152, 1209–1217.

Pike, C. J., Carroll, J. C., Rosario, E. R. and Barron, A. M. (2009). "Protective actions of sex steroid hormones in Alzheimer's disease," *Front Neuroendocr* 30, 239–258.

Plant, T. M. (2008). "Hypothalamic control of the pituitary-gonadal axis in higher primates: key advances over the last two decades," *J Neuroendocrinol* 20, 719–726.

Prossnitz, E. R. and Barton, M. (2011). "The G-protein-coupled estrogen receptor GPER in health and disease," *Nat Rev Endocrinol* 7, 715–726.

Puymirat, J., Miehe, M., Marchand, R., Sarlieve, L. and Dussault, J. H. (1991). "Immunocytochemical localization of thyroid hormone receptors in the adult rat brain," *Thyroid* 1, 173–184.

Rahman, F. and Christian, H. C. (2007). "Non-classical actions of testosterone: an update," *Trends Endocrinol Metab* 18, 371–378.

Rainbow, T. C., McGinnis, M. Y., Davis, P. G. and McEwen, B. S. (1982). "Application of anisomycin to the lateral ventromedial nucleus of the hypothalamus inhibits the activation of sexual behavior by estradiol and progesterone," *Brain Res* 233, 417–423.

Reynolds, R. M. (2010). "Corticosteroid-mediated programming and the pathogenesis of obesity and diabetes," *J Steroid Biochem Mol Biol* 122, 3–9.

Roselli, C. E., Liu, M. and Hurn, P. D. (2009). "Brain aromatization: classic roles and new perspectives," *Semin Reprod Med* 27, 207–217.

Rubinow, D. R. and Schmidt, P. J. (2006). "Gonadal steroid regulation of mood: the lessons of premenstrual syndrome," *Front Neuroendocr* 27, 210–216.

Rune, G. M. and Frotscher, M. (2005). "Neurosteroid synthesis in the hippocampus: role in synaptic plasticity," *Neurosci* 136, 833–842.

Sanchez, M. M., Young, L. J., Plotsky, P. M. and Insel, T. R. (2000). "Distribution of corticosteroid receptors in the rhesus brain: relative absence of glucocorticoid receptors in the hippocampal formation," *J Neurosci* 20, 4657–4668.

Schaafsma, S. M. and Pfaff, D. W. (2014). "Etiologies underlying sex differences in autism spectrum disorders," *Front Neuroendocr* 35, 255–271.

Scott, C. J., Tilbrook, A. J., Simmons, D. M., Rawson, J. A., Chu, S., Fuller, P. J. *et al.* (2000). "The distribution of cells containing estrogen receptor-alpha (ERalpha) and ERbeta messenger ribonucleic acid in the preoptic area and hypothalamus of the sheep: comparison of males and females," *Endocr* 141, 2951–2962.

Seckl, J. R., Dickson, K. L., Yates, C. and Fink, G. (1991). "Distribution of glucocorticoid and mineralocorticoid receptor messenger RNA expression in human postmortem hippocampus," *Brain Res* 561, 332–337.

Shughrue, P. J., Lane, M. V. and Merchenthaler, I. (1997). "Comparative distribution of estrogen receptor-alpha and -beta mRNA in the rat central nervous system," *J Comp Neurol* 388, 507–525.

Simerly, R. B. (2002). "Wired for reproduction: organization and development of sexually dimorphic circuits in the mammalian forebrain," *Annu Rev Neurosci* 25, 507–536.

Simerly, R. B., Carr, A. M., Zee, M. C. and Lorang, D. (1996). "Ovarian steroid regulation of estrogen and progesterone receptor messenger ribonucleic acid in the anteroventral periventricular nucleus of the rat," *J Neuroendocrinol* 8, 45–56.

Simerly, R. B., Chang, C., Muramatsu, M. and Swanson, L. W. (1990). "Distribution of androgen and estrogen receptor mRNA-containing cells in the rat brain: an in situ hybridization study," *J Comp Neurol* 294, 76–95.

Simpkins, J. W., Singh, M., Brock, C. and Etgen, A. M. (2012). "Neuroprotection and estrogen receptors," *Neuroendocr* 96, 119–130.

Sousa, N., Cerqueira, J. J. and Almeida, O. F. (2008). "Corticosteroid receptors and neuroplasticity," *Brain Res Rev* 57, 561–570.

Spary, E. J., Maqbool, A. and Batten, T. F. (2009). "Oestrogen receptors in the central nervous system and evidence for their role in the control of cardiovascular function," *J Chem Neuroanat* 38, 185–196.

Stein, D. G. (2008). "Progesterone exerts neuroprotective effects after brain injury," *Brain Res Rev* 57, 386–397.

Strauss, J. F. and Barbieri, R. L. (2004). *Yen and Jaffe's Reproductive Endocrinology*, 5th edn. (Philadelphia, PA: Elsevier Saunders).

Swaab, D. F., Chung, W. C., Kruijver, F. P., Hofman, M. A. and Hestiantoro, A. (2003). "Sex differences in the hypothalamus in the different stages of human life," *Neurobiol Aging* 24(Suppl. 1), S1–S16; discussion S17–S19.

Taves, M. D., Gomez-Sanchez, C. E. and Soma, K. K. (2011). "Extra-adrenal glucocorticoids and mineralocorticoids: evidence for local synthesis, regulation and function," *Am J Physiol* 301, E11–E24.

Tegethoff, M., Pryce, C. and Meinlschmidt, G. (2009). "Effects of intrauterine exposure to synthetic glucocorticoids on fetal, newborn, and infant hypothalamic-pituitary-adrenal axis function in humans: a systematic review," *Endocr Rev* 30, 753–789.

Tetel, M. J. and Pfaff, D. W. (2010). "Contributions of estrogen receptor-alpha and estrogen receptor-β to the regulation of behavior," *Biochim Biophys Acta* 1800, 1084–1089.

Thornton, J., Zehr, J. L. and Loose, M. D. (2009). "Effects of prenatal androgens on rhesus monkeys: a model system to explore the organizational hypothesis in primates," *Horm Behav* 55, 633–645.

Toran-Allerand, C. D. (1980). "Sex steroids and the development of the newborn mouse hypothalamus and preoptic area in vitro. II. Morphological correlates and hormonal specificity," *Brain Res* 189, 413–427.

Wagner, C. K. (2008). "Progesterone receptors and neural development: a gap between bench and bedside?" *Endocr* 149, 2743–2749.

Weigel, N. S. and Moore, N. L. (2007). www.nursa.org/article.cfm?doi=10.1621/nrs.05005.

Williams, T. J., Mitterling, K. L., Thompson, L. I., Torres-Reveron, A., Waters, E. M., McEwen, B. S. *et al.* (2011). "Age- and hormone-regulation of opioid peptides and synaptic proteins in the rat dorsal hippocampal formation," *Brain Res* 1379, 71–85.

Wintermantel, T. M., Campbell, R. E., Porteous, R., Bock, D., Grone, H. J., Todman, M. G. *et al.* (2006). "Definition of estrogen receptor pathway critical for estrogen positive feedback to gonadotropin-releasing hormone neurons and fertility," *Neuron* 52, 271–280.

Yang, J. H., Li, L. H., Lee, S., Jo, I. H., Lee, S. Y. and Ryu, P. D. (2007). "Effects of adrenalectomy on the excitability of neurosecretory parvocellular neurones in the hypothalamic paraventricular nucleus," *J Neuroendocrinol* 19, 293–301.

Yau, J. L. and Seckl, J. R. (2012). "Local amplification of glucocorticoids in the aging brain and impaired spatial memory," *Front Aging Neurosci* 4, 24.

Yu, S., Patchev, A. V., Wu, Y., Lu, J., Holsboer, F., Zhang, J. Z. *et al.* (2010). "Depletion of the neural precursor cell pool by glucocorticoids," *Ann Neurol* 67, 21–30.

Zheng, P. (2009). "Neuroactive steroid regulation of neurotransmitter release in the CNS: action, mechanism and possible significance," *Prog Neurobiol* 89, 134–152.

Zuloaga, D. G., Yahn, S. L., Pang, Y., Quihuis, A. M., Oyola, M. G., Reyna, A. *et al.* (2012). "Distribution and estrogen regulation of membrane progesterone receptor-β in the female rat brain," *Endocr* 153, 4432–4443.

10 Receptors for peptide hormones, neuropeptides and neurotransmitters

Peptide hormones, neuropeptides, neurotransmitters and other non-steroid chemical messengers regulate cellular and biochemical activity by binding to receptors located in the plasma membranes of their target cells. These chemical messengers are generally polar and water soluble and so cannot readily enter their target cells to influence the cell nucleus in the manner described for steroid and thyroid hormones (Chapter 9). In order to induce biochemical changes within the target cell, they act as first messengers to activate a second messenger, such as cAMP, within the cytoplasm of the target cell. The transduction of information from the first to the second messenger is accomplished through the activation of membrane protein transducers (G proteins) and enzymes, such as adenylate cyclase. This chapter discusses membrane receptors for peptide hormones and neurotransmitters, the mechanisms by which signal transduction across the cell membrane occurs, the role of G proteins and receptor tyrosine kinases in this signal transduction, the second messenger systems activated, and the actions of the second messengers in the target cells, with special emphasis on neural target cells.

Some of this material was introduced in Chapter 5 *and we will refer to relevant figures where appropriate.*

10.1 Membrane receptors

Membrane receptors are complex proteins embedded in the cell membrane. The function of these receptors is to recognize specific ligands in the blood (e.g. peptide hormones, neuropeptides) or in the synapse (e.g. neurotransmitters) and bind to them. Once this binding occurs, signal transduction across the cell membrane occurs as described in section 10.2 below. As noted in the description of steroid hormone receptors, modern techniques of immunohistochemistry (using antibodies to receptor proteins) and *in situ* hybridization (allowing visualization of peptide mRNA) permit the ready localization of membrane receptors in any tissue of interest. Gene cloning and sequencing techniques enabled molecular biologists to develop three-dimensional models of many membrane receptors. There are three distinct types of membrane receptor: (1) ligand-gated ion channel receptors; (2) guanine nucleotide binding protein (G-protein-) coupled receptors; and (3) transmembrane-regulated tyrosine kinases. All of these types of membrane receptors have three regions: an extracellular ligand binding domain, which protrudes from the cell surface and binds to the peptide, neurotransmitter or other ligand molecule; a transmembrane domain (or domains in the case of G protein receptors), which passes through the cell membrane; and an intracellular domain, which activates a second messenger system. Examples of these receptors are illustrated in Chapter 5 (Figures 5.3 and 5.13) and Figure 10.6 (below). Useful descriptions can be found in several textbooks (e.g. Boron and Boulpaep 2005; Squire *et al.* 2008).

10.1.1 Ligand-gated ion channel receptors

Some receptors (termed *ionotropic* receptors; see section 5.5) are ligand-gated ion channels; that is, when a neurotransmitter binds to the receptor, an ion channel is opened. These include the nicotinic acetylcholine receptor, the GABA/benzodiazepine receptor (Figure 5.2) and the glutamate receptor (Figure 5.3). The five subunits of the GABA receptor, for example, form the transmitter-gated ion channel. The GABA-activated membrane receptor opens chloride channels through the cell membrane. Some ion channels only open briefly, allowing a rapid influx of ions that cause a transient change in membrane potential. This occurs, for example, when acetylcholine binds to the nicotinic receptor to open sodium channels in neurons, resulting in an action potential (see Figure 5.10). Other ion channels may open for a longer duration, allowing a major influx of ions, such as calcium, into the cell.

10.1.2 G-protein-coupled receptors (GPCR)

This type of receptor (termed *metabotropic*; see section 5.5) is coupled to a G protein complex in the cell membrane as occurs in the dopamine receptor shown in Figure 5.13. Note that the receptor protein snakes in and out of the membrane so that there are seven transmembrane domains. All GPCRs have seven domains and this structure is attached to a G protein complex consisting of three subunits: $G\alpha$, $G\beta$ and $G\gamma$ (Figure 10.1) (Vassart and Costagliola 2011). G proteins couple the transduction of the neurotransmitter signal from the extracellular ligand-receptor complex to the intracellular second messenger system (Figure 10.1; see also section 10.2). They respond to a variety of stimuli ranging from photons, odorants, amino acids, neurotransmitters and neuropeptides. Eighty percent of all signal transduction across cell membranes involves these receptors (Millar and Newton 2010).

They are of great interest to pharmaceutical companies and Millar and Newton (2010) estimate more than 30 percent of all current therapeutics involve them. Table 10.1 lists several common drugs that bind to GPCRs. These authors also note that the approximately 800 known GPCRs offer considerable scope for the further development of new treatments for disease. There are also many examples of mutations in GPCRs causing endocrine diseases (for further reading, see Vassart and Costagliola 2011).

10.1.3 Transmembrane-regulated enzyme receptors

A third type of receptor, responsible for the activity of growth factors, neurotrophins (e.g. NGF), insulin and adipokines (e.g. leptin), is called a *receptor tyrosine kinase* or *trk*. These are single-stranded transmembrane proteins that join together (dimerize) when they bind the agonist, such as leptin (see Figure 10.6, below). A further interesting variation on the transmembrane structure is the insulin receptor that is essentially a pre-existing dimer; that is, the two transmembrane units are covalently connected (Nussey and Whitehead 2001).

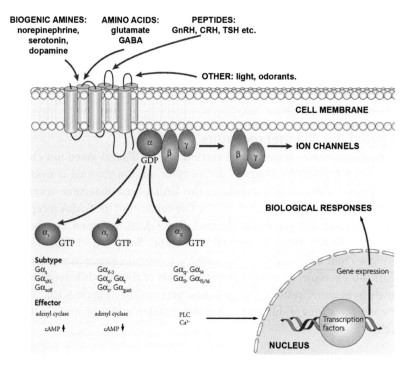

Figure 10.1 G-protein-coupled receptor signaling

As shown, various ligands use G-protein-coupled receptors (GPCRs) for cell stimulation. GPCRs interact with an intracellular trimer of G proteins composed of α, β and γ subunits. In the resting (unstimulated) state, the α subunits are bound to guanosine diphosphate (GDP). As soon as the GPCR binds a ligand, such as norepinephrine, GDP dissociates and is replaced with guanosine triphosphate (GTP). This facilitates the dissociation of Gα-GTP from the βγ subunit. The α subunits of G proteins are divided into three subfamilies: Gα$_s$, Gα$_i$ and Gα$_q$. Each G protein has several subtypes and activates several downstream effectors. For example, Gα$_s$ typically stimulates adenyl cyclase and increases levels of cyclic AMP (cAMP), whereas Gα$_i$ inhibits adenyl cyclase and lowers cAMP levels. Members of the Gα$_q$ family bind to and activate phospholipase C (PLC), which cleaves phosphatidylinositol bisphosphate (PIP2) into diacylglycerol and inositol triphosphate (IP3) (see also Figure 10.5). The Gβ subunits and Gγ subunits function as a dimer to activate, for example, ion channels. Ultimately, the integration of the functional activity of the G-protein-regulated signaling networks induces many biological and cellular functions.

Abbreviations: cAMP, cyclic adenosine monophosphate; CRH, corticotropin-releasing hormone; GABA, gamma-aminobutyric acid; GnRH, gonadotropin-releasing hormone; PLC, phospholipase C; TSH, thyroid-stimulating hormone. Reproduced with permission (Dorsam and Gutkind 2007).

Binding of insulin causes the two units to auto-phosphorylate, thus setting in motion the signaling cascade (see Figure 10.6 below).

10.1.4 Receptor regulation

The number of receptors in the cell membrane is not fixed, but can be *down-regulated* in order to change cellular response. For example, over-stimulation of some receptors will reduce the number of cell surface receptors in an attempt to reduce further stimulation. On the other hand, down-regulation/desensitization can be seen as a pathological response, as in the inability to respond to insulin in diabetics, or the reduced response to opiates with increasing dosages so that pain control is compromised. There are many other examples. Conversely, an absence of stimulation – for example, when dopamine levels are reduced in

Table 10.1 Some therapeutic drugs acting on G-protein-coupled receptors

Drug (brand name)	Receptor	Uses
Oxytocin (*Hospira*)	OT	Induction of labor
Lopressor (*Metoprolol*)	β_1 agonist	Lowers blood pressure
Salmeterol (*Serevent*)	β_2 antagonist	Opens airways
GH antagonist (*Octreotide*)	GH	Lowers GH secretion
Ranitidine (*Zantac*)	H2 antagonist	Counteracts stomach acid
Olanzapine (*Zyprexa*)	5-HT$_2$/ D2	Controls psychotic symptoms
GnRH agonists (e.g. *Buserilin; Leuprolide*)	GnRH	Fertility treatment; prostate cancer
Tetrahydrocannabinol (THC; *Dronabinol*)	CB1, CB2	Alzheimer's Disease*; Nausea
Naloxone (*Narcan*)	Opioid	Drug overdose

* Scotter *et al.* (2010). *Br J Pharmacol* 160, 480.

Parkinson's disease – leads to an *up-regulation* in dopamine receptors and consequent super-sensitivity (Ichise *et al.* 1999). Up-regulation can also be physiologically relevant, as in the increase in estradiol-induced pituitary GnRH receptors that are essential in producing the surge of LH midway through the menstrual cycle (Clayton 1988). The mechanism by which GPCR are down-regulated, or desensitized, is described below (section 10.1.5).

There are many examples of peptide hormones able to control their own receptor numbers. For example, growth hormone and insulin regulate their own receptor populations in several tissues (Flores-Morales *et al.* 2006; for further reading, see Grunberger *et al.* 1983). Receptor regulation can occur through changes in gene expression, to provide more receptor protein, or by removing/adding membrane receptors without a change in protein synthesis. For example, growth hormone receptor mRNA is reduced by fasting or low glucose levels, and increased by insulin, pregnancy or thyroid hormones (Flores-Morales *et al.* 2006). The mechanism by which receptors are reversibly removed from cell membranes is discussed below.

10.1.5 Receptor internalization

Several mechanisms exist to terminate, or at least modulate, neurotransmitter/neuropeptide signaling. One of these is called *desensitization* and is simply the uncoupling of the receptor from intracellular signaling pathways so that the strength of agonist stimulation is reduced. This will be described in the section 10.2. Another mechanism is to lower receptor abundance by removing, or sequestering, the receptor from the cell membrane (*internalization*). Such a mechanism probably exists for many hormones and neurotransmitters/ neuropeptides. In the present section, we will illustrate this concept using the β-adrenergic receptor as an example, although the principles can be applied to other GPCRs such as cannabinoid receptors (Smith *et al.* 2010). Figure 10.2 illustrates the mechanism. NE binds to β-adrenergic receptors, causing phosphorylation/desensitization to occur; that is, the G protein complex is uncoupled from the receptor (see section 10.2). This is followed

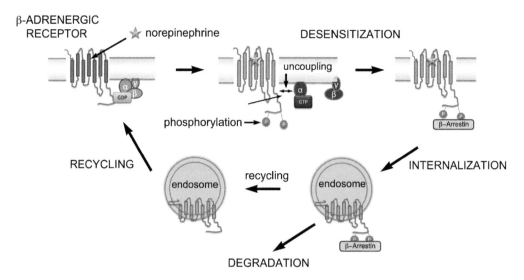

Figure 10.2 Desensitization, internalization and down-regulation of β-adrenergic receptors
Activation of the receptor by norepinephrine uncouples the receptor from the G protein complex, enabling a cellular response to occur as shown in Figure 10.1. The β-adrenergic receptor, free of the G proteins, is then phosphorylated and binds to a molecule called β-arrestin. This step effectively prevents the receptor from being reactivated and is therefore *desensitized* and will not respond to further stimulation. In addition, the receptor/ β-arrestin complex is removed from the cell membrane, internalized by endocytosis and then confined to an intracellular compartment called an *endosome*. Once internalized, β-adrenergic receptors may be recycled back to the cell surface following dephosphorylation, or alternatively can be degraded (*down-regulation*). Reproduced with permission, and redrawn (Smith *et al.* 2010).

immediately by the receptor being removed from the cell membrane (endocytosis) by a protein called β-arrestin. This internalization step sequesters the receptor into an endosomal compartment where it is either degraded or recycled back to the cell membrane. This recycling sequence has been described for several other receptors, including dopamine, opioids, LH and TSH (for further reading, see Hanyaloglu and von Zastrow 2008). Peptide hormones such as insulin and GH, which bind to tyrosine kinase receptors (e.g. see Figure 10.6), may also undergo receptor internalization, but desensitization of the intracellular signaling mechanism is the primary response to overstimulation. This will be discussed in section 10.3.5.

10.2 Signal transduction by G proteins

Receptors for most neurotransmitters, peptide hormones, neuropeptides and other chemical messengers are coupled to G proteins. As shown in Figure 10.1, G proteins are activated when the ligand binds to its membrane receptor. G protein subunits (e.g. G_s or G_i) are released from the αβγ subunit complex through binding to GTP. The GTP-activated G protein subunits can stimulate or inhibit the synthesis of second messengers and are, therefore, crucial in the regulation of the physiological effects of these ligands on their target cells (see section 10.3). The remaining βγ complex can also regulate cellular changes, usually via opening potassium ion channels. The cycle is reversed when GTP is hydrolyzed back to GDP.

Table 10.2 Examples of neurotransmitter/G protein subunit interactions

Ligand	Receptor	Subunit	Target
Norepinephrine	α_{1A}, α_{1B}	G_q	Heart, muscle, brain, lung
Norepinephrine	β_1, β_2	G_s	Heart, muscle, brain
Melatonin	MT_1, MT_2	G_i	Brain, pituitary
Serotonin	$5\text{-}HT_{1A,\ B,\ D}$	G_i	Brain
	$5\text{-}HT_{2A,\ B,\ C}$	G_q	Brain
	$5\text{-}HT_4$	G_s	Brain
Glutamate	$mGluR_1$, R_5	G_q	Brain
	$mGluR_{2,\ 3,\ 4}$	G_i	Brain
Cannabinoid	CB_1, CB_2	G_i	Brain
Prostaglandin E2	EP_1	G_q	Ubiquitous
ACTH	MC_2	G_s	Adrenal cortex
CRH	CRF_1, CRF_2	G_s	Pituitary corticotrophs
GnRH	GnRH	G_q	Pituitary gonadotrophs
Kisspeptin	GPR54	G_q	Brain
Opioids	μ, δ, κ	G_i	Brain
Somatostatin	SST_1–SST_5	G_i	Pituitary somatotrophs, pancreas

There are many types of G proteins, too many to be included in this text. Interested students are, however, referred elsewhere (Wettschureck and Offermanns 2005). Table 10.2 lists some examples of neurotransmitter/neuropeptide receptors that are linked to G protein subunits and shows that a given ligand, such as serotonin, can have different effects if its receptors are coupled to different G proteins in target cells. For example, 5HT-1A receptors are inhibitory and reduce cAMP levels, whereas 5HT-4 receptors are stimulatory. Likewise, one target cell may have receptors for a number of different hormones, each of which activates a different G protein and, through these G proteins, different second messenger systems and a corresponding different cellular response (Asa and Ezzat 2002).

In Chapter 6, we described how PRL secretion was regulated by TRH (stimulatory) and dopamine (inhibitory). Figure 10.3 illustrates how different G proteins mediate the intracellular responses of a single pituitary lactotroph to these two different first messengers. Dopamine exerts an inhibitory effect via a G_i subunit, and TRH stimulates PRL secretion via a G_s subunit acting through phospholipase C (PLC; see also Figure 10.5). There are many other examples of this type of dual regulation. One of them we have noted already: the dual control of pituitary GH secretion by GHRH and somatostatin (Figure 6.9; see also Figure 4.8). Note that a third receptor (ghrelin) also regulates GH secretion.

An additional control system that has a profound effect on G protein mechanisms is seen in the action of steroid hormones. As discussed in Chapter 9, steroid feedback is normally inhibitory. For example, cortisol feedback inhibits ACTH release by reducing CRH secretion from the hypothalamus, but also by preventing the stimulatory effect of CRH on ACTH secretion (Figure 6.4). In other words, the normal G protein mediated release of ACTH is prevented by cortisol binding to nuclear receptors. In addition, since steroid hormones are known to have rapid effects on G protein signaling via the cell membrane (Figure 9.9), this

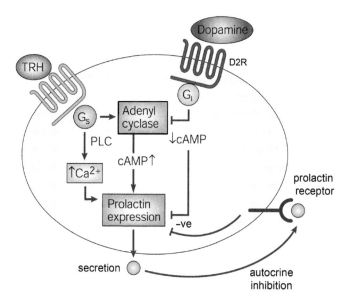

Figure 10.3 Prolactin secretion is controlled by two different G proteins
Lactotrophs express membrane receptors that induce prolactin synthesis and secretion. Thyrotropin-releasing hormone (TRH) signals through a G-protein-coupled receptor. Ligand binding activates the stimulatory G protein G_s, and induces cAMP formation, phospholipase C (PLC) activity and intracellular calcium ion flux. This leads to increased prolactin expression and secretion. Dopamine signaling, on the other hand, inhibits prolactin production. Dopamine signaling is mediated by the dopamine 2 receptor (D2R), which is coupled to the inhibitory G protein, G_i. Activation of this pathway leads to diminished cAMP production and reduced calcium ion (Ca^{2+}) flux, and, ultimately, reduced prolactin secretion. Prolactin also inhibits its own secretion through an autocrine mechanism. Reproduced with permission (Asa and Ezzat 2002).

provides another level of complexity in our understanding of G protein mechanisms. Gender differences in dopamine mechanisms have also been ascribed to the involvement of G proteins (Dluzen 2005). It should not be surprising that disorders of G protein receptors result in endocrine diseases and several other pathologies. Many receptors (e.g. muscarinic, glutamate, serotonin, CRH, somatostatin) have been implicated in Alzheimer's Disease (Thathiah and de Strooper 2011). GPCR signaling also appears to be important in treating some endocrine cancers (Dorsam and Gutkind 2007), and GPCR mutations seem to be responsible for endocrine diseases such as hypothyroidism, short stature, obesity and hypogonadism (Vassart and Costagliola 2011). The "GPCR Drug Pipeline" is an active source of new compounds designed to treat CNS disease (Lim 2010).

10.3 Second messenger systems

As shown in Figure 10.1, binding of a neurotransmitter, peptide hormone or neuropeptide to its receptor activates the G protein complex coupled to that receptor. $G\alpha$ can then regulate second messengers such as cAMP, cGMP, calcium ion and inositol phospholipids. Examples of the second messenger systems activated by a variety of peptide hormones, neurotransmitters and neuropeptides are shown in Table 10.3.

These second messengers stimulate biochemical changes in the cell by activating specific protein kinases (third messengers) such as protein kinase A (PKA). *Protein kinases are enzymes that activate (phosphorylate) proteins and they provide a link between membrane receptor stimulation and activation of nuclear signaling.* As shown in the example in Figure 10.4, α-MSH produces cAMP via a stimulatory G protein ($G\alpha s$), which in turn activates PKA. PKA subsequently phosphorylates a protein called CREB (*cAMP responsive element binding protein*) that acts as a transcription factor, binding to a response element

Table 10.3 Examples of second messenger systems activated by neurotransmitters and peptide hormones

First messenger	Target cell	Activity affected
A. Increased cAMP		
Catecholamines	cardiac cells: β-adrenergic receptors	Increase heart rate
Dopamine	D_1–D_5 receptors in brain	Neural activity
TRH	Pituitary thyrotroph cells	Secretion of TSH
Vasopressin	AVPR2 in kidney cells	Water reabsorption
TSH	Thyroid gland	Secretion of T3/T4
ACTH	Adrenal cortex	Glucocorticoid secretion
LH	Interstitial cells of testes	Testosterone secretion
	Luteal cells of ovary	Progesterone secretion
FSH	Ovarian granulosa cells	Estradiol secretion
B. Decreased cAMP		
Acetylcholine	Cardiac muscarinic receptors	Reduced heart rate
Catecholamines	α_2 adrenergic receptors	Smooth muscle relaxation
Dopamine	D_2 receptors in pituitary lactotrophs	Inhibit prolactin secretion
Opioid peptides	μ, δ, κ receptors in brain	Analgesia, euphoria, membrane potential
Somatostatin	SST_{1-5} receptors on pituitary somatotrophs	Inhibit GH release
Angiotensin II	AT_1 receptors kidney	Renal sodium retention
C. cGMP		
Atrial natriuretic peptide	Muscle ANP receptors A, B and C	Vasodilation
D. Ca^{++}– DAG		
Acetylcholine	M3 muscarinic receptors	Muscle contraction
Catecholamines	α_1-adrenergic receptors	Vasoconstriction
Oxytocin	Breast	Milk letdown
	Uterus	Muscle contraction
Vasopressin	AVPR 1 muscle	Vasoconstriction
TRH	Pituitary thyrotrophs	TSH secretion
GnRH	Pituitary gonadotrophs	LH and FSH secretion
Kisspeptin	GPR54 hypothalamus	GnRH secretion
Angiotensin II	AT_1 adrenal cortex	Release of aldosterone

on target genes. The phosphorylation step can be said to activate the CREB molecule. Protein phosphorylation plays an important role in the regulation of many physiological processes associated with signal transduction because it provides a common pathway through which neurotransmitters, neuropeptides, peptide hormones and steroid hormones (see also Figure 9.9) produce their biological effects on their target cells, even though these ligands activate different receptors, G proteins and second messengers (Purves *et al.* 2008).

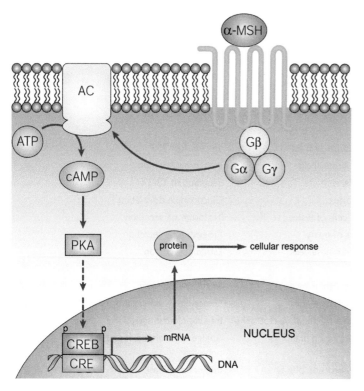

Figure 10.4 G protein receptors coupled to intracellular nuclear signaling

In this example, the neuropeptide α-MSH binds to its cell surface G-protein-coupled receptor and stimulates the formation of cAMP via activation of Gαs. Increased cAMP levels activate protein kinase A (PKA) which then phosphorylates (p) the transcription factor CREB (cAMP-responsive element binding protein) located in the cell nucleus. Phosphorylated CREB proteins activate genes that contain cAMP-responsive elements (CRE).

Abbreviations: AC, adenyl cyclase; ATP, adenosine triphosphate; mRNA, messenger RNA. Reproduced with permission (Chin 2003).

10.3.1 The cAMP second messenger system

cAMP was the earliest second messenger to be discovered and has been shown to regulate biochemical changes in a wide variety of target cells. Many first messengers act to stimulate cAMP synthesis, while others inhibit cAMP synthesis (see examples in Tables 10.2 and 10.3; and Figure 10.5). The stimulation of cAMP synthesis occurs via Gs and the inhibition of cAMP synthesis occurs via inhibitory G proteins (Gi). cAMP may have multiple actions once it is activated in a cell, for example: (1) alteration of the permeability of the cell membrane, via PKA, allowing ions such as Ca^{2+} to pass in and out of the cell; (2) changing the activity of enzymes by increasing their degree of phosphorylation; and (3) stimulation of protein synthesis through the actions of CREB in the cell nucleus. cAMP may also release calcium ions from intracellular stores and this calcium then acts as an intracellular messenger (see section 10.3.4) (for further reading, see Boron and Boulpaep 2005).

10.3.2 The cGMP second messenger system

cGMP is much less prevalent than cAMP, and is unevenly distributed throughout the body, with high levels in the brain, lung, heart, intestine and smooth muscle. cGMP acts as a second messenger for peptides such as ANP (Table 10.3). Also, as we saw in Chapter 5 (Figure 5.4), cGMP mediates the actions of nitric oxide (NO) in the brain. The activation of cGMP second messenger systems occurs in the same way as described for cAMP. When a ligand binds to its receptor, the G protein coupled to that receptor activates the enzyme guanylate cyclase that converts GTP to cyclic GMP. The cyclic GMP then activates protein kinase G which phosphorylates membrane or nuclear proteins. Alternatively, the guanylate

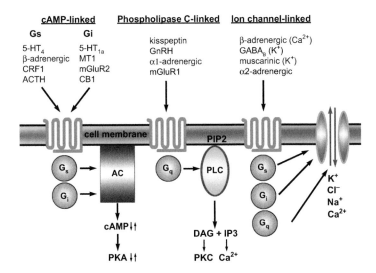

Figure 10.5 Specific examples of G-protein-coupled second messenger systems
A comparison of four different ways in which G protein receptors respond to various agonists to influence cell function. *The cAMP-linked system* responds to stimulatory signals (e.g. ACTH and CRH receptors) through the Gs subunit and elevates cAMP, which then activates protein kinase A (PKA). Conversely, inhibitory inputs such as cannabinoid (CB1) and serotonin (5-HT$_{1A}$) receptors inhibit cAMP and PKA via G$_i$ subunits. *The phospholipase C system (PLC)* is activated through stimulants such as kisspeptin and GnRH acting through Gq subunits. PLC subsequently liberates DAG and IP3 from the cell membrane, which then act as second messenger signals. Other G protein receptors (e.g. muscarinic and GABA) are *linked to ion channels* and these regulate entry of various ions such as K$^+$ and Cl$^-$.

Abbreviations: AC, adenyl cyclase; ACTH, adrenocorticotropin hormone; DAG, diacylglycerol; GABA, gamma-aminobutyric acid; IP3, inositol triphosphate; mGluR, metabotropic glutamate receptor; MT, melatonin receptor; PIP2, phosphatidylinositol diphosphate; PKC, protein kinase C; PLC, phospholipase C.

cyclase can be the intracellular domain of a receptor, as occurs in the atrial natriuretic factor receptor. Like cyclic cAMP, cGMP is deactivated by phosphodiesterase enzymes.

10.3.3 The inositol phospholipid (IP) – calcium second messenger system

Many peptides and neurotransmitters stimulate their target cells through the inositol phospholipid – calcium second messenger system (Figure 10.5). For example, receptors in target cells of the anterior pituitary activate the inositol phospholipid second messenger system in response to hypothalamic hormones such as TRH and CRH (Figures 10.3 and 10.7). Table 10.3 also shows that the IP system is involved in the secretion of aldosterone from the adrenal cortex in response to angiotensin II stimulation, the response of α_1-adrenergic receptors to catecholamines, and the response of the uterus to oxytocin. When one of these ligands binds to its receptor, a specific G protein (G$_q$) is activated. This protein is coupled to the enzyme phospholipase C (PLC) in the cell membrane and, when activated, PLC cleaves the membrane-bound lipid known as PIP2 (phosphatidylinositol diphosphate) into two components: inositol triphosphate (IP3) and diacylglycerol (DAG), each of which has distinct second messenger functions (Figure 10.5) (Purves *et al.* 2008). Diacyglycerol remains in the cell membrane and activates PKC, which then acts in a similar way to PKA by phosphorylating substrate proteins involved in a number of metabolic pathways. In the cell membrane, for example, it can phosphorylate peptide or neurotransmitter receptors, altering their affinity (and thus their sensitivity) to the peptides or neurotransmitters that bind to them. The other second messenger, IP3, enters the cytoplasm of the cell and releases Ca^{2+} from intracellular storage sites in the endoplasmic reticulum. Ca^{2+} ions then attach to their binding proteins (such as calmodulin) and act as intracellular messengers to facilitate secretion.

10.3.4 Calcium as a second messenger

Calcium ions are the most common intracellular messengers in neurons. They are involved in second messenger systems for a large number of neurotransmitters, peptides and growth factors, as shown by the examples in Table 10.3. Calcium is necessary for the action of cAMP, DAG and other second messengers and it functions as a second messenger in its own right. Calcium carries out its intracellular action by binding to calcium binding proteins, such as calmodulin, and it is this complex that activates protein kinases. Another protein, called calbindin, is present to maintain the levels of intracellular Ca^{2+} under tight control within physiological limits (Purves *et al.* 2008). Another critical protein that binds calcium ions is synaptotagmin. This is important for the secretion of neurotransmitters via the fusion of vesicles at the terminal membrane (see section 5.4; Figure 5.11). As a second messenger, calcium is important in metabolism, smooth muscle contraction, neurotransmitter release and hormone secretion. As noted above, norepinephrine, angiotensin II and a number of other ligands which activate the inositol phospholipid second messenger system stimulate the release of stored Ca^{2+} which then acts as a second messenger. Most cellular functions mediated by cAMP require the presence of calcium ions. For example, ACTH stimulation of the adrenal cortex requires calcium-bound calmodulin for adenylate cyclase activation of cAMP in order to cause the secretion of glucocorticoids.

The calcium ion second messenger system is additional to those for cAMP and cGMP and involves a variety of different calcium binding proteins including calmodulin, calbindin and parvalbumin. Calcium is stored inside the cell in calcium stores and is pumped out of the cell by calcium pumps. Thus, free Ca^{2+} levels are much lower inside the cell than outside, resulting in a calcium gradient across the cell membrane. The level of free (unbound) calcium inside the cell is regulated through two mechanisms. Calcium levels in the cell can be increased by opening the calcium channels in the cell membrane, allowing calcium to flow into the cell from the extracellular fluid, or by releasing intracellularly stored calcium. Finally, calcium channels in the cell membrane are also activated by GPCR, as are channels for Na^+, Cl^- and K^+ (Figure 10.5).

10.3.5 Tyrosine kinase second messenger systems

Many peptides and peptide hormones, such as insulin, growth hormone, leptin and insulin-like growth factor (IGF) use tyrosine kinases as second messengers. As examples, receptors for insulin and leptin are illustrated in Figure 10.6 (Schwartz *et al.* 2000). Note that the insulin receptor is a dimer, held together covalently, whereas the leptin receptor only exists as a dimer after leptin is bound in place. The bound ligands (insulin and leptin) activate

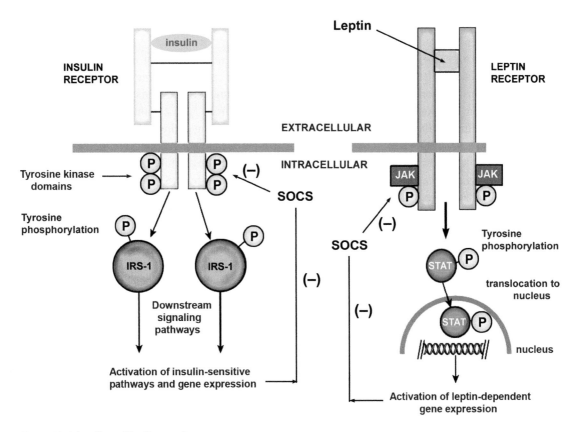

Figure 10.6 Insulin and leptin receptors

The insulin receptor comprises an extracellular subunit that enables binding of insulin, and an intracellular subunit that has intrinsic tyrosine kinase activity; that is, unlike the situation in Figure 10.4, where the protein kinase is located in the cytosol (PKA), the insulin receptor has a built-in kinase. Upon insulin binding, this kinase is activated and induces intracellular signaling proteins (IRS-1; insulin receptor substrate 1) by phosphorylating them on tyrosine residues. Activated IRS-1, in turn, regulates gene transcription, including the negative feedback regulator SOCS (*suppressor of cytokine signaling*) that prevents further stimulation by insulin. Unlike insulin receptors, however, *leptin receptors* do not have intrinsic tyrosine kinase activity. In contrast, the leptin receptor possesses docking sites for *janus kinases* (JAK), a family of tyrosine kinases. Thus, when leptin binds to its receptor, JAKS are bound to the intracellular receptor sites and these can then phosphorylate other chemical signals. The most important of these is STAT (*signal transduction and transcription*). Once phosphorylated, STAT travels into the cell nucleus and binds to response elements in target genes. One of these genes is *socs*, and its protein product SOCS exerts negative feedback on leptin stimulation at the receptor. Reproduced with permission (Schwartz *et al.* 2000).

the receptor tyrosine kinases. The difference between the receptors is that the insulin receptor kinase is part of the intracellular domain, and autophosphorylation occurs before phosphorylation of the signaling protein IRS-1 (*insulin receptor substrate 1*) takes place. IRS-1 is a second messenger. In contrast, the leptin receptor kinase exists as a separate entity called *janus kinase* (JAK) which is docked to the leptin receptor protein. JAK induces the phosphorylation of the intracellular second messenger STAT (*signal transduction and transcription*). Thus, just as we saw for PKA in Figure 10.4, here we see that the receptor kinases phosphorylate other intracellular messengers such as IRS-1 and STAT. It should go without saying that this is a relatively simple view of receptor tyrosine kinase action.

Interested students should look elsewhere for further details (for further reading, see Cohen 2006).

In an earlier section (10.1.5), we discussed the mechanisms of desensitization/down-regulation of G-protein-coupled receptors. In the case of tyrosine kinase receptors, the desensitization mechanism is particularly interesting because of its great clinical significance. For example, millions of patients suffering from Type 2 diabetes do not respond to insulin (*insulin resistance*) because their insulin signaling pathway is desensitized. In other words, insulin will bind to its receptor, but is unable to switch on intracellular signals. The reason for this is illustrated in Figure 10.6; a signaling inhibitor, SOCS (*suppressor of cytokine signaling*), is expressed following stimulation by IRS-1 and provides a negative feedback signal that prevents further activation of insulin receptors (Howard and Flier 2006). A similar explanation accounts for why obese patients do not respond to their own circulating leptin. Thus, *leptin resistance* is induced by negative feedback of SOCS that inhibits the leptin signaling pathway (Figure 10.6). This topic will be discussed in more detail in Chapter 12 (section 12.6.3; Figure 12.18).

10.3.6 Arachidonic acid derivatives as second messengers

In section 5.2.5, we introduced the idea that prostaglandins might function as neurotransmitters. These compounds are unusual in that they are not stored like neurotransmitters, in vesicles, but are produced on demand in almost every cell type, including neurons. There exists much better evidence that they act as intracellular signaling molecules. Their synthesis, like the inositol-phospholipid pathway (see section 10.3.3, above) depends on two enzymes – phospholipases C and A_2 – which act on cell membrane phospholipids to produce an intermediate compound called *arachidonic acid*. Arachidonic acid is further metabolized by enzymes such as COX-2 to prostaglandins PGE2 and PGF2α. A further pathway is mediated by the enzyme lipoxygenase, which yields another group of signaling molecules called *leukotrienes*. A detailed description of the involvement of prostaglandins and leukotrienes in neuroendocrine function is beyond the scope of this book, but they have been implicated in: (1) neurotransmitter secretion; (2) the production of fever; (3) pain pathways; and (4) neuroinflammation and neurodegenerative diseases (for further reading, see Phillis *et al.* 2006).

10.4 Interactions in second messenger systems

The second messenger systems described above may interact to regulate cellular activity. As an example, we saw in Figure 6.4 that CRH and ADH act simultaneously to induce the secretion of ACTH from the anterior pituitary gland. The mechanisms underlying this are shown in Figure 10.7. CRH activates cAMP and PKA, while vasopressin activates the inositol phospholipid-Ca^{2+} second messenger system, elevating intracellular Ca^{2+} levels and activating PKC. ACTH synthesis and release are then facilitated (Asa and Ezzat 2002). Another example is illustrated in Figure 9.9, where estradiol signals through nuclear receptors as well as through a G-protein-coupled receptor in the cell membrane that is itself coupled to two signaling mechanisms; namely a cAMP response and a PKC signal.

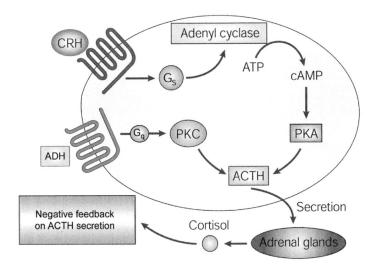

Figure 10.7 CRH and vasopressin (ADH) synergize in release of ACTH from anterior pituitary gland
In Figure 10.3, we showed how PRL secretion is controlled by two *opposing* receptor signals. This example illustrates the opposite: two receptors cooperating to increase ACTH secretion. The corticotroph expresses cell-surface receptors for corticotropin-releasing hormone (CRH) – a seven-transmembrane domain G-protein- (G_s-) coupled receptor. CRH binding activates Gs and leads to activation of adenyl cyclase and generation of cyclic AMP from ATP. In turn, cAMP activates intracellular protein kinase A (PKA). In contrast, signaling through the vasopressin receptor activates protein kinase C (PKC) via the Gq subunit. Signaling through both of these pathways induces secretion of adrenocorticotropic hormone (ACTH), resulting in cortisol production from the adrenal glands. This glucocorticoid binds to the glucocorticoid nuclear receptor to exert negative feedback on ACTH secretion. Reproduced with permission (Asa and Ezzat 2002).

10.5 Signal amplification

Although membrane receptors bind only to one ligand at a time, the biochemical activity stimulated in the target cell by that one ligand is amplified and prolonged through the second messenger cascade as shown in Figure 10.8. For example, a single molecule of NE binds to a β-adrenergic receptor that induces the formation of several molecules of cAMP via G_s and adenyl cyclase; that is, the original signal is amplified. Each molecule of cAMP activates one molecule of protein kinase A, but each activated protein kinase A molecule can phosphorylate many signal proteins, leading to changes in gene expression and cellular response. Thus, the second messenger system provides a mechanism for the rapid amplification of a singular signal from a hormone or neurotransmitter. In this way, a small quantity of the first messenger, which binds only briefly to its receptor, can lead to a high level of biochemical activity within a target cell and this biochemical activity can last for hours after the ligand has been deactivated. Purves *et al.* (2008) estimate that a single molecule of norepinephrine can generate thousands of molecules of cAMP.

10.6 Second messengers in the brain and neuroendocrine system

Each of the second messenger systems plays a role in signal transduction in the brain and neuroendocrine system. These signaling mechanisms have two major modulatory effects in

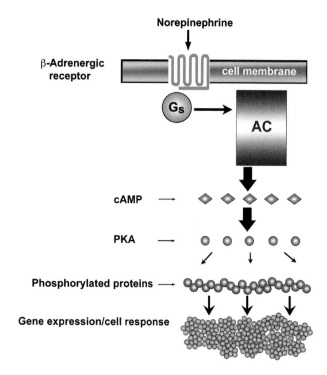

Norepinephrine

β-Adrenergic receptor

cell membrane

G$_S$

AC

cAMP

PKA

Phosphorylated proteins

Gene expression/cell response

Figure 10.8 Amplification in second messenger signaling

In this example, a single molecule of norepinephrine binds to the β-adrenergic receptor. This one-to-one binding generates the first amplification step by activating many G protein subunits that in turn allow adenyl cyclase to produce several molecules of cAMP. The next step – activation of protein kinase A (PKA) – is not amplified, but PKA will then phosphorylate and activate many signaling proteins. These signals in turn are further amplified to generate enhanced cellular responses.

Abbreviations: AC, adenyl cyclase; cAMP, cyclic adenosine monophosphate.

neural and neuroendocrine cells: they alter membrane permeability and regulate gene expression and protein synthesis. But above all, the sum total of the effects of all the signaling steps is to alter information processing by neurons.

10.6.1 Modulation of membrane permeability

When a neurotransmitter such as glutamate binds to its receptor at an *excitatory* synapse on a postsynaptic cell, the cell is depolarized as described in Chapter 5. The ability of the postsynaptic cell to generate action potentials in response to glutamate depends on the permeability of the cell membrane to Na^+ and K^+ ions (see Figure 5.10). This permeability is not just dependent on the incoming excitatory signal (glutamate), but is also controlled by other receptors, such as those for neuropeptides, peptide hormones and other neuromodulators such as GABA; that is, they can reduce the sensitivity to excitation by altering the permeability of ion channels so that depolarization is prevented or reduced. When such inhibitory receptors are present on presynaptic terminals, the net result is an inhibition of neurotransmitter secretion (Nestler *et al.* 2009). Other ion channel effects can readily be seen in tissues such as pituitary gonadotrophs. For example, GnRH induces secretion of LH and FSH via a G-protein-coupled receptor and a change in Ca^{2+} permeability. A further way of changing cell membrane ion permeability, and therefore neurotransmission, is to modify the number of receptors and ion channels that are present in the membrane. This is achieved by second messenger systems influencing gene expression so that fewer, or greater, numbers of channels are available. This is achieved by alterations in gene expression for the receptor/channel proteins.

10.6.2 Modulation of protein synthesis

The activation of second messengers can increase gene expression in neurons. For example, in neurosecretory cells in the hypothalamus and pineal gland the activation of cAMP by a neurotransmitter binding to its receptors stimulates the synthesis and release of oxytocin and melatonin, respectively. A non-neuronal example is the stimulation of hormone synthesis and release in the pituitary gland by hypothalamic hormones. Thus, CRH stimulates synthesis and release of ACTH (Figure 10.7) and TRH also regulates prolactin secretion (Figure 10.3).

At this point, it is worth emphasizing the different timescales of cellular responses to neurotransmitters or peptides binding to their membrane receptors. For example, muscle contraction, stimulated by the firing of nerves at the neuromuscular junction, is a very rapid response triggered by ion channel gated receptors (for acetylcholine) that takes only a few milliseconds to occur. Similar fast responses occur in the brain following glutamate-mediated opening of sodium channels (Figure 5.3). Single action potentials occur over a timescale of about 5 milliseconds. Subsequent changes in membrane permeability, which depend on the activation of cAMP or PKC (Figure 10.5), and which result in prolonged cell firing or in hormone release, may take minutes to occur. For example, as we previously noted, Figure 6.12 shows that in oxytocin neurons bursts of action potentials last from 5 to 15 seconds. This is sufficient to release oxytocin in sufficient quantities to reach the mammary glands to induce milk ejection. Long-term changes, such as hormone synthesis, learning and memory formation, all of which involve protein synthesis, take much longer to occur. The amplification of neurotransmitter signals by second messengers thus provides the cellular mechanism for the transformation of rapid synaptic activity into the long-term biochemical changes necessary for neural growth and differentiation and for habituation, sensitization and other forms of learning and memory to occur (for further reading, see Nicholls *et al.* 2001).

10.6.3 Drug action

Synthetic drugs that act as neurotransmitter agonists or antagonists (see Table 5.5 for examples) bind to receptors and act through the same second messenger systems as the neurotransmitters that they are mimicking or blocking. These drugs thus alter cellular activity by their effects on the second messenger cascade or by altering the level of enzymes such as phosphodiesterase, which deactivate second messengers. Through this mechanism, these drugs can alter the synthesis and release of neurohormones from neuroendocrine cells. Remember also, as discussed in section 10.1, that some G-protein-coupled receptors are therapeutic targets for several drugs (Table 10.1).

10.6.4 Receptor dysfunction

Disorders of receptors, G proteins and/or second messenger systems disrupt cellular activity, causing physiological and psychological disorders. For example, insulin and leptin resistance (receptor desensitization) occur in obesity and diabetes. Another neuroendocrine example is the case of down-regulated pituitary GnRH receptors following overstimulation (see section 7.4), resulting in abnormally low LH secretion (Figure 7.4). There are several examples of genetic mutations in G protein receptors; disabled GHRH receptors result in dwarfism because patients cannot respond to GHRH and therefore they have very low GH levels. Similarly, mutations in the GH receptor also result in dwarfism (Brooks and Waters 2010). Mutations in TRH receptors are known to cause hypothyroidism (for further reading, see Lania *et al.* 2006; and Engel and Schally 2007). Alzheimer's Disease may involve changes in protein kinase C, such that activators of PKC could be useful as cognition-enhancing therapeutics (Sun and Alkon 2010). Although schizophrenia is commonly thought to involve defects in dopamine function, new evidence implies that a reduction in mGluR3 receptors may account for some symptoms (Wroblewska and Lewis 2009). Finally, abnormalities in receptor-G protein coupling have been suggested as an underlying cause of mood disorders such as depression (for further reading, see Schreiber and Avissar 2007).

10.7 Comparison of neurotransmitter/neuropeptide and steroid hormone action at their target cells

As described in Chapter 9, steroid hormones enter their target cells and form steroid-receptor complexes. These then travel to the nucleus as a dimer, and bind to hormone response elements on target genes. In contrast, neurotransmitters and peptide hormones modify cell function via membrane receptors and the activation of second messenger systems. As shown previously in Figure 9.9, the cellular effects of these two types of signals

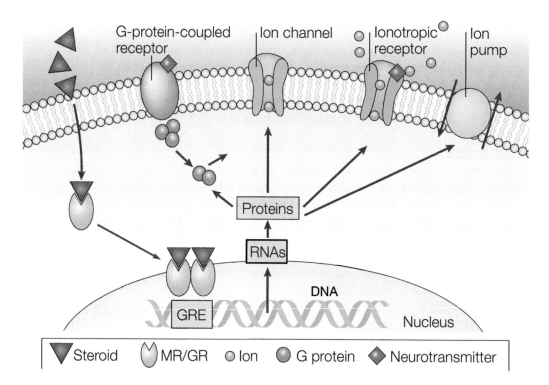

Figure 10.9 Intracellular steroid receptors regulate membrane permeability to ions

Adrenal steroids such as cortisol cross the cell membrane (red triangles, left), bind to intracellular receptors that dimerize and influence gene transcription via glucocorticoid response elements (GREs) on target genes (see also Figure 9.1). Cortisol stimulates the synthesis of several species of mRNA that encode different proteins such as receptors, ion channels, etc. In this way, cortisol regulates the membrane responsiveness of cells to stimulation.

Abbreviations: mRNA, messenger RNA; MR/GR, mineralocorticoid/glucocorticoid receptors. Reproduced with permission (de Kloet *et al.* 2005).

can be complementary; that is, they both control the responses of their target cells to neurotransmitters and neuropeptides by regulating the synthesis of receptors, G proteins and ion channels (genomic effects) or by altering membrane permeability (non-genomic effects). Another, more specific, example of such an arrangement is described by de Kloet *et al.* (2005). Figure 10.9 illustrates that binding of cortisol to intracellular MR and GR affects membrane permeability. This is accomplished by changes in gene transcription that result in the synthesis of G-protein-coupled receptors, ion channels, ionotropic receptors and ion pumps. Such alterations contribute to the sensitivity of the cell to stimulation with a variety of chemical signals.

10.8 Summary

This chapter examined the mechanisms by which peptide hormones, neurotransmitters and neuropeptides stimulate biochemical changes in their target cells by binding to receptors located in the cell membrane. Receptors are complex proteins, often having a number of subunits. The number of receptors that a cell has for each hormone, such as insulin, is not constant, but can be regulated homospecifically by insulin levels in the circulation, by other hormones or by non-hormonal factors. Ligand-bound cell surface receptors can be taken into the cell by endocytosis, where they are deactivated or recycled to the cell membrane. Once a ligand binds to its receptor, the signal is transduced by G proteins or, in the case of tyrosine kinase receptors, by autophosphorylation. G proteins activate enzymes that produce various second messengers within the cell. These second messengers amplify the original signal, so that a few first messengers can promote high levels of biochemical activity within the target cell. The best-known second messenger is cAMP, but cGMP, diacylglycerol, inositol triphosphate and calcium ions all act as second messengers, singly or in combination. Calcium ions, for example, whether free or bound to calcium-binding proteins, such as calmodulin, interact with the other second messengers to activate protein kinase enzymes in the cell membrane and cytoplasm. Protein kinase enzymes, such as PKC, phosphorylate substrate proteins to promote changes in cell membrane permeability and phosphorylate nuclear proteins to activate mRNA and protein synthesis. The second messenger systems described in this chapter function in the anterior pituitary (target cells of hypothalamic hormones), in the endocrine and non-endocrine target cells of pituitary hormones, and the target cells for insulin, glucagon and other peptide hormones in both the body and the brain. In addition, neurotransmitters and neuropeptides act at synaptic and non-synaptic receptors through the same second messenger systems that regulate neural growth, behavioral changes, emotional and motivational arousal, and short- and long-term memory. Steroid hormones act at intracellular and extracellular sites to modulate the responses of their target cells to neurotransmitters and peptides by regulating the number of receptors and G proteins in these cells. Disorders of receptors and second messenger systems result in physiological dysfunction that may underlie disorders such as depression, schizophrenia and Alzheimer's Disease.

FURTHER READING

Boron, W. F. and Boulpaep, E. L. (2005). *Medical Physiology*, updated edn. (Philadelphia, PA: Elsevier Saunders).

Cohen, P. (2006). "The twentieth century struggle to decipher insulin signalling," *Nat Rev Mol Cell Biol* 7, 867–873.

Engel, J. B. and Schally, A. V. (2007). "Drug insight: clinical use of agonists and antagonists of luteinizing-hormone-releasing hormone," *Nat Clin Pract Endocrinol Metab* 3, 157–167.

Grunberger, G., Taylor, S. I., Dons, R. F. and Gorden, P. (1983). "Insulin receptors in normal and disease states," *Clin Endocrinol Metab* 12, 191–219.

Hanyaloglu, A. C. and von Zastrow, M. (2008). "Regulation of GPCRs by endocytic membrane trafficking and its potential implications," *Annu Rev Pharmacol Toxicol* 48, 537–568.

Lania, A. G., Mantovani, G. and Spada, A. (2006). "Mechanisms of disease: mutations of G proteins and G-protein-coupled receptors in endocrine diseases," *Nat Clin Pract Endocrinol Metab* 2, 681–693.

Nestler, E. J., Hyman, S. E. and Malenka, R. C. (2009). *Molecular Neuropharmacology: A Foundation for Clinical Neuroscience*, 2nd edn. (New York: McGraw-Hill).

Nicholls, J. G., Martin, A. R., Wallace, B. G. and Fuchs, P. A. (2001). *From Neuron to Brain*, 4th edn. (Sunderland: Sinauer Associates).

Phillis, J. W., Horrocks, L. A. and Farooqui, A. A. (2006). "Cyclooxygenases, lipoxygenases, and epoxygenases in CNS: their role and involvement in neurological disorders," *Brain Res Rev* 52, 201–243.

Schreiber, G. and Avissar, S. (2007). "Regulators of G-protein-coupled receptor-G-protein coupling: antidepressants mechanism of action," *Expert Rev Neurother* 7, 75–84.

Vassart, G. and Costagliola, S. (2011). "G protein-coupled receptors: mutations and endocrine diseases," *Nat Rev Endocr* 7, 362–372.

REVIEW QUESTIONS

10.1   Cyclic AMP is a well-described second messenger involved in the activation of many cell types, including neurons. Name *three* other second messengers.

10.2   Steroid and non-steroid peptide hormones can both regulate gene expression, but they do so by different pathways. What is the difference?

10.3   Neuropeptides and neurotransmitters bind to receptors located in the neuron membrane. Based on structure alone, there are two major membrane receptor types. What are they?

10.4   Describe, in one sentence, how membrane receptor numbers are down-regulated.

10.5   Name two pathways that are activated following ACTH binding to a membrane ACTH receptor.

10.6   Which enzyme is activated by cyclic AMP to signal the nucleus to begin gene expression?

10.7   Which are the two ways in which G proteins can influence cyclic AMP synthesis?

10.8   In order to act as a second messenger, calcium ions must bind to a calcium __ __ within the cytoplasm of the cell.

10.9   Dopamine and TRH bind to the same cell (a lactotroph) to regulate PRL secretion. How do they do this?

10.10   The α subunits of G proteins stimulate/inhibit second messenger production. What role do the β and γ subunits serve?

10.11   One of the functions of G proteins is to control the production of intracellular cAMP. They also influence some components of the neuron cell membrane. Provide one example.

ESSAY QUESTIONS

10.1   Describe the significance of G proteins in the transduction of membrane signals in neurons.

10.2   Using the leptin, or insulin, receptor as an example, describe how membrane receptor tyrosine kinases communicate with the cell nucleus.

10.3   Outline the functions of calcium ions and calmodulin as second messengers in peptide hormone target cells.

10.4   Compare and contrast the mechanisms through which steroid and peptide hormones activate genomic responses in their target cells.

10.5   "Pituitary GnRH receptors are rapidly down-regulated by inappropriate stimulation with GnRH." Describe some of the experiments that confirm this statement, and provide one example of how this effect can be put to clinical use.

10.6   Describe how the inositol phospholipid second messenger system interacts with the cyclic AMP second messenger system to regulate physiological changes, such as secretion, in cells.

REFERENCES

Asa, S. L. and Ezzat, S. (2002). "The pathogenesis of pituitary tumours," *Nat Rev Cancer* 2, 836–849.

Boron, W. F. and Boulpaep, E. L. (2005). *Medical Physiology*, updated edn. (Philadelphia, PA: Elsevier Saunders).

Brooks, A. J. and Waters, M. J. (2010). "The growth hormone receptor: mechanism of activation and clinical implications," *Nat Rev Endocrinol* 6, 515–525.

Chin, L. (2003). "The genetics of malignant melanoma: lessons from mouse and man," *Nat Rev Cancer* 3, 559–570.

Clayton, R. N. (1988). "Mechanism of GnRH action in gonadotrophs," *Hum Reprod* 3, 479–483.

de Kloet, E. R., Joels, M. and Holsboer, F. (2005). "Stress and the brain: from adaptation to disease," *Nat Rev Neurosci* 6, 463–475.

Dluzen, D. E. (2005). "Unconventional effects of estrogen uncovered," *Trends Pharmacol Sci* 26, 485–487.

Dorsam, R. T. and Gutkind, J. S. (2007). "G-protein-coupled receptors and cancer," *Nat Rev Cancer* 7, 79–94.

Flores-Morales, A., Greenhalgh, C. J., Norstedt, G. and Rico-Bautista, E. (2006). "Negative regulation of growth hormone receptor signaling," *Mol Endocrinol* 20, 241–253.

Howard, J. K. and Flier, J. S. (2006). "Attenuation of leptin and insulin signaling by SOCS proteins," *Trends Endocrinol Metab* 17, 365–371.

Ichise, M., Kim, Y. J., Ballinger, J. R., Vines, D., Erami, S. S., Tanaka, F. *et al.* (1999). "SPECT imaging of pre- and postsynaptic dopaminergic alterations in L-dopa-untreated PD," *Neurol* 52, 1206–1214.

Lim, W. K. (2010). "GPCR drug pipeline: new compounds for CNS diseases," *Frontiers in CNS Drug Discovery* 1, 400–412.

Millar, R. P. and Newton, C. L. (2010). "The year in G protein-coupled receptor research," *Mol Endocrinol* 24, 261–274.

Nussey, S. S. and Whitehead, S. A. (2001). *Endocrinology: An Integrated Approach* (Oxford: Bios Scientific Publishers), www.ncbi.nlm.nih.gov/books/NBK20.

Purves, D., Augustine, G. J., Fitzpatrick, D., Hall, W. C., LaMantia, A.-S., McNamara, J. O. *et al.* (2008). *Neuroscience*, 4th edn. (Sunderland: Sinauer Associates).

Schwartz, M. W., Woods, S. C., Porte, D., Jr., Seeley, R. J. and Baskin, D. G. (2000). "Central nervous system control of food intake," *Nature* 404, 661–671.

Scotter, E. L., Abood, M. E. and Glass, M. (2010). "The endocannabinoid system as a target for the treatment of neurodegenerative disease," *Br J Pharmacol* 160, 480–498.

Smith, T. H., Sim-Selley, L. J. and Selley, D. E. (2010). "Cannabinoid CB1 receptor-interacting proteins: novel targets for central nervous system drug discovery?" *Br J Pharmacol* 160, 454–466.

Squire, L. R., Berg, D. E., Bloom, F. E., du Lac, S., Ghosh, A. and Spitzer, N. C. (2008). *Fundamental Neuroscience*, 3rd edn. (London: Academic Press).

Sun, M. K. and Alkon, D. L. (2010). "Pharmacology of protein kinase C activators: cognition-enhancing and antidementic therapeutics," *Pharmacol Ther* 127, 66–77.

Thathiah, A. and de Strooper, B. (2011). "The role of G protein-coupled receptors in the pathology of Alzheimer's disease," *Nat Rev Neurosci* 12, 73–87.

Vassart, G. and Costagliola, S. (2011). "G protein-coupled receptors: mutations and endocrine diseases," *Nat Rev Endocrinol* 7, 362–372.

Wettschureck, N. and Offermanns, S. (2005). "Mammalian G proteins and their cell type specific functions," *Physiol Rev* 85, 1159–1204.

Wroblewska, B. and Lewis, D. A. (2009). "Validating novel targets for pharmacological interventions in schizophrenia," *Am J Psych* 166, 753–756.

Neuropeptides I: classification, synthesis and co-localization with classical neurotransmitters

11

Many chemical messengers regulate neural activity, including neurotransmitters (Chapter 5), steroid hormones (Chapter 9) and peptide hormones (Chapter 10). This chapter, and Chapter 12, examines how the class of chemical messengers termed *neuropeptides* regulates neural activity. The topic of neuropeptides is now an extensive one and we divide the coverage into two parts. This chapter will describe the classification and synthesis of neuropeptides and their co-localization with classical neurotransmitters. Chapter 12 examines the functions of neuropeptides in the brain and neuroendocrine system.

11.1 Classification of neuropeptides

The realization that neuropeptides can act as neurotransmitters took place after the discovery of most of the "classical," or small molecule, neurotransmitters, such as NE, glutamate and ACh. Perhaps the simplest and obvious difference between these two classes of neurotransmitter is the size of the molecules; neuropeptides range in size from 2 to at least 40 amino acids, whereas a molecule such as NE is derived from a single amino acid (tyrosine) (see Table 7.1). Another difference is that neuropeptides are more versatile in their range of biological activities. For example, some peptide hormones are synthesized in endocrine glands, in fat cells and in the GI tract, but are also produced in the brain, where they act as neurotransmitters or neuromodulators. Another significant distinguishing characteristic of neuropeptides is their mode of synthesis. Classical neurotransmitters such as amino acids or catecholamines are formed by two or three enzymatic steps, often in the nerve terminal. Neuropeptides, on the other hand, are synthesized from large prepropeptides in the neuronal cell body (e.g. see section 7.2.1; Figure 7.2).

In terms of nomenclature, peptide hormones were traditionally and sensibly named after the first function they were known to serve. Thus, the pituitary hormones (ACTH, TSH, FSH, GH, etc.) were named for their actions at their target cells; for example, TSH stimulates the thyroid gland. Hypothalamic-releasing hormones (CRH, TRH, GnRH, GHRH, etc.) were named for their functions at pituitary target cells; and the hormones of the gastrointestinal (GI) tract (CCK, VIP, gastrin, etc.) were named based on their gastrointestinal functions. Nevertheless, when the GI peptides were discovered in the brain, this naming process resulted in considerable confusion, made worse when pituitary hormones were also found there. Thus, the name of a neuropeptide (as shown in

Table 11.1 Categories of mammalian brain peptides

Hypothalamic-releasing hormones	**Growth factors**
Thyrotropin-releasing hormone	Nerve growth factor
Gonadotropin-releasing hormone	Epidermal growth factor
Somatostatin	Fibroblast growth factors
Corticotropin-releasing hormone	Endothelial growth factor
Growth hormone releasing hormone	Brain-derived neurotrophic factor
Posterior pituitary	**Opioid peptides**
Vasopressin	β-endorphin
Oxytocin	Enkephalins
	Dynorphins
Anterior pituitary	Nociceptin (orphanin FQ)
Adrenocorticotropic hormone	
α-melanocyte-stimulating hormone	**Others**
Prolactin	Kisspeptin
Luteinizing hormone	Angiotensin II
Follicle-stimulating hormone	Bombesin
Growth hormone	Bradykinin
Thyroid-stimulating hormone	Calcitonin
β-endorphin	δ-sleep-inducing peptide
	Neuropeptide Y
Gastrointestinal peptides	Neurotensin
Vasoactive intestinal peptide	Thymosin
Cholecystokinin	Atrial natriuretic factor
Gastrin	Brain natriuretic factor
Substance P	
Neuropeptide Y	
Insulin	
Glucagon	
Secretin	
Motilin	
Pancreatic polypeptide	
Galanin	

Table 11.1) may indicate its hormonal function, but may not give any clues as to its role as a neuropeptide.

The human genome contains approximately 90 genes that encode neuropeptide precursors, and the *Neuropeptides Database* (www.neuropeptides.nl/) lists more than 100 neuropeptides and their parent gene sequences (for further reading, see Nestler *et al.* 2009; and Iversen *et al.* 2009). This means that each gene product – *a pre-propeptide* – is processed into several smaller peptides, each with individual biological activity (see section 11.2).

Table 11.2 **Endogenous opioid peptides**	
Precursor	Opioid peptide product
Proenkephalin	[Met]-enkephalin
	[Leu]-enkephalin
	Peptide E
	BAM 22P
	Metorphamide
Pro-opiomelanocortin	β-endorphin
Prodynorphin	Dynorphin A
	Dynorphin A (1–8)
	Dynorphin B
	α-neoendorphin
	β-neoendorphin
Pronociceptin/orphanin-FQ	Nociceptin/orphanin-FQ
	Endomorphin-1
	Endomorphin-2
Prodermorphin and	Dermorphin
Prodeltorphin	Deltorphin
	Deltorphin I
	Deltorphin II

11.2 Synthesis, storage, release and deactivation of neuropeptides

A significant portion of this topic was covered in earlier sections. Section 5.2.4 pointed out the differences between neuropeptide and classical neurotransmitter biosynthesis (see Figure 5.8). Neuropeptide and peptide hormone processing was covered in detail in section 7.2.1 (Figures 7.1 and 7.2). An additional useful example is that of the endogenous opioid peptides. Table 11.2 illustrates the complexity of the five known families of endogenous opioids: the endorphins, enkephalins, dynorphins, endomorphins and deltorphins (Imura *et al.* 1985; Corbett *et al.* 2006).

Three of these families are shown in Figure 11.1. Each family is derived from a different precursor protein: pro-opiomelanocortin (POMC), proenkephalin, prodynorphin, pronociceptin/orphanin-FQ and prodermorphin/prodeltorphin, respectively. α-MSH, ACTH and β-endorphin are derived from the POMC molecule (section 3.2.2; Figure 3.4). Similarly, proenkephalin contains four copies of leu-enkephalin and one of met-enkephalin. Prodynorphin also contains leu-enkephalin, but not met-enkephalin, and also encodes dynorphin A, dynorphin B and the neo-endorphins. All of the propeptides are cleaved to smaller peptides by *prohormone convertases* as described in section 3.2.2. This principle, of neuropeptides being derived from larger (pre-)propeptides, can be applied to any of the brain peptides (Nestler *et al.* 2009).

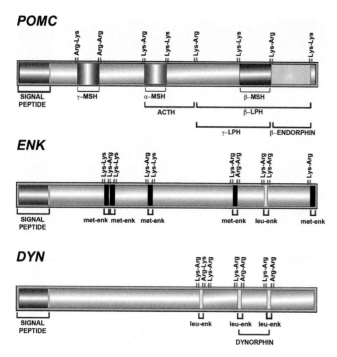

Figure 11.1 Schematic diagram of three opioid pre-propeptides

Pro-opiomelanocortin (*pomc*), enkephalin (*enk*) and dynorphin (*dyn*) genes encode large pre-propeptides that are proteolytically cleaved to a series of smaller, biologically active neuropeptides. The cleavage sites are pairs of amino acids as shown, either arginine-lysine (Arg-Lys), or Arg-Arg, or Lys-Lys junctions. These junctions are recognized by enzymes called pro-protein convertases that cleave the junctions to produce smaller peptides, as illustrated. The signal peptides are removed in the endoplasmic reticulum to produce the propeptides (see section 7.2.1). Note that knowledge of the pre-propeptide amino acid sequences allows us to predict which of the smaller peptides will be produced.

11.2.1 Storage and release

Neuropeptides are stored in large, dense-core secretory vesicles in nerve endings and they are released when the neuron is depolarized. In contrast, classical neurotransmitters such as ACh and GABA are usually stored in small synaptic vesicles. An important characteristic of neuro-peptide neurons is that they often contain both types of secretory vesicles, so some neurons actually have two or more neurotransmitters in their terminals. Table 11.3 lists some examples of co-localization of neuropeptides and small molecule neurotransmitters. In addition, there are other examples where two neuropeptides are co-localized; that is, some propeptides give rise to multiple, smaller peptides. This means that, for example, leu-enkephalin will be co-localized with dynorphin A and B in the same vesicles; and α-MSH will be stored and released along with β-endorphin. There are also many instances of classical neurotransmitters being co-localized, without a neuropeptide, for example: NE and ACh; glutamate and GABA; glutamate and DA (Trudeau and Gutierrez 2007; El Mestikawy *et al.* 2011).

What could be the purpose of more than one neurotransmitter being released from the same neuron simultaneously? First, we have to assume that receptors for both neurotrans-mitters/neuropeptides are localized postsynaptically. Second, if this is so, then perhaps the receptors exert opposite or even synergistic effects on postsynaptic neurons, thereby altering the excitability of that cell. There is evidence for such interactions (Nestler *et al.* 2009), and an example is shown in Figure 11.2 (Nusbaum *et al.* 2001; Burnstock 2004). This neuron terminal arrangement shows the co-localization of a small molecule neurotrans-mitter with a neuropeptide. When the presynaptic neuron is firing at low frequency, only the neurotransmitter is released and each action potential is followed by an excitatory

Table 11.3 Co-localization of neuropeptides and classical (small molecule) neurotransmitters

Neuropeptide	Neurotransmitter	Co-localization site
NPY	NE	Neurons of locus ceruleus; sympathetic neurons
Dynorphin	GABA	Striatal neurons
Substance P		
VIP	ACh	Parasympathetic neurons
Neurotensin	DA	Substantia nigra
CCK		
Galanin	GABA	Spinal cord interneurons
Somatostatin	GABA	Amygdala
Substance P	GABA	Amygdala
Orexin A	Glutamate	Hypothalamus
β-endorphin	GABA	POMC neurons
GnRH	Glutamate	Hypothalamus
CRH	"	"
TRH	"	"
Somatostatin	"	"
GHRH	GABA	Hypothalamus

A more complete list can be found in Torrealba and Carrasco (2004).

potential in the postsynaptic neuron (Figure 11.2A). However, an increase in the stimulation frequency induces the secretion of *both* the neuropeptide and the neurotransmitter. Thus, when the neuropeptide binds to its receptor, it prevents the excitatory effect of the neurotransmitter and induces a marked *hyperpolarization* in the postsynaptic neuron (Figure 11.2B).

11.2.2 Deactivation

Neurotransmitters such as NE are deactivated by specific enzymes and by reuptake into the presynaptic terminal. These mechanisms are illustrated in Figure 5.14 and discussed in section 5.6. Neuropeptides, on the other hand, are deactivated through degradation by relatively non-specific peptidase enzymes that exist throughout the brain and nervous system. For example, the enkephalins are degraded by "enkephalinases," but these same enzymes are active against many peptides. Another difference between neuropeptides and small molecule neurotransmitters is that the former are not inactivated via a reuptake mechanism.

11.3 Exploring the relationships among neuropeptides, neurotransmitters and hormones

Neuropeptides are closely and functionally related to both "true" hormones (such as insulin) and neurotransmitters (such as NE) and, as noted previously, the terminology used to describe these chemical messengers is often confusing (refer back to Table 1.1,

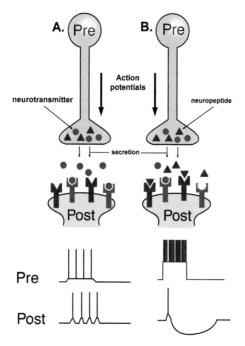

Figure 11.2 An example of co-localization and co-transmission with a neurotransmitter and a neuropeptide

On the left side of this figure (A), a presynaptic neuron fires action potentials at a relatively low frequency so that only the small molecule neurotransmitter (blue) is released. This is represented by the rapid onset, short-duration postsynaptic spikes at "Pre." Postsynaptically, the result is a succession of excitatory postsynaptic potentials ("Post"), time-locked to each presynaptic action potential. On the right (B), the firing frequency of the presynaptic neuron is increased, now inducing co-release of the neurotransmitter and the neuropeptide (red) and causing a qualitative change in the postsynaptic response. The high frequency train of presynaptic action potentials induces an initial, single spike followed by a smooth and prolonged hyperpolarization as a result of the inhibitory action of the co-released neuropeptide; that is, release of the neuropeptide overcomes the excitatory effect of the fast-acting neurotransmitter. Blue structures represent neurotransmitter receptors; red structures represent neuropeptide receptors. Reproduced with permission (Nusbaum *et al.* 2001).

for example). The relationships among these different neuroregulators can be examined in three ways: by looking at their common biosynthetic pathways, their common evolutionary pathways or their common embryological origins.

11.3.1 Common biosynthetic pathways

As we have seen, neuropeptides usually occur in families that are based on common pre-propeptide precursors. For example, ACTH, α-MSH and β-endorphin belong to the same family, because they are all derived from pro-opiomelanocortin (Figures 11.1 and 7.2). Likewise, the enkephalins are synthesized from proenkephalin (see section 11.2, above) and the tachykinins (Substance P, Neurokinin A, Neuropeptide K) are synthesized from pre-protachykinin (Nestler *et al.* 2009). As discussed in section 7.2.1, neuropeptide precursors undergo cleavage at different locations, often at lysine-arginine junctions, to give rise to unique peptides in different cells. For example, pre-procholecystokinin contains the sequences for all forms of CCK. The endocrine glands of the intestine produce the hormone CCK-33, while the neurons of the brain produce CCK-8 and the gastrointestinal nerves innervating the pancreas produce the small CCK-4 fragment. The catecholamines, indoleamines and steroid hormones also form families of chemical messengers in which different molecules are synthesized through common biosynthetic pathways in different cells (see Figures 5.8 and 7.3).

11.3.2 Common evolutionary pathways

The same families of hormones, neurotransmitters, peptides and cytokines are found in all vertebrates, and there are vertebrate-like hormones and neurotransmitters in

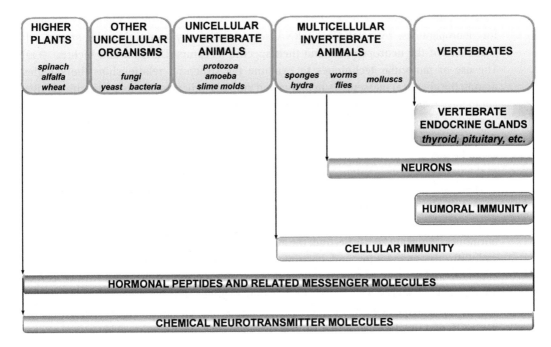

Figure 11.3 Evolutionary origins of chemical messengers
Neurotransmitter and neuropeptide molecules are present in higher plants and unicellular invertebrates. Specialized neurons evolved in the multicellular invertebrates, while endocrine glands did not exist until the vertebrates evolved. Cellular immunity may also have evolved in the multicellular invertebrates and humoral immunity in the vertebrates. The information in this figure is adapted from Roth *et al.* (1985), with permission.

multicellular and unicellular invertebrates and in the higher plants, as illustrated in Figure 11.3 (Le Roith *et al.* 1986; Roth *et al.* 1986). For example, substances similar to insulin, glucagon, somatostatin, substance P and ACTH are found in insects, crustaceans and molluscs, and vertebrate-like peptides are found in plants such as alfalfa (TRH), tobacco (somatostatin and interferon), spinach (insulin) and wheat (opioids). The chemicals that act as neurotransmitters and neuropeptides are thus phylogenetically very old, but specific neurons did not exist until the higher invertebrates evolved, and endocrine glands did not appear until the vertebrates evolved (see Figure 11.3). These findings led to the hypothesis that the neural, endocrine and immune systems evolved phylogenetically in a Darwinian fashion from a common neuropeptide system which exists in unicellular organisms. Thus, the chemical signals used in mammalian cellular communication may have evolved in different ways, some becoming neurotransmitters and others neuropeptides.

The phylogenetic relationships among the neuropeptides in some families are well known, as shown for the neurohypophyseal hormones oxytocin and vasopressin (Mohr *et al.* 1995). Likewise, the GnRH family consists of multiple neuropeptides found in the brains of fish, birds and mammals (Roch *et al.* 2011) and the opioid precursors, such as POMC, show phylogenetic relationships throughout the vertebrates (Cruz-Gordillo *et al.* 2010). The gut peptides (CCK, gastrin, glucagon, secretin and VIP) may all have evolved from a single VIP-like peptide (Dimaline and Dockray 1994; Baldwin *et al.* 2010).

Neuropeptide evolution can be studied by examining the structure of the genes that code for neuropeptides, by analyzing mRNA sequences, or by examining the amino acid sequences of the neuropeptides and their specific receptors (Darlison and Richter 1999). The use of molecular techniques to examine the structural relationships (homologies) among the chemical messengers in plants and animals indicates that neuropeptide families could have evolved from mutations or substitutions in the amino acid sequences of a small number of original biologically active peptides. There is recent molecular genetic evidence that evolutionary changes continue to occur in modern humans and that some of these could be linked to neuropsychiatric disorders (Moalic *et al.* 2010). For example, mutations in the vasopressin gene might be associated with alcoholism.

There are a number of reasons for thinking that chemical messengers evolved via changes in their amino acid sequences. First, the evolution of individual hormones can be traced through the vertebrates as illustrated by the evolution of the neurohypophyseal hormones (Minakata 2010). Second, the evolution of families of related peptides such as GH and PRL, the tachykinins and gastrointestinal hormones can be traced through their common amino acid sequences (Stewart and Channabasavaiah 1979). Third, certain amino acid sequences are shared by different peptides. For example, CCK and gastrin share peptide sequences, as do glucagon, secretin and VIP, even though these peptides have different functions. Fourth, different peptides with similar amino acid sequences have similar functions in different species. For example, secretin stimulates pancreatic function in mammals, but not birds, and VIP stimulates pancreatic function in birds, but not mammals (Krieger 1984). Details of the mechanisms involved in the molecular evolution of the peptides are discussed by Acher (1993).

11.3.3 Common embryological origins

As noted already (see section 2.2.7), some neuropeptides are biosynthesized in peripheral tissues as well as in the brain. For example, VIP, substance P, somatostatin, bombesin, neurotensin and NPY are GI hormones, but their genes are also expressed in brain cells. In addition, there are several cell types, usually referred to as *neuroendocrine cells*, which occur outside the brain, but behave in some ways just like neurons. Such cells, scattered throughout the body, have been called the "Diffuse Neuroendocrine System" (DNES) (for further reading, see Toni 2004). These cells are defined by the following criteria: (1) neuroendocrine cells produce a neurotransmitter and/or neuromodulator or a neuropeptide hormone; (2) these substances are localized within membrane-bound granules or vesicles to enable secretion by exocytosis; (3) neuroendocrine cells differ from neurons in that they lack axons and specialized nerve terminals; their mode of action is therefore likely to be autocrine/paracrine (Day and Salzet 2002). Knowledge of the embryological origin of the DNES makes clear that the cells are neural in nature. The developing embryo has three layers: the ectoderm, mesoderm and endoderm, and two of these are shown in Figure 11.4. The brain and spinal cord develop from the neural ectoderm, via the neural plate and neural tube. A separate group of cells, those of the neural crest, is also derived from the neural ectoderm. Neural crest cells differentiate into peripheral sensory neurons, neurons of the sympathetic nervous system, melanocytes, endocrine cells and myriad other cell types (Table 11.4; Crane and Trainor 2006; for further reading, see Gammill and Bronner-Fraser 2003). Some of the neurons listed in Table 11.4 are able to synthesize both

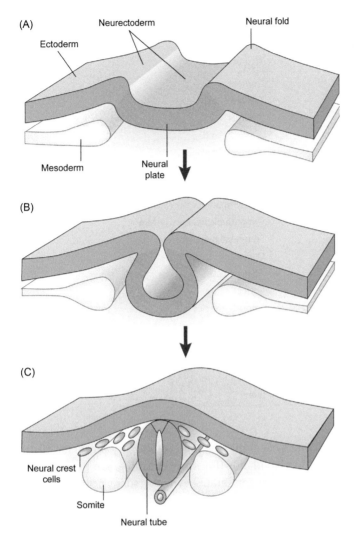

(A)

Ectoderm
Neurectoderm
Neural fold
Mesoderm
Neural plate

(B)

(C)

Neural crest cells
Somite
Neural tube

Figure 11.4 Embryological origins of the nervous system
The vertebrate embryo is shown as having two distinct layers (A): an outer ectoderm and an inner mesoderm. A third inner layer, the endoderm, is omitted for clarity. The central part of the ectoderm gives rise to the nervous system (see below), whereas the outer region develops into the epidermis (plus nails, hair and teeth). The mesoderm develops into somites, which are the precursors of muscles and bone. The endoderm gives rise to, for example, glands such as liver and the lining of the digestive tract. The neural ectoderm and neural plate undergo a physical transformation into the neural tube (B and C), with specialized cells called *neural crest cells* developing from the neural fold. The neural tube is transformed into the brain and spinal cord, whereas neural crest cells migrate away from the neural tube and give rise to several progeny, including neurons and glia of the autonomic ganglia, chromaffin cells of the adrenal medulla and neurons of the enteric nervous system. Reproduced with permission (Gammill and Bronner-Fraser 2003).

neurotransmitters and neuropeptides. A typical example is the chromaffin cell of the adrenal medulla, which secretes epinephrine. Chromaffin cells are innervated by preganglionic cholinergic axons (section 2.2.9 and Figure 5.20) and behave as postganglionic neurons even though they do not possess axons and nerve terminals. Depolarization of these cells induces the secretion of epinephrine and NE. However, chromaffin cells also express neuropeptides such as enkephalin and neuropeptide Y (Fraser *et al.* 1997; Henion and Landis 1992), and are also capable of expressing a variety of growth factors, including fibroblast growth factor (FGF-2) and TGF-β (Unsicker 1993). Thus, chromaffin cells conform to the criteria suggested for DNES cells, as outlined above.

Two further examples are: (1) the *β-cells of the pancreas* – their major product is the peptide insulin (see section 2.2.8), but serotonin is also biosynthesized in these cells and may regulate β-cell mass during pregnancy (Kim *et al.* 2010); and (2) *C-cells of the thyroid gland* that express calcitonin as well as serotonin (Russo *et al.* 1996). Pearse suggested that

Table 11.4 Cells and tissues derived from the neural crest

Cell types	Tissues/organs
Sensory neurons	Cornea
Cholinergic neurons	Teeth
Adrenergic neurons	Thymus
Glial cells	Thyroid gland
Schwann cells	Parathyroid gland
Chromaffin cells	Adrenal gland
Calcitonin-producing cells	Spinal ganglia
Melanocytes	Parasympathetic nervous system
Chondroblasts, chondrocytes	Sympathetic nervous system
Osteoblasts, osteocytes	Peripheral nervous system
Fibroblasts	Connective tissue
Striated myoblasts	Adipose tissue
Smooth myoblasts	Cardiac septa
Adipocytes	Smooth muscles
Mesenchymal cells	Pancreas
Enteric ganglionic neurons	Blood vessels
Pancreatic β cells	Endothelia
	Heart
	Brain

the cells of the diffuse neuroendocrine system may be the evolutionary precursor of the more specialized nerve cells and endocrine glands (for further reading, see Pearse 1983; Modlin *et al.* 2006). As the multicellular invertebrates evolved, some neuroendocrine precursor cells may have evolved into neurons that synthesized primarily neurotransmitters (Figure 11.3). In vertebrates, some neuroendocrine precursor cells may have evolved into neurons and some evolved into the peptide-secreting cells of the endocrine system. Still other neuroendocrine cells retained the ability to synthesize both amines and peptides (Le Roith *et al.* 1982). The hypothesis that distinct neural and endocrine systems evolved from the diffuse neuroendocrine system suggests that neurosecretory cells represent an intermediary cell type between the primitive neuroendocrine cell and the specialized neuron.

11.4 Coexistence (co-localization) of neurotransmitters and neuropeptides

Section 11.3.1 and Table 11.3 discussed the co-localization of neurotransmitters and neuropeptides. Proof of co-localization awaited the development of sophisticated double (and sometimes triple) immunohistochemical labeling using specific antibodies and/or *in situ hybridization*. For example, Landry *et al.* (1997) demonstrated co-localization of galanin and oxytocin/vasopressin and Meister and Hokfelt (1988) performed multiple labelings with GABAergic, catecholaminergic, cholinergic and neuropeptidergic neurons in the hypothalamus.

11.4.1 Synthesis, storage and release of co-transmitters

In order for a neuron to synthesize and release both neurotransmitters and neuropeptides, that cell must have both of the biosynthetic pathways illustrated in Figure 5.7. In such a cell, neuropeptide synthesis will begin with mRNA being translated to the pre-propeptide on the ribosomes and endoplasmic reticulum. The propeptide is formed after removal of the signal peptide (see Figure 7.1) and the completed peptides are formed inside the secretory vesicles by convertases while being transported down the axons. In contrast, neurotransmitter synthesis will be completed in the secretory vesicles at the nerve terminals through the action of the enzymes packaged into the synaptic vesicles along with the neurotransmitter transmitter precursors (Figures 5.7 and 5.8). Small molecule neurotransmitters tend to be localized to small vesicles, whereas neuropeptides are stored in the large vesicles. Thus, some neuron terminals will have both types of vesicle, and each will be released depending on the frequency of stimulation (see Figure 11.2).

11.4.2 Dale's Principle and the problem of co-localization

It was originally thought that each neuron had the ability to synthesize and release only one neurotransmitter. This was loosely termed "Dale's Principle" or "Dale's Law." As discussed above (section 11.4.1), the discovery of the co-localization of classical neurotransmitters and neuropeptides in the same neuronal terminals has forced a necessary re-examination of this principle (Burnstock 1976; Osborne 1981). At face value, Dale's Principle might seem to be incorrect. However, Dale's Principle stated that: "any one class of nerve cells operates at all of its synapses by the same chemical transmission *mechanism*." This is not the same thing as saying that a neuron uses only one neurotransmitter; that is, the mechanism may in fact be the release of two, or more, neurotransmitters. This means that the co-localization of classical neurotransmitters and neuropeptides within the same neuron does not violate this principle.

11.4.3 Which neurotransmitters and neuropeptides are co-localized?

There are a number of combinations of neurotransmitters and neuropeptides that can be co-localized in the same neuron. Table 11.3 reveals some of these combinations. It is important to emphasize here that co-localization means only that combinations of neurotransmitters and neuropeptides can be released from certain neurons. It does not mean that they are necessarily localized to the same secretory vesicle (but see below). We outlined various combinations in section 11.2.1. In most cases, neurotransmitters and neuropeptides are stored in separate vesicles. It is not difficult to see why this is so. For example, each neurotransmitter is derived from a distinct precursor, such as tyrosine (NE) or choline (ACh). These precursors are transported into neurons and their terminals using specific transporters. In addition, secretory vesicles also possess specific transporters for particular neurotransmitters and each contain the necessary enzymes for neurotransmitter biosynthesis. So each vesicle type manufactures and stores a specific neurotransmitter. In the case of neuropeptides, each one is derived from a specific gene and is localized, again, to a specific vesicle. An obvious exception to this rule is when more than one neuropeptide is derived from a single gene; for example, β-endorphin and α-MSH are produced from the pro-opiomelanocortin pre-propeptide (Figure 7.2), or leu-enkephalin and dynorphin are cleaved from the prodynorphin peptide (Figure 11.1). In other words, these pairs of peptides are made in the same vesicles and are therefore secreted together. A well-described example of the co-localization of a neuropeptide and a neurotransmitter is reviewed by Hrabovszky and Liposits (2008). These authors discuss the evidence for glutamate in GnRH neuron terminals (i.e. the terminals in the median eminence close to the portal vessels; see Figure 6.7): (1) GnRH and glutamate appear to be contained in separate vesicles; (2) GnRH neurons express the gene for the glutamate transporter vGluT2; that is, the transporter necessary for glutamate to be taken up into glutamate secretory vesicles; (3) vGluT2 and GnRH immunofluorescence were co-localized to the same neuron terminals; and (4) the terminals also possessed ionotropic glutamate receptors (Figure 5.3). The secretion of glutamate in the vicinity of presynaptic glutamate receptors is thought to regulate the release of GnRH via an autocrine effect (Figure 11.5). The review also provides evidence for glutamate co-localized with CRH, TRH, somatostatin, oxytocin and vasopressin, suggesting that glutamate co-localization may be a general phenomenon in hypothalamic neurosecretory cells (Hrabovszky and Liposits 2008).

However, there are some reports of co-localization of neurotransmitters and/or neuropeptides *in the same vesicle*. Whether these are rare occurrences needs to be determined, but the evidence for such co-localization exists for GnRH/galanin in hypothalamic neurons (Liposits *et al.* 1995), DA and CCK in striatal neurons (Llona *et al.* 1994) and dynorphin/oxytocin in hypothalamic neurohypophyseal neurons (Douglas and Russell 2001). These data imply that the pre-prohormones for GnRH and galanin are somehow processed synchronously in the cell body and then the propeptides are sorted and packaged into the same vesicle (Liposits *et al.* 1995; for further reading, see Merighi 2002). In the example of DA and CCK, it may be that the CCK-containing vesicles also possess a dopamine transporter for the uptake of dopamine. More recently, it seems that glutamate should be

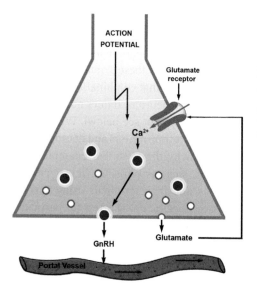

Figure 11.5 Autocrine regulation of GnRH secretion by glutamate

Gonadotropin-releasing hormone (GnRH) and glutamate are co-localized to GnRH neuron terminals. Action potential-induced depolarization of the terminal causes secretion of both GnRH and glutamate. Glutamate then acts in an autocrine fashion to release internal supplies of Ca^{2+} that induces further secretion of GnRH.

recognized as a co-transmitter, from the same vesicles, in cholinergic, 5-HT, GABA and catecholamine neurons (for further reading, see El Mestikawy *et al.* 2011).

Although no rules have been established to predict which chemical messengers will be co-localized, it appears that monoamines coexist most frequently with the gastrointestinal peptides. The neuropeptides derived from pro-opiomelanocortin (e.g. β-endorphin) are not co-localized with any of the classical neurotransmitters or neuropeptides derived from other precursors. In contrast, neuropeptides derived from proenkephalin and prodynorphin can be co-localized with classical neurotransmitters and other peptides (Kupfermann 1991).

11.4.4 Co-localization in the pituitary gland

The localization of neuropeptides in the *posterior pituitary* reflects the neural nature of this part of the gland; that is, neuropeptides are found in nerve terminals whose cell bodies are located in the PVN and SON (see, e.g., Figure 3.1). These neurons are rich in neuropeptides: oxytocin neurons co-localize oxytocin, CCK, dynorphin, met-enkephalin and CRH, whereas vasopressin neurons contain ADH, dynorphin, galanin and leu-enkephalin (Meister *et al.* 1990; Landry *et al.* 1997; for further reading, see: Meister 1993; and van Leeuwen *et al.* 1998). Oxytocin and vasopressin can also be localized to the same neurons (Telleria-Diaz *et al.* 2001). Note also that like releasing hormone neurons (section 11.4.3), oxytocin and vasopressin neuron terminals also contain glutamate (Hrabovszky *et al.* 2007). There is also evidence for co-localization in cells of the *anterior pituitary*. We already know that ACTH and β-endorphin are co-localized in corticotrophs because they are encoded by the same gene, propiomelanocortin (POMC; see Figure 3.4). Other examples include lactotroph cells (prolactin) and thyrotroph (TSH) cells that both synthesize VIP; gonadotroph cells (LH and FSH) also synthesize neuropeptide Y, angiotensin II, dynorphin and enkephalin; and somatotroph cells (GH) also synthesize neuropeptide Y (Houben and Denef 1990). Co-localization was also reported in human pituitary tumor tissue; neurotensin in TSH cells and gonadotrophs; VIP in gonadotrophs; and substance P in corticotrophs (Reyes *et al.* 2007). Also, as we saw for hypothalamic neurons, gonadotrophs, thyrotrophs and corticotrophs co-localize glutamate and express glutamate transporters (Hrabovszky *et al.* 2006; Kocsis *et al.* 2010). In human pituitary and pituitary tumor tissue, GABA has been localized to GH cells (End *et al.* 2005). Little is yet known about the functional consequences of these co-localizations.

11.5 Localization of neuropeptide cell bodies and pathways in the brain

Neuropeptidergic neurons are located throughout the brain. The study of these neurons is now such a huge research endeavor that it would be impractical to outline here the localization of all known neuropeptides. As described in earlier sections (e.g. Figure 9.11), powerful techniques are available to specifically localize any neurochemical of interest. Cloning of neuropeptide genes permits detection and localization of the mRNA by *in situ hybridization* (Figure 9.5), usually to the cell body. In addition, specific antibodies allow the visualization of neuropeptides by immunohistochemical and immunofluorescence methods. This latter

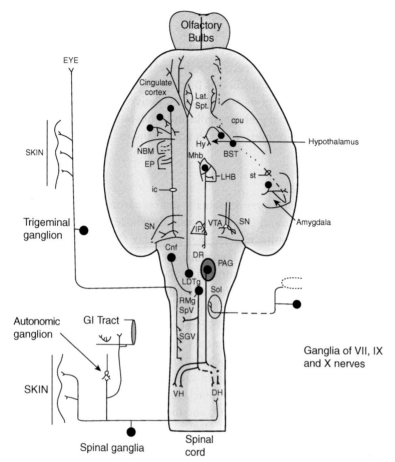

Figure 11.6 Pathways containing Substance P (SP) in the rat nervous system

The brain is shown in horizontal view. A detailed description of these pathways can be found in Ribeiro-da-Silva and Hokfelt (2000), but this information is unnecessary for our purposes in this book. It is sufficient to point out that SP is found in nerves throughout the nervous system, including neuroendocrine pathways such as those in hypothalamus, amygdala, autonomic ganglia and GI tract. Reproduced with permission (Ribeiro-da-Silva and Hokfelt 2000).

approach is capable of revealing detailed pathways of neuropeptidergic innervation throughout the brain and spinal cord. Pioneers in this field were T. Hokfelt and his co-workers (1980), who mapped the brain in terms of neuropeptides. For example, Figure 11.6 illustrates a typical detailed outline of the neuroanatomical localization of Substance P in the central and peripheral nervous systems (Ribeiro-da-Silva and Hokfelt 2000). Their paper also includes a discussion of co-localization of SP with serotonin.

Another example is the immunohistochemical labeling of hypothalamic oxytocin neurons. These cells were discussed earlier and their location in PVN and SON can be seen in Figure 6.10. The important point to remember is that each brain nucleus (PVN and SON) contains both types of neuron; that is, oxytocin neurons and vasopressin neurons are localized side by side and the individual peptides can be visualized using double-labeling (Ludwig and Leng 2006). Other neuropeptides present in PVN neurons include CRH (see section 4.3.2) and glucagon-like peptide (GLP-1), normally found in the GI tract (see section 2.2.7). Figure 11.7 illustrates the immunofluorescent localization of these three neuropeptides in PVN neurons (Tauchi *et al.* 2008). Note how this technique is sensitive enough to reveal GLP-1 neurons making contact with oxytocin neurons.

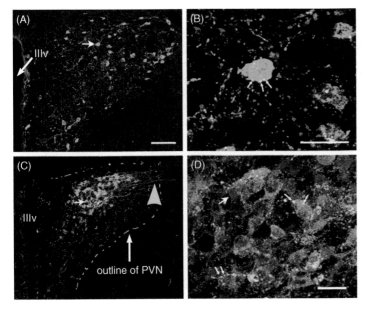

Figure 11.7 Immunofluorescent localization of oxytocin (OT), corticotropin-releasing hormone (CRH) and GLP-1 in the paraventricular nucleus (PVN)

Oxytocin-expressing cells in the PVN are shown in green, and GLP-1 fibers in orange (A). The oxytocin cell marked with an arrow in (A) is magnified in (B) and shows that OT neurons were apposed by GLP-1-immunoreactive boutons, suggesting synaptic contacts. GLP-1-OT appositions (yellow) are indicated by small arrows. (C) CRH immunopositive cells localized in the PVN, corresponding to the region of densest GLP-1 labeling. Note the CRH fibers coursing away from the PVN (arrowhead). In (D), the magnified image in (C) (small arrow) shows evidence for GLP-1-CRH appositions on numerous cells in this region. GLP-1-CRH appositions are indicated by small arrows.

Abbreviation: IIIv, third ventricle.
Reproduced with permission (Tauchi *et al.* 2008).

These examples are typical of many investigations into the localization of neuropeptides in the brain and spinal cord. The localization of several hypothalamic neuropeptide genes, by *in situ hybridization*, in the *human brain* is illustrated in Figure 11.8. This report by Krolewski *et al.* (2010) shows the localization of the expression patterns for vasopressin, CRH, MCH (melanin concentrating hormone) and orexin genes.

While neuropeptides are found throughout the brain in varying concentrations, and very few brain areas lack neuropeptides, the regions which have the highest concentration of neuropeptides are the neuroendocrine regions of the hypothalamus and median eminence, which contain nearly 40 neuropeptides (Merchenthaler 1991). The amygdala, hippocampus, other areas of the limbic system and the medulla and pons of the brain stem also have high concentrations of neuropeptides. Nieuwenhuys (1985) has pointed out that brain areas with the highest concentrations of neuropeptides are often coincident with those that have high densities of steroid hormone receptors (Figures 9.11 to 9.13). It is not difficult to find examples of such interactions for any neuropeptide of interest. For example, steroid hormone/neuropeptide interactions are important in attempts to understand the neural basis of social anxiety (Choleris *et al.* 2008) and even in human social-emotional interactions (Bos *et al.* 2012). In addition, inspection of Figure 5.15 reveals that amine neurotransmitters are also well represented in these same brain areas.

11.5.1 Species and sex differences in neuropeptide localization

As noted, it is beyond the scope of this book to itemize the locations of all neuropeptides in several species. It is a general rule that neuropeptides occur in the same brain areas in all mammals, but there may be species differences in the exact neural locations of both the

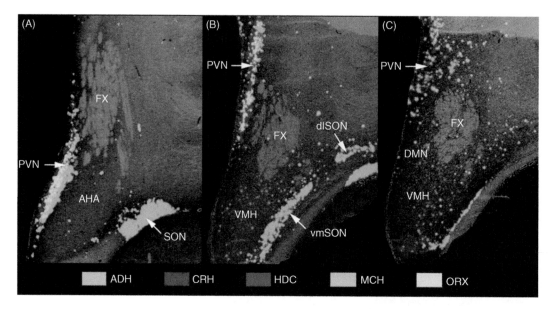

Figure 11.8 Localization of neuropeptide genes by *in situ* hybridization in human brain tissue
The three brain sections in this figure are from the region of the paraventricular and supraoptic nuclei (PVN and SON). The midline (third ventricle) is to the left. Each gene product is color-coded as indicated at the bottom of the figure. In (A), PVN neurons express ADH (vasopressin), CRH and orexin. In contrast, SON neurons express only ADH. In (B), at a slightly different hypothalamic location, PVN and SON labeling remains the same as in (A), but some isolated neurons are also labeled for MCH (melanin concentrating hormone; blue). In (C), MCH neurons are more numerous and now we can also observe labeling for histidine decarboxylase (purple; HDC), which is a marker for the neurotransmitter histamine.
   Abbreviations: AHA, anterior hypothalamic area; dlSON, dorsolateral supraoptic nucleus; DMN, dorsomedial hypothalamic nucleus; FX, fornix; PVN, paraventricular nucleus; SON, supraoptic nucleus; VMH, ventromedial hypothalamic nucleus; vmSON, ventromedial supraoptic nucleus. Reproduced with permission (Krolewski *et al.* 2010).

neuropeptide cell bodies and the receptors for these neuropeptides. Students interested in particular neuropeptides are referred to PubMed (www.ncbi.nlm.nih.gov/pubmed/). Of more interest are sex differences in the neural distribution of neuropeptides. For example, females have higher levels of brain oxytocin than males and have oxytocin in some neural areas in which it is absent in males (Häussler *et al.* 1990). There are also sex differences in vasopressin distribution (Semaan and Kauffman 2010; de Vries and Panzica 2006). A further striking example is the sex difference in hypothalamic *kiss1* expression (female >> male) reported by Kauffman (2010) (see Figure 11.9). This difference is thought to account for sex differences in puberty onset and in the ability of women to generate a positive feedback of estradiol on luteinizing hormone (LH) secretion (see Figure 12.11). In general, however, differences in neuropeptide levels/distribution are due to the organizational effects of gonadal steroids on neural growth and differentiation during prenatal development (see section 9.9.5).

11.6 Neuropeptide receptors and second messenger systems

This topic was covered in detail in Chapter 10. Many neuropeptides bind to G-protein-coupled membrane receptors. For example, Figure 10.3 illustrates the binding of TRH to

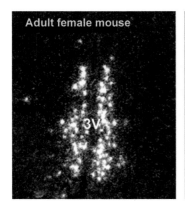
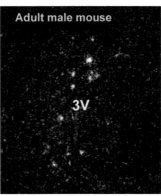

Figure 11.9 Sex difference in *kiss1* gene expression in mouse hypothalamus
kiss1 mRNA was detected by *in situ* hybridization. These two brain sections, from adult male and female mice, were obtained from a region of the hypothalamus called the anteroventral periventricular nucleus–periventricular nucleus continuum (AVPV/PeN). Females obviously have significantly more *kiss1*-positive neurons (denoted by clusters of white silver grains) in the AVPV/PeN than do males.
Abbreviation: 3V, third ventricle.
Reproduced with permission (Kauffman 2010).

G-protein-coupled receptors in pituitary thyrotrophs. Figures 10.7 and 12.4 show similar binding for CRH and kisspeptin, respectively. Other neuropeptides, such as leptin and growth factors, interact with tyrosine kinase receptors (Figure 10.6). The second messengers involved may be any of the ones covered in Chapter 10; that is, cAMP, ions, STAT proteins, etc. Some neuropeptides can bind to synaptic receptors, where they act as neurotransmitters or co-transmitters, while others, such as opioids, may bind to distant non-synaptic receptors, where they act as neuromodulators. Through the second messenger cascade, neuropeptides regulate receptor levels and alter membrane permeability, increasing or decreasing the sensitivity of the cell to neurotransmitter stimulation as shown in Figure 10.9. Neuropeptides can also regulate protein synthesis in their target cells, altering the synthesis, storage and release of neurotransmitters, neuropeptides or other proteins.

11.6.1 The importance of neuropeptide agonists and antagonists

The identification and classification of receptors for the small-molecule neurotransmitters listed in Table 5.3 relied on the use of synthetic agonist and antagonist drugs which selectively bind to specific membrane receptor proteins as discussed in section 5.8. Section 10.1 discussed the clinical importance of drugs acting at G-protein-coupled receptors (Table 10.1). Once the effects of such drugs are known, they can be used to treat clinical disorders of neurotransmitter action. In order to investigate neuropeptide receptors, to determine their actions on second messenger systems and target cell activity, and to develop clinical treatments for neuropeptide-related disorders, it is important to develop synthetic neuropeptide agonists and antagonists. Natural neuropeptides are rapidly metabolized, poorly transported from the blood to the brain and cannot be given orally. *Synthetic* neuropeptide agonists and antagonists have been successfully synthesized for a number of neuropeptides, with promising therapeutic application. A selection of these is shown in Table 11.5. Important advances have also been made in developing *non-peptide drugs* for neuropeptide receptors. For example, non-peptide vasopressin antagonists (*Vaptans*) may be useful in treating hypervolemia (elevated blood volume) resulting from heart failure or kidney failure (Decaux *et al.* 2008; Fields and Bhardwaj 2009).

Table 11.5 **Some neuropeptide drugs and their therapeutic applications**

Receptor	Drug and clinical application	Reference
Somatostatin	Octreotide (agonist) (*acromegaly*)	Heaney and Melmed 2004
GH	Pegvisomant (antagonist) (*acromegaly*)	Manjila *et al.* 2010
NPY	Synthetic peptides (agonist and antagonist) (*obesity*)	Sato *et al.* 2009
Somatostatin	Lanreotide (agonist) (*GI tract tumors*)	Appetecchia and Baldelli 2010
GnRH	Buserelin; Cetrorelix (agonist and antagonist) (*prostate cancer; ovarian cancer; breast cancer*)	Engel and Schally 2007
CRH	Synthetic peptides (antagonist) (*alcoholism*)	Lowery and Thiele 2010
Oxytocin	Synthetic peptides (agonist and antagonist) (*autism; osteoporosis; diabetes*)	Viero *et al.* 2010
Orexin	Synthetic peptides (antagonist) (*sleep promotion*)	Coleman and Renger 2010
Nociceptin/Orphanin FQ	Synthetic peptides (agonist and antagonist) (*pain*)	Largent-Milnes and Vanderah 2010
Substance P	Casopitant (antagonist) (*depression*)	Di Fabio *et al.* 2011

11.6.2 Identification and localization of neuropeptide target cells in the brain

At the beginning of this chapter, the identification and localization of neuropeptide cell bodies and pathways was discussed. To fully understand neuropeptidergic function, it is necessary to identify their target cells; that is, those cells that express neuropeptide receptors. The older techniques of receptor binding and autoradiography have been superseded by techniques of immunohistochemistry (for receptor proteins) and *in situ hybridization* (for receptor mRNA) as was described for the identification of steroid hormone receptors in section 9.2. For many neuropeptide receptors, the distribution does not always match the distribution of the cells that make the peptide, simply because the target cells may be located some distance away. This is especially true for neuropeptides, which diffuse away from the terminals to act in a neuromodulatory fashion, rather than to remain in the synaptic cleft where postsynaptic receptors are usually located. The cloning of neuropeptide receptor genes has allowed the precise localization of many receptor mRNAs in the brain. Using opioid receptors as an example, Table 5.3 revealed that there are several opioid receptor subtypes that appear to "prefer" particular opioid peptides, although there is some overlap in these preferences (Julius 1997). For example, β-endorphin binds mostly to μ- and δ-receptors; enkephalins are fairly specific to δ-receptors, but with some binding to μ-receptors; dynorphins bind to κ- and μ-receptors. The newer opioid peptides such as endomorphins and deltorphins (see Table 11.2) bind specifically to μ-receptors and δ-receptors, respectively. A fourth receptor, ORL1 (*opioid receptor-like receptor*) binds nociceptin/orphanin-FQ

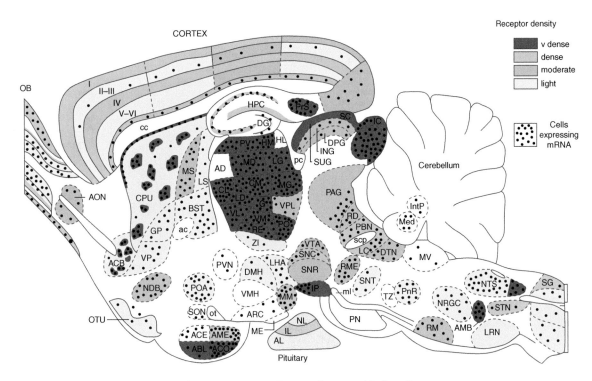

Figure 11.10 Distribution of μ-opioid receptor mRNA and μ-opioid receptor binding sites
μ-opioid receptors are found throughout the brain. This map also reveals that some brain areas, such as the thalamus (e.g. CM, CL and MD), have a high density of μ-opioid receptor mRNA coincident with high levels of receptor sites. Other regions, such as layers I and IV of the cortex, have dense binding (i.e. many receptor sites), but no expression of receptor mRNA. In this case, the cell bodies that express the receptor mRNA could be located in different layers, such as II and III. These cells would then transport the receptor proteins to their terminals in layers I and IV.
    Abbreviations: CL, centrolateral thalamus; CM, centromedial thalamus; MD, dorsomedial thalamus; OB, olfactory bulb; POA, preoptic area; PVN, paraventricular nucleus. A full list of abbreviations can be found in Mansour *et al.* (1995b). Reproduced with permission (Mansour *et al.* 1995b).

(Table 11.2; Mollereau and Mouledous 2000). The cloning of these receptor genes allows the localization of their mRNAs and Figure 11.10 illustrates just such a map for μ-opioid receptors (Mansour *et al.* 1995b; for further reading, see Mansour *et al.* 1995a). The figure also shows that the cells that express the receptor gene (i.e. mRNA) may not always coincide with the location of the receptor protein (detected by receptor binding). This implies that the receptor protein is transported away from the cell body, often over long distances. With this information on receptor distribution, it becomes possible to investigate the many neural systems in which opioids are implicated; for example, pain pathways; reproduction; water balance. This approach can be applied to any receptor for any neuropeptide of interest.

11.7 Neuropeptides and the blood-brain barrier

The cells and extracellular fluid of the brain are efficiently quarantined from the main circulatory system by the blood-brain barrier (BBB). This barrier helps to homeostatically

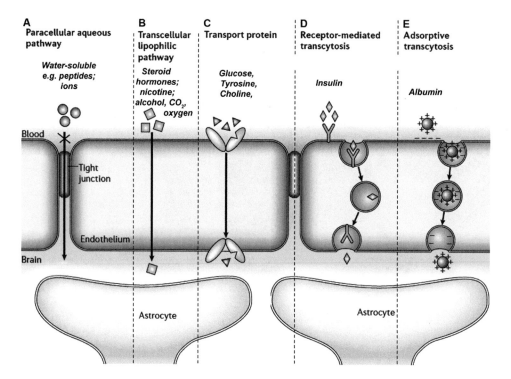

Figure 11.11 Entry sites for various ions, molecules, hormones, etc. to cross the blood-brain barrier
Schematic representation of the endothelial cells, and the tight junctions that connect them, forming the blood-brain barrier (BBB). There are several routes for molecular traffic across the BBB: (A) the tight junctions severely restrict penetration of water-soluble compounds, including peptides; (B) in contrast, the lipid membranes of the endothelium readily allow passage of lipid-soluble molecules such as steroid hormones; (C) some substances, such as tyrosine, choline and glucose, are actively transported across the BBB by specific transporters; (D) certain proteins, such as insulin, are taken up by specific receptor-mediated endocytosis and transcytosis as a means to cross the BBB; (E) native plasma proteins such as albumin are poorly transported, but cationization (acquiring a positive charge) can increase their uptake by endocytosis and transcytosis. Most CNS drugs (e.g. nicotine, narcotics, alcohol) enter via route (B). Reproduced with permission (Abbott *et al.* 2006).

maintain the ionic and chemical environment of the brain. For example, blood levels of amino acids and peptides can change dramatically following a meal; blood K^+ and Na^+ levels can vary widely with exercise; toxins may flood the system in illness, and hormone levels can fluctuate with sleep and age. The brain has to be protected from these changes that might alter neurotransmission and neuronal excitability. The BBB consists of tight junctions between the endothelial cells of blood capillaries that prevent blood components from entering the brain (Figure 11.11).

The figure illustrates that: (A) water-soluble peptides, ions and water-soluble drugs cannot enter the brain through the tight junctions; on the other hand, (B) lipid-soluble molecules such as steroid hormones, alcohol, heroin and nicotine, and gases such as oxygen, can readily enter the brain by diffusing across the endothelial cell membranes; (C) some compounds, such as glucose, tyrosine and choline, which would normally be excluded from the brain are actively pumped into the brain via specific transporters (e.g. Figure 5.8); also, (D) large molecules such as insulin require a receptor-mediated transport mechanism in order to enter the brain; (E) large, blood plasma proteins such as albumin are normally restricted from entering the brain, but small amounts can gain access via adsorptive transcytosis. Thus, we can see that the BBB is a selective barrier that acts to protect the brain (Abbott *et al.* 2006).

In contrast, there are some areas of the brain where the BBB is absent. These regions are called *circumventricular organs* (CVO) and their blood capillaries do not have tight junctions, thus permitting entry of peptides into the brain (Boron and Boulpaep 2005). Such capillaries are said to be *fenestrated* ("having perforations"). The location of circumventricular organs is shown in Figure 11.12 (Dantzer *et al.* 2008). As described in Chapter 12, one of these, the median eminence, is particularly important for peptide feedback. For example, fat cells and GI tract cells respond to food intake by secreting peptides such as ghrelin and leptin, which then convey information into the brain via the median eminence.

A major consequence of an intact BBB is the difficulty of treating CNS diseases through use of therapeutic drugs, especially peptides, which have difficulty entering the brain. This problem is being attacked in various ways, one of which is to design pro-drugs that readily cross the BBB before being enzymatically converted to the active drug. A good example is the use of morphine to treat chronic pain. Morphine is a polar molecule that does not readily enter the brain. Heroin, however, is much more lipid soluble and can freely enter the brain, where it is then converted back to morphine (Begley 2004; for further reading, see Pardridge 2007; and Pasha and Gupta 2010). Another example is the use of L-Dopa to treat Parkinson's disease. These patients have abnormally low levels of brain DA because of the

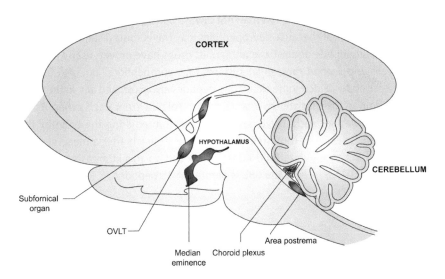

Figure 11.12 Location of the brain circumventricular organs
Regions of the brain where the blood-brain barrier (BBB) is absent or leaky are called circumventricular organs. This schematic diagram illustrates these sites. Notice that one of them is the hypothalamic median eminence, an important conduit for many peptide and protein hormones that enter the brain from the circulation.

Abbreviation: OVLT, organum vasculosum of the laminae terminalis. Reproduced with permission (Dantzer *et al.* 2008).

loss of dopaminergic neurons. DA will not cross the BBB, but its precursor, L-Dopa, can do so (see Figure 5.8). This means that patients can be treated with L-Dopa. An alternative and controversial viewpoint on the ability of peptides to cross the BBB is provided by Kastin and Pan (2010). These authors have evidence that circulating peptides can, and do, enter the brain. Whether these studies will lead to advances in therapeutics remains unknown.

11.8 Summary

This chapter describes some of the different classes of neuropeptides and compares the mechanisms for their synthesis, storage and release with those for the classical, small molecule, neurotransmitters. Like peptide hormones, neuropeptides are synthesized as pre-prohormones on the ribosomes and stored in secretory granules where they undergo further modification, to smaller fragments, by proprotein convertases. The inter-relationships among the neuropeptides, neurotransmitters and hormones are examined through their common biosynthetic pathways, common evolutionary pathways and common embryological origins. Certain peripheral neuroendocrine cells are able to produce both monoamine neurotransmitters and neuropeptides and show plasticity during development, with some becoming neurons and others becoming endocrine cells. In the brain, many neurons synthesize and store both monoamine neurotransmitters and neuropeptides. They may not necessarily be stored in the same vesicles, but two, or more, neurotransmitters/neuropeptides can be co-released when the cell is stimulated. Co-localization of peptides is also observed in the pituitary gland. All areas of the brain synthesize neuropeptides, and the hypothalamus is a particularly rich source. The areas of maximal neuropeptide receptor concentration may not always correspond to the areas of the highest density of cell bodies, simply because neuropeptides may travel some distance from their release sites. Synthetic neuropeptide agonist and antagonist drugs have proven useful, not just in identifying target cells and second messenger systems for neuropeptides, but also for the treatment of some disease states. Since some circulating peptides also exist, and act, in the brain as neuropeptides, it is important that the two systems are isolated from each other. Thus, peripheral peptides, and molecules such as NE, are prevented from entering the brain by the BBB. However, other peptides, such as those released from the stomach after a meal, need to reach brain receptors in order to regulate food intake. They can do this in areas such as the median eminence and other CVOs, where the BBB is permeable to peptides.

FURTHER READING

El Mestikawy, S., Wallen-Mackenzie, A., Fortin, G. M., Descarries, L. and Trudeau, L. E. (2011). "From glutamate co-release to vesicular synergy: vesicular glutamate transporters," *Nat Rev Neurosci* 12, 204–216.

Gammill, L. S. and Bronner-Fraser, M. (2003). "Neural crest specification: migrating into genomics," *Nat Rev Neurosci* 4, 795–805.

Iversen, L. L., Iversen, S. D., Bloom, D. E. and Roth, R. H. (2009). *Introduction to Neuropsychopharmacology* (New York: Oxford University Press).

Mansour, A., Fox, C. A., Burke, S., Akil, H. and Watson, S. J. (1995a). "Immunohistochemical localization of the cloned mu opioid receptor in the rat CNS," *J Chem Neuroanat* 8, 283–305.

Meister, B. (1993). "Gene expression and chemical diversity in hypothalamic neurosecretory neurons," *Mol Neurobiol* 7, 87–110.

Merighi, A. (2002). "Costorage and coexistence of neuropeptides in the mammalian CNS," *Prog Neurobiol* 66, 161–190.

Modlin, I. M., Champaneria, M. C., Bornschein, J. and Kidd, M. (2006). "Evolution of the diffuse neuroendocrine system – clear cells and cloudy origins," *Neuroendocrinology* 84, 69–82.

Nestler, E. J., Hyman, S. E. and Malenka, R. C. (2009). *Molecular Neuropharmacology: A Foundation for Clinical Neuroscience*, 2nd edn. (New York: McGraw-Hill).

Pardridge, W. M. (2007). "Blood-brain barrier delivery," *Drug Discov Today* 12, 54–61.

Pasha, S. and Gupta, K. (2010). "Various drug delivery approaches to the central nervous system," *Expert Opin Drug Deliv* 7, 113–135.

Pearse, A. G. (1983). "The neuroendocrine division of the nervous system: APUD cell as neurons or paraneurons" in N. N. Osborne (ed.), *Dale's Principle and Communication between Neurons* (Oxford: Pergamon Press), pp. 37–48.

Toni, R. (2004). "The neuroendocrine system: organization and homeostatic role," *J Endocrinol Invest* 27, 35–47.

van Leeuwen, F. W., Verwer, R. W., Spence, H., Evans, D. A. and Burbach, J. P. (1998). "The magnocellular neurons of the hypothalamo-neurohypophyseal system display remarkable neuropeptidergic phenotypes leading to novel insights in neuronal cell biology," *Prog Brain Res* 119, 115–126.

REVIEW QUESTIONS

11.1   Describe two differences between neuropeptides and so-called "classical neurotransmitters."

11.2   What is the primary difference in the synthetic pathways of classical neurotransmitters versus neuropeptides?

11.3   Which important process takes place inside secretory vesicles in order that neuropeptides and neurotransmitters can be secreted?

11.4   Some neurons biosynthesize both monoamine transmitters and neuropeptides. Which are stored in small vesicles and which are stored in large vesicles in the nerve terminal?

11.5   In general, each neurotransmitter (such as norepinephrine or serotonin) is stored in a separate secretory vesicle. Why do some vesicles contain more than one neuropeptide?

11.6   Give an example of a neuron that secretes two different neuropeptides from the same secretory vesicle.

11.7   A commonly accepted version of Dale's Principle is that one neuron releases one neurotransmitter. In terms of co-localization of neurotransmitters, what is a more accurate version of this principle?

11.8   What is the name of the propeptide for ACTH and β-endorphin?

11.9 During evolution, neurotransmitter-like molecules first occurred in ____, while endocrine glands first appeared in the _____.

11.10 Which layer of the early embryo forms the brain, spinal cord and neural crest cells of the developing embryo?

11.11 Chromaffin cells of the adrenal medulla are part of the "Diffuse Neuroendocrine System." Name two properties of these cells that show them to behave like central neurons.

11.12 Which neurohypophyseal (posterior pituitary) hormones are co-localized and co-released with dynorphin?

11.13 Which area of the brain has the highest concentration of neuropeptides?

11.14 Peptides, such as insulin or leptin, are large molecules that do not easily enter the brain from the circulation. How does insulin breach the blood-brain barrier (BBB) to reach areas of the brain that contain insulin receptors?

ESSAY QUESTIONS

11.1 Describe the application of immunocytochemistry and *in situ* hybridization techniques to the discovery and localization of neuropeptides in the brain. Your answer should focus on a single example; for example, kisspeptin, GnRH, etc.

11.2 Compare and contrast the neuronal mechanisms for the biosynthesis, storage and release of classical neurotransmitters versus neuropeptides.

11.3 Discuss the biosynthesis of the opioid family of neuropeptides in the brain. Include a description of the cells that produce these neuropeptides, and the different functions of each family member.

11.4 Discuss the theory of the evolution of the neuroendocrine system as proposed in Figure 11.3.

11.5 What could be the purpose of more than one neurotransmitter or neuropeptide being released from the same neuron simultaneously?

11.6 Describe the co-localization of neuropeptides with posterior pituitary hormones. Comment on the functions of these neuropeptides in the regulation of pituitary hormone secretion.

11.7 Discuss the importance of developing neuropeptide agonists and antagonists for use as therapeutic drugs.

11.8 Describe the mechanisms by which peptide hormones and neuropeptides can cross the BBB.

11.9 Using vasopressin and kisspeptin as examples of central neuropeptides, describe the sex differences known to exist in their brain localization.

11.10 Most neuropeptides, for example, the endogenous opioid peptides, are localized to discrete areas of the brain. However, the receptors for these neuropeptides may also be found elsewhere in the brain. Discuss the reasons for this apparent discrepancy.

REFERENCES

Abbott, N. J., Ronnback, L. and Hansson, E. (2006). "Astrocyte-endothelial interactions at the blood-brain barrier," *Nat Rev Neurosci* 7, 41–53.

Acher, R. (1993). "Neurohypophysial peptide systems: processing machinery, hydroosmotic regulation, adaptation and evolution," *Regul Pept* 45, 1–13.

Appetecchia, M. and Baldelli, R. (2010). "Somatostatin analogues in the treatment of gastroenteropancreatic neuroendocrine tumours, current aspects and new perspectives," *J Exp Clin Cancer Res* 29, 19.

Baldwin, G. S., Patel, O. and Shulkes, A. (2010). "Evolution of gastrointestinal hormones: the cholecystokinin/gastrin family," *Curr Opin Endocrinol Diabetes Obes* 17, 77–88.

Begley, D. J. (2004). "Delivery of therapeutic agents to the central nervous system: the problems and the possibilities," *Pharmacol Ther* 104, 29–45.

Boron, W. F. and Boulpaep, E. L. (2005). *Medical Physiology*, updated edn. (Philadelphia, PA: Elsevier Saunders).

Bos, P. A., Panksepp, J., Bluthe, R. M. and Honk, J. V. (2012). "Acute effects of steroid hormones and neuropeptides on human social-emotional behavior: a review of single administration studies," *Front Neuroendocr* 33, 17–35.

Burnstock, G. (1976). "Do some nerve cells release more than one transmitter?" *Neurosci* 1, 239–248.

Burnstock, G. (2004). "Cotransmission," *Curr Opin Pharmacol* 4, 47–52.

Choleris, E., Devidze, N., Kavaliers, M. and Pfaff, D. W. (2008). "Steroidal/neuropeptide interactions in hypothalamus and amygdala related to social anxiety," *Prog Brain Res* 170, 291–303.

Coleman, P. J. and Renger, J. J. (2010). "Orexin receptor antagonists: a review of promising compounds patented since 2006," *Expert Opin Ther Pat* 20, 307–324.

Corbett, A. D., Henderson, G., McKnight, A. T. and Paterson, S. J. (2006). "75 years of opioid research: the exciting but vain quest for the Holy Grail," *Br J Pharmacol* 147(Suppl. 1), S153–S162.

Crane, J. F. and Trainor, P. A. (2006). "Neural crest stem and progenitor cells," *Annu Rev Cell Dev Biol* 22, 267–286.

Cruz-Gordillo, P., Fedrigo, O., Wray, G. A. and Babbitt, C. C. (2010). "Extensive changes in the expression of the opioid genes between humans and chimpanzees," *Brain Behav Evol* 76, 154–162.

Dantzer, R., O'Connor, J. C., Freund, G. G., Johnson, R. W. and Kelley, K. W. (2008). "From inflammation to sickness and depression: when the immune system subjugates the brain," *Nat Rev Neurosci* 9, 46–56.

Darlison, M. G. and Richter, D. (1999). "Multiple genes for neuropeptides and their receptors: co-evolution and physiology," *Trends Neurosci* 22, 81–88.

Day, R. and Salzet, M. (2002). "The neuroendocrine phenotype, cellular plasticity, and the search for genetic switches: redefining the diffuse neuroendocrine system," *Neuro Endocrinol Lett* 23, 447–451.

De Vries, G. J. and Panzica, G. C. (2006). "Sexual differentiation of central vasopressin and vasotocin systems in vertebrates: different mechanisms, similar endpoints," *Neurosci* 138, 947–955.

Decaux, G., Soupart, A. and Vassart, G. (2008). "Non-peptide arginine-vasopressin antagonists: the vaptans," *Lancet* 371, 1624–1632.

Di Fabio, R., Alvaro, G., Griffante, C., Pizzi, D. A., Donati, D., Mattioli, M. *et al.* (2011). "Discovery and biological characterization of (2R,4 S)-1'-acetyl-N-{(1R)-1-[3,5-bis(trifluoromethyl)phenyl]ethyl}-2-(4-fl uoro-2-methylphenyl)-N-methyl-4,4'-bipiperidine-1-carboxamide as a new potent and selective neurokinin 1 (NK1) receptor antagonist clinical candidate," *J Med Chem* 54, 1071–1079.

Dimaline, R. and Dockray, G. J. (1994). "Evolution of the gastrointestinal endocrine system (with special reference to gastrin and CCK)," *Baillieres Clin Endocrinol Metab* 8, 1–24.

Douglas, A. J. and Russell, J. A. (2001). "Endogenous opioid regulation of oxytocin and ACTH secretion during pregnancy and parturition," *Prog Brain Res* 133, 67–82.

El Mestikawy, S., Wallen-Mackenzie, A., Fortin, G. M., Descarries, L. and Trudeau, L. E. (2011). "From glutamate co-release to vesicular synergy: vesicular glutamate transporters," *Nat Rev Neurosci* 12, 204–216.

End, K., Gamel-Didelon, K., Jung, H., Tolnay, M., Ludecke, D., Gratzl, M. *et al.* (2005). "Receptors and sites of synthesis and storage of gamma-aminobutyric acid in human pituitary glands and in growth hormone adenomas," *Am J Clin Pathol* 124, 550–558.

Engel, J. B. and Schally, A. V. (2007). "Drug insight: clinical use of agonists and antagonists of luteinizing-hormone-releasing hormone," *Nat Clin Pract Endocrinol Metab* 3, 157–167.

Fields, J. D. and Bhardwaj, A. (2009). "Non-peptide arginine-vasopressin antagonists (vaptans) for the treatment of hyponatremia in neurocritical care: a new alternative?" *Neurocrit Care* 11, 1–4.

Fraser, M., Matthews, S. G., Braems, G., Jeffray, T. and Challis, J. R. (1997). "Developmental regulation of preproenkephalin (PENK) gene expression in the adrenal gland of the ovine fetus and newborn lamb: effects of hypoxemia and exogenous cortisol infusion," *J Endocrinol* 155, 143–149.

Häussler, H. U., Jirikowski, G. F. and Caldwell, J. D. (1990). "Sex differences among oxytocin-immunoreactive neuronal systems in the mouse hypothalamus," *J Chem Neuroanat* 3, 271–276.

Heaney, A. P. and Melmed, S. (2004). "Molecular targets in pituitary tumours," *Nat Rev Cancer* 4, 285–295.

Henion, P. D. and Landis, S. C. (1992). "Developmental regulation of leucine-enkephalin expression in adrenal chromaffin cells by glucocorticoids and innervation," *J Neurosci* 12, 3818–3827.

Hokfelt, T., Johansson, O., Ljungdahl, A., Lundberg, J. M. and Schultzberg, M. (1980). "Peptidergic neurones," *Nature* 284, 515–521.

Houben, H. and Denef, C. (1990). "Regulatory peptides produced in the anterior pituitary," *Trends Endocrinol Metab* 1, 398–403.

Hrabovszky, E., Deli, L., Turi, G. F., Kallo, I. and Liposits, Z. (2007). "Glutamatergic innervation of the hypothalamic median eminence and posterior pituitary of the rat," *Neurosci* 144, 1383–1392.

Hrabovszky, E., Kallo, I., Turi, G. F., May, K., Wittmann, G., Fekete, C. *et al.* (2006). "Expression of vesicular glutamate transporter-2 in gonadotrope and thyrotrope cells of the rat pituitary. Regulation by estrogen and thyroid hormone status," *Endocr* 147, 3818–3825.

Hrabovszky, E. and Liposits, Z. (2008). "Novel aspects of glutamatergic signalling in the neuroendocrine system," *J Neuroendocrinol* 20, 743–751.

Imura, H., Kato, Y., Nakai, Y., Nakao, K., Tanaka, I., Jingami, H. *et al.* (1985). "Endogenous opioids and related peptides: from molecular biology to clinical medicine. The Sir Henry Dale lecture for 1985," *J Endocrinol* 107, 147–157.

Julius, D. (1997). "Another opiate for the masses?" *Nature* 386, 442.

Kastin, A. J. and Pan, W. (2010). "Concepts for biologically active peptides," *Curr Pharm Des* 16, 3390–3400.

Kauffman, A. S. (2010). "Gonadal and nongonadal regulation of sex differences in hypothalamic Kiss1 neurones," *J Neuroendocrinol* 22, 682–691.

Kim, H., Toyofuku, Y., Lynn, F. C., Chak, E., Uchida, T., Mizukami, H. *et al.* (2010). "Serotonin regulates pancreatic beta cell mass during pregnancy," *Nat Med* 16, 804–808.

Kocsis, Z. S., Molnar, C. S., Watanabe, M., Daneels, G., Moechars, D., Liposits, Z. *et al.* (2010). "Demonstration of vesicular glutamate transporter-1 in corticotroph cells in the anterior pituitary of the rat," *Neurochem Int* 56, 479–486.

Krieger, D. T. (1984). "Brain peptides," *Vitam Horm* 41, 1–50, 275–281.

Krolewski, D. M., Medina, A., Kerman, I. A., Bernard, R., Burke, S., Thompson, R. C. *et al.* (2010). "Expression patterns of corticotropin-releasing factor, arginine vasopressin, histidine decarboxylase, melanin-concentrating hormone, and orexin genes in the human hypothalamus," *J Comp Neurol* 518, 4591–4611.

Kupfermann, I. (1991). "Functional studies of cotransmission," *Physiol Rev* 71, 683–732.

Landry, M., Roche, D., Angelova, E. and Calas, A. (1997). "Expression of galanin in hypothalamic magnocellular neurones of lactating rats: co-existence with vasopressin and oxytocin," *J Endocrinol* 155, 467–481.

Largent-Milnes, T. M. and Vanderah, T. W. (2010). "Recently patented and promising ORL-1 ligands: where have we been and where are we going?" *Expert Opin Ther Pat* 20, 291–305.

Le Roith, D., Shiloach, J. and Roth, J. (1982). "Is there an earlier phylogenetic precursor that is common to both the nervous and endocrine systems?" *Peptides* 3, 211–215.

Le Roith, D., Delahunty, G., Wilson, G. L., Roberts, C. T., Jr., Shemer, J., Hart, C. *et al.* (1986). "Evolutionary aspects of the endocrine and nervous systems," *Recent Prog Horm Res* 42, 549–587.

Liposits, Z., Reid, J. J., Negro-Vilar, A. and Merchenthaler, I. (1995). "Sexual dimorphism in copackaging of luteinizing hormone-releasing hormone and galanin into neurosecretory vesicles of hypophysiotrophic neurons: estrogen dependency," *Endocr* 136, 1987–1992.

Llona, I., Annaert, W. G., Jacob, W. and De Potter, W. P. (1994). "Co-storage in large 'dense-core' vesicles of dopamine and cholecystokinin in rat striatum," *Neurochem Int* 25, 573–581.

Lowery, E. G. and Thiele, T. E. (2010). "Pre-clinical evidence that corticotropin-releasing factor (CRF) receptor antagonists are promising targets for pharmacological treatment of alcoholism," *CNS Neurol Disord Drug Targets* 9, 77–86.

Ludwig, M. and Leng, G. (2006). "Dendritic peptide release and peptide-dependent behaviours," *Nat Rev Neurosci* 7, 126–136.

Manjila, S., Wu, O. C., Khan, F. R., Khan, M. M., Arafah, B. M. and Selman, W. R. (2010). "Pharmacological management of acromegaly: a current perspective," *Neurosurg Focus* 29, E14.

Mansour, A., Fox, C. A., Akil, H. and Watson, S. J. (1995b). "Opioid-receptor mRNA expression in the rat CNS: anatomical and functional implications," *Trends Neurosci* 18, 22–29.

Meister, B., Cortes, R., Villar, M. J., Schalling, M. and Hokfelt, T. (1990). "Peptides and transmitter enzymes in hypothalamic magnocellular neurons after administration of hyperosmotic stimuli: comparison between messenger RNA and peptide/protein levels," *Cell Tissue Res* 260, 279–297.

Meister, B. and Hokfelt, T. (1988). "Peptide- and transmitter-containing neurons in the medio-basal hypothalamus and their relation to GABAergic systems: possible roles in control of prolactin and growth hormone secretion," *Synapse* 2, 585–605.

Merchenthaler, I. (1991). "Current status of brain hypophysiotropic factors: morphologic aspects," *Trends Endocrinol Metab* 2, 219–226.

Minakata, H. (2010). "Oxytocin/vasopressin and gonadotropin-releasing hormone from cephalopods to vertebrates," *Ann NY Acad Sci* 1200, 33–42.

Moalic, J. M., Le Strat, Y., Lepagnol-Bestel, A. M., Ramoz, N., Loe-Mie, Y., Maussion, G. *et al.* (2010). "Primate-accelerated evolutionary genes: novel routes to drug discovery in psychiatric disorders," *Curr Med Chem* 17, 1300–1316.

Mohr, E., Meyerhof, W. and Richter, D. (1995). "Vasopressin and oxytocin: molecular biology and evolution of the peptide hormones and their receptors," *Vitam Horm* 51, 235–266.

Mollereau, C. and Mouledous, L. (2000). "Tissue distribution of the opioid receptor-like (ORL1) receptor," *Peptides* 21, 907–917.

Nestler, E. J., Hyman, S. E. and Malenka, R. C. (2009). *Molecular Neuropharmacology: A Foundation for Clinical Neuroscience*, 2nd edn. (New York: McGraw Hill).

Nieuwenhuys, R. (1985). *Chemoarchitecture of the Brain* (Berlin: Springer Verlag).

Nusbaum, M. P., Blitz, D. M., Swensen, A. M., Wood, D. and Marder, E. (2001). "The roles of co-transmission in neural network modulation," *Trends Neurosci* 24, 146–154.

Osborne, N. N. (1981). "Communication between neurones: current concepts," *Neurochem Int* 3, 3–16.

Reyes, R., Valladares, F., Diaz-Flores, L., Feria, L., Alonso, R., Tramu, G. *et al.* (2007). "Immunohistochemical localization of hormones and peptides in the human pituitary cells in a case of hypercortisolism by ACTH secreting microadenoma," *Histol Histopathol* 22, 709–717.

Ribeiro-da-Silva, A. and Hokfelt, T. (2000). "Neuroanatomical localisation of Substance P in the CNS and sensory neurons," *Neuropeptides* 34, 256–271.

Roch, G. J., Busby, E. R. and Sherwood, N. M. (2011). "Evolution of GnRH: diving deeper," *Gen Comp Endocrinol* 171, 1–16.

Roth, J., Le Roith, D., Collier, E. S., Weaver, N. R., Watkinson, A., Cleland, C. F. *et al.* (1985). "Evolutionary origins of neuropeptides, hormones and receptors: possible applications to immunology," *J Immunol* 135, 816s–819s.

Roth, J., Le Roith, D., Lesniak, M. A., de Pablo, F., Bassas, L. and Collier, E. (1986). "Molecules of intercellular communication in vertebrates, invertebrates and microbes: do they share common origins?" *Prog Brain Res* 68, 71–79.

Russo, A. F., Clark, M. S. and Durham, P. L. (1996). "Thyroid parafollicular cells. An accessible model for the study of serotonergic neurons," *Mol Neurobiol* 13, 257–276.

Sato, N., Ogino, Y., Mashiko, S. and Ando, M. (2009). "Modulation of neuropeptide Y receptors for the treatment of obesity," *Expert Opin Ther Pat* 19, 1401–1415.

Semaan, S. J. and Kauffman, A. S. (2010). "Sexual differentiation and development of forebrain reproductive circuits," *Curr Opin Neurobiol* 20, 424–431.

Stewart, J. M. and Channabasavaiah, K. (1979). "Evolutionary aspects of some neuropeptides," *Fed Proc* 38, 2302–2308.

Tauchi, M., Zhang, R., D'Alessio, D. A., Stern, J. E. and Herman, J. P. (2008). "Distribution of glucagon-like peptide-1 immunoreactivity in the hypothalamic paraventricular and supraoptic nuclei," *J Chem Neuroanat* 36, 144–149.

Telleria-Diaz, A., Grinevich, V. V. and Jirikowski, G. F. (2001). "Co-localization of vasopressin and oxytocin in hypothalamic magnocellular neurons in water-deprived rats," *Neuropeptides* 35, 162–167.

Toni, R. (2004). "The neuroendocrine system: organization and homeostatic role," *J Endocrinol Invest* 27, 35–47.

Torrealba, F. and Carrasco, M. A. (2004). "A review on electron microscopy and neurotransmitter systems," *Brain Res Rev* 47, 5–17.

Trudeau, L. E. and Gutierrez, R. (2007). "On cotransmission and neurotransmitter phenotype plasticity," *Mol Interv* 7, 138–146.

Unsicker, K. (1993). "The trophic cocktail made by adrenal chromaffin cells," *Exp Neurol* 123, 167–173.

Viero, C., Shibuya, I., Kitamura, N., Verkhratsky, A., Fujihara, H., Katoh, A. *et al.* (2010). "REVIEW: Oxytocin: crossing the bridge between basic science and pharmacotherapy," *CNS Neurosci Ther* 16, e138–e156.

12 Neuropeptides II: function

As noted in Chapter 11, the human genome contains about 90 genes that encode neuropeptide precursors (pre-propeptides). The biologically active neuropeptide products of the pre-propeptides number at least 100, and there are likely to be many more waiting to be discovered. Neuropeptides are synthesized in a wide variety of neurons in many brain regions and more often than not are co-localized and co-released with classical neurotransmitters. Should neuropeptides therefore be categorized as neurotransmitters? An alternative description is that of *neuromodulator*, since in many instances they modify the neural effects of classical neurotransmitters. A good example of this is shown in Figure 11.2, where co-release of a neuropeptide totally modifies the influence of the co-localized neurotransmitter on a postsynaptic neuron. This chapter will illustrate the neurotransmitter and neuromodulator actions of neuropeptides on the neuroendocrine system, the autonomic nervous system (ANS) and the central nervous system. First, however, we will explore whether neuropeptides are best described as neurotransmitters or neuromodulators, or both.

12.1 Neurotransmitter and neuromodulator actions of neuropeptides: a dichotomy or a continuum?

An initial useful exercise is to establish criteria by which a neuromodulator could be defined. Recall that in Chapter 5 several criteria were established to ascertain whether a neurochemical might be considered to be a neurotransmitter (Table 5.1). In brief, these are: (1) synthesized in neurons; (2) present in the presynaptic nerve terminals, usually contained in vesicles, and released into the synapse in amounts sufficient to stimulate the postsynaptic cell; (3) whether endogenously released or applied exogenously, a neurotransmitter should have the same effect on the postsynaptic cell (i.e. it activates the same ion channels or second messenger pathways); (4) receptors specific to the neurotransmitter should be present postsynaptically; (5) receptor antagonists should prevent #3; and (6) a specific deactivating mechanism should exist in the synapse.

Strictly speaking, all of these criteria apply equally well to neuropeptides, with some qualifications. For example, classical neurotransmitters are made by enzymatic transformation (criterion #1) of a single amino acid transported into the neuron from the circulation, whereas neuropeptides are produced via changes in gene expression. However, subsequent proteolytic cleavage into smaller peptides can generate several different neuropeptides, each acting at different receptors. Nonetheless, even though the mode of synthesis differs, the final product (classical neurotransmitter versus neuropeptide) acts similarly at postsynaptic receptors. Another apparent difference is the way in which their action is terminated: classical neurotransmitters via enzymatic degradation or reuptake into the nerve terminal; neuropeptides by non-specific peptidases. Neurons do not have reuptake mechanisms for neuropeptides. Thus, the two types of molecule are deactivated (criterion #6), although by different mechanisms. Criteria # 2, 3 and 5 apply equally well to both classical neurotransmitters and neuropeptides, although for #2 it is obvious that classical neurotransmitters are stored in small vesicles compared to the large vesicles containing neuropeptides. Criterion #4 (postsynaptic receptors) is the one that represents the biggest difference between the two molecular entities. Because classical neurotransmitters tend to be removed quickly from the synapse, their receptors are normally in close proximity on the postsynaptic cell. So the signaling is very "private" and localized. In contrast, neuropeptides are almost hormone-like in their ability to travel some distance from the synapse before binding to a receptor. As a general rule, neuropeptide receptors are located away from the immediate postsynaptic site (Torrealba and Carrasco 2004). This concept is illustrated in Figure 12.1, which shows how a neuropeptide acts at long range to influence the activity of many neurons (*neuromodulation*).

Figure 12.2 also illustrates how neuropeptides and neurotransmitters are secreted from nerve terminals before binding to a receptor (Zoli *et al.* 1999). In Figure 12.2A, neurotransmitters bind to receptors in the synapse, but can spill over to bind to extra-synaptic sites (see also Figure 11.5). In contrast, neuropeptides have the ability to travel over long distances to bind to non-synaptic receptors (Figure 12.2B).

The concept that some neurotransmitters *and* neuropeptides act away from the synaptic cleft has been formalized as *volume transmission* rather than neuromodulation (Zoli *et al.* 1999; for further reading, see Agnati *et al.* 2006). In fact, neuropeptides may be preferentially released, not from nerve terminals at all, but from dendrites (see below, section 12.3.2). The concept is well described for secretion of oxytocin and vasopressin from hypothalamic neurons (Landgraf and Neumann 2004; Ludwig and Leng 2006). This represents the biggest difference between classical neurotransmitters and neuropeptides. In summary, Table 12.1 is an informative comparison of the major characteristics of classical neurotransmitters and neuropeptides (Landgraf and Neumann 2004).

12.2 Neurotransmitter actions of neuropeptides

As outlined in section 12.1, the criteria that establish whether a substance is a neurotransmitter or not apply equally well to neuropeptides. Indeed, there is a wealth of evidence to suggest that many neuropeptides, including Substance P, neurotensin, CCK, VIP, the enkephalins and others, meet most of these criteria. Some classical neurotransmitters such as GABA and glutamate bind to ligand-gated ion channel receptors (Figures 5.2 and 5.3), and trigger rapid onset (milliseconds), short-duration opening of ion channels. Other neurotransmitters, such as the muscarinic cholinergic and

A. Neurotransmitter secretion: 30–50µm maximum diffusion distance

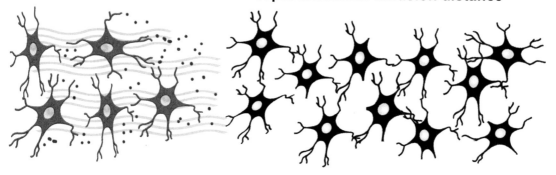

• = neurotransmitter

B. Neuropeptide secretion >1,000µm diffusion distance

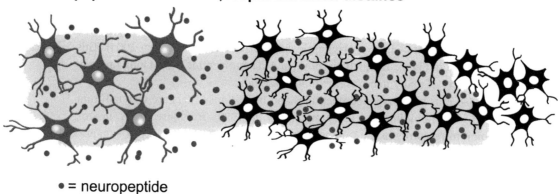

• = neuropeptide

Diffusion ⟶

Figure 12.1 Comparison of diffusion distances for classical neurotransmitter versus neuropeptide release
In A, the release of a classical neurotransmitter, such as DA, is highly localized (blue lines). Diffusion is minimal largely because of a very efficient removal of the neurotransmitter from the synapse, either by enzymatic degradation or by reuptake into the nerve terminals (see Chapter 5). In contrast, B shows that neuropeptides diffuse long distances away from the synapse (yellow area). This occurs partly because of limited enzymatic degradation of the neuropeptide following release, but also because there is no reuptake mechanism for neuropeptides. This means the neuropeptide molecules remain intact and diffuse through the extracellular fluid. Reproduced with permission and redrawn (Fuxe *et al.* 2010).

β-adrenergic agonists, act via G-protein-coupled receptors to trigger relatively slow (seconds) intracellular signaling (see Chapter 10). Neuropeptides also bind to G-protein-coupled receptors and trigger slow onset, long-duration changes in membrane potential (see Figure 11.2). They do this by affecting ion channels in the membrane of the post-synaptic cell, usually by activating second messenger systems (see Figure 10.5). A third mechanism by which neuropeptides influence neurotransmission is via tyrosine kinase receptors (see Figure 10.6). In this case, large neuropeptides such as NGF and BDNF, normally considered to be neurotrophins, are released as synaptic messengers to bind to tyrosine kinase receptors (Merighi 2002).

Table 12.1 **Basic characteristics of neuropeptides compared to classical (small molecule) neurotransmitters**

Number	– Higher (many neuropeptides)
Molecule size	– Larger
Molecules released	– Fewer (lower synaptic concentration)
Frequency of discharge	– Higher (impulse trains)
Concentration in extracellular fluid	– Lower
Chemical information	– Higher (complex molecules)
Sites of action	– Remote (diffusion away from synapse)
Receptor location	– Inside and outside of synapse
Receptor recognition sites	– More
Binding affinity	– Higher
Reuptake mechanism	– None
Replenishment in terminal	– Slow (needs pre-propeptide synthesis)
Selectivity	– Higher

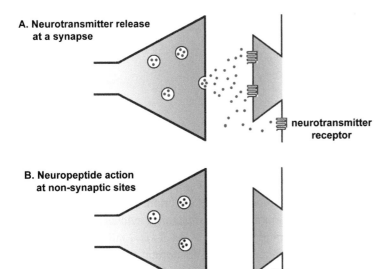

A. Neurotransmitter release at a synapse

neurotransmitter receptor

B. Neuropeptide action at non-synaptic sites

neuropeptide receptors

Figure 12.2 Synaptic and non-synaptic binding of neurotransmitters and neuropeptides
This figure complements Figure 12.1 by emphasizing the localized nature of the receptors for neurotransmitters (A) compared to neuropeptide receptors that are often found some distance from the release site (B). Reproduced with permission and redrawn (Zoli *et al.* 1999).

12.2.1 How well do neuropeptides meet the criteria defining a neurotransmitter?

In this section, we will use a newly discovered neuropeptide, *kisspeptin*, as a prototypical neuropeptide and consider whether it meets the criteria necessary to establish it as a neurotransmitter. Kisspeptin is the product of *kiss1* gene expression and is a critical hypothalamic neuropeptide in the neuroendocrine regulation of puberty and the reproductive system (Oakley *et al.* 2009). Using the criteria listed in Table 5.1.

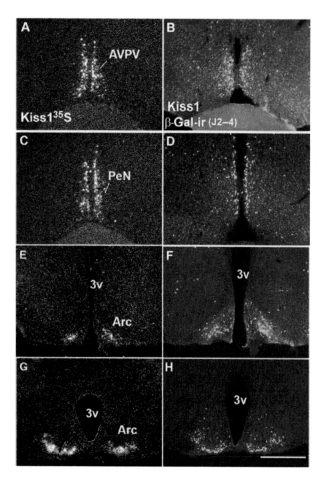

Figure 12.3 Distribution of hypothalamic neurons that express the *kiss1* gene
The figure shows the use of two different techniques to identify neurons that express *kiss1* mRNA. The left panels (A, C, E and G) show four different levels (frontal planes) of the mouse hypothalamus, including the arcuate nucleus (Arc). Intense labeling of *kiss1* mRNA with a radioactive probe (labeled with the isotope ^{35}S) is shown in white and indicates clusters of neurons that express kisspeptin as a neurotransmitter. The right-hand panel (B, D, F and H) illustrates a similar distribution of *kiss1* mRNA, but these are obtained from a mutant mouse expressing a genetically engineered *kiss1* gene. The *kiss1* mRNA carries a fluorescent label (β-Gal) that is easily detected with simple histological techniques (i.e. non-radioactively).

Abbreviations: 3v, third ventricle; ARC, arcuate nucleus; AVPV, anteroventral periventricular nucleus; β-Gal-ir, β-Galactosidase immunoreactivity; PeN, anterior periventricular nucleus. Reproduced with permission (Cravo *et al.* 2011).

Criterion 1 *Is it made in the brain?* This question is important because kisspeptin is present in peripheral blood and could theoretically reach the brain. However, *kiss1* mRNA is readily visualized by *in situ* hybridization in the hypothalamus, evidence that the *kiss1* gene is expressed there. Figure 12.3 illustrates the localization of *kiss1* mRNA in two ways: *first*, by *in situ* hybridization with a radioactive label in normal mouse brain sections (left panel); and *second*, in mice genetically engineered to produce neuronal *kiss1* mRNA that has an easily detected non-radioactive label (right panel) (Cravo *et al.* 2011). Both techniques provide very similar localizations of *kiss1*-positive neurons. In addition, the neuropeptide product of *kiss1*, kisspeptin, is detectable in mouse and human brain using immunohistochemistry (Clarkson *et al.* 2009; Hrabovszky *et al.* 2010). Also, as in the *in situ* hybridization data, localization is in specific brain areas. So, together with the mRNA data, the first criterion is firmly satisfied.

Criterion 2 *Is kisspeptin contained in secretory vesicles in nerve terminals?* Strong evidence for this is currently lacking, but Oakley *et al.* (2009) describe preliminary findings that kisspeptin immunoreactivity is detectable at synaptic contacts in the arcuate nucleus of the hypothalamus. Clarkson and Herbison (2006) also demonstrated that GnRH neurons are contacted by kisspeptin-positive fibers. In addition, kisspeptin-positive axons terminate close to POMC neurons in the arcuate nucleus (Fu and van den Pol 2010). These data suggest a nerve terminal location for kisspeptin, but electron microscopic studies are required to confirm a vesicular site for this neuropeptide.

Criterion 3 *Does kisspeptin induce electrophysiological changes in target neurons?* Kisspeptin induces secretion of GnRH from hypothalamic neurons and several groups have demonstrated that kisspeptin depolarizes and excites GnRH neurons (Oakley *et al.* 2009; Zhang *et al.* 2008). In addition, kisspeptin increases the firing rate of POMC and GnRH neurons (Fu and van den Pol 2010; Han *et al.* 2005). These findings confirm that kisspeptin behaves as a neurotransmitter by directly depolarizing hypothalamic neurons.

Criterion 4 *Does direct application of kisspeptin into the hypothalamus induce physiological changes?* Kisspeptin causes secretion of GnRH, and subsequently LH, when injected peripherally or when injected directly into the brain (Navarro *et al.* 2005; Messager *et al.* 2005).

Criterion 5 *Are receptors for kisspeptin present in hypothalamus?* Kisspeptin signals through a G-protein-coupled receptor sometimes called GPR54, but recently christened kiss1r (Oakley *et al.* 2009). This receptor uses a Gq-PLC-phosphoinositol second messenger system (see section 10.3.3) and Figure 12.4 illustrates how GnRH secretion is regulated through the kisspeptin receptor.

A detailed *in situ* hybridization study in the mouse brain revealed that *kiss1r* mRNA is present in high abundance in the POA of the hypothalamus (Herbison *et al.* 2010). Of more significance, most GnRH neurons expressed *kiss1r* mRNA and therefore possess kisspeptin receptors. This is consistent with the evidence given for Criterion #4; that is, kisspeptin induces secretion of GnRH. In addition, mouse mutants that do not express *kiss1r* mRNA,

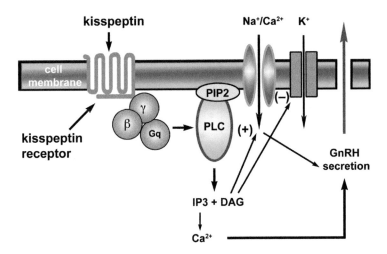

Figure 12.4 Kisspeptin signals through a G-protein-coupled receptor to induce secretion of gonadotropin-releasing hormone (GnRH)
Kisspeptin binding activates the Gq protein that in turn stimulates activity of phospholipase C (PLC). PLC cleaves PIP2 (phosphatidylinositol diphosphate) into two signaling molecules, DAG (diacylglycerol) and IP3 (inositol triphosphate) (see section 10.3.3). DAG regulates ion channels to regulate depolarization of the nerve terminal whereas IP3 liberates intracellular stores of Ca^{2+} ions. The latter is important for the secretion of GnRH.

and which therefore have no kisspeptin receptors, do not respond to kisspeptin with release of GnRH (Lapatto *et al.* 2007).

Criterion 6 *Kisspeptin receptor antagonists should block the effects of kisspeptin.* It is already known that when kisspeptin receptors are absent, or are ineffective, the reproductive system does not work. For example, in humans (and in mice), with a loss-of-function mutation in KISS1R, puberty is absent and these patients are infertile (Gianetti and Seminara 2008). By implication, the blockade of kisspeptin receptors with a synthetic antagonist should also disrupt reproduction in normal animals and humans. Pineda *et al.* (2010) demonstrated that a new kisspeptin antagonist, P-234, blocked puberty in mice and prevented an ovulatory surge of LH and FSH during a normal estrous cycle. The antagonist also prevented kisspeptin-induced increases in inositol phosphates (Figure 12.4) and the increase in cell firing (see criterion #3) (Millar *et al.* 2010). Figure 12.5 illustrates that when the antagonist is injected directly into the brain of sheep or rats, there is an inhibition

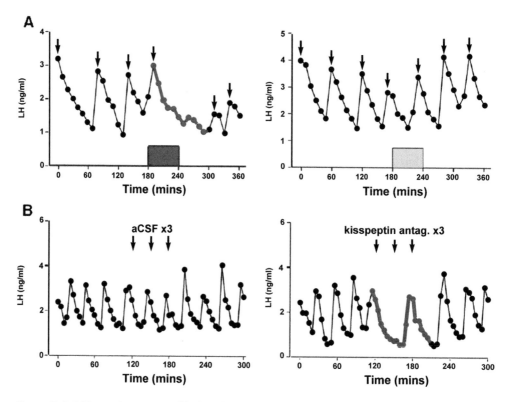

Figure 12.5 A kisspeptin antagonist blocks pulsatile secretion of luteinizing hormone (LH)
In A, the kisspeptin antagonist P-234 inhibits LH secretion and pulsatility (red bar) in a sheep model. Antagonist P-234 was infused directly into the cerebrospinal fluid (CSF) of a female sheep and LH levels were determined in peripheral blood. Pulses are indicated by arrows; note the elimination of pulses by the antagonist. In contrast, a control infusion (green bar) had no effect on LH pulses. Results in B represent the effects of injection of the antagonist directly into the hypothalamic median eminence of rat brain. 500 nanoliters of artificial CSF (control; left-hand panel; three injections) had no effect on LH pulses, whereas 50 picomoles of kisspeptin antagonist (right-hand panel; three injections at 30-minute intervals) markedly slowed the frequency of LH pulses. Reproduced with permission (Millar *et al.* 2010).

of pulsatile secretion of LH, strongly implicating kisspeptin in this phenomenon. Taken all together, these data confirm that endogenous kisspeptin acts through specific receptors in the brain to regulate physiological systems and does so acting as a neurotransmitter.

Criterion 7 *Specific mechanism for termination of action.* In the case of small molecule, classical neurotransmitters, there are two main mechanisms for terminating neurotransmitter action; that is, degradation by inactivating enzymes or reuptake into the nerve terminal by specific reuptake mechanisms (Figure 5.14). In the case of neuropeptides, such as kisspeptin, there does not appear to be a reuptake mechanism. In addition, although kisspeptin is sensitive to peptidase-induced degradation, like all neuropeptides, little is known concerning the specificity of this action. However, some synthetic analogs of kisspeptin have been designed to resist degradation and one of these was shown to be fourfold more potent than the parent kisspeptin in stimulating LH secretion (Curtis *et al.* 2010). This observation suggests that kisspeptin is normally degraded in the brain as a way to terminate its activity.

Summary

With the exception of criterion #2 (localization to vesicles), #5 (precise whereabouts of receptors; synaptic or extra-synaptic) and #7 (no specific degradation mechanism), kisspeptin appears to act as a neurotransmitter. Criteria #2 and #5 must await further detailed studies on localization, and #7 is consistent with other known neuropeptides.

12.3 Neuromodulator actions of neuropeptides

Section 12.2 introduced the idea that neuropeptides behave differently from small molecule, classical neurotransmitters. It seems likely that most, if not all, neuropeptides are *co-localized* with at least one classical neurotransmitter (Hokfelt *et al.* 2000). For this reason alone, it appears possible that the co-release of a neuropeptide will serve to modulate the action of the neurotransmitter. A good example is shown in Figure 11.2. What are the mechanisms by which neuropeptides influence the neurochemical/neurophysiological actions of classical neurotransmitters? There are at least three possible sites where neuropeptides may modulate the actions of the primary neurotransmitters in postsynaptic cells: (1) effects on neurotransmitter receptors; (2) changing the second messenger systems; and (3) modifying ion channels. In this section, bear in mind that the examples given may not always include co-release. In other words, neuropeptides are able to modify classical neurotransmitter action following release from other nerve terminals.

12.3.1 Neuropeptide regulation of postsynaptic receptor density

Neuropeptides can modulate the sensitivity of postsynaptic cells to classical neurotransmitters by up- or down-regulating their receptors. An example of this is seen in the opioid system. Opioid peptides and glutamate are major regulators in the endocrine and reproductive axes (Vuong *et al.* 2010; Maffucci and Gore 2009). Much research has taken place in this area, but one example will serve to show how an opiate, morphine, can acutely regulate glutamate receptor abundance. A single injection of morphine in rats significantly reduced glutamate receptor mRNA in hypothalamus and in hippocampus, but had no effect in spinal cord (Le Greves *et al.* 1998). Thus, the inhibitory influence of morphine or heroin on fertility may be partially mediated through changes in glutamate receptors.

Second-messenger systems may also be modulated by neuropeptides such as VIP (see section 2.2.7). VIP binds to a G-protein-coupled receptor and increases cAMP levels. In the cerebral cortex, VIP interacts with norepinephrine (acting at α1-adrenergic receptors) to potentiate an increase in cAMP production (Schaad *et al.* 1989). A similar synergy is also seen with VIP and glutamate in the production of arachidonic acid as second messenger (Stella and Magistretti 1996). An example using calcium ions as second messenger is the synergistic action of norepinephrine and a peptide called *pituitary adenylate cyclase-activating polypeptide* (PACAP) on secretion of vasopressin from hypothalamic neurons (Shioda *et al.* 2000).

12.3.2 Electrophysiological responses to neuropeptides

Neuropeptides may alter the permeability of ion channels in both pre- and postsynaptic cells. Such changes modify the electrophysiological activity of neurons. An example of this was shown in Figure 11.2, where a neuropeptide induced a profound hyperpolarization in the postsynaptic cell. The pattern of presynaptic neuron firing determines whether small or large synaptic vesicles will be released from the nerve terminal. Low frequency stimulation releases small vesicles (neurotransmitters only) and this induces corresponding excitatory postsynaptic potentials. In contrast, high frequency stimulation releases large vesicles containing neuropeptides that hyperpolarize postsynaptic neurons. Thus, a change in presynaptic firing pattern can determine the electrophysiological response of the target cells.

The study of how *specific* neuropeptides influence electrophysiological properties of neurons is a vast area of neuroscientific endeavor and is beyond the scope of this book. Nonetheless, we will provide several examples within the neuroendocrine field. The techniques employed are applicable to any investigation of neuropeptide effects in the brain. For example, Chapter 6 introduced two new neuropeptides, kisspeptin and GnIH, which are critical components of the reproductive neuroendocrine system (section 6.3.3; Figure 6.7). The physiological effects of kisspeptin are already noted (see criterion #3, above, in section 12.2.1). A further illustration of how kisspeptin excites GnRH neurons in the reproductive hypothalamus is shown in Figure 12.6. Note how a single application of kisspeptin (labeled as Kiss-1) induces a prolonged excitation of GnRH neurons. In contrast, GnIH rapidly and reversibly inhibits this excitation (Wu *et al.* 2009). The authors speculate that the opposing effects of these neuropeptides may be important in the control of puberty, ovulation and sex behavior.

Electrophysiological techniques have also succeeded in clarifying the mechanism by which DA regulates prolactin (PRL) secretion. We saw in Chapter 6 (Figure 6.8) that DA acts as an inhibitory factor for PRL release. Under some circumstances, however, PRL secretion is *stimulated* by TRH, the releasing hormone that controls thyroid hormone output (see Figure 4.9). It had been assumed that TRH countered the inhibitory effect of DA by acting directly on pituitary lactotrophs to induce PRL release. Lyons *et al.* (2010) revealed that TRH modifies the firing of DA neurons to allow PRL secretion to occur. This is another example of how electrophysiological techniques, applied to neuropeptide effects, are able to clarify difficult physiological questions.

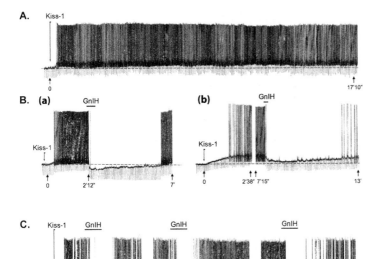

Figure 12.6 Activation of neurons by kisspeptin is blocked by gonadotropin inhibitory hormone (GnIH)

These electrophysiological experiments were conducted in slices of mouse brain maintained in tissue culture to keep neurons alive. In panel A, a brief application (5 seconds) of kisspeptin (Kiss-1) induced a fast and prolonged neuronal excitation that lasted > 17 minutes. In panel B *(a and b)* are two more neurons that respond similarly to that in A, but the neuronal firing is abruptly interrupted by application of GnIH. In panel C, a single Kiss-1 activated neuron was repeatedly inhibited by GnIH. Each inhibitory period was followed by renewed excitation, suggesting that a single application of Kiss-1 was able to continue to fire action potentials for up to 23 minutes. Reproduced with permission (Wu *et al.* 2009).

An electrophysiological approach to the study of neuroendocrine neurons is a difficult undertaking. GnRH neurons, for example, are scattered through the hypothalamus and are therefore hard to locate. Together with their small size, this makes it very difficult to electrophysiologically record from them. However, with new transgenic techniques, it is feasible to breed experimental animals, like mice and rats, which express so-called *fusion genes* (*transgenes*) that are labeled with green fluorescent protein (GFP) (see Figure 12.3 for a further example). In other words, it is possible to create an animal whose GnRH neurons will glow green under a fluorescence microscope; that is, only cells that express the GnRH-GFP fusion gene will be visible. This approach has been exploited for the electrophysiological study of GnRH neurons (Suter *et al.* 2000a; Suter *et al.* 2000b; for further reading, see Moenter 2010). The technique has also been applied to hypothalamic neurons expressing a vasopressin-GFP fusion gene. Recall that vasopressin neurons are located in the SON/PVN regions of the hypothalamus and project their axons down to the posterior lobe of the pituitary (see Figure 6.10). Ueta *et al.* (2008) describe the production of a transgenic rat that expresses the vasopressin-GFP fusion gene. The SON, PVN and posterior pituitary all glow green under fluorescence microscopy and live, individual neurons can be readily obtained for recording purposes in real time. Figure 12.7 illustrates some of their data, including an acutely isolated vasopressin neuron (A) and a whole posterior pituitary (B). This technique allows electrical recording from an identified vasopressin neuron (D). The recording electrode is visible and a typical recording of a vasopressin neuron firing is seen in (E). With this approach, it is possible to study oxytocin and vasopressin neurons, each labeled with a different color, in the same transgenic rat (Katoh *et al.* 2011). In summary, this is a powerful technique for investigating individual living neurons, either electrophysiologically or biochemically.

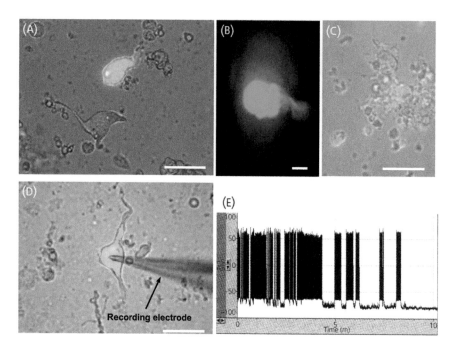

Figure 12.7 The use of green fluorescent protein (GFP) to identify and isolate vasopressin neurons from rat supraoptic nucleus (SON)

The figure shows isolated SON neurons (A and D) that glow green because of the presence of vasopressin-GFP neuropeptide. Panel C illustrates that vasopressin neuron terminals, obtained from the posterior pituitary, also glow green because of vasopressin-GFP stored in these terminals. An intact whole posterior pituitary gland (B) is labeled for the same reason. Panel D reveals the great utility of the technique, allowing recordings to be made from single living vasopressin neurons. In this panel, the recording electrode is clearly visible and panel E shows a representative train of action potentials obtained from a recording experiment. Reproduced with permission (Ueta *et al.* 2008).

In a final example, electrical activity of neuroendocrine neurons can be inferred from their biological response; that is, the pattern of their secretory activity. An earlier example described how the firing pattern of oxytocin neurons could be correlated with the feeding behavior of suckling rat pups (Figure 6.12). This study was greatly extended by Lincoln and Wakerley (1974), who demonstrated that electrical activity of identified SON neurons in rat mothers was correlated with oxytocin release, milk ejection from the nipples and subsequent feeding of the pups. These pioneering experiments led, more than 30 years later, to a clearer understanding of how oxytocin neurons are also able to time the occurrence of birth (parturition) by releasing oxytocin (Brunton and Russell 2008). Figure 12.8 reveals that, close to the time of birth, the pregnant uterus signals to oxytocin neurons, via NE and noradrenergic receptors, to release oxytocin from dendrites. This secretion causes electrical coupling of oxytocin neurons, via oxytocin autoreceptors, so that many neurons fire simultaneously, driving an increased release of oxytocin from the terminals in the posterior pituitary. Oxytocin then enters the circulation and causes contraction of the uterus to assist in the birth process.

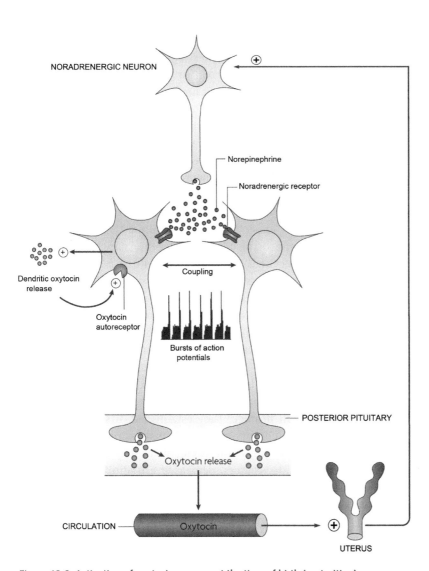

Figure 12.8 Activation of oxytocin neurons at the time of birth (parturition)
Parturition begins with uterine contractions. These stimulate (+) noradrenergic neurons in the brain that project
to oxytocin neurons. The noradrenergic input releases norepinephrine that activates bursts of action potentials
in the oxytocin neurons, and this activation is further enhanced by coupling between the oxytocin neurons.
Simultaneously, dendritic release of oxytocin, and binding of oxytocin to autoreceptors, is critically important in
driving burst-firing. The coordinated burst-firing of oxytocin neurons leads to pulsatile oxytocin secretion from
the posterior pituitary gland into the circulation, thus promoting uterine contractions. Reproduced with
permission (Brunton and Russell 2008).

12.3.3 Neuropeptide modulation of neurotransmitter release from presynaptic neurons

Chapter 11 introduced co-release of neurotransmitters and neuropeptides as having significant effects on postsynaptic cells (e.g. see Figure 11.2). An additional critical function of

neuropeptides, and neurotransmitters as well, is their ability to exert an autocrine negative feedback on the presynaptic terminal. This is called *presynaptic inhibition*. Bear in mind, however, that some presynaptic receptors, such as those for glutamate, also exert *positive* feedback on presynaptic terminals (e.g. see Figure 11.5; for further reading, see Deng *et al.* 2010). One example of presynaptic inhibition, although not for a neuropetide, is in the cannabinoid system (Figure 5.5), where endocannabinoids exert a negative feedback effect on neurotransmitter release from presynaptic terminals via a G-protein-coupled receptor and changes in Ca^{2+} and K^+ channels. These receptors are also called *autoreceptors*. Such a mechanism is operative in controlling the strength of neurotransmission (Wu and Saggau 1997; for further reading, see Miller 1998).

An example from the neuroendocrine system is conveniently taken from the regulation of oxytocin neurons already described in section 12.3.2 (Figure 12.8). These neurons co-localize oxytocin and dynorphin, and kappa opioid autoreceptors are localized to the neuron terminals (Figure 12.9) (Brown *et al.* 2000; Douglas and Russell 2001). Oxytocin secretion is restrained by the secreted dynorphin exerting negative feedback via kappa opioid receptors. This is particularly important during pregnancy and ensures that oxytocin release does not cause uterine contractions before the end of pregnancy. As parturition approaches, the ability of dynorphin to prevent oxytocin secretion is down-regulated, thus permitting oxytocin secretion, uterine contractions and induction of labor.

Another example of presynaptic inhibition, again using the opioid system, is described by Devidze *et al.* (2008). This group investigated how opioid peptides were able to decrease sex behavior (arousal) in mice. They established, first, that estradiol increased sex behavior by inducing excitatory synaptic transmission in the VMH of ovariectomized female mice. Using electrophysiological methods, they then showed that opioids, acting at presynaptic μ-opioid receptors, decreased estradiol-induced sexual arousal by inhibiting synaptic transmission. Devidze *et al.* (2008) concluded that the VMH contains excitatory glutametergic terminals that are regulated by presynaptic μ-opioid receptors.

In summary, there is an extensive literature on presynaptic neuropeptide receptors and their ability to reduce synaptic transmission (for further reading, see Schlicker and Kathmann 2008). This review includes, for example, receptors for neuropeptide NPY, ACTH and orexins, as well as opioids. Presynaptic receptors are coupled to inhibitory G proteins (G_i) and are able to block voltage-dependent Ca^{2+} channels and open voltage-dependent K^+ channels. Both of these actions prevent secretion of neuro-transmitters. In addition, presynaptic inhibition involves a mechanism that interferes with the fusion of secretory vesicles with the nerve terminal membrane. Recall that Chapter 5 discussed the process by which secretory vesicles fuse with nerve terminals to facilitate secretion of neurotransmitters. The fusion machinery includes a vesicle membrane protein called *synaptotagmin* (Figure 5.11). This protein appears to be the target of presynaptic inhibition; that is, G-protein-coupled receptors respond to, for example, opioids by releasing the Gβγ subunit from the G protein complex (see Figure 10.1), which then binds to synaptotagmin to prevent vesicle fusion and secretion (Blackmer *et al.* 2005).

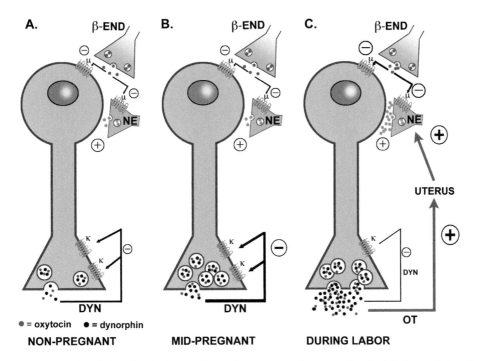

Figure 12.9 Regulation of oxytocin secretion during pregnancy from nerve terminals in the posterior pituitary through presynaptic inhibition by opioid peptides

In *non-pregnant* animals (A), oxytocin (OT) cells express μ-opioid receptors at the level of the cell body and κ-receptors are located on oxytocin nerve terminals in the posterior pituitary gland where feedback from dynorphin (DYN), co-released with oxytocin, can inhibit further oxytocin release. In *mid-pregnancy* (B), the auto-inhibitory mechanisms are increased in the posterior pituitary gland, restraining the release of oxytocin and hence increasing the accumulation of pituitary oxytocin stores. By the end of pregnancy and during labor (C), the auto-inhibitory mechanisms acting on the posterior pituitary gland are down-regulated, allowing the release of large amounts of OT that stimulate uterine contractions and childbirth. Oxytocin secretion is further enhanced by sensory feedback from the uterus through norepinephrine (NE). The stimulatory effect of NE overcomes an inhibitory effect of β-endorphin (β-END) at the cell body and NE terminals. Reproduced with permission and redrawn (Brown *et al.* 2000).

12.4 Regulatory effects of neuropeptides on the neuroendocrine system

As discussed in Chapter 11, many types of neuropeptide cell bodies are found in the hypothalamus, an area that is critical for the regulation of the neuroendocrine system. Neuropeptides can alter hormone secretion either by modulating the release of classical neurotransmitters that in turn regulate neuroendocrine neurons, or by direct action on neuroendocrine neurons. Several examples of the involvement of neuropeptides in hypothalamic regulatory systems are known. For example, in section 12.2.1, a case was made for hypothalamic kisspeptin acting as an important neurotransmitter/neuromodulator in the reproductive system, controlling the secretion of another neuropeptide, GnRH.

Table 12.2 Effects of selected neuropeptides on the neuroendocrine system

	GH	PRL	ACTH	TSH	LH	FSH	OXY	VP
Endogenous opioids	↑	↑	↑ (acute) ↓(chronic)	↓	↓	↓↑↓?	↓	↓
Substance P								
Central	↓	0?	0↑?	0	0↓?	0	–	–
Peripheral	↑	↑	0	0	↓	↓	–	–
Neurotensin								
Central	↓	↓	–	↓	↑↓?	0	–	–
Peripheral	↑	↑	–	↑	↑↓?	0	–	–
CCK	↑	↑	↑	↓	↓	↓	–	–
VIP	↑	↑	↑	0	↑↓	0	↑	↑
NPY								
Central	↓	0	↑	–	↑↓*	0	–	–
Angiotensin	↓	↓	↑	–	↑	0	–	–
Bombesin	↑	↑	0	↓	0	0	–	–

Notes:
↑ = Stimulation of hormone release; ↓ = inhibition of hormone release;
0 = no effect; ? = contradictory effects; – = unknown;
* = ↑ with estradiol, ↓ without estradiol.
Source: Ottlecz 1987; Müller and Nistico 1989; McCann 1991; Dumont *et al.* 1992.

Also, one class of neuropeptide, the opioids, modulated the secretion of another neuropeptide, oxytocin (see Figure 12.9). Chapter 6 detailed many examples of how hypothalamic neuropeptides controlled pituitary hormone secretion and some of these are listed in Table 12.2.

As emphasized previously, this book is meant to provide an introduction to the neuroendocrine system rather than a detailed list of neuropeptides implicated in hypothalamic function. Therefore, this section will outline two physiological systems to serve as examples of the importance of neuropeptides: (1) the reproductive system as exemplified by kisspeptin and GnRH; and (2) the control of body weight and energy intake.

12.5 Kisspeptin and GnRH as hypothalamic regulators of fertility

In section 12.2.1, the neuropeptide kisspeptin was shown to satisfy most of the criteria necessary to establish it as a neurotransmitter/neuromodulator that acts within the hypothalamus to regulate the secretion of GnRH (Figure 6.7). As noted already, neuropeptides are always formed as pre-propeptides (e.g. see Figures 7.2 and 11.1) and kisspeptin is no exception. The human *kiss1* gene encodes a 145 amino acid pre-propeptide (pre-prokisspeptin) that contains the sequence for kisspeptin-54 (also known as *metastin*)

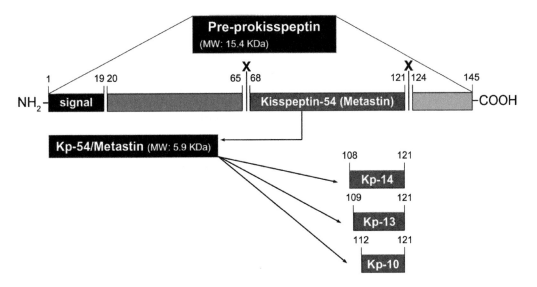

Figure 12.10 Structure and cleavage of human pre-prokisspeptin
Pre-prokisspeptin is a 145 amino acid peptide that contains a 19 amino acid signal peptide and a central 54 amino acid region, flanked by two cleavage sites (denoted by X). Proteolytic processing of pre-prokisspeptin by the enzyme furin generates kisspeptin-54 (also called *metastin*) (Harihar *et al.* 2014). Further cleavage of kisspeptin-54 gives rise to kisspeptins of lower molecular weight: kisspeptin-14 (Kp-14), Kp-13 and Kp-10. Reproduced with permission (Roa *et al.* 2008).

(Figure 12.10; Roa *et al.* 2008). The bioactive neuropeptide (kisspeptin-54) is formed by cleavage as shown, and three other smaller peptides have been isolated. However, although all four peptides appear to bind to the kisspeptin receptor, it remains unknown whether the smaller peptides are produced physiologically (Oakley *et al.* 2009).

Also in keeping with many neuropeptides, kisspeptin is expressed in several other tissue sites. It was originally discovered as a metastasis-inhibiting substance (*metastin*) in several tumors – such as those in pancreas, ovary, thyroid and breast – and is expressed in placenta, ovary, pancreatic islets, cardiovascular system, fat tissue and several brain areas outside of the hypothalamus (Roa *et al.* 2008; Oakley *et al.* 2009; Brown *et al.* 2008).

How is a neuropeptidergic interplay between kisspeptin and GnRH important for ensuring fertility? In experiments in mice with genetically targeted gene deletions, the lack of kisspeptin receptors, or *kiss1* gene expression, induces infertility. In other words, without kisspeptin or its receptor, the reproductive system fails to function. Female mutant mice do not reach puberty and are unable to ovulate, and male mice are also immature with very small testes, low testosterone and low pituitary secretion of LH/FSH. Human patients with inactivating mutations in the kisspeptin receptor (KISS1 R) are similarly infertile (Chan *et al.* 2009; Seminara and Crowley 2008). There are two major hypothalamic sites for kisspeptin neurons; these are the *arcuate nucleus* and the *anteroventral periventricular nucleus* (AVPV; Figure 12.3). The work of many laboratories led to the suggested neuronal wiring shown in Figure 12.11. Part of the negative feedback effect of estradiol (see Figures 8.8 and 8.9) is exerted through inhibition of *kiss1* expression in the hypothalamic arcuate nucleus, with a consequent reduction in GnRH secretion. Conversely, *positive*

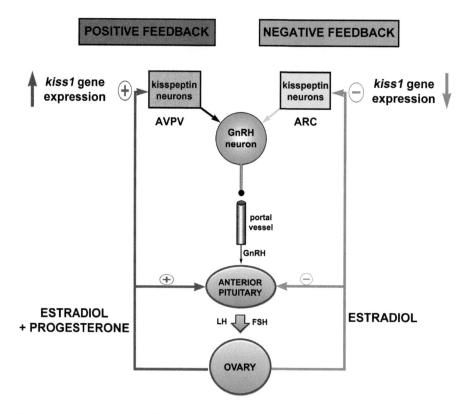

Figure 12.11 Schematic model for the role of *kiss1* in the generation of the pre-ovulatory surge of LH and the neuroendocrine control of ovulation

Estradiol (E2) and progesterone (P) are mandatory for induction of the pre-ovulatory surge of luteinizing hormone (LH) (*positive feedback*); a phenomenon that involves: (i) activation of kisspeptin (Kp) neurons at the anteroventral periventricular nucleus (AVPV) at the time of the surge; and (ii) a state of enhanced responsiveness to kisspeptin, likely at the level of GnRH neurons, during the peri-ovulatory period. In addition, E2 and P are essential for the modulation of (iii) increased pituitary responsiveness to GnRH. In contrast, the *negative feedback* of E2 on LH secretion is exerted at the arcuate nucleus (ARC), where *kiss1* gene expression and Kp levels are reduced. This decrease effectively stops GnRH secretion and therefore LH release. Reproduced with permission and redrawn (Roa *et al.* 2008).

feedback of estradiol (and often progesterone as well) *increases kiss1* expression in AVPV and stimulates kisspeptin to induce secretion of GnRH (Roa *et al.* 2008). This stimulation of GnRH secretion is mediated by kisspeptin receptors as described earlier (section 12.2.1). In summary, kisspeptin is a key neuropeptide in the hypothalamic control of the reproductive system. It seems likely that kisspeptin may be clinically invaluable in the treatment of human reproductive disorders (Hameed *et al.* 2011).

12.6 Neuropeptides and the regulation of food intake and body weight

Data provided by the World Health Organization reveal that more than 700 million people worldwide will be obese by 2015. In addition, at least 2.5 billion will be classified as being overweight (Suzuki *et al.* 2010; for further reading, see Guyenet and Schwartz 2012). In Canada, 60 percent of adults and 25 percent of children are overweight or obese and these figures represent an annual health cost burden of approximately $6 billion (www.obesity network.ca). In the United Kingdom, 25 percent of all adults are obese (Zaninotto *et al.* 2009). Since obesity is firmly associated with a decrease in the quality of life and an escalating incidence of cardiovascular disease, Type 2 diabetes, osteoarthritis, cancer, hypertension, stroke and Alzheimer's Disease, it is imperative that major efforts be made to understand the neuroendocrinology of appetite regulation. The major neural centers for appetite regulation are located in the hypothalamus and brain stem and this chapter will focus primarily on the former.

Chapter 2 introduced the topic of peptides released as hormones from a variety of peripheral tissues such as the GI tract (Figure 2.3) and fat (adipose tissue; Figure 2.8). A critical fat-derived hormone is leptin and we will discuss this below (see also Figures 1.4 and 2.8). The present chapter will also show that circulating peptides such as leptin (from fat) and ghrelin (from stomach) can reach the brain, where they regulate some aspects of food intake and metabolism. Note that many of the GI peptides are also produced in the brain where they act as neuropeptides.

Students are encouraged to consult an excellent online poster, provided by Drs. Dietrich and Horvath (2010), which illustrates much of the information to follow (www.nature.com/ nrn/posters/feeding).

12.6.1 Gut peptides and the regulation of appetite

Peptide hormones of the GI tract were introduced in Chapter 2 and many of them are shown in Figure 2.3. GI tissue represents a large endocrine organ and it expresses at least 30 peptide genes that in turn produce more than 100 bioactive peptides (Suzuki *et al.* 2010). Primary functions within the GI tract include digestion, absorption of nutrients, contraction/relaxation of smooth muscle walls and sphincters, and secretion of fluids and electrolytes. However, some of these peptides also target the brain to control both hunger and satiety. For example, even anticipation of a meal causes secretion of some hormones. In contrast, the mechanical and chemical stimulation of the stomach and intestine, caused by the presence of food, induces release of other hormones from the GI tract. How do these peptide hormones affect appetite? They do so by targeting the brain in two distinct ways: (1) two groups of neurons in the hypothalamic arcuate nucleus respond to peptide hormones with opposing effects on food intake; and (2) neurons in the brainstem receive sensory input from the GI tract via the vagus nerve, such that information such as gut distension and gut hormone levels are detected. Figure 12.12 summarizes these mechanisms (Murphy and Bloom 2006; see also Steinert *et al.* 2012). The two groups of arcuate neurons referred to in (1), above, are first, neurons expressing the *POMC* gene and second, those producing the co-localized neuropeptides NPY and *agouti-related protein (AgRP)*. The latter is responsible for stimulating appetite (termed *orexigenic*), whereas POMC neurons reduce food intake by suppressing appetite (*anorexigenic*).

Although not covered in this book, note that these neurons are part of a complex hypothalamic circuitry that also sends information to higher brain centers, to the sympathetic nervous system and to the neuroendocrine thyroid axis (for further reading, see Morton *et al.* 2006; and www.nature.com/nrn/posters/feeding).

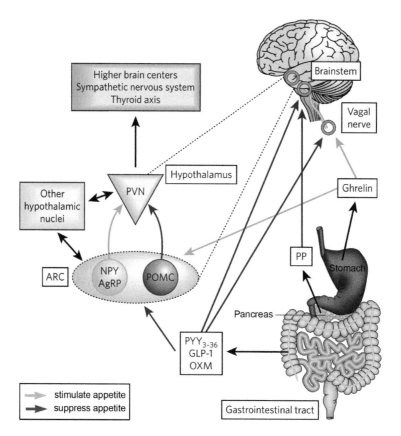

Figure 12.12 Regulation of appetite by gut hormones

Three hormones – PYY$_{3-36}$, GLP-1 and oxyntomodulin (OXM) – are released from the gastrointestinal tract following a meal. They can directly stimulate anorectic pathways in the hypothalamus and brainstem to reduce appetite, and may also act through the vagus nerve. Pancreatic polypeptide (PP) is secreted from the pancreas after a meal and is thought to reduce appetite by directly signaling to neurons in the brainstem. Ghrelin is released from the stomach when we are hungry/fasting and might signal directly to the hypothalamus, or through the vagus nerve, to stimulate food intake. The arcuate nucleus (ARC) is important in integrating gut hormone energy homeostasis signals via NPY/AgRP neurons and POMC neurons. These neurons signal to the PVN and other hypothalamic nuclei to increase or decrease appetite, respectively (see also Figures 12.13 and 12.16). Green arrows indicate orexigenic signals and red arrows indicate anorectic signals.

Abbreviations: pro-opiomelanocortin (POMC); paraventricular nucleus (PVN). Reproduced with permission (Murphy and Bloom 2006).

12.6.2 Gut peptides

How do hypothalamic neurons respond to GI peptide hormones in the circulation? Figure 12.12 illustrates that all except ghrelin inhibit appetite. *Ghrelin* is therefore *orexigenic*; that is, it stimulates food intake. Ghrelin is released from the stomach and stimulates the firing rate of NPY/AgRP neurons in the arcuate nucleus. These in turn induce food intake. Since ghrelin levels in the blood are elevated before meals, when we are hungry, and are reduced after eating, ghrelin has been called the "hunger hormone." Injection of rats with ghrelin causes an increase in food intake, and continued treatment induced weight gain (Gardiner *et al.* 2008).

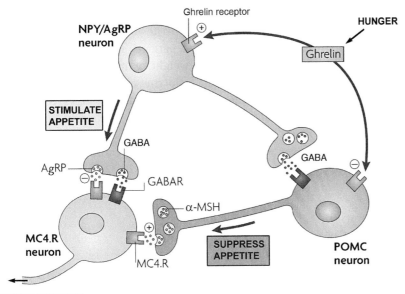

Figure 12.13 Hunger-induced ghrelin signaling in the arcuate nucleus
Two groups of neurons in the arcuate nucleus have opposite effects on feeding: (1) *Pro-opiomelanocortin* (POMC) neurons inhibit feeding behavior; and (2) NPY/AgRP (*neuropeptide Y and agouti-related peptide*) neurons which inhibit POMC neurons and activate feeding. POMC neurons secrete α-MSH, an agonist of MC4.R receptors (melanocortin4 receptor), which activates neuronal pathways that inhibit feeding. In contrast, NPY/AgRP neurons are stimulated by ghrelin and respond by releasing GABA to inhibit POMC neurons. However, they also inhibit downstream (MC4.R) neurons by secreting AgRP and GABA; GABA is inhibitory and AgRP competes with α-MSH for the MC4.R, thereby acting as an antagonist to induce food-seeking behavior. There is evidence that ghrelin also inhibits POMC neurons directly. The involvement of other hormones in this system is shown in Figure 12.16. Reproduced with the generous permission of Dr. T. L. Horvath.

Investigations in human volunteers revealed similar data. Ghrelin increased food intake in both obese and lean subjects (Druce *et al.* 2005; Druce *et al.* 2006). Figure 12.13 illustrates in more detail how the two populations of arcuate neurons (i.e. POMC and NPY/AgRP) respond to an increase in ghrelin. POMC neurons normally release α-MSH that binds to melanocortin receptors (MC4.R) to activate neurons that inhibit feeding. However, when we are hungry, the stomach releases ghrelin that in turn activates NPY/AgRP neurons. These neurons release GABA and AgRP that inhibit POMC neurons and their targets. AgRP acts as an antagonist at the MC4.R. Elevated ghrelin levels may also directly inhibit POMC neurons (Figure 12.13). *The net result of ghrelin action* is to induce feeding by counteracting POMC-induced appetite suppression.

These studies led to the suggestion that ghrelin might be useful in treating patients with *cachexia*. Cachexia is a physical wasting condition associated with diseases such as cancer and AIDS, where there is a marked loss of body weight. Is it possible that ghrelin treatment could counter the wasting effects of such diseases? The answer seems to be yes, and DeBoer (2011) reviewed several clinical trials in which ghrelin treatment induced significant improvements in appetite and weight gain. Another interesting effect of ghrelin is in the treatment of patients suffering from anorexia nervosa. Such patients have a fear of gaining weight, exhibit starvation-induced abnormal behavior and several biochemical and endocrine abnormalities due to malnutrition. In a pilot study, patients treated with ghrelin for 14 days reported an increase in feelings of hunger, an increase in their energy intake and a continued increase in their body weight up to six months post-treatment (Hotta *et al.* 2009). Ghrelin, or its synthetic analogs, could therefore be a useful treatment for some types of eating disorder (for further reading, see Castaneda *et al.* 2010; and Tong and D'Alessio 2011).

PYY$_{3-36}$ (peptide tyrosine tyrosine)

PYY$_{3-36}$ is released from the GI tract in response to eating, and is particularly sensitive to high fat intake. In contrast, in obese subjects, there is an attenuated increase after eating. This indicates an impaired satiety response that might be responsible for increased food intake (le Roux *et al.* 2006). Treatment of rats with PYY$_{3-36}$ inhibits food intake (Gardiner *et al.* 2008; for further reading, see Chaudhri *et al.* 2008). This peptide also has an anorectic effect in humans and can reduce calories consumed in both lean and obese individuals (Batterham *et al.* 2003a; Steinert *et al.* 2010). Of great interest are the observations that when PYY$_{3-36}$ is combined with two other GI peptides (GLP-1 or oxyntomodulin; see below) the effect on food intake in humans is additive, suggesting that these peptides could be useful in the treatment of obesity (Field *et al.* 2010a; Neary *et al.* 2005). PYY$_{3-36}$ probably acts on the same neurons as those described above for ghrelin; that is, it is likely to inhibit NPY/AgRP neurons to reduce appetite and to stimulate POMC neurons to promote satiety. However, and as shown in Figure 12.12, PYY$_{3-36}$ also exerts effects via the brainstem and vagus nerve.

Glucagon-like peptide-1 (GLP-1)

Chapter 2 described how the peptide hormone glucagon was produced in the pancreas. In the GI tract, however (and the brain), the pre-proglucagon peptide is cleaved, not to glucagon, but to GLP-1 and oxyntomodulin (OXM; see below) (Field *et al.* 2010b). Like PYY$_{3-36}$, GLP-1 is released from the GI tract following a meal and GLP-1 administration inhibits food intake in laboratory animals and in humans (Gardiner *et al.* 2008; van Bloemendaal *et al.* 2014). For example, even in very hungry food-deprived rats, GLP-1 potently prevents food intake, and chronic treatment of normal rats induces weight loss. As we noted above, combined treatment with GLP-1 and PYY$_{3-36}$ is very effective in reducing appetite in humans. Unfortunately, GLP-1 is rapidly degraded in the bloodstream, but newer, long-acting synthetic analogs appear promising in their ability to reduce body weight (Hayes *et al.* 2011; for further reading, see Barrera *et al.* 2011). These analogs may also be useful in the treatment of Type 2 diabetes as well as Alzheimer's Disease (Tong and Sandoval 2011; McClean *et al.* 2011). GLP-1, unlike PYY$_{3-36}$, appears to mainly affect the brainstem in its effect on food intake.

Oxyntomodulin (OXM)

OXM is also a product of the pre-proglucagon peptide, and, like GLP-1 and PYY$_{3-36}$, is released from GI cells postprandially in proportion to caloric intake. OXM reduces food intake in animals and in humans and continued treatment is able to decrease body weight, especially in obese subjects (Gardiner *et al.* 2008; Field *et al.* 2009; Pocai 2012). Just as for GLP-1, synthetic analogs of OXM are required to fully exploit its promise for weight control, but as stated already it may also be particularly effective when combined with PYY$_{3-36}$. OXM appears to bind to the same receptor as GLP-1 and to target the arcuate nucleus in its effects on food intake (for further reading, see Druce and Bloom 2006).

Pancreatic polypeptide (PP)

PP is secreted from the Islets of Langerhans in the pancreas (see section 2.2.8) and blood levels remain elevated up to 6 hours after meals, in proportion to the number of calories consumed (Simpson and Bloom 2010). Treatment of dogs, mice and humans with PP leads to a reduction in food intake (Batterham *et al.* 2003b). Mice engineered to over-express the PP gene, producing high levels of PP, are lean and show reduced food intake, and obese mice given repeated injections of PP lose weight (Chaudhri *et al.* 2008). In humans, the effect of an infusion of PP persisted for 24 hours. This suggests that PP may be another peptide useful in the treatment of obesity. Figure 12.12 indicates that the main site of action is at the brainstem, but the hypothalamus is also a likely target (Lin *et al.* 2009).

To summarize, the role of GI peptides in the control of appetite and body weight is complex and likely to become more so with the isolation of additional gene products. Nevertheless, such knowledge is invaluable in attempts to control the obesity epidemic (for further reading, see Moran and Dailey 2009). Another aspect of this regulatory system involves hormones derived from fat tissue (e.g. leptin) and this will be discussed in the next section. But why are so many different peptide hormones involved in inhibiting food intake? It may be that their secretion is differentially sensitive to various stimuli, and they may have different targets in the brain. Also, some of them act additively. Much more information is needed, perhaps in terms of kinetics, to elucidate specific individual effects. On the other hand, the regulation of food intake is so crucial to survival that the control systems probably include built-in redundancies. For example, mice lacking the GLP-1 receptor (and which are therefore non-responsive to GLP-1) have normal body weight and food intake, suggesting that compensatory mechanisms have taken over elsewhere in the appetite-regulating system (for further reading, see Seeley *et al.* 2000).

12.6.3 Leptin: a fat peptide hormone

This section will focus on leptin as the prototypical adipose tissue hormone (for further reading, see Li 2011). This is because the field of fat hormones is now so extensive that it would be impossible to do it justice in a textbook like this one. For example, since its discovery in 1994, leptin has generated over 25,000 publications (approximately three per day!) and is now implicated in many neural, reproductive, metabolic, immune, endocrine and neuroendocrine systems (for further reading, see Gautron and Elmquist 2011).

Section 2.2.13 introduced the concept that adipose tissue functions as a large endocrine organ that secretes a new family of hormones called *adipokines*. Figure 2.8 illustrated some of the known substances secreted from fat, which now number in excess of 50. What is the

Figure 12.14 Leptin-deficient mouse and lean control
On the left is an obese mouse, designated *ob/ob*, which completely lacks leptin production by fat cells. On the right is a normal, lean mouse of the same age that has normal levels of blood leptin. Note that *ob/ob* mice respond to leptin treatment by losing weight and transforming their appearance to that of control, lean mice. Reproduced with permission (Chicurel 2000).

role of adipokines in the control of energy balance? We know most about leptin, and its mode of action is shown in Figure 2.9. Leptin is secreted from fat cells in order to reduce food intake and to maintain body fat levels. It does this by instructing the hypothalamus, which contains leptin receptors, to reduce appetite and increase energy expenditure. This description is very similar to that for gut peptides, discussed above (section 12.6.2). So it is logical that there exists some overlap between the actions of leptin and those for GI peptide hormones. It is important to remember, however, that signals from the GI tract occur around the time of meals, whereas those from fat are concerned with long-term maintenance of adipose tissue levels. The crucial importance of leptin in the homeostatic control of fat levels can best be appreciated in laboratory animals carrying a defective leptin gene that results in an absence of leptin. Animals and humans lacking leptin are obese (*humans*: Figure 1.4; *mice*: Figure 12.14). Such a mutation is rare in humans, but is common in animals. Note that mice with mutations in the *leptin receptor* gene have high blood levels of leptin, but they are unable to respond to it, so they are also obese.

The remarkable slimming effect of leptin treatment in obese patients with complete leptin deficiency is seen in children (Figure 1.4) and in adults (Figure 12.15; Licinio *et al.* 2004).

In contrast, although leptin can reduce food intake in all species examined to date, including human, monkey and sheep (Henry and Clarke 2008), this strategy does not work in obese humans. Paradoxically, such people have a high fat mass and corresponding elevated levels of leptin. But they are unresponsive to leptin and this is called *leptin resistance* (see later in this section). Note, however, that in both normal weight and obese humans leptin blood levels are markedly reduced following food restriction (Weigle *et al.* 1997; Wisse *et al.* 1999).

Leptin feedback effects on the brain

As previously emphasized for gut peptides, the hypothalamus is a primary target of circulating leptin. In fact, the same group of neurons in the arcuate nucleus that responds to, for example, ghrelin (Figure 12.13), is also sensitive to leptin. Figure 12.16 reveals that leptin, along with insulin, stimulates POMC neurons to induce satiety and prevent further

Figure 12.15 Leptin treatment reduces body weight in obese patients lacking leptin gene expression
Three patients are shown here at baseline (A) and after leptin replacement (B). They received subcutaneous injections of human leptin once daily in the evening (18:00–20:00) at low physiological replacement doses in the range of 0.01–0.04 mg/kg for 18 months. For comparison purposes, two research nurses whose weights were stable during this period are shown. Patients' faces were blurred to maintain confidentiality. Arrows indicate the striking leptin-induced loss in body weight. Reproduced with permission (Licinio *et al.* 2004).

food intake. In addition, they inhibit NPY/AgRP neurons to prevent a hunger/feeding response (Schwartz and Morton 2002; Henry and Clarke 2008). Notice also from Figure 12.16 that when body fat levels fall, and therefore leptin and insulin secretion decrease: (1) NPY/AgRP neurons are relieved from inhibition and begin to drive the eating response; and (2) POMC neurons are no longer stimulated to yield satiety.

Leptin feedback into the brain is mediated by leptin receptors. As discussed in Chapter 10, leptin and insulin receptors are of the *receptor tyrosine kinase* family (Figure 10.6). Leptin receptors are not restricted to the arcuate nucleus and are found in many extra-hypothalamic brain areas (Caron *et al.* 2010; Scott *et al.* 2009), including hippocampus and DA neurons of the midbrain. In rodents, leptin stimulates energy expenditure via the sympathetic nervous system, although this is not yet confirmed in human studies (Dardeno *et al.* 2010). And, as discussed below, leptin's action in the midbrain is related to the rewarding aspects of eating behavior.

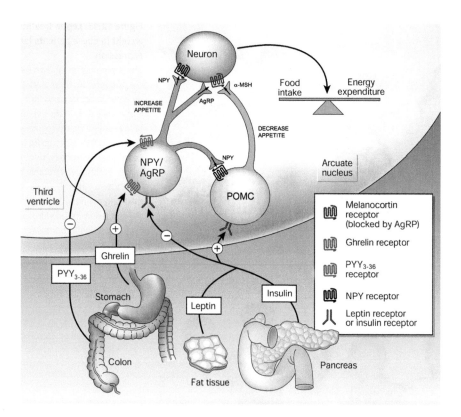

Figure 12.16 Hormones that control eating

This figure effectively combines most of the information shown in Figures 12.12 and 12.13. Leptin and insulin (lower part of the figure) circulate in the blood at concentrations proportional to body fat mass. They decrease appetite by *inhibiting* neurons (center) that produce the molecules NPY and AgRP, while *stimulating* α-MSH (POMC) neurons in the arcuate nucleus region of the hypothalamus. NPY and AgRP stimulate eating, and α-MSH inhibits eating, via other neurons (top). Activation of NPY/AgRP-expressing neurons inhibits POMC neurons. The hormone ghrelin stimulates appetite by activating the NPY/AgRP-expressing neurons. PYY$_{3-36}$, released from the colon, inhibits these neurons and thereby decreases appetite for up to 12 hours. Reproduced with permission (Schwartz and Morton 2002).

Leptin resistance

A complete absence of leptin in experimental animals, and in patients with a mutated leptin gene, leads to obesity. In contrast, the obesity in people that we see all around us is paradoxically associated with high blood levels of leptin secreted from the large fat mass. This implies that obese people are resistant to their own leptin, and strongly suggests that leptin treatment cannot be a therapeutic option for obesity (see below). This was briefly discussed in Chapter 10 (section 10.3.5), and Figure 10.6 revealed that chronic overstimulation of leptin and insulin receptors produces desensitization of the signaling pathway through production of SOCS proteins. This is illustrated in slightly more detail in Figure 12.17 (Munzberg and Myers 2005; for further reading, see Myers *et al.* 2010). On the left is the normal signaling pathway for leptin: binding of leptin to its receptor induces

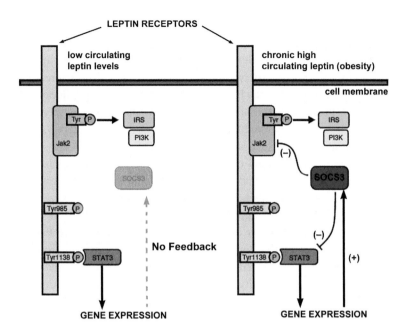

Figure 12.17 Signaling through the leptin receptor and induction of leptin resistance
At low, physiological levels, leptin binding to its receptor activates the Jak2 tyrosine kinase (*janus kinase*), leading to the autophosphorylation of tyrosine residues on Jak2 and the phosphorylation of Tyr985 and Tyr1138 on the intracellular tail of the receptor (labeled as blue P). Jak2 also activates the IRS-PI3K signaling pathway. Phosphorylation of Tyr1138 mediates the activation of the transcription factor STAT3 (*signal transduction and transcription* 3) that further regulates gene expression (see Figure 10.6). One of these genes, *socs3*, expresses the SOCS3 protein and when leptin signaling increases the levels of this protein, it initiates an inhibitory feedback action shown on the right. SOCS3 binds to Jak2 and STAT3 and prevents further signaling through the leptin receptor. This explains the diminishing effectiveness of increasing leptin concentrations in obesity: at low concentrations of leptin, where baseline STAT3 activation is modest, induction of SOCS3 would also be modest, and incremental changes in leptin would be almost fully translated into changes in leptin receptor signaling. In contrast, at high levels of circulating leptin (as in obesity), the increased baseline STAT3 activation would result in increased expression of SOCS3, which would mitigate most of the expected increase in leptin receptor signaling. Reproduced with permission (Munzberg and Myers 2005).

phosphorylation (activation; see Figure 10.8 for another illustration of this principle) of various tyrosine residues in the receptor protein. These phosphorylation sites activate transcription factors such as IRS/PI3K (*insulin receptor substrate 1/phosphoinositide 3 kinase*) and STAT3 (see section 10.3.5) that are translocated to the nucleus where they regulate target gene transcription (e.g. the *pomc* gene in Figure 12.16). Another gene stimulated by leptin is that for *socs3*, a negative feedback regulator of the leptin signaling cascade. When leptin levels are low this gene remains inactivated. However, on the right of Figure 12.17, note that high, chronic levels of leptin increase the expression of *socs3* leading to negative feedback on the signaling pathway, and the induction of leptin resistance. In other words, even high levels of leptin do not stimulate the leptin receptor.

Leptin and reward

The control of body weight is a homeostatic system that depends on many hormonal signals, including leptin. In spite of this complex hormonal feedback, the brain is capable of completely overriding any feelings of satiety; that is, obesity is caused by overeating even in the face of a diminished need for energy intake. We sometimes eat simply because it is pleasurable to do so and for some people this becomes compulsive, leading to obesity. The mesolimbic DA system (e.g. the ventral tegmentum and nucleus accumbens; Figure 5.15) – strongly associated with hedonic stimuli such as alcohol, drugs of abuse and sex – also provides the incentive for the reward of food consumption. In fact, DA neurons in this brain region possess leptin receptors and animal studies indicate that leptin injection into the mesolimbic DA system is able to diminish food intake. This suggests the interesting possibility that obesity-related leptin resistance leads to enhanced craving for food because of reduced leptin signaling (Davis *et al.* 2010; for further reading, see Pandit *et al.* 2011). At least one study revealed that leptin treatment of leptin-deficient patients strongly reduced feelings of hunger *and* reduced food intake, and this was correlated with a leptin-induced reduction in neuronal activity in the ventral striatum as measured by functional MRI (magnetic resonance imaging). In other words, in the non-treated patients, the sight of food activated striatal neurons, whereas in leptin-treated patients it did not (Farooqi *et al.* 2007; for further reading, see Rosenbaum *et al.* 2008; Frank *et al.* 2011). This result suggests that in an obesity-induced, leptin-resistant state mesolimbic DA neurons are not inhibited by leptin and remain active, thereby producing feelings of hunger. Volkow and Wise have drawn some interesting parallels between obesity and addiction to drugs such as cocaine: "The guidelines for prevention and treatment of the two disorders are remarkably similar, and some of the same pharmacological interventions that are promising for the control of drug intake are also promising for controlling the intake of food. Few fields seem to offer as much potential for cross-fertilization as the fields of addiction and obesity research" (Volkow and Wise 2005).

Therapeutic uses of leptin

The powerful slimming effect of leptin treatment in those obese patients who completely lack circulating leptin (see Figure 1.4) was noted above. Such patients seem not to develop leptin resistance, probably because they received only single injections of leptin each day (Paz-Filho *et al.* 2011). In addition, and of great interest in these studies, was the realization that leptin was important not only for regulating food intake and body weight, but also had a profound beneficial influence on other physiological systems. Table 12.3 lists some of these effects.

There is also evidence that leptin treatment of certain women with amenhorrea (i.e. no menstrual cycles) leads to restored fertility. The women tested had a normal puberty, but they eventually stopped menstruating because of strenuous physical activity and/or low body weight (Chou *et al.* 2011). Their leptin levels were below normal and daily injections of leptin over a period of 36 weeks were successful in normalizing leptin levels and in restoring menstrual cycles. Such patients also see benefits in terms of increased bone density (for further reading, see Sienkiewicz *et al.* 2011).

What are the possibilities of treating obesity with leptin? People, and children, who are obese have high circulating levels of leptin and are leptin resistant. Clinical trials of high

Table 12.3 **Some effects of leptin replacement therapy in leptin-deficient patients**	
Endocrine effects	*Reversal* of type 2 diabetes
	Increase in insulin sensitivity and decrease in insulin secretion
	Appearance of secondary sexual characteristics
	Menstrual periods established
	Testosterone levels become normal (men)
	Increase in 24-hour cortisol secretion with a more regular circadian pattern
	Increase in insulin-like growth factor binding protein (IGFBP1)
	Maintenance of adequate growth velocity
	Regulation of the thyroid-stimulating hormone (TSH) rhythmicity
Body composition	*Weight loss*, mostly fat, up to 54 percent of initial body weight
Brain and behavior	*Decrease* in caloric intake
	Increase in physical activity
	Increase in gray matter concentration*
	Activation of brain areas involved with satiety and *inhibition* of areas involved with hunger
	Increase in cognitive development
	Changes from infantile to adult-like behavior

Data from Paz-Filho *et al.* 2011 and *London *et al.* 2011.

dose leptin as a treatment for obesity have been largely unsuccessful (Dardeno *et al.* 2010). Since there are other hormones capable of reducing appetite, such as PYY_{3-36} and GLP-1 (see Figure 12.12), perhaps a combination of leptin with these peptides might be efficacious in reducing body weight. Little evidence for this is yet available, but the pancreatic peptide hormone *amylin* (Figure 2.3) is effective when combined with leptin (Ravussin *et al.* 2009; for further reading, see Tam *et al.* 2011). Figure 12.18 shows that a leptin/amylin combination (using synthetic analogs of each drug), given to overweight/obese patients reduced body weight by 12 kg over a period of 20 weeks. The authors speculate that amylin may act to reverse leptin resistance by stimulating the STAT3 signaling pathway shown in Figure 12.17 (for further reading, see Roth *et al.* 2008). In summary, leptin, perhaps in combination with other peptides, remains a possible therapeutic option in attempts to deal with the obesity epidemic.

12.7 Visceral, cognitive and behavioral effects of neuropeptides

In addition to their effects on the neuroendocrine system, neuropeptides modulate the visceral functions of the ANS and the cognitive and behavioral functions of the central nervous system (McCall and Singer 2012). By way of example, all three of these effects are closely related as shown for oxytocin and vasopressin (Figure 12.19).

In the hypothalamus, central amygdala and brainstem nuclei, neuropeptides regulate visceral functions concerned with basic bodily needs. These include hunger and feeding,

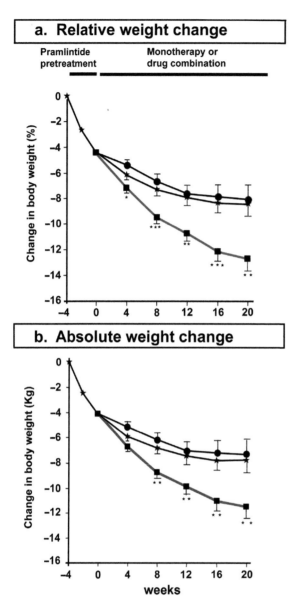

Figure 12.18 Effect of leptin and amylin combination treatment on weight loss in obese/overweight patients
Patients were treated twice per day with the leptin analog *metreleptin* and the amylin analog *pramlintide*. The data show (A) percent change and (B) absolute change in body weight from the time of enrollment, during the lead-in and treatment phases (metreleptin $n = 19$; pramlintide $n = 38$; pramlintide/metreleptin $n = 36$).
     * $p < 0.05$; ** $p < 0.01$; *** $p < 0.001$. Reproduced with permission and redrawn (Ravussin *et al.* 2009)

fluid regulation and thirst, pain perception and avoidance responses, autonomic and behavioral aspects of temperature regulation and sexual arousal, as well as the functions of the gastrointestinal and circulatory systems. In the hippocampus, basolateral amygdala and other forebrain and limbic structures, neuropeptides regulate emotional and motivational states associated with these basic bodily needs and coordinate the adaptive behavioral responses and general arousal levels correlated with these affective states. Learning and memory are important aspects of adaptive behavior that are regulated by neuropeptide action in the neocortex as well as in the hippocampus and other limbic areas (Figure 12.20).

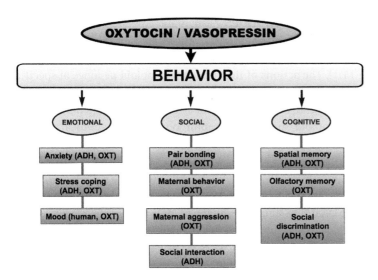

Figure 12.19 Involvement of brain oxytocin and vasopressin in a variety of behaviors
A wide variety of behaviors are modulated through central release of oxytocin and vasopressin. For example, following birth and the beginning of pup suckling, oxytocin is released to stimulate maternal behavior and oxytocin and vasopressin are involved in pair bonding.

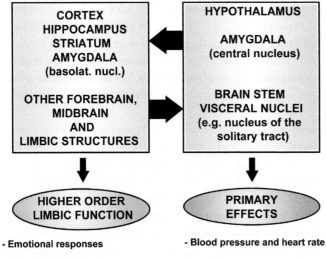

Figure 12.20 Schematic view of neuropeptidergic brain regions that regulate visceral, cognitive and behavioral responses
Brain stem nuclei contain neuropeptide receptors that regulate visceral functions. Higher order structures such as cortical, limbic and hypothalamic nuclei modulate the cognitive and behavioral correlates of these affective functions.

This section reviews the versatility of neuropeptides on these adaptive behaviors. As we have noted previously, however, this book is not designed to attempt an exhaustive coverage of such a vast field of information. For this reason, we have chosen representative neuropeptides to illustrate particular points and to provide a template by which the actions of other neuropeptides can readily be understood.

12.7.1 The endogenous opioid peptides

These neuropeptides were introduced in Chapter 11. There are several known opioid families (see Table 11.2) and much is known about the genes that express the opioid neuropeptides (see Figure 11.1). The actions of the endogenous opioid peptides at specific receptors in the brain and spinal cord have been extensively studied. This section briefly reviews the analgesic, visceral, cognitive and behavioral effects of the endogenous opioids and the correlations between opioid functions, receptor types and anatomical sites of action.

Analgesic effects

It has been known for centuries that opiate agonist drugs, such as morphine, can efficiently relieve pain (analgesia). The anatomy and physiology of pain pathways is adequately covered elsewhere (Purves *et al.* 2008; Berne and Levy 2000) and the multiple types of opiate receptor have been covered (see section 11.6.2). Briefly, the μ-opiate receptor appears to be crucial for analgesic drugs – acting in the spinal cord, brain stem and forebrain pathways – to prevent transmission of painful stimuli. It is true that other opiate receptors are found in the same pathways, but only the μ-opiate receptor mediates the analgesic effect of morphine. This was elegantly demonstrated in mice lacking the μ-opiate receptor (*μ-receptor knockouts*). These mice failed to respond to morphine in pain perception tests; that is, they continued to feel pain even though they were given large amounts of morphine (Matthes *et al.* 1996; Iversen 1996). The result shows conclusively that the μ-opiate receptor is the site for analgesic actions of opiates (for further reading on *opioid knockouts*, see Kieffer and Gaveriaux-Ruff 2002; and on *opiate receptors and analgesia*, see Bodnar 2010).

The discovery of the endogenous opioid peptides, such as enkephalins and β-endorphin, held out the promise of newer and more efficient analgesic drugs free of addictive side effects. Unfortunately, the results of extensive research in this field have been disappointing and have not delivered on this promise (Iversen *et al.* 2009). On the other hand, several powerful synthetic opiates (e.g. fentanyl) are very effective in reducing severe pain, although all opiates suffer from their propensity to induce dependence and abuse (Iversen *et al.* 2009; Corbett *et al.* 2006). It is also worth remembering that the *neuroendocrine system* is especially vulnerable to chronic exposure to opiates, particularly in patients being treated for chronic pain. Table 12.4 outlines a brief summary of these effects, with a comparison to animal studies (Vuong *et al.* 2010). Thus, some patients suffer from decreased libido, erectile dysfunction, amenorrhea, bone loss and impaired insulin secretion.

Effects on the ANS

Within the ANS, stimulation of opiate receptors depresses respiration, reduces gastrointestinal motility, inhibits catecholamine secretion from the adrenal medulla, modulates heart rate, reduces blood pressure and lowers body temperature. Opiates and opioid peptides depress respiration through their action at μ- and δ-opiate receptors in the respiratory centers of the brainstem. The anti-diarrheal effect of opiates occurs through their action at μ-receptors in the gastrointestinal tract, where they inhibit gastrointestinal motility and the peristaltic reflex. Within the cardiovascular system, stimulation of μ-, δ- and κ-receptors

Table 12.4 **Opiate effects on the endocrine systems of animals and humans**

Hormone	Acute		Chronic	
	Animals	Human	Animals	Human
GH	↑	↑	=	no data
PRL	↑	↑	↑	↑/=
TSH	↓	↑	no data	no data
ACTH	↑	↓	?	↓/=
LH	↓	↓↓	↓	↓↓
FSH	=	=	=	=
Estradiol	↓	↓↓	=	↓/=
Testosterone	↓	↓↓	↓	↓↓
ADH	?	?	?	?
OT	↓	↓	↓/=	↓/=

Key: ↑, stimulation; ↓, inhibition; ?, conflicting; =, no change

inhibits sympathetic stimulation of the heart, reducing stress-induced increases in heart rate. Opiates and opioid peptides reduce blood pressure through their action in the brain and by inhibiting norepinephrine release from sympathetic nerves. Opioids, especially dynorphin, act at presynaptic κ-receptors on posterior pituitary nerve terminals to regulate fluid balance by inhibiting vasopressin release, just as we saw previously for oxytocin secretion (see section 12.3.3; and Figure 12.9).

Cognitive and behavioral effects

The behavioral effects of endogenous opioid peptides have been the focus of intensive research for some time and are the topic of annual reviews over the past 32 years (Bodnar 2010). These reviews outline in detail the behavioral impact of molecular, pharmacological and genetic manipulation of opioid peptides, opioid receptors, opioid agonists and opioid antagonists. As such, they represent an invaluable source of information in this field. This section will merely touch on some aspects of the widespread involvement of opioids in behavioral research. Details of the euphoric and addictive/dependence aspects of opiate use can be conveniently found elsewhere (Iversen *et al.* 2009).

Opioids and binge eating

Animal experiments revealed that food intake, and especially palatable food in binge eating models, could be reduced by blocking μ-opiate receptors. The results implicate endogenous opioids in the drive to over-consume highly palatable food. This work led directly to the question of whether similar drugs might be effective in humans. As many as 6.6 percent of the population indulge in binge eating and this is strongly associated with obesity. Single, acute treatment with opioid receptor antagonists such as naloxone and naltrexone were effective in decreasing short-term food intake (decreases between 11 and 29 percent) (Nathan and Bullmore 2009). Similarly, naltrexone reduced the frequency of binge eating and also reduced binge duration in bulimic patients and obese binge eaters (Marrazzi *et al.* 1995). Although in animal experiments drugs such as naltrexone could reduce body weight, similar studies in humans have been largely negative. Nathan and Bullmore (2009) have suggested that success may depend on the use of more specific antagonists and to focus on specific groups of patients such as binge eaters (Pecina and Smith 2010; Ziauddeen *et al.* 2013).

Opioids and alcoholism

Once again, animal experiments led to the conclusion that alcoholics may have abnormal endogenous opioid systems. Opioid receptor antagonists, such as naloxone or naltrexone, decreased ethanol intake in animals. More specific receptor blockade with κ-opiate receptor antagonists also attenuated ethanol consumption. In humans, naltrexone, a non-specific opioid receptor antagonist, decreased alcohol consumption by alcoholics and was also effective in reducing relapse rates, as well as craving and ethanol intake (Gianoulakis 2009). Naltrexone is now approved in Canada and the USA for the treatment of alcoholism.

Opioids and sex behavior

In both men and women, opiates such as heroin and morphine, acting on opioid receptors, produce high-risk sexual behaviors. Multiple sexual partners, avoidance of condom use and high incidence of STD/HIV diagnosis are commonplace (Frohmader *et al.* 2010). However, although low doses of heroin can increase sexual desire, chronic intake of such drugs leads not only to addiction but to blunted sexual desire and sexual response together with erectile and ejaculatory dysfunction (Meston and Frohlich 2000). These effects of chronic opiate use are not surprising given that opiates and opioid peptides decrease sex hormone secretion by a profound inhibitory action on GnRH neurons in the hypothalamus (Dudas and Merchenthaler 2004). Based on these studies, we can speculate that if endogenous opioid peptides are involved in sex behavior, then opioid antagonists such as naltrexone should have the opposite effect. Several reports indicate that this is so. For example, a group of 19 male patients, displaying compulsive sexual behavior, were treated with naltrexone (Raymond *et al.* 2010). Seventeen of 19 patients reported a reduction in problematic sexual behavior for periods ranging from two months to two years. In a study on adolescent sex offenders, 15 of 21 patients responded positively to naltrexone with a decrease in sexual fantasies (Ryback 2004). These representative examples indicate that endogenous opioid peptides can be implicated in human sex behavior (for further reading, see Bostwick and Bucci 2008).

In summary, there is abundant evidence that endogenous opioid peptides play a significant role in regulating some aspects of human behavior. Interested students should consult Sauriyal *et al.* (2011) for a discussion of the putative role of opioids in depression, anxiety, epilepsy and stress.

12.7.2 Posterior pituitary peptides

Some aspects of the release of oxytocin and vasopressin (ADH) from the posterior pituitary gland have been noted. Acting as pituitary hormones, they are critical in childbirth, in milk letdown when babies suckle at the breast and in the regulation of water balance (e.g. see

Chapter 3 and Figure 8.3). However, earlier in the present chapter, the idea that oxytocin was secreted from dendrites as well as from nerve terminals in the posterior pituitary was introduced (Figure 12.8). Further, oxytocin and ADH nerve fibers are found outside the hypothalamus, especially in the limbic system and spinal cord (Buijs *et al.* 1978; de Vries 2008; Rood and de Vries 2011). Receptors for these two neuropeptides are also localized to many brain regions in several species (for references, see Schorscher-Petcu *et al.* 2009; and Lee *et al.* 2009). This suggests that the release of oxytocin and ADH, *outside the hypothalamus*, subserves a possible neurotransmitter function unrelated to their hormonal action. Note, however, that oxytocin and ADH, released from dendrites of the neurons that project to the posterior pituitary, can also diffuse for large distances in the brain (see Figure 12.1) (Ludwig and Leng 2006). Under some circumstances, it is possible to induce dendritic release within the rat brain in the absence of a concomitant secretion from the posterior pituitary; that is, dendritic secretion is controlled differently from axonal release. For example, in a forced swimming procedure – an emotional and physical stressor – both ADH and oxytocin are released from dendrites, but only oxytocin is secreted from the pituitary (Landgraf and Neumann 2004). Social defeat, on the other hand, induces an increase in brain (dendritic) secretion of oxytocin, but not ADH, with blood levels of both neuropeptides remaining unchanged. This review lists a wide variety of chemical and behavioral stimuli that provoke brain release of ADH and oxytocin and it is now clear that both neuropeptides have profound effects on behavior (Landgraf and Neumann 2004). Figure 12.19 outlines some of these.

Oxytocin

A concise review by Neumann (2008) serves as a convenient introduction to brain oxytocin acting as a regulator of emotional and social behavior (for further reading and more detailed descriptions, see Lee *et al.* 2009; and Veenema and Neumann 2008). Brain oxytocin is implicated in the neural control of *anxiety*. Acute or chronic brain injection of oxytocin is anxiolytic (i.e. prevents or reduces anxiety) in rats and mice of both sexes. However, blocking endogenous oxytocin with an oxytocin receptor antagonist increased anxiety only in pregnant or lactating females. This suggests the importance of sex hormones in oxytocin's actions. Since oxytocin also inhibits the *stress response* (the hypothalamic-pituitary-adrenal axis) to a variety of physical and emotional stressors, and since oxytocin levels are high during pregnancy, this suggests that it might play a critical role in calming the mother around the time of birth. In addition, oxytocin facilitates *maternal behavior*, an extra benefit to the offspring. This powerful effect on maternal behavior is also seen when oxytocin is injected into the brains of virgin rats. Newborn animals receive additional benefits from their mother's oxytocin through an increase in *maternal aggression* against potential threats. A direct role for oxytocin is not as clear here, but injection of an oxytocin receptor antagonist directly into the brain (especially in the amygdala and in the paraventricular nucleus (PVN)) significantly reduced the level of aggression. Oxytocin is also firmly implicated in controlling *sexual behavior*: in female rats, oxytocin elicits lordosis behavior, whereas in males it is released in the PVN during successful mating. In several species, including monkey, oxytocin promotes erectile function. Finally, oxytocin is implicated in *social memory and recognition*. For example, brain oxytocin release in the olfactory bulb is related to newborn lamb recognition and bonding with the mother. In mice, oxytocin plays an important role in the Bruce Effect; that is, the ability of a pregnant mouse to recognize her mate and remain pregnant. Recognition of a novel mate results in loss of the pregnancy. Mice with a disabled oxytocin gene (oxytocin knockouts) could not tell the difference between mate and non-mate, with consequent loss of pregnancy. Is there evidence for behavioral effects of oxytocin in *humans*? (for further reading, see Meyer-Lindenberg *et al.* 2011). Brain injections in humans are impractical, and oxytocin given systemically will not cross the blood-brain barrier. However, neuropeptides given *intranasally* are able to reach the brain (Born *et al.* 2002) and it is therefore possible to conduct experiments in humans in which oxytocin is delivered directly to the brain. For example, oxytocin seems to promote face recognition and to reduce levels of anxiety to psychosocial stress. In both men and women, blood oxytocin levels are increased during sexual arousal and orgasm, although it is unclear whether intranasal oxytocin can induce or enhance arousal. Oxytocin acting at oxytocin receptors in the human brain have been implicated in the neurochemistry of romantic love (Zeki 2007; Carter and Porges 2013). In terms of maternal behavior, there is, as in animal experiments, some evidence for an influence of oxytocin, but these conclusions were based on blood, rather than brain, levels of oxytocin (Galbally *et al.* 2011). It is generally believed that oxytocin can facilitate feelings of attachment and trust, and intranasal oxytocin appears to generate feelings of generosity and an appreciation of others' feelings (empathy). Conversely, there are suggestions that disruptions in oxytocin signaling may lead to psychiatric consequences such as schizophrenia and autism (Kuehn 2011; Yamasue *et al.* 2012; Miller 2012).

Vasopressin (ADH)

Interested students will find that the behavioral effects of vasopressin are covered in detail by Caldwell *et al.* (2008). In this section, the discussion will be restricted to just a few specific points. For example, ADH is strongly implicated in the *regulation of aggression*, particularly in male Syrian hamsters. Microinjection of ADH into the anterior hypothalamus facilitates aggression and is testosterone-dependent. The effect can be blocked by ADH receptor antagonists, specifically the subtype designated Avpr1b. Thus, oral treatment with an Avpr1b antagonist reduces aggression in mice and hamsters. Female transgenic mice lacking the *Avpr1b* gene show reduced maternal aggression. In humans, there is a positive correlation between high levels of ADH in the cerebrospinal fluid (CSF) and aggression towards individuals (Coccaro *et al.* 1998). An interesting correlative study suggests that some individuals with conduct disorder, as well as prisoners with a history of violence, have reduced levels of autoantibodies to ADH; that is, this allows elevated levels of ADH to interact with receptors (Fetissov *et al.* 2006). There is a large literature on the influence of ADH in *learning and memory*. ADH receptor antagonists, and the use of ADH receptor knockout mice, reveal a positive correlation between ADH and memory. Human investigations are less conclusive. Healthy elderly volunteers given repeated intranasal ADH showed no improvement in long-term memory, although both acute and chronic treatment of boys with learning disorders improved recall of stories. A final clinical point is that there is good evidence for abnormal ADH levels in patients with *anxiety disorders and major depression*. These studies depend on measuring blood levels of ADH and assume that brain ADH levels are also abnormal. However, there are also reports that brain *ADH* mRNA is elevated in PVN and SON particularly in depressed patients of the melancholic subtype. Similarly, an increase in ADH immunoreactivity (ADH peptide level) was found in the PVN of people who had committed suicide. Neumann and Landgraf (2012) outline a theory that a balance between oxytocin and vasopressin neuropeptide systems is critical for normal emotional behavior; for example, oxytocin exerts antidepressive effects and vasopressin has an opposite effect. Several clinical trials have attempted to target ADH receptors for potential treatments of *autism and schizophrenia* (Ryckmans 2010). Some characteristics of autism, such as atypical social behavior and increased anxiety and emotionality, are affected by ADH in experimental animals and especially in males. It is interesting, therefore, that autism may be more prevalent in boys (Pfaff *et al.* 2011). Genetic analysis is proving to be helpful, with the possibility that the *Avpr1a* receptor gene is an autism susceptibility gene (Yirmiya *et al.* 2006; Yang *et al.* 2010).

12.7.3 Hypothalamic hypophysiotropic peptide hormones

These hormones – often called hypothalamic releasing or inhibiting hormones – were first introduced in section 4.3. However, in addition to their classical effects on cells of the anterior pituitary, these hypothalamic hormones behave as neuropeptides and have a wide range of central effects through their release in many extra-hypothalamic brain regions. Some of these effects are summarized here.

Somatostatin

Somatostatin exerts inhibitory control of GH secretion from the anterior pituitary (Figures 4.8 and 4.11) and regulates insulin release from the pancreas (section 2.2.8). In addition, somatostatin receptors are found throughout the brain and there are six known subtypes: sst1, sst2A and 2B, and sst3–sst5. The distribution of the mRNAs for each receptor is described for the rat brain (Schindler *et al.* 1996; Dournaud *et al.* 2011). There are distinct patterns for *sst1–sst4* mRNAs, with specific localization in cerebral cortex, hippocampus, amygdala and cerebellum. *Sst5* mRNA is found only in the pituitary gland. The distribution of the corresponding receptor proteins was determined by immunohistochemistry (Helboe and Moller 2000). These data imply that somatostatin exerts effects throughout the brain and not just in the neuroendocrine system. Therefore, an important question is: what is the source of the somatostatin that binds to these receptors? The rat brain and human brain have been mapped for somatostatin mRNA and there are many neuronal groups that express somatostatin, including neocortex, amygdala, basal ganglia, midbrain, hippocampus and medulla (Mengod *et al.* 1992; Priestley *et al.* 1991). Somatostatin immunoreactivity has been localized to many of the same sites, confirming the likely role of somatostatin as a neuropeptide in many brain regions (Viollet *et al.* 2008). In keeping with this wide distribution, there is evidence for a regulatory/neuromodulatory function of somatostatin in motor activity, cognition, sleep and sensory processing. It is now known, for example, that a derangement in brain somatostatin in human temporal lobe may be involved in the defects associated with Alzheimer's Disease (Gahete *et al.* 2010; Epelbaum *et al.* 2009).

Growth hormone releasing hormone (GHRH)

GHRH stimulates GH secretion from the anterior pituitary (section 4.3.4; Figure 4.11). However, the mRNA for both GHRH and the GHRH receptor were found in cerebral cortex and brain stem, suggesting a non-neuroendocrine role for GHRH (Matsubara *et al.* 1995; Szentirmai *et al.* 2007). Cortical GHRH appears to regulate some aspects of sleep (Liao *et al.* 2010), and *ghrh* is reported to be a susceptibility gene for anxiety and panic disorder (Gratacos *et al.* 2009). An influence of GHRH on food intake is known. Injection of GHRH into the brain stimulates feeding, but has no other behavioral effect, nor is it dependent on GH release from the pituitary gland (Feifel and Vaccarino 1994). Further work by this group showed that an injection of GHRH specifically into the suprachiasmatic nuclei (SCN; see section 8.2.1; Figure 8.5) not only increased feeding, but also caused a phase advance of running activity in hamsters; that is, it altered a circadian rhythm (Vaccarino *et al.* 1995). This suggests that GHRH is an important component of the brain clock mechanism. Finally, in a recent review, there is good evidence for widespread effects of GHRH throughout the body. These include a stimulatory/supportive influence on pancreatic β-cells and a restorative effect on cardiac tissue following a heart attack (infarction) (Kiaris *et al.* 2011).

Corticotropin-releasing hormone (CRH)

As seen earlier, CRH, synthesized in PVN neurons, is responsible for the secretion of ACTH from the anterior pituitary (section 6.3.1; Figures 1.1 and 6.4). CRH is also expressed in neurons in many areas of the brain. Alon *et al.* (2009) used a new transgenic mouse that expressed a fluorescent CRH peptide to demonstrate the presence of CRH in numerous nuclei throughout the brain and especially in cortex and amygdala. As expected, there were high levels of CRH in the PVN. They also localized CRH mRNA by *in situ* hybridization and showed that the distribution was almost identical to that for the peptide. This confirms that CRH is expressed in neurons outside the hypothalamus. CRH binds predominantly to the CRF1 receptor and this subtype is found in all brain regions shown to be CRH-positive. However, the receptor is localized to many other areas as well, indicating that CRH probably diffuses away from its release sites and acts as a neuromodulator/neurotransmitter in the brain (Chen *et al.* 2000; van Pett *et al.* 2000). It is not surprising therefore that when CRH is injected directly into the brain, it evokes a variety of responses, including arousal, anxiety, changes in the ANS, inhibition of the reproductive system, inhibition of the digestive system and anorexia (Richard *et al.* 2002). More specifically, injection of CRH into the brains of experimental animals produces behavior reminiscent of *human depression*: that is, increased anxiety, reduced slow wave sleep, psychomotor alterations, anhedonia, and decreased appetite and libido. In human beings, increased cerebrospinal fluid CRH concentrations are observed in major depression and suicide victims, and several postmortem studies revealed evidence for increased CRH neurotransmission in major depression and suicide. CRH expression levels (mRNA and peptide) are elevated in hypothalamus, cortical areas and the locus coeruleus (Binder and Nemeroff 2010; for further reading, see Clark and Kaiyala 2003). In terms of *food intake*, CRH is a powerful anorectic agent and it reduces feeding in a variety of animals, including rats, mice, fish and sheep. CRH treatment also prevents weight gain in genetically obese rats (Richard *et al.* 2002). Whether the CRH system is a useful target for treatment of human obesity remains to be seen, but the discovery of other CRH-like peptides (e.g. the

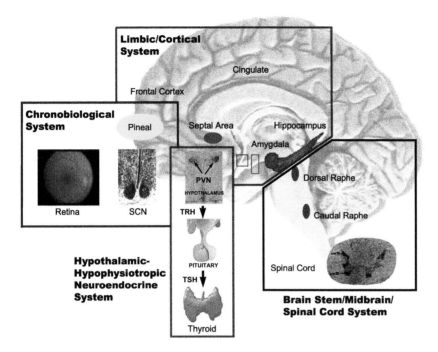

Figure 12.21 The thyroid hormone releasing hormone (TRH) homeostatic modulatory system
This important system comprises four distinct, functionally integrated components, including: (a) the hypothalamic-pituitary-thyroid neuroendocrine system (green box); (b) the brainstem-midbrain-spinal cord system (black box); (c) the limbic-cortical system (red box); and (d) the chronobiological system (purple box). Note that boxes (a), (b) and (d) include brain histological sections illustrating anatomical locations; e.g. paraventricular nucleus (PVN) in (a).
   Abbreviations: SCN, suprachiasmatic nuclei; TSH, thyroid-stimulating hormone. Reproduced with permission (Gary *et al.* 2003).

urocortins), and a second CRH receptor (CRF2), suggests some promise for such an approach (for further reading, see Doyon *et al.* 2004; and Zorrilla *et al.* 2003).

Thyroid hormone releasing hormone (TRH)

In Chapter 4 (section 4.3.1), we introduced TRH as a hypothalamic-releasing hormone that controls the thyroid system (Figure 6.6) and, under some circumstances, prolactin release as well (Figure 4.9). However, TRH peptide, *trh* mRNA and TRH receptors are found in the hypothalamus and throughout the CNS as well, suggesting that TRH acts as a neuromodulator/neurotransmitter. For example, TRH gene expression was found in hypothalamus, cortex, amygdala, hippocampus and septum (Knoblach and Kubek 1997; Fliers *et al.* 1998). G-protein-coupled TRH receptors, designated TRH-R1 and TRH-R2, are localized in a large number of brain regions, including the spinal cord (O'Dowd *et al.* 2000; Heuer *et al.* 2000). Thousands of publications describe the myriad actions of TRH in the CNS (Yarbrough *et al.* 2007). In an effort to provide a conceptual framework for these effects, Gary *et al.* (2003) proposed that these functions are regulated by four anatomically distinct brain systems (Figure 12.21). The *neuroendocrine system* (also illustrated in Figure 6.6) is responsible for regulation of metabolism and thermogenesis. The *limbic/cortical system* is associated with

Table 12.5 Some therapeutic applications of TRH

- Antidepressant effects in major depression
- Therapeutic effects in amyotrophic lateral sclerosis/motor neuron disease
- Anticonvulsant actions in epilepsies
- Therapeutic effects in Alzheimer's Disease
- Attenuation of memory impairment
- Decrease in schizophrenic psychotic symptoms
- Antagonism of alcohol effects
- Neurological improvements post-stroke and head trauma
- Reversal of benzodiazepine-induced sedation
- Improved cognition in short-duration alcoholism
- Stimulates respiration post-general anesthesia
- Increases cerebral blood flow
- Attenuates mania and alcohol withdrawal effects

behavioral functions such as arousal, cognition, mood, anxiety and seizure modulation. The *brain stem/spinal cord* system has effects on GI function, pain, respiratory drive and thermoregulation. The *chronobiological system*, consisting of the suprachiasmatic nuclei (SCN)/retina/pineal gland (Figure 8.5), is responsive to TRH in modifying circadian variations of some physiological functions. TRH receptors were detected in the SCN (Manaker *et al.* 1985) and in the retina (Satoh *et al.* 1993).

The surprisingly wide range of physiological effects exerted by TRH has led to possible therapeutic applications. Some of these are listed in Table 12.5 (data from Yarbrough *et al.* 2007). Note, however, that TRH has a short half-life in blood (minutes) and may also have difficulty crossing the blood-brain barrier. For these reasons, metabolically stable TRH analogs have been developed to further extend their therapeutic potential in areas such as depression, chronic fatigue syndrome, recovery from anesthesia, attention deficit hyperactivity disorder (ADHD), anxiety disorders and disturbed circadian rhythms (e.g. jet lag) (Gary *et al.* 2003) (for further reading, see Lechan and Fekete 2006).

Gonadotropin-releasing hormone (GnRH)

A description of the critical importance of *hypothalamic* GnRH in the reproductive system was provided in Chapter 4 (section 4.3.3; see Figure 6.7). GnRH is, however, widely distributed throughout the CNS, elegantly demonstrated using transgenic mice that express GnRH coupled to a fluorescent marker protein such as GFP (see also section 12.3.2, above, and Figures 12.3 and 12.7 as examples of this approach) (Spergel *et al.* 2001; Herbison *et al.* 2001). This transgenic technique allows GnRH-positive neurons and fibers to be easily detected by fluorescence microscopy. Thus, GnRH fibers were also observed in the median eminence and in the amygdala, hippocampus, septum and midbrain. Neurons were identified in the hypothalamus, as anticipated, but in olfactory bulb and septum and in spinal cord (Dolan *et al.* 2003). The distribution of GnRH *receptors* is even more widespread and this suggests that GnRH acts as a neurotransmitter and a neuromodulator by diffusing away from nerve terminals. One estimate suggests that about 800 GnRH neurons communicate with approximately 30,000 neurons in 34 different brain areas (Boehm *et al.* 2005), presumably via GnRH receptors. Localization of GnRH receptors was accomplished by standard immunohistofluorescence microscopy with antibodies to the GnRH receptor peptide (Albertson *et al.* 2008), but more recently by the transgenic technique already mentioned; that is, the GnRH receptor gene was modified to express a GnRH receptor protein coupled to a marker protein such as GFP (Schang *et al.* 2011; Wen *et al.* 2011). Wen *et al.* (2011) also demonstrated that the GnRH receptors that they detected were functional; that is, neurons responded to stimulation with GnRH. In addition, they listed the many brain areas that contain GnRH receptors (see also Table 12.6; Skinner *et al.* 2009) and point out that several of these areas are associated with control of odor and pheromone processing, sexual behavior, appetite, defensive behavior, motor programs and the relay of information to higher cortical areas (for further reading, see Skinner *et al.* 2009). This would be consistent with known effects of GnRH injected into the brain. For example, it has been known for some time that GnRH facilitates sexual behavior in rodents of both sexes (Schiml and Rissman 2000). In an interesting recent study, an *antagonist* of GnRH had significant anti-anxiety and antidepressive effects and was able to prevent memory impairment in rats treated with β-amyloid, suggesting a possible role for GnRH in Alzheimer's Disease (Telegdy *et al.* 2010; for further reading, see Wang *et al.* 2010).

12.7.4 Anterior pituitary hormones

All of the neuropeptides covered so far in this section are of brain origin, including the posterior pituitary hormones oxytocin and vasopressin. *Anterior pituitary hormones*, on the other hand, are expressed in endocrine cells such as gonadotrophs and somatotrophs before entering the circulation to act as hormones. There is some evidence, however, that these hormones do have behavioral effects when injected into the brain. The question of whether they may be physiologically relevant depends on two issues: first, is it possible that hormones such as LH or GH can enter the brain from the peripheral circulation (i.e. following release from the anterior pituitary); and, second, is there evidence for biosynthesis in brain cells, and for the presence of receptors in the brain? All of the anterior pituitary hormones are very large molecules that appear unlikely to cross the blood-brain barrier; GH, for example, is a single polypeptide chain of 191 amino acids (molecular weight approx. 22 kDa). Nonetheless, there is some evidence that this might occur (see discussion in Barron *et al.* 2006; for further reading, see Banks 2008). For example, LH and GH polypeptides have been localized in rat and primate brain (Hojvat *et al.* 1985; Hojvat *et al.* 1982), specifically in amygdala, hippocampus and cortex.

Luteinizing hormone (LH) and follicle-stimulating hormone (FSH)

LH and FSH peptides have been identified in the brain, plus gene expression for FSH, the FSH receptor and the LH receptor (Lathe 2001; Chu *et al.* 2008; Lei *et al.* 1993; Pandolfi *et al.* 2009). The spinal cord also expresses LH receptor mRNA (Rao *et al.* 2003). Intracerebroventricular injection of HCG (human chorionic gonadotropin; an LH-like hormone) in rats affects feeding behavior, sleep-wake cycles and exploratory activity (Barron *et al.* 2006). The hippocampus, a brain region critical for memory, appears to be a primary target region for LH and FSH. Epidemiological studies have correlated aging-related elevations in LH secretion (e.g. during menopause) with cognitive decline and an

Table 12.6 GnRH target sites outside the hypothalamic-pituitary-gonadal axis

Tissue	Evidence	Species
Nervous tissue		
Retina	a, b, c	Mouse, rat, vole
Olfactory bulb	a, c	Mouse, rat
Cortex, especially piriform	a, c	Mouse, rat
Lateral septum	a, b, c	Mouse, rat
Preoptic area	b, c, d	Mouse, sheep
Arcuate nucleus	b, c, d	Mouse, sheep
Hippocampus	a, c	Mouse, rat, sheep
Amygdala	a, c	Mouse, rat, sheep
Central gray	a, c	Mouse, rat, sheep
Cerebellum	b, c, d	Mouse, rat
Spinal cord	b	Sheep
Pituitary		
Growth hormone	a, d	Rat, human
Prolactin	d	Rat, human
Thyroid-stimulating hormone		
Adrenocorticotrophic hormone	d	Rat
Other		
Kidney	a, b, d	Mouse, rat, human
Liver	a, b	Mouse, rat, human
Heart	a, b, c, d	Mouse, human
Bladder	a, b	Mouse, dog, human
Tooth	b, c	Mouse
Adrenal	a, b	Mouse, rat, human, cow
Skin	a, b	Mouse, rat, dog
Skeletal muscle	b	Rat, human
Spleen	a, d	Mouse, rat
Lymphocytes	a, c, d	Mouse, rat, human

a. detection of receptors by receptor binding methods;
b. detection of GnRH mRNA;
c. detection of GnRH receptor protein by immunohistochemistry;
d. detection of GnRH responses by electrophysiology.
Reproduced with permission (Skinner *et al.* 2009)

increase in prevalence of Alzheimer's Disease (Barron *et al.* 2006; Meethal *et al.* 2005). This correlation does not necessarily implicate cognitive decline with elevated *brain* LH levels, but significant increases in the neuronal content of LH have been reported in hippocampus from Alzheimer's patients (Bowen *et al.* 2002). These same authors used a mouse model of Alzheimer's Disease to show that *inhibition* of LH secretion improved cognitive performance and decreased amyloid-β deposition in brain tissue (Meethal *et al.* 2005).

Growth hormone

GH gene expression is detectable in many non-pituitary tissues, including immune, reproductive, alimentary, respiratory, muscular, skeletal and cardiovascular systems. In these tissues, it is probable that GH acts locally as an autocrine/paracrine factor, rather than as a circulating endocrine hormone. The brain is also a major site of GH expression (Harvey 2010; Alba-Betancourt *et al.* 2011; for further reading, see Harvey and Hull 2003) and GH receptor protein and mRNA are present in cerebral cortex, hypothalamus, thalamus, hippocampus and cerebellum (Lobie *et al.* 1993; Castro *et al.* 2000). Brain-derived GH may be especially important in the developing brain, including the retina (Baudet *et al.* 2009). In hippocampus, *gh* mRNA levels are higher in adults, and levels are sexually dimorphic (females > males). Hippocampal *gh* mRNA is also stimulated by estradiol treatment in females and by stress in males (Donahue *et al.* 2006). This group also reported that hippocampal-dependent learning stimulated *gh* mRNA levels almost sixfold in the hippocampus in male rats (Donahue *et al.* 2002). It seems possible, therefore, that some aspects of mood and cognition are controlled by brain-derived GH and this may be useful in understanding, and treating, the aging brain (Creyghton *et al.* 2004).

Prolactin (PRL)

This multitasking hormone exerts at least 300 different effects in many tissues throughout the body (Bole-Feysot *et al.* 1998). However, although the primary site of PRL synthesis is the anterior pituitary gland, *prl* gene expression is also present in several brain areas. These include hypothalamus (PVN, POA, arcuate), septum, thalamus and amygdala (Emanuele *et al.* 1992; Roselli *et al.* 2008). Brain *prl* gene expression is up-regulated by estradiol treatment (DeVito *et al.* 1992), by stress (Torner *et al.* 2002) and by physiological stimuli such as suckling (Torner *et al.* 2004) and the estrous cycle (Torner *et al.* 1999). As we shall see, PRL influences multiple brain functions, including maternal behavior, fertility, food intake, the stress response, circadian rhythms and the control of its own secretion through hypothalamic negative feedback. Prolactin receptors that mediate these effects have been localized in rodent brain (Brown *et al.* 2010; Mann and Bridges 2002; Bakowska and Morrell 2003). A helpful review of prolactin's effects in the brain is provided by Grattan and Kokay (2008): (1) PRL regulates its own secretion by binding to PRL receptors on hypothalamic DA neurons (see Figure 6.8); (2) PRL injection into the brain potently induces maternal behavior that can be blocked with a PRL receptor antagonist (Bridges *et al.* 2001); (3) PRL stimulates food intake via PRL receptors in the PVN (Sauve and Woodside 2000; Naef and Woodside 2007); (4) brain PRL exerts a profound anxiolytic and anti-stress influence in female rats, and this may be particularly important in the period immediately after giving birth (Donner *et al.* 2007; Slattery and Neumann 2008); and (5) high PRL secretion (*hyperprolactinemia*) causes infertility in both men and women. The mechanism for this is an inhibition of pulsatile secretion of GnRH (see section 6.3.3; Figure 6.7) probably via reduction of kisspeptin release or via GABA inhibition of GnRH neurons (Kokay *et al.* 2011); (6) in rodents, a successful pregnancy is dependent upon daily surges of PRL (at 0300 and 1700) that recur for 10 to 12 days. This rhythm is generated by PRL acting in the hypothalamus and can be induced by injection of PRL into the brain of virgin rats (Helena *et al.* 2009; for further reading, see Bertram *et al.* 2010).

Thyroid-stimulating hormone (TSH)

The possibility that TSH may have effects on brain function has received little attention compared to other anterior pituitary hormones. The TSH protein was localized to several rat brain regions (DeVito 1989) and more recently in the embryonic chick brain (Murphy and Harvey 2002). Immunoreactive brain TSH levels are influenced by endocrine manipulations such as thyroidectomy, and they occur in the absence of the pituitary (Ottenweller and Hedge 1982; Hojvat *et al.* 1985). Brain content of TSH can therefore be regulated and it is clearly not of pituitary origin. Evidence that TSH acts in the brain is strongly suggested by the presence of TSH receptors and TSH mRNA in brain regions such as hippocampus, cortex, hypothalamus, pineal and retina (Grommen *et al.* 2006; Crisanti *et al.* 2001; Bockmann *et al.* 1997). At present, there is little evidence to suggest that brain TSH/TSH receptors regulate any brain function. Nevertheless, TSH levels and TSH receptor expression appear to be elevated in cerebral cortex and cerebellum of patients with Down's Syndrome or Alzheimer's Disease (Labudova *et al.* 1999). The physiological and pathological relevance of central TSH/TSH receptor complexes to brain function remains to be determined.

Adrenocorticotropic hormone (ACTH)

It has been known for some time that ACTH, and related peptides such as α-MSH, acts in the brain to produce behavioral changes (De Wied and Jolles 1982). Recall that ACTH is a product of the pro-opiomelanocortin gene (*pomc*; Figure 11.1) and is secreted from the pituitary gland under the control of CRH and cortisol (Figure 6.4). The POMC prohormone (propeptide) is expressed in neurons, however, and, as can be seen in Figure 11.1, may generate several different neuropeptides, including ACTH (Raffin-Sanson *et al.* 2003; Bicknell 2008). Which of these peptides is produced depends on the tissue of origin. For example, in *anterior pituitary*, POMC is processed largely to ACTH and β-endorphin, with little processing to the smaller peptides. *In the brain* – mainly in the arcuate nucleus of the hypothalamus – the accepted view is that POMC is processed mostly to α-MSH (see Figure 12.13) and β-endorphin. However, there is evidence that ACTH is also produced (Schulz *et al.* 2010; Pritchard *et al.* 2003), suggesting that ACTH might behave as a neuropeptide in the brain in addition to its role in regulating the adrenal gland. This is consistent with the known influence of ACTH when injected into the brain; for example, ACTH and α-MSH are equipotent in inhibiting food intake when applied directly into the hypothalamus (Schulz *et al.* 2010). ACTH and α-MSH both bind to the MC4.R melanocortin receptor (see Figure 12.13) which is widely expressed in the brain (Mountjoy 2010). Thus, the septum, which expresses MC4.R receptors, responds to injected ACTH by inducing anxiety and a reduction in social interactions in rats, whereas injection into the substantia nigra or periaqueductal gray produces excessive grooming (Mountjoy 2010). A complete list of behavioral effects would include: increased stretching and yawning; increased penile erections; increased female sexual behavior (lordosis); effects on memory; and changes in pain sensitivity. There is also some evidence that ACTH, or its analogs, could be effective in promoting recovery from nerve damage and possibly preventing some aspects of neurodegeneration (Mountjoy 2010; Bertolini *et al.* 2009). In summary, ACTH – like the other anterior pituitary hormones – is a potent neuropeptide, as well as a circulating endocrine hormone. It seems likely that central effects of ACTH will have therapeutic relevance in the future.

12.7.5 Gastrointestinal peptides

Earlier chapters (see Chapters 2 and 11) emphasized the diversity of peptides originating in the GI tract (Figure 2.3 and Table 11.1). This section will not deal with all of these peptides, but it is probably true to say that their genes are expressed in the nervous system, as well as in the GI tract. As already outlined above, for pituitary hormones, the test of whether GI peptides act as brain neuropeptides/neuromodulators depends on whether: (a) they are expressed in brain tissue; and (b) their receptors are present in the brain. In other words, we have seen already that peptides such as PYY_{3-36} and ghrelin are released from GI tissue and reach the brain, via the bloodstream, where they modify food intake (section 12.6). This section deals with a brief selection of *brain-derived* GI peptides.

Glucagon-like peptide-1 (GLP-1)

As noted in section 12.6.2, GLP-1 is released from the GI tract and inhibits food intake. In the brain, GLP-1 is synthesized in neurons of the solitary tract and the brainstem and nerve fibers extend to the hypothalamus, where GLP-1 receptors are located (Tauchi *et al.* 2008; Llewellyn-Smith *et al.* 2011). Release of GLP-1 in the PVN regulates CRH secretion in the stress response, and infusion of GLP-1 into the brain reduces food intake via an effect in the arcuate nucleus. Tauchi *et al.* (2008) also showed that GLP-1 innervates oxytocin neurons in the SON and perhaps participates in the processes of childbirth and lactation. Brain GLP-1 also elevates blood pressure and heart rate, and stimulates brainstem neurons to activate the sympathetic nervous system (Trapp and Hisadome 2011).

Cholecystokinin (CCK)

CCK and CCK receptors are expressed in many regions of the mammalian brain, including the human brain (Noble *et al.* 1999; Inui 2003; for further reading, see Beinfeld 2001). In general terms, brain CCK is a regulator of multiple behaviors, especially of cognitive function and food intake (Rotzinger *et al.* 2010). For example, in mice lacking the CCK gene (CCK knockouts; CCK-KO), food intake was increased in the light period, but decreased in the dark, compared to controls (Lo *et al.* 2008). CCK-KO mice also showed increased anxiety and impaired memory. Consistent with data obtained from animal experiments, humans with a defective *cck* gene have problems with panic disorder, schizophrenia, alcoholism and hallucinations, perhaps because of reductions in brain levels of CCK (Inui 2003). Attempts have been made to treat anxiety with anti-CCK drugs, so far without success (Steckler 2010). CCK and CCK receptors are co-localized with endogenous opioid systems and there is evidence that CCK receptors may be involved in pain processing. This has led to the development of drugs designed to treat neuropathic pain (Hanlon *et al.* 2011).

Vasoactive intestinal polypeptide (VIP)

VIP is an important gut hormone involved in gut motility and digestion, but this peptide is also present throughout the brain, with significant gene expression in the cerebral cortex, hippocampus, amygdala, SCN and hypothalamus (Dickson and Finlayson 2009). VIP receptors (called $VPAC_1$ and $VPAC_2$) are distributed in most parts of the brain, indicating an important role for this peptide in neuronal function (Joo *et al.* 2004). For example, transgenic mice deficient in *vip* gene expression showed significant deficits in recall of learned behavior (Chaudhury *et al.* 2008) and in spatial memory (Gozes 2010). Gozes *et al.*

(1996) demonstrated that treatment of a mouse model of Alzheimer's Disease with VIP produced an amelioration of memory loss. Use of the VIP-knockout mouse also demonstrated that VIP is critical for transmitting circadian information from the SCN (Loh *et al.* 2008). The circadian rhythm in corticosterone and ACTH was completely lost in these animals, even though the adrenal response to stress remained intact; that is, a footshock generated a surge of corticosterone in mutant and control mice (for further reading, see Aton *et al.* 2005). Current research indicates that VIP and its receptors are important therapeutic targets for treating several neurological conditions, such as Alzheimer's Disease, Parkinson's disease, and autism spectrum disorders (White *et al.* 2010).

Insulin

Insulin is a vital hormone in terms of glucose homeostasis, and facilitates glucose and amino acid uptake into muscle, fat and liver tissue (Figure 8.6; Nussey and Whitehead 2001). Insulin receptors are, however, expressed throughout the brain, suggesting that the brain is an insulin target organ (Marks *et al.* 1990; Schulingkamp *et al.* 2000). The primary receptor sites are hippocampus, hypothalamus, cerebellum, cortex and olfactory bulb. The central effects of insulin are thought to occur after this large polypeptide crosses the blood-brain barrier (Banks 2004; Banks 2008). Nonetheless, there is also good evidence that the insulin gene is expressed by brain cells (Schechter and Abboud 2001).

The effect of insulin on *hypothalamic* insulin receptors is to inhibit food intake and to induce a permanent loss of body weight (Gerozissis 2004). Insulin receptors in other brain regions, however, especially in hippocampus and cortex, appear to have little connection to the regulation of glucose and body weight, but instead mediate improvements in learning and memory (Stockhorst *et al.* 2004; Zhao *et al.* 2004). There is also evidence that brain insulin expression, and insulin signaling, may be defective in Alzheimer's Disease (de la Monte and Wands 2005). Using postmortem brain samples, gene expression levels for insulin and insulin receptor were reduced by four- to tenfold in hippocampal tissue from Alzheimer patients compared to controls (Steen *et al.* 2005). These authors suggest that a reduction in insulin-insulin receptor signaling could be part of the disease process and therefore open to therapeutic intervention (de la Monte *et al.* 2009).

Ghrelin

There is evidence for ghrelin acting as a hunger-induced circulating orexigenic hormone after release from the stomach (section 12.6.2; Figure 12.13). This effect is mediated by hypothalamic receptors, but ghrelin receptors are found in many other brain areas, including cortex, hippocampus, thalamus, Raphé nuclei and spinal cord (Ferrini *et al.* 2009). This strongly implies that ghrelin should have effects in the CNS distinct from those involving hunger. Whether ghrelin is made in the brain is controversial. There is evidence for ghrelin mRNA in the hypothalamus, suggesting a putative role for ghrelin as a neurotransmitter/neuromodulator, but others have failed to find ghrelin peptide in neurons (Furness *et al.* 2011). There is little information on other brain areas as sites for ghrelin expression (Hou *et al.* 2006; Ferrini *et al.* 2009). Nevertheless, ghrelin has well-described effects on CNS mechanisms, including inhibition of pain (Sibilia *et al.* 2006) and modification of circadian rhythms (Yi *et al.* 2008). Also of interest is the ability of ghrelin *antagonists* to suppress CNS-mediated intake of palatable food and to reduce alcohol, cocaine and amphetamine

Table 12.7 Summary of behavioral/visceral/cognitive effects of neuropeptides

Neuropeptide	Effect
Opioid peptides (e.g. β-endorphin, dynorphin)	Pain Narcotic dependence; and abuse Sex behavior Body temperature Respiration Binge eating Blood pressure
Oxytocin; vasopressin	Emotional, social and cognitive Fluid balance Sex behavior; aggression
Leptin	Food intake; appetite Food reward
Hypothalamic-releasing hormones (e.g. CRH, GnRH)	Cognition; sleep; brain aging Food intake; anxiety Arousal Depression; body temperature
Anterior pituitary hormones (e.g. LH, prolactin, GH)	Feeding; sleep; brain aging; mood; maternal behavior; stress; circadian rhythms; anxiety; sex behavior; memory; pain
Gut peptides (e.g. CCK, VIP, GLP-1)	Appetite; blood pressure; heart rate; pain; memory; brain aging; sleep; circadian rhythms; depression; maternal behavior

consumption (Dickson *et al.* 2011). In their review on ghrelin and mental health, Steiger *et al.* (2011) provide evidence for ghrelin's role in sleep-wake regulation, memory and in affective and other CNS disorders. A fuller appreciation of the effects of ghrelin in the brain could therefore assist in developing new therapies for sleep, psychiatric disorders and drug addiction.

Miscellaneous: peptide YY; secretin; amylin

Three more of the peptides listed in Figure 2.3 will briefly complete this section. PYY_{3-36} is a gut-derived appetite suppressor (Figures 12.12 and 12.16). PYY_{3-36} also: (a) increases spontaneous physical activity when injected into the brain during the dark phase of the light-dark cycle (Pfluger *et al.* 2011); (b) reduces depression-related behavior (Painsipp *et al.* 2011); and (c) decreases the motivation to eat (Chandarana and Batterham 2008). *Secretin* is a peptide released from the duodenum following food intake. It inhibits gastric acid secretion and induces bicarbonate release from the pancreas in response to an acid load from the stomach (Boron and Boulpaep 2005). In keeping with other gut peptides, however, secretin and secretin receptors are expressed in several brain regions (Cheng *et al.* 2011; Yuan *et al.* 2011). Injection of secretin, both peripherally and centrally, inhibits food intake (Cheng *et al.* 2011). Investigation of mouse mutants that lack secretin receptors revealed significant behavioral effects of secretin. For example, receptor-deficient mice were hyperactive, displayed abnormal social recognition behavior and had poor motor coordination and learning skills (Nishijima *et al.* 2006). *Amylin* is co-released from the pancreas with insulin. When given peripherally, it can significantly reduce food intake and body weight by acting on brain sites such as the area postrema (Moran and Dailey 2009). Amylin receptors are found throughout the brain and spinal cord (Paxinos *et al.* 2004; Huang *et al.* 2010). In contrast, amylin mRNA was detected only in the preoptic area of the hypothalamus and then only in rat mothers (Dobolyi 2009). No expression was detectable in males or in virgin females. This suggests that brain amylin may contribute to regulating maternal behavior. Other effects in the brain include suppression of visceral pain (Huang *et al.* 2010), depression of motor activity (Baldo and Kelley 2001) and inhibition of sexual activity (Clementi *et al.* 1999).

12.7.6 Summary
Neuropeptides play a critical role in the regulation of virtually all autonomic and central nervous system functions. This section focused on the individual effects of only a selection of peptides. It is worth remembering that many peptides interact to impose further regulation of these systems. Table 12.7 summarizes much of the information provided in this section.

12.8 Summary

Neuropeptides can function as neurotransmitters and neuromodulators. Many neuropeptides meet the majority of the seven criteria used to define neurotransmitters. Both neurotransmitters and neuropeptides have multiple actions at pre- and postsynaptic receptors. Neuropeptides bind to G-protein-coupled receptors, often found in non-synaptic locations, and modify second messenger systems such as cAMP and ion channels. Some neuropeptides bind to a second type of receptor called a tyrosine kinase receptor. Differences between neurotransmitters and neuromodulators are listed in Table 12.1. As neuromodulators, neuropeptides exert their effects by regulating postsynaptic receptor density, G protein coupling and the activation of second messenger systems in postsynaptic target cells, or by altering the electrophysiological responses of neurons by modulating ion channels and membrane potential. The co-transmitter actions of neuropeptides and neurotransmitters result in additive or subtractive effects on the amount of second messenger production in the target cell. An important function of neuropeptides is to exert negative feedback on the presynaptic terminal to alter the release of neurotransmitters and neuropeptides from nerve terminals. Neuropeptides regulate Ca^{2+} and K^+ channels, thus depolarizing or hyperpolarizing the cell, regulating its sensitivity to neurotransmitter stimulation and regulating release of neurotransmitters.

In the neuroendocrine system, neuropeptides regulate many aspects of the secretion of posterior and anterior pituitary hormones, as well as hormones secreted from peripheral endocrine glands. The endogenous opioids, adipokines, such as leptin, and the gastrointestinal peptides are particularly potent regulators of the neuroendocrine system. Through their action on neurons of the hypothalamus, brain stem and limbic system, neuropeptides also regulate some aspects of the ANS and CNS. Neuropeptides regulate integrated adaptive responses to bodily needs such as food, water, pain and temperature regulation and the visceral, affective and behavioral expressions of these bodily needs. The GI tract, for example, is the largest endocrine organ and produces approximately 30 peptides, some of which act as neuropeptides to control food intake and appetite. Hypothalamic, pituitary and gastrointestinal peptides also have visceral, cognitive and behavioral effects via their receptors in the central and peripheral components of the ANS and CNS. Neuropeptides have been implicated in the etiology of psychiatric disorders such as depression, schizophrenia and dementia, and in neurological disorders, such as Alzheimer's, Huntington's and Parkinson's diseases. How useful this knowledge might be in therapeutic terms remains to be established.

FURTHER READING

Agnati, L. F., Leo, G., Zanardi, A., Genedani, S., Rivera, A., Fuxe, K. *et al.* (2006). "Volume transmission and wiring transmission from cellular to molecular networks: history and perspectives," *Acta Physiol (Oxf)* 187, 329–344.

Aton, S. J., Colwell, C. S., Harmar, A. J., Waschek, J. and Herzog, E. D. (2005). "Vasoactive intestinal polypeptide mediates circadian rhythmicity and synchrony in mammalian clock neurons," *Nat Neurosci* 8, 476–483.

Banks, W. A. (2008). "The blood-brain barrier: connecting the gut and the brain," *Regul Pept* 149, 11–14.

Barrera, J. G., Sandoval, D. A., D'Alessio, D. A. and Seeley, R. J. (2011). "GLP-1 and energy balance: an integrated model of short-term and long-term control," *Nat Rev Endocr* 7, 507–516.

Beinfeld, M. C. (2001). "An introduction to neuronal cholecystokinin," *Peptides* 22, 1197–1200.

Bertram, R., Helena, C. V., Gonzalez-Iglesias, A. E., Tabak, J. and Freeman, M. E. (2010). "A tale of two rhythms: the emerging roles of oxytocin in rhythmic prolactin release," *J Neuroendocrinol* 22, 778–784.

Bodnar, R. J. (2010). "Endogenous opiates and behavior: 2009," *Peptides* 31, 2325–2359.

Bostwick, J. M. and Bucci, J. A. (2008). "Internet sex addiction treated with naltrexone," *Mayo Clin Proc* 83, 226–230.

Caldwell, H. K., Lee, H. J., Macbeth, A. H. and Young, W. S., 3rd. (2008). "Vasopressin: behavioral roles of an 'original' neuropeptide," *Prog Neurobiol* 84, 1–24.

Castaneda, T. R., Tong, J., Datta, R., Culler, M. and Tschop, M. H. (2010). "Ghrelin in the regulation of body weight and metabolism," *Front Neuroendocr* 31, 44–60.

Chaudhri, O. B., Salem, V., Murphy, K. G. and Bloom, S. R. (2008). "Gastrointestinal satiety signals," *Annu Rev Physiol* 70, 239–255.

Clark, M. S. and Kaiyala, K. J. (2003). "Role of corticotropin-releasing factor family peptides and receptors in stress-related psychiatric disorders," *Semin Clin Neuropsychiatry* 8, 119–136.

Deng, P. Y., Xiao, Z., Jha, A., Ramonet, D., Matsui, T., Leitges, M. *et al.* (2010). "Cholecystokinin facilitates glutamate release by increasing the number of readily releasable vesicles and releasing probability," *J Neurosci* 30, 5136–5148.

Doyon, C., Moraru, A. and Richard, D. (2004). "The corticotropin-releasing factor system as a potential target for antiobesity drugs," *Drug News Perspect* 17, 505–517.

Druce, M. R. and Bloom, S. R. (2006). "Oxyntomodulin: a novel potential treatment for obesity," *Treat Endocrinol* 5, 265–272.

Frank, S., Heni, M., Moss, A., von Schnurbein, J., Fritsche, A., Häring, H. U. *et al.* (2011). "Leptin therapy in a congenital leptin-deficient patient leads to acute and long-term changes in homeostatic, reward, and food-related brain areas," *J Clin Endocrinol Metab* 96, E1283–E1287.

Gautron, L. and Elmquist, J. K. (2011). "Sixteen years and counting: an update on leptin in energy balance," *J Clin Invest* 121, 2087–2093.

Guyenet, S. J. and Schwartz, M. W. (2012). "Clinical review: regulation of food intake, energy balance, and body fat mass: implications for the pathogenesis and treatment of obesity," *J Clin Endocrinol Metab* 97, 745–755.

Harvey, S. and Hull, K. (2003). "Neural growth hormone: an update," *J Mol Neurosci* 20, 1–14.

Kieffer, B. L. and Gaveriaux-Ruff, C. (2002). "Exploring the opioid system by gene knockout," *Prog Neurobiol* 66, 285–306.

Lechan, R. M. and Fekete, C. (2006). "The TRH neuron: a hypothalamic integrator of energy metabolism," *Prog Brain Res* 153, 209–235.

Lee, H. J., Macbeth, A. H., Pagani, J. H. and Young, W. S., 3rd. (2009). "Oxytocin: the great facilitator of life," *Prog Neurobiol* 88, 127–151.

Li, M. D. (2011). "Leptin and beyond: an odyssey to the central control of body weight," *Yale J Biol Med* 84, 1–7.

Meyer-Lindenberg, A., Domes, G., Kirsch, P. and Heinrichs, M. (2011). "Oxytocin and vasopressin in the human brain: social neuropeptides for translational medicine," *Nat Rev Neurosci* 12, 524–538.

Miller, R. J. (1998). "Presynaptic receptors," *Annu Rev Pharmacol Toxicol* 38, 201–227.

Moenter, S. M. (2010). "Identified GnRH neuron electrophysiology: a decade of study," *Brain Res* 1364, 10–24.

Moran, T. H. and Dailey, M. J. (2009). "Minireview: gut peptides: targets for antiobesity drug development?" *Endocr* 150, 2526–2530.

Myers, M. G., Jr., Leibel, R. L., Seeley, R. J. and Schwartz, M. W. (2010). "Obesity and leptin resistance: distinguishing cause from effect," *Trends Endocrinol Metab* 21, 643–651.

Pandit, R., de Jong, J. W., Vanderschuren, L. J. and Adan, R. A. (2011). "Neurobiology of overeating and obesity: the role of melanocortins and beyond," *Eur J Pharmacol* 660, 28–42.

Rosenbaum, M., Sy, M., Pavlovich, K., Leibel, R. L. and Hirsch, J. (2008). "Leptin reverses weight loss-induced changes in regional neural activity responses to visual food stimuli," *J Clin Invest* 118, 2583–2591.

Roth, J. D., Roland, B. L., Cole, R. L., Trevaskis, J. L., Weyer, C., Koda, J. E. *et al.* (2008). "Leptin responsiveness restored by amylin agonism in diet-induced obesity: evidence from nonclinical and clinical studies," *Proc Natl Acad Sci USA* 105, 7257–7262.

Schlicker, E. and Kathmann, M. (2008). "Presynaptic neuropeptide receptors," *Handb Exp Pharmacol* 184, 409–434.

Seeley, R. J., Woods, S. C. and D'Alessio, D. (2000). "Targeted gene disruption in endocrine research – the case of glucagon-like peptide-1 and neuroendocrine function," *Endocr* 141, 473–475.

Sienkiewicz, E., Magkos, F., Aronis, K. N., Brinkoetter, M., Chamberland, J. P., Chou, S. *et al.* (2011). "Long-term metreleptin treatment increases bone mineral density and content at the lumbar spine of lean hypoleptinemic women," *Metabolism* 60, 1211–1221.

Skinner, D. C., Albertson, A. J., Navratil, A., Smith, A., Mignot, M., Talbott, H. *et al.* (2009). "Effects of gonadotrophin-releasing hormone outside the hypothalamic-pituitary-reproductive axis," *J Neuroendocrinol* 21, 282–292.

Tam, C. S., Lecoultre, V. and Ravussin, E. (2011). "Novel strategy for the use of leptin for obesity therapy," *Exp Opin Biol Therap* 11, 1677–1685.

Tong, J. and D'Alessio, D. (2011). "Eating disorders and gastrointestinal peptides," *Curr Opin Endocrinol Diabetes Obes* 18, 42–49.

Veenema, A. H. and Neumann, I. D. (2008). "Central vasopressin and oxytocin release: regulation of complex social behaviours," *Prog Brain Res* 170, 261–276.

Wang, L., Chadwick, W., Park, S. S., Zhou, Y., Silver, N., Martin, B. *et al.* (2010). "Gonadotropin-releasing hormone receptor system: modulatory role in aging and neurodegeneration," *CNS Neurol Disord Drug Targets* 9, 651–660.

Williams, K. W. and Elmquist, J. K. (2012). "From neuroanatomy to behavior: central integration of peripheral signals regulating feeding behavior," *Nat Neurosci* 15, 1350–1355.

Zorrilla, E. P., Tache, Y. and Koob, G. F. (2003). "Nibbling at CRF receptor control of feeding and gastrocolonic motility," *Trends Pharmacol Sci* 24, 421–427.

REVIEW QUESTIONS

12.1   What is the one criterion used to define a classical neurotransmitter that all neuropeptides *fail* to meet?

12.2   Receptors for neuropeptides, unlike those for classical neurotransmitters, are often located where?

12.3   The neuropeptide kisspeptin binds to a G-protein-coupled receptor which stimulates which second messenger system?

12.4   Name three sites at which neuropeptides can act as co-transmitters to modulate the effects of co-released neurotransmitters at postsynaptic cells.

12.5   How does dynorphin control the release of oxytocin from pitiuitary nerve terminals?

12.6   Name one *anorexigenic* hypothalamic neuronal system, and one *orexigenic* neuron system.

12.7   What is the "hunger hormone"?

12.8   Name two hormones released from the GI tract.

12.9   In mice and humans, what happens to body weight in the absence of leptin?

12.10   What is the intracellular negative feedback mechanism that induces leptin resistance?

12.11   Oxytocin regulates uterine contractions. It is also a powerful regulator of behavior. Name three types of behavior controlled by this hormone.

12.12   Endogenous opioid peptides (EOP), such as β-endorphin, are implicated in pain mechanisms. They also regulate neuroendocrine secretion. Name three pituitary hormones affected by EOP.

12.13   Hypothalamic-releasing hormones, in addition to stimulating secretion of pituitary hormones, have profound behavioral effects. Give one example of a behavioral effect of: (a) GHRH; (b) CRH; and (c) TRH.

ESSAY QUESTIONS

12.1   Discuss the criteria that should be satisfied in order that a newly discovered brain neuropeptide can be described as a neurotransmitter. Emphasize the experimental techniques used in these investigations.

12.2   The discovery of leptin raised the exciting possibility that leptin treatment of obese patients would lead to a reduction in body weight. With the exception of one group of patients, this approach has not worked. Discuss the reasons for this failure.

12.3   In the absence of brain kisspeptin, or its receptor, the reproductive system fails to function. Discuss this statement in terms of both animal and human studies.

12.4   Ghrelin has been christened the "hunger hormone." Discuss this statement. Describe both animal and human experiments designed to probe its mechanism of action.

12.5   Oxytocin is a powerful modulator of emotional and social behavior. Discuss the behavioral effects of oxytocin in humans, paying special attention to whether oxytocin can reach the brain following treatment. Is oxytocin part of the neurochemical process governing falling in love?

12.6   The receptors for gonadotropin-releasing hormone (GnRH) are expressed widely in the body, indicating that this neuropeptide may have many actions distinct from its effect on the release of LH and FSH from the piuitary. Describe its role in the control of sexual behavior. Is there evidence for an effect in humans?

REFERENCES

Alba-Betancourt, C., Aramburo, C., Avila-Mendoza, J., Ahumada-Solorzano, S. M., Carranza, M., Rodriguez-Mendez, A. J. et al. (2011). "Expression, cellular distribution, and heterogeneity of growth hormone in the chicken cerebellum during development," Gen Comp Endocrinol 170, 528–540.

Albertson, A. J., Navratil, A., Mignot, M., Dufourny, L., Cherrington, B. and Skinner, D. C. (2008). "Immunoreactive GnRH type I receptors in the mouse and sheep brain," J Chem Neuroanat 35, 326–333.

Alon, T., Zhou, L., Perez, C. A., Garfield, A. S., Friedman, J. M. and Heisler, L. K. (2009). "Transgenic mice expressing green fluorescent protein under the control of the corticotropin-releasing hormone promoter," Endocr 150, 5626–5632.

Bakowska, J. C. and Morrell, J. I. (2003). "The distribution of mRNA for the short form of the prolactin receptor in the forebrain of the female rat," Brain Res Mol Brain Res 116, 50–58.

Baldo, B. A. and Kelley, A. E. (2001). "Amylin infusion into rat nucleus accumbens potently depresses motor activity and ingestive behavior," Am J Physiol Regul Integr Comp Physiol 281, R1232–R1242.

Banks, W. A. (2004). "The source of cerebral insulin," Eur J Pharmacol 490, 5–12.

Banks, W. A. (2008). "The blood-brain barrier: connecting the gut and the brain," *Regul Pept* 149, 11–14.

Barron, A. M., Verdile, G. and Martins, R. N. (2006). "The role of gonadotropins in Alzheimer's disease: potential neurodegenerative mechanisms," *Endocrine* 29, 257–269.

Batterham, R. L., Cohen, M. A., Ellis, S. M., Le Roux, C. W., Withers, D. J., Frost, G. S. *et al.* (2003a). "Inhibition of food intake in obese subjects by peptide YY3–36," *N Engl J Med* 349, 941–948.

Batterham, R. L., Le Roux, C. W., Cohen, M. A., Park, A. J., Ellis, S. M., Patterson, M. *et al.* (2003b). "Pancreatic polypeptide reduces appetite and food intake in humans," *J Clin Endocrinol Metab* 88, 3989–3992.

Baudet, M. L., Rattray, D., Martin, B. T. and Harvey, S. (2009). "Growth hormone promotes axon growth in the developing nervous system," *Endocr* 150, 2758–2766.

Berne, R. M. and Levy, M. N. (2000). *Principles of Physiology,* 3rd edn. (St. Louis, MO: Mosby).

Bertolini, A., Tacchi, R. and Vergoni, A. V. (2009). "Brain effects of melanocortins," *Pharmacol Res* 59, 13–47.

Bicknell, A. B. (2008). "The tissue-specific processing of pro-opiomelanocortin," *J Neuroendocrinol* 20, 692–699.

Binder, E. B. and Nemeroff, C. B. (2010). "The CRF system, stress, depression and anxiety-insights from human genetic studies," *Mol Psychiatry* 15, 574–588.

Blackmer, T., Larsen, E. C., Bartleson, C., Kowalchyk, J. A., Yoon, E. J., Preininger, A. M. *et al.* (2005). "G protein betagamma directly regulates SNARE protein fusion machinery for secretory granule exocytosis," *Nat Neurosci* 8, 421–425.

Bockmann, J., Winter, C., Wittkowski, W., Kreutz, M. R. and Bockers, T. M. (1997). "Cloning and expression of a brain-derived TSH receptor," *Biochem Biophys Res Commun* 238, 173–178.

Bodnar, R. J. (2010). "Endogenous opiates and behavior: 2009," *Peptides* 31, 2325–2359.

Boehm, U., Zou, Z. and Buck, L. B. (2005). "Feedback loops link odor and pheromone signaling with reproduction," *Cell* 123, 683–695.

Bole-Feysot, C., Goffin, V., Edery, M., Binart, N. and Kelly, P. A. (1998). "Prolactin (PRL) and its receptor: actions, signal transduction pathways and phenotypes observed in PRL receptor knockout mice," *Endocr Rev* 19, 225–268.

Born, J., Lange, T., Kern, W., McGregor, G. P., Bickel, U. and Fehm, H. L. (2002). "Sniffing neuropeptides: a transnasal approach to the human brain," *Nat Neurosci* 5, 514–516.

Boron, W. F. and Boulpaep, E. L. (2005). *Medical Physiology*, updated edn. (Philadelphia, PA: Elsevier Saunders).

Bowen, R. L., Smith, M. A., Harris, P. L., Kubat, Z., Martins, R. N., Castellani, R. J. *et al.* (2002). "Elevated luteinizing hormone expression colocalizes with neurons vulnerable to Alzheimer's disease pathology," *J Neurosci Res* 70, 514–518.

Bridges, R., Rigero, B., Byrnes, E., Yang, L. and Walker, A. (2001). "Central infusions of the recombinant human prolactin receptor antagonist, S179D-PRL, delay the onset of maternal behavior in steroid-primed, nulliparous female rats," *Endocr* 142, 730–739.

Brown, C. H., Russell, J. A. and Leng, G. (2000). "Opioid modulation of magnocellular neurosecretory cell activity," *Neurosci Res* 36, 97–120.

Brown, R. E., Imran, S. A., Ur, E. and Wilkinson, M. (2008). "Kiss-1 mRNA in adipose tissue is regulated by sex hormones and food intake," *Mol Cell Endocrinol* 281, 64–72.

Brown, R. S., Kokay, I. C., Herbison, A. E. and Grattan, D. R. (2010). "Distribution of prolactin-responsive neurons in the mouse forebrain," *J Comp Neurol* 518, 92–102.

Brunton, P. J. and Russell, J. A. (2008). "The expectant brain: adapting for motherhood," *Nat Rev Neurosci* 9, 11–25.

Buijs, R. M., Swaab, D. F., Dogterom, J. and van Leeuwen, F. W. (1978). "Intra- and extrahy-pothalamic vasopressin and oxytocin pathways in the rat," *Cell Tissue Res* 186, 423–433.

Caron, E., Sachot, C., Prevot, V. and Bouret, S. G. (2010). "Distribution of leptin-sensitive cells in the postnatal and adult mouse brain," *J Comp Neurol* 518, 459–476.

Carter, C. S. and Porges, S. W. (2013). "The biochemistry of love: an oxytocin hypothesis," *EMBO Reports* 14, 12–16.

Castro, J. R., Costoya, J. A., Gallego, R., Prieto, A., Arce, V. M. and Senaris, R. (2000). "Expression of growth hormone receptor in the human brain," *Neurosci Lett* 281, 147–150.

Chan, Y. M., Broder-Fingert, S. and Seminara, S. B. (2009). "Reproductive functions of kisspeptin and Gpr54 across the life cycle of mice and men," *Peptides* 30, 42–48.

Chandarana, K. and Batterham, R. (2008). "Peptide YY," *Curr Opin Endocrinol Diabetes Obes* 15, 65–72.

Chaudhri, O. B., Salem, V., Murphy, K. G. and Bloom, S. R. (2008). "Gastrointestinal satiety signals," *Annu Rev Physiol* 70, 239–255.

Chaudhury, D., Loh, D. H., Dragich, J. M., Hagopian, A. and Colwell, C. S. (2008). "Select cognitive deficits in vasoactive intestinal peptide deficient mice," *BMC Neurosci* 9, 63.

Chen, Y., Brunson, K. L., Muller, M. B., Cariaga, W. and Baram, T. Z. (2000). "Immunocytochemical distribution of corticotropin-releasing hormone receptor type-1 (CRF (1))-like immunoreactivity in the mouse brain: light microscopy analysis using an antibody directed against the C-terminus," *J Comp Neurol* 420, 305–323.

Cheng, C. Y., Chu, J. Y. and Chow, B. K. (2011). "Central and peripheral administration of secretin inhibits food intake in mice through the activation of the melanocortin system," *Neuropsychopharmacology* 36, 459–471.

Chicurel, M. (2000). "Whatever happened to leptin?" *Nature* 404, 538–540.

Chou, S. H., Chamberland, J. P., Liu, X., Matarese, G., Gao, C., Stefanakis, R. *et al.* (2011). "Leptin is an effective treatment for hypothalamic amenorrhea," *Proc Natl Acad Sci USA* 108, 6585–6590.

Chu, C., Gao, G. and Huang, W. (2008). "A study on co-localization of FSH and its receptor in rat hippocampus," *J Mol Histol* 39, 49–55.

Clarkson, J., d'Anglemont de Tassigny, X., Colledge, W. H., Caraty, A. and Herbison, A. E. (2009). "Distribution of kisspeptin neurones in the adult female mouse brain," *J Neuroendocrinol* 21, 673–682.

Clarkson, J. and Herbison, A. E. (2006). "Postnatal development of kisspeptin neurons in mouse hypothalamus; sexual dimorphism and projections to gonadotropin-releasing hormone neurons," *Endocr* 147, 5817–5825.

Clementi, G., Busa, L., de Bernardis, E., Prato, A. and Drago, F. (1999). "Effects of centrally injected amylin on sexual behavior of male rats," *Peptides* 20, 379–382.

Coccaro, E. F., Kavoussi, R. J., Hauger, R. L., Cooper, T. B. and Ferris, C. F. (1998). "Cerebrospinal fluid vasopressin levels: correlates with aggression and serotonin function in personality-disordered subjects," *Arch Gen Psychiatry* 55, 708–714.

Corbett, A. D., Henderson, G., McKnight, A. T. and Paterson, S. J. (2006). "75 years of opioid research: the exciting but vain quest for the Holy Grail," *Br J Pharmacol* 147(Suppl. 1), S153–S162.

Cravo, R. M., Margatho, L. O., Osborne-Lawrence, S., Donato, J., Jr., Atkin, S., Bookout, A. L. *et al.* (2011). "Characterization of Kiss1 neurons using transgenic mouse models," *Neurosci* 173, 37–56.

Creyghton, W. M., van Dam, P. S. and Koppeschaar, H. P. (2004). "The role of the somatotropic system in cognition and other cerebral functions," *Semin Vasc Med* 4, 167–172.

Crisanti, P., Omri, B., Hughes, E., Meduri, G., Hery, C., Clauser, E. *et al.* (2001). "The expression of thyrotropin receptor in the brain," *Endocr* 142, 812–822.

Curtis, A. E., Cooke, J. H., Baxter, J. E., Parkinson, J. R., Bataveljic, A., Ghatei, M. A. *et al.* (2010). "A kisspeptin-10 analog with greater in vivo bioactivity than kisspeptin-10," *Am J Physiol Endocrinol Metab* 298, E296–E303.

Dardeno, T. A., Chou, S. H., Moon, H. S., Chamberland, J. P., Fiorenza, C. G. and Mantzoros, C. S. (2010). "Leptin in human physiology and therapeutics," *Front Neuroendocr* 31, 377–393.

Davis, J. F., Choi, D. L. and Benoit, S. C. (2010). "Insulin, leptin and reward," *Trends Endocrinol Metab* 21, 68–74.

de la Monte, S. M., Longato, L., Tong, M. and Wands, J. R. (2009). "Insulin resistance and neurodegeneration: roles of obesity, type 2 diabetes mellitus and non-alcoholic steatohepatitis," *Curr Opin Investig Drugs* 10, 1049–1060.

de la Monte, S. M. and Wands, J. R. (2005). "Review of insulin and insulin-like growth factor expression, signaling, and malfunction in the central nervous system: relevance to Alzheimer's disease," *J Alzheimers Dis* 7, 45–61.

de Vries, G. J. (2008). "Sex differences in vasopressin and oxytocin innervation of the brain," *Prog Brain Res* 170, 17–27.

De Wied, D. and Jolles, J. (1982). "Neuropeptides derived from pro-opiocortin: behavioral, physiological, and neurochemical effects," *Physiol Rev* 62, 976–1059.

DeBoer, M. D. (2011). "Ghrelin and cachexia: will treatment with GHSR-1a agonists make a difference for patients suffering from chronic wasting syndromes?" *Mol Cell Endocrinol* 340, 97–105.

Deng, P. Y., Xiao, Z., Jha, A., Ramonet, D., Matsui, T., Leitges, M. *et al.* (2010). "Cholecystokinin facilitates glutamate release by increasing the number of readily releasable vesicles and releasing probability," *J Neurosci* 30, 5136–5148.

Devidze, N., Zhang, Q., Zhou, J., Lee, A. W., Pataky, S., Kow, L. M. *et al.* (2008). "Presynaptic actions of opioid receptor agonists in ventromedial hypothalamic neurons in estrogen- and oil-treated female mice," *Neurosci* 152, 942–949.

DeVito, W. J. (1989). "Thyroid hormone regulation of hypothalamic immunoreactive thyrotropin," *Endocr* 125, 1219–1223.

DeVito, W. J., Avakian, C., Stone, S. and Ace, C. I. (1992). "Estradiol increases prolactin synthesis and prolactin messenger ribonucleic acid in selected brain regions in the hypophysectomized female rat," *Endocr* 131, 2154–2160.

Dickson, L. and Finlayson, K. (2009). "VPAC and PAC receptors: from ligands to function," *Pharmacol Ther* 121, 294–316.

Dickson, S. L., Egecioglu, E., Landgren, S., Skibicka, K. P., Engel, J. A. and Jerlhag, E. (2011). "The role of the central ghrelin system in reward from food and chemical drugs," *Mol Cell Endocrinol* 340, 80–87.

Dobolyi, A. (2009). "Central amylin expression and its induction in rat dams," *J Neurochem* 111, 1490–1500.

Dolan, S., Evans, N. P., Richter, T. A. and Nolan, A. M. (2003). "Expression of gonadotropin-releasing hormone and gonadotropin-releasing hormone receptor in sheep spinal cord," *Neurosci Lett* 346, 120–122.

Donahue, C. P., Jensen, R. V., Ochiishi, T., Eisenstein, I., Zhao, M., Shors, T. *et al.* (2002). "Transcriptional profiling reveals regulated genes in the hippocampus during memory formation," *Hippocampus* 12, 821–833.

Donahue, C. P., Kosik, K. S. and Shors, T. J. (2006). "Growth hormone is produced within the hippocampus where it responds to age, sex, and stress," *Proc Natl Acad Sci USA* 103, 6031–6036.

Donner, N., Bredewold, R., Maloumby, R. and Neumann, I. D. (2007). "Chronic intracerebral prolactin attenuates neuronal stress circuitries in virgin rats," *Eur J Neurosci* 25, 1804–1814.

Douglas, A. J. and Russell, J. A. (2001). "Endogenous opioid regulation of oxytocin and ACTH secretion during pregnancy and parturition," *Prog Brain Res* 133, 67–82.

Dournaud, P., Slama, A., Beaudet, J. M. and Epelbaum, J. (2011). "Somatostatin receptors" in R. Quirion, A.Björklund and T. Hökfelt (eds.), *Handbook of Chemical Neuroanatomy* (Boston, MA: Elsevier), vol. 16, pp. 1–43.

Druce, M. R., Neary, N. M., Small, C. J., Milton, J., Monteiro, M., Patterson, M. *et al.*(2006). "Subcutaneous administration of ghrelin stimulates energy intake in healthy lean human volunteers," *Int J Obes (Lond)* 30, 293–296.

Druce, M. R., Wren, A. M., Park, A. J., Milton, J. E., Patterson, M., Frost, G. *et al.*(2005). "Ghrelin increases food intake in obese as well as lean subjects," *Int J Obes (Lond)* 29, 1130–1136.

Dudas, B. and Merchenthaler, I. (2004). "Close anatomical associations between beta-endorphin and luteinizing hormone-releasing hormone neuronal systems in the human diencephalon," *Neurosci* 124, 221–229.

Dumont, Y., Martel, J. C., Fournier, A., St-Pierre, S. and Quirion, R. (1992). "Neuropeptide Y and neuropeptide Y receptor subtypes in brain and peripheral tissues," *Prog Neurobiol* 38, 125–167.

Emanuele, N. V., Jurgens, J. K., Halloran, M. M., Tentler, J. J., Lawrence, A. M. and Kelley, M. R. (1992). "The rat prolactin gene is expressed in brain tissue: detection of normal and alternatively spliced prolactin messenger RNA," *Mol Endocrinol* 6, 35–42.

Epelbaum, J., Guillou, J. L., Gastambide, F., Hoyer, D., Duron, E. and Viollet, C. (2009). "Somatostatin, Alzheimer's disease and cognition: an old story coming of age?" *Prog Neurobiol* 89, 153–161.

Farooqi, I. S., Bullmore, E., Keogh, J., Gillard, J., O'Rahilly, S. and Fletcher, P. C. (2007). "Leptin regulates striatal regions and human eating behavior," *Science* 317, 1355.

Feifel, D. and Vaccarino, F. J. (1994). "Growth hormone-regulatory peptides (GHRH and somatostatin) and feeding: a model for the integration of central and peripheral function," *Neurosci Biobehav Rev* 18, 421–433.

Ferrini, F., Salio, C., Lossi, L. and Merighi, A. (2009). "Ghrelin in central neurons," *Curr Neuropharmacol* 7, 37–49.

Fetissov, S. O., Hallman, J., Nilsson, I., Lefvert, A. K., Oreland, L. and Hokfelt, T. (2006). "Aggressive behavior linked to corticotropin-reactive autoantibodies," *Biol Psychiatry* 60, 799–802.

Field, B. C., Chaudhri, O. B. and Bloom, S. R. (2009). "Obesity treatment: novel peripheral targets," *Br J Clin Pharmacol* 68, 830–843.

Field, B. C., Chaudhri, O. B. and Bloom, S. R. (2010a). "Bowels control brain: gut hormones and obesity," *Nat Rev Endocrinol* 6, 444–453.

Field, B. C., Wren, A. M., Peters, V., Baynes, K. C., Martin, N. M., Patterson, M. *et al.* (2010b). "PYY3–36 and oxyntomodulin can be additive in their effect on food intake in overweight and obese humans," *Diabetes* 59, 1635–1639.

Fliers, E., Wiersinga, W. M. and Swaab, D. F. (1998). "Physiological and pathophysiological aspects of thyrotropin-releasing hormone gene expression in the human hypothalamus," *Thyroid* 8, 921–928.

Frohmader, K. S., Pitchers, K. K., Balfour, M. E. and Coolen, L. M. (2010). "Mixing pleasures: review of the effects of drugs on sex behavior in humans and animal models," *Horm Behav* 58, 149–162.

Fu, L. Y. and van den Pol, A. N. (2010). "Kisspeptin directly excites anorexigenic proopiomelanocortin neurons but inhibits orexigenic neuropeptide Y cells by an indirect synaptic mechanism," *J Neurosci* 30, 10205–10219.

Furness, J. B., Hunne, B., Matsuda, N., Yin, L., Russo, D., Kato, I. *et al.* (2011). "Investigation of the presence of ghrelin in the central nervous system of the rat and mouse," *Neurosci* 193, 1–9.

Fuxe, K., Dahlstrom, A. B., Jonsson, G., Marcellino, D., Guescini, M., Dame, M. *et al.* (2010). "The discovery of central monoamine neurons gave volume transmission to the wired brain," *Prog Neurobiol* 90, 82–100.

Gahete, M. D., Rubio, A., Duran-Prado, M., Avila, J., Luque, R. M. and Castano, J. P. (2010). "Expression of somatostatin, cortistatin, and their receptors, as well as DA receptors, but not of neprilysin, are reduced in the temporal lobe of Alzheimer's disease patients," *J Alzheimers Dis* 20, 465–475.

Galbally, M., Lewis, A. J., Ijzendoorn, M. and Permezel, M. (2011). "The role of oxytocin in mother-infant relations: a systematic review of human studies," *Harv Rev Psychiatry* 19, 1–14.

Gardiner, J. V., Jayasena, C. N. and Bloom, S. R. (2008). "Gut hormones: a weight off your mind," *J Neuroendocrinol* 20, 834–841.

Gary, K. A., Sevarino, K. A., Yarbrough, G. G., Prange, A. J., Jr. and Winokur, A. (2003). "The thyrotropin-releasing hormone (TRH) hypothesis of homeostatic regulation: implications for TRH-based therapeutics," *J Pharmacol Exp Ther* 305, 410–416.

Gerozissis, K. (2004). "Brain insulin and feeding: a bi-directional communication," *Eur J Pharmacol* 490, 59–70.

Gianetti, E. and Seminara, S. (2008). "Kisspeptin and KISS1R: a critical pathway in the reproductive system," *Reprod* 136, 295–301.

Gianoulakis, C. (2009). "Endogenous opioids and addiction to alcohol and other drugs of abuse," *Curr Top Med Chem* 9, 999–1015.

Gozes, I. (2010). "VIP-PACAP 2010: my own perspective on modulation of cognitive and emotional behavior," *J Mol Neurosci* 42, 261–263.

Gozes, I., Bardea, A., Reshef, A., Zamostiano, R., Zhukovsky, S., Rubinraut, S. *et al.* (1996). "Neuroprotective strategy for Alzheimer disease: intranasal administration of a fatty neuropeptide," *Proc Natl Acad Sci USA* 93, 427–432.

Gratacos, M., Costas, J., de Cid, R., Bayes, M., Gonzalez, J. R., Baca-Garcia, E., de Diego, Y. *et al.* (2009). "Identification of new putative susceptibility genes for several psychiatric disorders by association analysis of regulatory and non-synonymous SNPs of 306 genes involved in neuro-transmission and neurodevelopment," *Am J Med Genet B Neuropsychiatr Genet* 150B, 808–816.

Grattan, D. R. and Kokay, I. C. (2008). "Prolactin: a pleiotropic neuroendocrine hormone," *J Neuroendocrinol* 20, 752–763.

Grommen, S. V., Taniuchi, S., Janssen, T., Schoofs, L., Takahashi, S., Takeuchi, S. *et al.* (2006). "Molecular cloning, tissue distribution, and ontogenic thyroidal expression of the chicken thyrotropin receptor," *Endocr* 147, 3943–3951.

Hameed, S., Jayasena, C. N. and Dhillo, W. S. (2011). "Kisspeptin and fertility," *J Endocrinol* 208, 97–105.

Han, S. K., Gottsch, M. L., Lee, K. J., Popa, S. M., Smith, J. T., Jakawich, S. K. *et al.* (2005). "Activation of gonadotropin-releasing hormone neurons by kisspeptin as a neuroendocrine switch for the onset of puberty," *J Neurosci* 25, 11349–11356.

Hanlon, K. E., Herman, D. S., Agnes, R. S., Largent-Milnes, T. M., Kumarasinghe, I. R., Ma, S. W. *et al.* (2011). "Novel peptide ligands with dual acting pharmacophores designed for the pathophysiology of neuropathic pain," *Brain Res* 1395, 1–11.

Harihar, S., Pounds, K. M., Iwakuma, T., Seidah, N. G. and Welch, D. R. (2014). "Furin is the major proprotein convertase required for KISS-1-to-Kisspeptin processing," *PLoS One* 9, e84958.

Harvey, S. (2010). "Extrapituitary growth hormone," *Endocrine* 38, 335–359.

Hayes, M. R., Kanoski, S. E., Alhadeff, A. L. and Grill, H. J. (2011). "Comparative effects of the long-acting GLP-1 receptor ligands, liraglutide and exendin-4, on food intake and body weight suppression in rats," *Obesity (Silver Spring)* 19, 1342–1349.

Helboe, L. and Moller, M. (2000). "Localization of somatostatin receptors at the light and electron microscopial level by using antibodies raised against fusion proteins," *Prog Histochem Cytochem* 35, 3–64.

Helena, C. V., McKee, D. T., Bertram, R., Walker, A. M. and Freeman, M. E. (2009). "The rhythmic secretion of mating-induced prolactin secretion is controlled by prolactin acting centrally," *Endocr* 150, 3245–3251.

Henry, B. A. and Clarke, I. J. (2008). "Adipose tissue hormones and the regulation of food intake," *J Neuroendocrinol* 20, 842–849.

Herbison, A. E., de Tassigny, X., Doran, J. and Colledge, W. H. (2010). "Distribution and postnatal development of GPR54 gene expression in mouse brain and gonadotropin-releasing hormone neurons," *Endocr* 151, 312–321.

Herbison, A. E., Pape, J. R., Simonian, S. X., Skynner, M. J. and Sim, J. A. (2001). "Molecular and cellular properties of GnRH neurons revealed through transgenics in the mouse," *Mol Cell Endocrinol* 185, 185–194.

Heuer, H., Schafer, M. K., O'Donnell, D., Walker, P. and Bauer, K. (2000). "Expression of thyrotropin-releasing hormone receptor 2 (TRH-R2) in the central nervous system of rats," *J Comp Neurol* 428, 319–336.

Hojvat, S., Baker, G., Kirsteins, L. and Lawrence, A. M. (1982). "Growth hormone (GH) immunoreactivity in the rodent and primate CNS: distribution, characterization and presence posthypophysectomy," *Brain Res* 239, 543–557.

Hojvat, S., Emanuele, N., Baker, G., Kirsteins, L. and Lawrence, A. M. (1985). "Brain thyroid-stimulating hormone: effects of endocrine manipulations," *Brain Res* 360, 257–263.

Hokfelt, T., Broberger, C., Xu, Z. Q., Sergeyev, V., Ubink, R. and Diez, M. (2000). "Neuropeptides – an overview," *Neuropharmacol* 39, 1337–1356.

Hotta, M., Ohwada, R., Akamizu, T., Shibasaki, T., Takano, K. and Kangawa, K. (2009). "Ghrelin increases hunger and food intake in patients with restricting-type anorexia nervosa: a pilot study," *Endocr J* 56, 1119–1128.

Hou, Z., Miao, Y., Gao, L., Pan, H. and Zhu, S. (2006). "Ghrelin-containing neuron in cerebral cortex and hypothalamus linked with the DVC of brainstem in rat," *Regul Pept* 134, 126–131.

Hrabovszky, E., Ciofi, P., Vida, B., Horvath, M. C., Keller, E., Caraty, A. *et al.* (2010). "The kisspeptin system of the human hypothalamus: sexual dimorphism and relationship with gonadotropin-releasing hormone and neurokinin B neurons," *Eur J Neurosci* 31, 1984–1998.

Huang, X., Yang, J., Chang, J. K. and Dun, N. J. (2010). "Amylin suppresses acetic acid-induced visceral pain and spinal c-fos expression in the mouse," *Neurosci* 165, 1429–1438.

Inui, A. (2003). "Neuropeptide gene polymorphisms and human behavioural disorders," *Nat Rev Drug Discov* 2, 986–998.

Iversen, L. L. (1996). "How does morphine work?" *Nature* 383, 759–760.

Iversen, L. L., Iversen, S. D., Bloom, D. E. and Roth, R. H. (2009). *Introduction to Neuropsychopharmacology* (New York: Oxford University Press).

Joo, K. M., Chung, Y. H., Kim, M. K., Nam, R. H., Lee, B. L., Lee, K. H. *et al.* (2004). "Distribution of vasoactive intestinal peptide and pituitary adenylate cyclase-activating polypeptide receptors (VPAC1, VPAC2, and PAC1 receptor) in the rat brain," *J Comp Neurol* 476, 388–413.

Katoh, A., Fujihara, H., Ohbuchi, T., Onaka, T., Hashimoto, T., Kawata, M. *et al.* (2011). "Highly visible expression of an oxytocin-monomeric red fluorescent protein 1 fusion gene in the hypothalamus and posterior pituitary of transgenic rats," *Endocr* 152, 2768–2774.

Kiaris, H., Chatzistamou, I., Papavassiliou, A. G. and Schally, A. V. (2011). "Growth hormone-releasing hormone: not only a neurohormone," *Trends Endocrinol Metab* 22, 311–317.

Knoblach, S. M. and Kubek, M. J. (1997). "Changes in thyrotropin-releasing hormone levels in hippocampal subregions induced by a model of human temporal lobe epilepsy: effect of partial and complete kindling," *Neurosci* 76, 97–104.

Kokay, I. C., Petersen, S. L. and Grattan, D. R. (2011). "Identification of prolactin-sensitive GABA and kisspeptin neurons in regions of the rat hypothalamus involved in the control of fertility," *Endocr* 152, 526–535.

Kuehn, B. M. (2011). "Scientists probe oxytocin therapy for social deficits in autism, schizophrenia," *JAMA* 305, 659–661.

Labudova, O., Cairns, N., Koeck, T., Kitzmueller, E., Rink, H. and Lubec, G. (1999). "Thyroid stimulating hormone-receptor overexpression in brain of patients with Down syndrome and Alzheimer's disease," *Life Sci* 64, 1037–1044.

Landgraf, R. and Neumann, I. D. (2004). "Vasopressin and oxytocin release within the brain: a dynamic concept of multiple and variable modes of neuropeptide communication," *Front Neuroendocr* 25, 150–176.

Lapatto, R., Pallais, J. C., Zhang, D., Chan, Y. M., Mahan, A., Cerrato, F. *et al.* (2007). "Kiss1-/- mice exhibit more variable hypogonadism than Gpr54-/- mice," *Endocr* 148, 4927–4936.

Lathe, R. (2001). "Hormones and the hippocampus," *J Endocrinol* 169, 205–231.

Lee, H. J., Macbeth, A. H., Pagani, J. H. and Young, W. S., 3rd. (2009). "Oxytocin: the great facilitator of life," *Prog Neurobiol* 88, 127–151.

Le Greves, P., Huang, W., Zhou, Q., Thornwall, M. and Nyberg, F. (1998). "Acute effects of morphine on the expression of mRNAs for NMDA receptor subunits in the rat hippocampus, hypothalamus and spinal cord," *Eur J Pharmacol* 341, 161–164.

le Roux, C. W., Batterham, R. L., Aylwin, S. J., Patterson, M., Borg, C. M., Wynne, K. J. *et al.* (2006). "Attenuated peptide YY release in obese subjects is associated with reduced satiety," *Endocr* 147, 3–8.

Lei, Z. M., Rao, C. V., Kornyei, J. L., Licht, P. and Hiatt, E. S. (1993). "Novel expression of human chorionic gonadotropin/luteinizing hormone receptor gene in brain," *Endocr* 132, 2262–2270.

Liao, F., Taishi, P., Churchill, L., Urza, M. J. and Krueger, J. M. (2010). "Localized suppression of cortical growth hormone-releasing hormone receptors state-specifically attenuates electroencephalographic delta waves," *J Neurosci* 30, 4151–4159.

Licinio, J., Caglayan, S., Ozata, M., Yildiz, B. O., de Miranda, P. B., O'Kirwan, F., *et al.* (2004). "Phenotypic effects of leptin replacement on morbid obesity, diabetes mellitus, hypogonadism, and behavior in leptin-deficient adults," *Proc Natl Acad Sci USA* 101, 4531–4536.

Lin, S., Shi, Y. C., Yulyaningsih, E., Aljanova, A., Zhang, L., Macia, L., *et al.* (2009). "Critical role of arcuate Y4 receptors and the melanocortin system in pancreatic polypeptide-induced reduction in food intake in mice," *PLoS One* 4, e8488.

Lincoln, D. W. and Wakerley, J. B. (1974). "Electrophysiological evidence for the activation of supraoptic neurones during the release of oxytocin," *J Physiol* 242, 533–554.

Llewellyn-Smith, I. J., Reimann, F., Gribble, F. M. and Trapp, S. (2011). "Preproglucagon neurons project widely to autonomic control areas in the mouse brain," *Neurosci* 180, 111–121.

Lo, C. M., Samuelson, L. C., Chambers, J. B., King, A., Heiman, J., Jandacek, R. J. *et al.* (2008). "Characterization of mice lacking the gene for cholecystokinin," *Am J Physiol Regul Integr Comp Physiol* 294, R803–R810.

Lobie, P. E., Garcia-Aragon, J., Lincoln, D. T., Barnard, R., Wilcox, J. N. and Waters, M. J. (1993). "Localization and ontogeny of growth hormone receptor gene expression in the central nervous system," *Brain Res Dev Brain Res* 74, 225–233.

Loh, D. H., Abad, C., Colwell, C. S. and Waschek, J. A. (2008). "Vasoactive intestinal peptide is critical for circadian regulation of glucocorticoids," *Neuroendocrinology* 88, 246–255.

London, E. D., Berman, S. M., Chakrapani, S., Delibasi, T., Monterosso, J., Erol, H. K. *et al.* (2011). "Short-term plasticity of gray matter associated with leptin deficiency and replacement," *J Clin Endocr Metab* 96, E1212–E1220.

Ludwig, M. and Leng, G. (2006). "Dendritic peptide release and peptide-dependent behaviours," *Nat Rev Neurosci* 7, 126–136.

Lyons, D. J., Horjales-Araujo, E. and Broberger, C. (2010). "Synchronized network oscillations in rat tuberoinfundibular DA neurons: switch to tonic discharge by thyrotropin-releasing hormone," *Neuron* 65, 217–229.

Maffucci, J. A. and Gore, A. C. (2009). "Chapter 2: hypothalamic neural systems controlling the female reproductive life cycle gonadotropin-releasing hormone, glutamate, and GABA," *Int Rev Cell Mol Biol* 274, 69–127.

Manaker, S., Winokur, A., Rostene, W. H. and Rainbow, T. C. (1985). "Autoradiographic localization of thyrotropin-releasing hormone receptors in the rat central nervous system," *J Neurosci* 5, 167–174.

Mann, P. E. and Bridges, R. S. (2002). "Prolactin receptor gene expression in the forebrain of pregnant and lactating rats," *Brain Res Mol Brain Res* 105, 136–145.

Marks, J. L., Porte, D., Jr., Stahl, W. L. and Baskin, D. G. (1990). "Localization of insulin receptor mRNA in rat brain by in situ hybridization," *Endocr* 127, 3234–3236.

Marrazzi, M. A., Kinzie, J. and Luby, E. D. (1995). "A detailed longitudinal analysis on the use of naltrexone in the treatment of bulimia," *Int Clin Psychopharmacol* 10, 173–176.

Matsubara, S., Sato, M., Mizobuchi, M., Niimi, M. and Takahara, J. (1995). "Differential gene expression of growth hormone (GH)-releasing hormone (GRH) and GRH receptor in various rat tissues," *Endocr* 136, 4147–4150.

Matthes, H. W., Maldonado, R., Simonin, F., Valverde, O., Slowe, S., Kitchen, I. *et al.* (1996). "Loss of morphine-induced analgesia, reward effect and withdrawal symptoms in mice lacking the mu-opioid-receptor gene," *Nature* 383, 819–823.

McCall, C. and Singer, T. (2012). "The animal and human neuroendocrinology of social cognition, motivation and behavior," *Nat Neurosci* 15, 681–688.

McCann, S. M. (1991). "Neuroregulatory peptides" in M. Motta (ed.), *Brain Endocrinology*, 2nd edn. (New York: Raven Press), pp. 1–30.

McClean, P. L., Parthsarathy, V., Faivre, E. and Holscher, C. (2011). "The diabetes drug liraglutide prevents degenerative processes in a mouse model of Alzheimer's disease," *J Neurosci* 31, 6587–6594.

Meethal, S. V., Smith, M. A., Bowen, R. L. and Atwood, C. S. (2005). "The gonadotropin connection in Alzheimer's disease," *Endocrine* 26, 317–326.

Mengod, G., Rigo, M., Savasta, M., Probst, A. and Palacios, J. M. (1992). "Regional distribution of neuropeptide somatostatin gene expression in the human brain," *Synapse* 12, 62–74.

Merighi, A. (2002). "Costorage and coexistence of neuropeptides in the mammalian CNS," *Prog Neurobiol* 66, 161–190.

Messager, S., Chatzidaki, E. E., Ma, D., Hendrick, A. G., Zahn, D., Dixon, J. *et al.* (2005). "Kisspeptin directly stimulates gonadotropin-releasing hormone release via G protein-coupled receptor 54," *Proc Natl Acad Sci USA* 102, 1761–1766.

Meston, C. M. and Frohlich, P. F. (2000). "The neurobiology of sexual function," *Arch Gen Psychiatry* 57, 1012–1030.

Millar, R. P., Roseweir, A. K., Tello, J. A., Anderson, R. A., George, J. T., Morgan, K. *et al.* (2010). "Kisspeptin antagonists: unraveling the role of kisspeptin in reproductive physiology," *Brain Res* 1364, 81–89.

Miller, G. (2012). "The promise and perils of oxytocin," *Science* 339, 267–269.

Moran, T. H. and Dailey, M. J. (2009). "Minireview: gut peptides: targets for antiobesity drug development?" *Endocr* 150, 2526–2530.

Morton, G. J., Cummings, D. E., Baskin, D. G., Barsh, G. S. and Schwartz, M. W. (2006). "Central nervous system control of food intake and body weight," *Nature* 443, 289–295.

Mountjoy, K. G. (2010). "Distribution and function of melanocortin receptors within the brain," *Adv Exp Med Biol* 681, 29–48.

Müller, E. E. and Nistico, G. (1989). *Brain Messengers and the Pituitary* (San Diego, CA: Academic Press).

Munzberg, H. and Myers, M. G., Jr. (2005). "Molecular and anatomical determinants of central leptin resistance," *Nat Neurosci* 8, 566–570.

Murphy, A. E. and Harvey, S. (2002). "Extrapituitary TSH in early chick embryos: pit-1 dependence?" *J Mol Neurosci* 18, 77–87.

Murphy, K. G. and Bloom, S. R. (2006). "Gut hormones and the regulation of energy homeostasis," *Nature* 444, 854–859.

Naef, L. and Woodside, B. (2007). "Prolactin/leptin interactions in the control of food intake in rats," *Endocr* 148, 5977–5983.

Nathan, P. J. and Bullmore, E. T. (2009). "From taste hedonics to motivational drive: central mu-opioid receptors and binge-eating behaviour," *Int J Neuropsychopharmacol* 12, 995–1008.

Navarro, V. M., Castellano, J. M., Fernandez-Fernandez, R., Tovar, S., Roa, J., Mayen, A. *et al.* (2005). "Characterization of the potent luteinizing hormone-releasing activity of Kiss-1 peptide, the natural ligand of GPR54," *Endocr* 146, 156–163.

Neary, N. M., Small, C. J., Druce, M. R., Park, A. J., Ellis, S. M., Semjonous, N. M. *et al.* (2005). "Peptide YY3–36 and glucagon-like peptide-17–36 inhibit food intake additively," *Endocr* 146, 5120–5127.

Neumann, I. D. (2008). "Brain oxytocin: a key regulator of emotional and social behaviours in both females and males," *J Neuroendocrinol* 20, 858–865.

Neumann, I. D. and Landgraf, R. (2012). "Balance of brain oxytocin and vasopressin: implications for anxiety, depression, and social behaviors," *Trends Neurosci* 35, 649–659.

Nishijima, I., Yamagata, T., Spencer, C. M., Weeber, E. J., Alekseyenko, O., Sweatt, J. D. *et al.* (2006). "Secretin receptor-deficient mice exhibit impaired synaptic plasticity and social behavior," *Hum Mol Genet* 15, 3241–3250.

Noble, F., Wank, S. A., Crawley, J. N., Bradwejn, J., Seroogy, K. B., Hamon, M. *et al.* (1999). "International Union of Pharmacology. XXI. Structure, distribution, and functions of cholecystokinin receptors," *Pharmacol Rev* 51, 745–781.

Nussey, S. S. and Whitehead, S. A. (2001). *Endocrinology: An Integrated Approach* (Oxford: Bios Scientific Publishers), www.ncbi.nlm.nih.gov/books/NBK20.

O'Dowd, B. F., Lee, D. K., Huang, W., Nguyen, T., Cheng, R., Liu, Y. *et al.* (2000). "TRH-R2 exhibits similar binding and acute signaling but distinct regulation and anatomic distribution compared with TRH-R1," *Mol Endocrinol* 14, 183–193.

Oakley, A. E., Clifton, D. K. and Steiner, R. A. (2009). "Kisspeptin signaling in the brain," *Endocr Rev* 30, 713–743.

Ottenweller, J. E. and Hedge, G. A. (1982). "Thyrotropin-like immunoreactivity in the pituitary and three brain regions of the female rat: diurnal variations and the effect of thyroidectomy," *Endocr* 111, 515–521.

Ottlecz, A. (1987). "Action of gastrointestinal polypeptide hormones on pituitary anterior lobe function," *Front Horm Res* 15, 282–298.

Painsipp, E., Herzog, H., Sperk, G. and Holzer, P. (2011). "Sex-dependent control of murine emotional-affective behaviour in health and colitis by peptide YY and neuropeptide Y," *Br J Pharmacol* 163, 1302–1314.

Pandolfi, M., Pozzi, A. G., Canepa, M., Vissio, P. G., Shimizu, A., Maggese, M. C. *et al.* (2009). "Presence of beta-follicle-stimulating hormone and beta-luteinizing hormone transcripts in the brain of *Cichlasoma dimerus* (Perciformes: Cichlidae): effect of brain-derived gonadotropins on pituitary hormone release," *Neuroendocrinology* 89, 27–37.

Paxinos, G., Chai, S. Y., Christopoulos, G., Huang, X. F., Toga, A. W., Wang, H. Q. *et al.* (2004). "In vitro autoradiographic localization of calcitonin and amylin binding sites in monkey brain," *J Chem Neuroanat* 27, 217–236.

Paz-Filho, G., Wong, M. L. and Licinio, J. (2011). "Ten years of leptin replacement therapy," *Obes Rev* 12, e315–e323.

Pecina, S. and Smith, K. S. (2010). "Hedonic and motivational roles of opioids in food reward: implications for overeating disorders," *Pharmacol Biochem Behav* 97, 34–46.

Pfaff, D. W., Rapin, I. and Goldman, S. (2011). "Male predominance in autism: neuroendocrine influences on arousal and social anxiety," *Autism Res* 4, 163–176.

Pfluger, P. T., Castaneda, T. R., Heppner, K. M., Strassburg, S., Kruthaupt, T., Chaudhary, N. *et al.* (2011). "Ghrelin, peptide YY and their hypothalamic targets differentially regulate spontaneous physical activity," *Physiol Behav* 105, 52–61.

Pineda, R., Garcia-Galiano, D., Roseweir, A., Romero, M., Sanchez-Garrido, M. A., Ruiz-Pino, F. *et al.* (2010). "Critical roles of kisspeptins in female puberty and preovulatory gonadotropin surges as revealed by a novel antagonist," *Endocr* 151, 722–730.

Pocai, A. (2012). "Unraveling oxyntomodulin, GLP-1's enigmatic brother," *J Endocrinol* 215, 335–346.

Priestley, J. V., Rethelyi, M. and Lund, P. K. (1991). "Semi-quantitative analysis of somatostatin mRNA distribution in the rat central nervous system using in situ hybridization," *J Chem Neuroanat* 4, 131–153.

Pritchard, L. E., Oliver, R. L., McLoughlin, J. D., Birtles, S., Lawrence, C. B., Turnbull, A. V. *et al.* (2003). "Proopiomelanocortin-derived peptides in rat cerebrospinal fluid and hypothalamic extracts: evidence that secretion is regulated with respect to energy balance," *Endocr* 144, 760–766.

Purves, D., Augustine, G. J., Fitzpatrick, D., Hall, W. C., LaMantia, A.-S., McNamara, J. O. *et al.* (2008). *Neuroscience*, 4th edn. (Sunderland: Sinauer Associates).

Raffin-Sanson, M. L., de Keyzer, Y. and Bertagna, X. (2003). "Proopiomelanocortin, a polypeptide precursor with multiple functions: from physiology to pathological conditions," *Eur J Endocrinol* 149, 79–90.

Rao, S. C., Li, X., Rao Ch, V. and Magnuson, D. S. (2003). "Human chorionic gonadotropin/luteinizing hormone receptor expression in the adult rat spinal cord," *Neurosci Lett* 336, 135–138.

Ravussin, E., Smith, S. R., Mitchell, J. A., Shringarpure, R., Shan, K., Maier, H. *et al*. (2009). "Enhanced weight loss with pramlintide/metreleptin: an integrated neurohormonal approach to obesity pharmacotherapy," *Obesity (Silver Spring)* 17, 1736–1743.

Raymond, N. C., Grant, J. E. and Coleman, E. (2010). "Augmentation with naltrexone to treat compulsive sexual behavior: a case series," *Ann Clin Psychiatry* 22, 56–62.

Richard, D., Lin, Q. and Timofeeva, E. (2002). "The corticotropin-releasing factor family of peptides and CRF receptors: their roles in the regulation of energy balance," *Eur J Pharmacol* 440, 189–197.

Roa, J., Aguilar, E., Dieguez, C., Pinilla, L. and Tena-Sempere, M. (2008). "New frontiers in kisspeptin/GPR54 physiology as fundamental gatekeepers of reproductive function," *Front Neuroendocr* 29, 48–69.

Rood, B. D. and De Vries, G. J. (2011). "Vasopressin innervation of the mouse (*Mus musculus*) brain and spinal cord," *J Comp Neurol* 519, 2434–2474.

Roselli, C. E., Bocklandt, S., Stadelman, H. L., Wadsworth, T., Vilain, E. and Stormshak, F. (2008). "Prolactin expression in the sheep brain," *Neuroendocrinology* 87, 206–215.

Rotzinger, S., Lovejoy, D. A. and Tan, L. A. (2010). "Behavioral effects of neuropeptides in rodent models of depression and anxiety," *Peptides* 31, 736–756.

Ryback, R. S. (2004). "Naltrexone in the treatment of adolescent sexual offenders," *J Clin Psychiatry* 65, 982–986.

Ryckmans, T. (2010). "Modulation of the vasopressin system for the treatment of CNS diseases," *Curr Opin Drug Discov Devel* 13, 538–547.

Satoh, T., Feng, P., Kim, U. J. and Wilber, J. F. (1993). "Identification of thyrotropin-releasing hormone receptor messenger RNA in the rat central nervous system and eye," *Brain Res Mol Brain Res* 19, 175–178.

Sauriyal, D. S., Jaggi, A. S. and Singh, N. (2011). "Extending pharmacological spectrum of opioids beyond analgesia: multifunctional aspects in different pathophysiological states," *Neuropeptides* 45, 175–188.

Sauve, D. and Woodside, B. (2000). "Neuroanatomical specificity of prolactin-induced hyperphagia in virgin female rats," *Brain Res* 868, 306–314.

Schaad, N. C., Schorderet, M. and Magistretti, P. J. (1989). "Accumulation of cyclic AMP elicited by vasoactive intestinal peptide is potentiated by noradrenaline, histamine, adenosine, baclofen, phorbol esters, and ouabain in mouse cerebral cortical slices: studies on the role of arachidonic acid metabolites and protein kinase C," *J Neurochem* 53, 1941–1951.

Schang, A. L., Counis, R., Magre, S., Bleux, C., Granger, A., Ngo-Muller, V. *et al*. (2011). "Reporter transgenic mouse models highlight the dual endocrine and neural facet of GnRH receptor function," *Ann NY Acad Sci* 1220, 16–22.

Schechter, R. and Abboud, M. (2001). "Neuronal synthesized insulin roles on neural differentiation within fetal rat neuron cell cultures," *Brain Res Dev Brain Res* 127, 41–49.

Schiml, P. A. and Rissman, E. F. (2000). "Effects of gonadotropin-releasing hormones, corticotropin-releasing hormone, and vasopressin on female sexual behavior," *Horm Behav* 37, 212–220.

Schindler, M., Humphrey, P. P. and Emson, P. C. (1996). "Somatostatin receptors in the central nervous system," *Prog Neurobiol* 50, 9–47.

Schorscher-Petcu, A., Dupre, A. and Tribollet, E. (2009). "Distribution of vasopressin and oxytocin binding sites in the brain and upper spinal cord of the common marmoset," *Neurosci Lett* 461, 217–222.

Schulingkamp, R. J., Pagano, T. C., Hung, D. and Raffa, R. B. (2000). "Insulin receptors and insulin action in the brain: review and clinical implications," *Neurosci Biobehav Rev* 24, 855–872.

Schulz, C., Paulus, K., Lobmann, R., Dallman, M. and Lehnert, H. (2010). "Endogenous ACTH, not only alpha-melanocyte-stimulating hormone, reduces food intake mediated by hypothalamic mechanisms," *Am J Physiol Endocrinol Metab* 298, E237–E244.

Schwartz, M. W. and Morton, G. J. (2002). "Obesity: keeping hunger at bay," *Nature* 418, 595–597.

Scott, M. M., Lachey, J. L., Sternson, S. M., Lee, C. E., Elias, C. F., Friedman, J. M. *et al*. (2009). "Leptin targets in the mouse brain," *J Comp Neurol* 514, 518–532.

Seminara, S. B. and Crowley, W. F., Jr. (2008). "Kisspeptin and GPR54: discovery of a novel pathway in reproduction," *J Neuroendocrinol* 20, 727–731.

Shioda, S., Yada, T., Muroya, S., Uramura, S., Nakajo, S., Ohtaki, H. *et al*. "Functional significance of colocalization of PACAP and catecholamine in nerve terminals," *Ann NY Acad Sci* 921, 211–217.

Sibilia, V., Lattuada, N., Rapetti, D., Pagani, F., Vincenza, D., Bulgarelli, I. *et al*. (2006). "Ghrelin inhibits inflammatory pain in rats: involvement of the opioid system," *Neuropharmacol* 51, 497–505.

Simpson, K. A. and Bloom, S. R. (2010). "Appetite and hedonism: gut hormones and the brain," *Endocrinol Metab Clin North Am* 39, 729–743.

Skinner, D. C., Albertson, A. J., Navratil, A., Smith, A., Mignot, M., Talbott, H. *et al*. (2009). "Effects of gonadotrophin-releasing hormone outside the hypothalamic-pituitary-reproductive axis," *J Neuroendocrinol* 21, 282–292.

Slattery, D. A. and Neumann, I. D. (2008). "No stress please! Mechanisms of stress hyporesponsiveness of the maternal brain," *J Physiol* 586, 377–385.

Spergel, D. J., Kruth, U., Shimshek, D. R., Sprengel, R. and Seeburg, P. H. (2001). "Using reporter genes to label selected neuronal populations in transgenic mice for gene promoter, anatomical, and physiological studies," *Prog Neurobiol* 63, 673–686.

Steckler, T. (2010). "Developing small molecule nonpeptidergic drugs for the treatment of anxiety disorders: is the challenge still ahead?" *Curr Top Behav Neurosci* 2, 415–428.

Steen, E., Terry, B. M., Rivera, E. J., Cannon, J. L., Neely, T. R., Tavares, R. *et al*. (2005). "Impaired insulin and insulin-like growth factor expression and signaling mechanisms in Alzheimer's disease – is this type 3 diabetes?" *J Alzheimers Dis* 7, 63–80.

Steiger, A., Dresler, M., Schussler, P. and Kluge, M. (2011). "Ghrelin in mental health, sleep, memory," *Mol Cell Endocrinol* 340, 88–96.

Steinert, R. E., Poller, B., Castelli, M. C., Drewe, J. and Beglinger, C. (2010). "Oral administration of glucagon-like peptide 1 or peptide YY 3–36 affects food intake in healthy male subjects," *Am J Clin Nutr* 92, 810–817.

Steinert, R. E., Meyer-Gerspach, A. C. and Beglinger, C. (2012). "The role of the stomach in the control of appetite and the secretion of satiation peptides," *Am J Physiol* 302, E666–E673.

Stella, N. and Magistretti, P. J. (1996). "Vasoactive intestinal peptide (VIP) and pituitary adenylate cyclase-activating polypeptide (PACAP) potentiate the glutamate-evoked release of arachidonic acid from mouse cortical neurons. Evidence for a cAMP-independent mechanism," *J Biol Chem* 271, 23705–23710.

Stockhorst, U., de Fries, D., Steingrueber, H. J. and Scherbaum, W. A. (2004). "Insulin and the CNS: effects on food intake, memory, and endocrine parameters and the role of intranasal insulin administration in humans," *Physiol Behav* 83, 47–54.

Suter, K. J., Song, W. J., Sampson, T. L., Wuarin, J. P., Saunders, J. T., Dudek, F. E. *et al*. (2000a). "Genetic targeting of green fluorescent protein to gonadotropin-releasing hormone neurons: characterization of whole-cell electrophysiological properties and morphology," *Endocr* 141, 412–419.

Suter, K. J., Wuarin, J. P., Smith, B. N., Dudek, F. E. and Moenter, S. M. (2000b). "Whole-cell recordings from preoptic/hypothalamic slices reveal burst firing in gonadotropin-releasing hormone neurons identified with green fluorescent protein in transgenic mice," *Endocr* 141, 3731–3736.

Suzuki, K., Simpson, K. A., Minnion, J. S., Shillito, J. C. and Bloom, S. R. (2010). "The role of gut hormones and the hypothalamus in appetite regulation," *Endocr J* 57, 359–372.

Szentirmai, E., Yasuda, T., Taishi, P., Wang, M., Churchill, L., Bohnet, S. *et al.* (2007). "Growth hormone-releasing hormone: cerebral cortical sleep-related EEG actions and expression," *Am J Physiol Regul Integr Comp Physiol* 293, R922–R930.

Tauchi, M., Zhang, R., D'Alessio, D. A., Stern, J. E. and Herman, J. P. (2008). "Distribution of glucagon-like peptide-1 immunoreactivity in the hypothalamic paraventricular and supraoptic nuclei," *J Chem Neuroanat* 36, 144–149.

Telegdy, G., Adamik, A., Tanaka, M. and Schally, A. V. (2010). "Effects of the LHRH antagonist Cetrorelix on affective and cognitive functions in rats," *Regul Pept* 159, 142–147.

Tong, J. and Sandoval, D. A. (2011). "Is the GLP-1 system a viable therapeutic target for weight reduction?" *Rev Endocr Metab Disord* 12, 187–195.

Torner, L., Maloumby, R., Nava, G., Aranda, J., Clapp, C. and Neumann, I. D. (2004). "In vivo release and gene upregulation of brain prolactin in response to physiological stimuli," *Eur J Neurosci* 19, 1601–1608.

Torner, L., Nava, G., Duenas, Z., Corbacho, A., Mejia, S., Lopez, F. *et al.* (1999). "Changes in the expression of neurohypophyseal prolactins during the estrous cycle and after estrogen treatment," *J Endocrinol* 161, 423–432.

Torner, L., Toschi, N., Nava, G., Clapp, C. and Neumann, I. D. (2002). "Increased hypothalamic expression of prolactin in lactation: involvement in behavioural and neuroendocrine stress responses," *Eur J Neurosci* 15, 1381–1389.

Torrealba, F. and Carrasco, M. A. (2004). "A review on electron microscopy and neurotransmitter systems," *Brain Res Rev* 47, 5–17.

Trapp, S. and Hisadome, K. (2011). "Glucagon-like peptide 1 and the brain: central actions-central sources?" *Auton Neurosci* 161, 14–19.

Ueta, Y., Fujihara, H., Dayanithi, G., Kawata, M. and Murphy, D. (2008). "Specific expression of optically active reporter gene in arginine vasopressin-secreting neurosecretory cells in the hypothalamic-neurohypophyseal system," *J Neuroendocrinol* 20, 660–664.

Vaccarino, F. J., Sovran, P., Baird, J. P. and Ralph, M. R. (1995). "Growth hormone-releasing hormone mediates feeding-specific feedback to the suprachiasmatic circadian clock," *Peptides* 16, 595–598.

Van Bloemendaal, L., ten Kulve, J. S., la Fleur, S. E., Ijzerman, R. G. and Diamant, M. (2014). "Effects of glucagon-like peptide 1 on appetite and body weight: focus on the CNS," *J Endocrinol* 221, T1–T16.

Van Pett, K., Viau, V., Bittencourt, J. C., Chan, R. K., Li, H. Y., Arias, C. *et al.* (2000). "Distribution of mRNAs encoding CRF receptors in brain and pituitary of rat and mouse," *J Comp Neurol* 428, 191–212.

Viollet, C., Lepousez, G., Loudes, C., Videau, C., Simon, A. and Epelbaum, J. (2008). "Somatostatinergic systems in brain: networks and functions," *Mol Cell Endocrinol* 286, 75–87.

Volkow, N. D. and Wise, R. A. (2005). "How can drug addiction help us understand obesity?" *Nat Neurosci* 8, 555–560.

Vuong, C., Van Uum, S. H., O'Dell, L. E., Lutfy, K. and Friedman, T. C. (2010). "The effects of opioids and opioid analogs on animal and human endocrine systems," *Endocr Rev* 31, 98–132.

Weigle, D. S., Duell, P. B., Connor, W. E., Steiner, R. A., Soules, M. R. and Kuijper, J. L. (1997). "Effect of fasting, refeeding, and dietary fat restriction on plasma leptin levels," *J Clin Endocrinol Metab* 82, 561–565.

Wen, S., Gotze, I. N., Mai, O., Schauer, C., Leinders-Zufall, T. and Boehm, U. (2011). "Genetic identification of GnRH receptor neurons: a new model for studying neural circuits underlying reproductive physiology in the mouse brain," *Endocr* 152, 1515–1526.

White, C. M., Ji, S., Cai, H., Maudsley, S. and Martin, B. (2010). "Therapeutic potential of vasoactive intestinal peptide and its receptors in neurological disorders," *CNS Neurol Disord Drug Targets* 9, 661–666.

Wisse, B. E., Campfield, L. A., Marliss, E. B., Morais, J. A., Tenenbaum, R. and Gougeon, R. (1999). "Effect of prolonged moderate and severe energy restriction and refeeding on plasma leptin concentrations in obese women," *Am J Clin Nutr* 70, 321–330.

Wu, L. G. and Saggau, P. (1997). "Presynaptic inhibition of elicited neurotransmitter release," *Trends Neurosci* 20, 204–212.

Wu, M., Dumalska, I., Morozova, E., van den Pol, A. N. and Alreja, M. (2009). "Gonadotropin inhibitory hormone inhibits basal forebrain vGluT2-gonadotropin-releasing hormone neurons via a direct postsynaptic mechanism," *J Physiol* 587, 1401–1411.

Yang, S. Y., Cho, S. C., Yoo, H. J., Cho, I. H., Park, M., Yoe, J. *et al.* (2010). "Family-based association study of microsatellites in the 5' flanking region of AVPR1A with autism spectrum disorder in the Korean population," *Psychiatry Res* 178, 199–201.

Yamasue, H., Yee, J. R., Hurlemann, R., Rilling, J. K., Chen, F. S., Meyer-Lindenberg, A. *et al.* (2012). "Integrative approaches utilizing oxytocin to enhance prosocial behavior: from animal and human social behavior to autistic social dysfunction," *J Neurosci* 32, 14109–14117.

Yarbrough, G. G., Kamath, J., Winokur, A. and Prange, A. J., Jr. (2007). "Thyrotropin-releasing hormone (TRH) in the neuroaxis: therapeutic effects reflect physiological functions and molecular actions," *Med Hypotheses* 69, 1249–1256.

Yi, C. X., Challet, E., Pevet, P., Kalsbeek, A., Escobar, C. and Buijs, R. M. (2008). "A circulating ghrelin mimetic attenuates light-induced phase delay of mice and light-induced Fos expression in the suprachiasmatic nucleus of rats," *Eur J Neurosci* 27, 1965–1972.

Yirmiya, N., Rosenberg, C., Levi, S., Salomon, S., Shulman, C., Nemanov, L. *et al.* (2006). "Association between the arginine vasopressin 1a receptor (AVPR1a) gene and autism in a family-based study: mediation by socialization skills," *Mol Psychiatry* 11, 488–494.

Yuan, Y., Lee, L. T., Ng, S. S. and Chow, B. K. (2011). "Extragastrointestinal functions and transcriptional regulation of secretin and secretin receptors," *Ann NY Acad Sci* 1220, 23–33.

Zaninotto, P., Head, J., Stamatakis, E., Wardle, H. and Mindell, J. (2009). "Trends in obesity among adults in England from 1993 to 2004 by age and social class and projections of prevalence to 2012," *J Epidemiol Community Health* 63, 140–146.

Zeki, S. (2007). "The neurobiology of love," *FEBS Letters* 581, 2575–2579.

Zhang, C., Roepke, T. A., Kelly, M. J. and Ronnekleiv, O. K. (2008). "Kisspeptin depolarizes gonadotropin-releasing hormone neurons through activation of TRPC-like cationic channels," *J Neurosci* 28, 4423–4434.

Zhao, W. Q., Chen, H., Quon, M. J. and Alkon, D. L. (2004). "Insulin and the insulin receptor in experimental models of learning and memory," *Eur J Pharmacol* 490, 71–81.

Ziauddeen, H., Chamberlain, S. R., Nathan, P. J., Koch, A., Maltby, K., Bush, M. *et al.* (2013). "Effects of mu-opioid receptor antagonist GSK1521498 on hedonic and consummatory eating behavior: a proof of mechanism study in binge-eating obese subjects," *Mol Psych* 18, 1287–1293.

Zoli, M., Jansson, A., Sykova, E., Agnati, L. F. and Fuxe, K. (1999). "Volume transmission in the CNS and its relevance for neuropsychopharmacology," *Trends Pharmacol Sci* 20, 142–150.

Cytokines and the interaction between the neuroendocrine and immune systems

Cytokines were introduced in Chapter 1 (section 1.4.9) as signaling molecules secreted by cells of the immune system. They are important components of the interconnected neural, endocrine and immune systems (see Figure 1.2). For example, various types of stress, including academic examinations, influence the immune system via an activation of the hypothalamic-pituitary-adrenal axis and the secretion of glucocorticoids (see Figures 6.1 and 6.4). Likewise, an immune response that stimulates white blood cells to produce cytokines, for example following an infection, has effects on the hypothalamic control of various hormones such as ACTH, GH and PRL. Since most immune cells have receptors for these hormones, the immune system is affected directly by pituitary hormones as well as by adrenal output of glucocorticoids or catecholamines (Figure 1.2). The immune system, therefore, via secretion of cytokines, participates in a classic neuroendocrine feedback system. There are also profound sex differences in immune responses and some of this variation is due to the effects of sex steroids, such as testosterone and estradiol. For example, 80 percent of patients with autoimmune diseases (e.g. rheumatoid arthritis; multiple sclerosis) are women; 60 percent of adult asthma cases are women; and men are at least 1.6-fold more likely than women to die from malignant cancers (Klein 2012).

This chapter begins with an overview of those cells of the immune system that secrete cytokines, and then discusses the immune functions of the thymus gland and its hormones. The roles of cytokines in the immune response to antigens (i.e. substances that cause an immune response) and in the development of blood cells are then summarized and the neuromodulatory effects of cytokines on the brain and neuroendocrine system are examined. This is followed by a discussion of the neural and endocrine regulation of the immune system and the hypothalamic integration of neural, endocrine and immune systems.

13.1 The cells of the immune system

The immune system consists of many different cell types, including several that secrete cytokines; i.e. monocytes, macrophages, T lymphocytes (T cells), B lymphocytes (B cells) and natural killer (NK) cells (see Figure 13.1). The primary role of the immune system is the protection of the body from foreign invaders such as bacteria and viruses and from abnormal cellular development as occurs in tumor cells. To do this, the immune system must be able to discriminate foreign (non-self) cells from the body's own cells (self). Almost all substances have regions called antigenic determinants or *epitopes* that can stimulate an immune response. The term *antigen* is used to refer to the whole molecule that carries the epitope(s). All cells (both foreign and self) carry many epitope recognition sites (antibodies) on their surfaces. When a foreign antigen is detected, an immune response is activated.

The immune system has two branches, each of which plays a different, but complementary, role in the protection against infection. The *humoral immune system* protects the body from extracellular antigens (e.g. pollen, bacteria, toxins, etc.). When an antigen is detected in the circulation or other extracellular fluids, the humoral immune response is activated and antibodies (immunoglobulins) are secreted from B cells to inactivate the foreign material. The *cell-mediated immune system* protects the body from intracellular antigens (e.g. viruses), as well as from foreign (transplanted) tissue, tumor cells and protozoa. If one of the body's cells is infected by a virus or a tumor develops, then the cell-mediated immune response is activated and the infected cell or tumor cell is destroyed by cytotoxic T cells or NK cells.

13.1.1 Monocytes and macrophages

Monocytes develop from precursor cells in the bone marrow and travel through the bloodstream to various tissues, where they mature into macrophages (see Figure 13.7). Macrophages ingest (phagocytose) and break down foreign antigens and present epitopes of these antigens to T cells. Macrophages synthesize and release cytokines, including several interleukins and interferons.

13.1.2 T lymphocytes (T cells)

The precursor cells for T cells are produced in the bone marrow and migrate to the thymus gland through the circulatory system (see section 13.2; Figure 13.1). In the thymus gland, T cell precursors differentiate to form helper T cells (T_H), cytotoxic T cells (T_c) and suppressor T cells (T_s). Cytotoxic T cells destroy body cells that have been infected by viruses, tumor cells and foreign cells (e.g. tissue transplants). Helper T cells release cytokines that stimulate the immune functions of other lymphocytes in a number of ways as discussed in section 13.3. Suppressor T cells are able to suppress immune responses by inhibiting the actions of T_H and T_C cells. The maturation, differentiation and proliferation of T cells in the thymus gland is regulated by thymic hormones and the thymic microenvironment (see section 13.2). After the T cells mature in the thymus gland, they are stored in the secondary lymph organs (e.g. the spleen, lymph nodes and Peyer's patches, the gut-associated lymph tissue) before being released to circulate through the body (see Figure 13.1). If the thymus gland is absent, the T cells fail to mature and cell-mediated immunity cannot occur. Because the cytokines released by the T_H cells (e.g. interleukins, interferons, transforming growth factor) mediate antibody production by B cells, the lack of T cells also inhibits humoral immunity.

13.1.3 B lymphocytes (B cells)

As with T cells, the precursors of the B cells are produced in the bone marrow. In mammals, B cells mature in the bone marrow and then migrate to the secondary lymph organs, where they are stored and released into the circulation. In birds, B cells mature in the bursa of Fabricius before being stored in the secondary lymph organs (Figure 13.1). The committed B cells are antigen specific and when they detect "their" antigen in the body, they produce antigen-specific antibodies (immunoglobulins) that bind to the antigens and inactivate

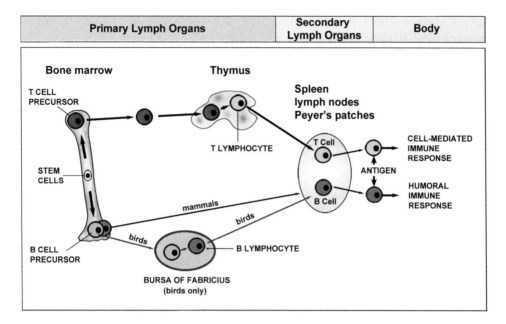

Figure 13.1 Origin and development of T and B lymphocytes
Multipotential hematopoietic stem cells in the bone marrow produce the precursors for both T and B cells. In mammals and birds, T cell precursors migrate from the bone marrow through the bloodstream to the thymus gland, where they differentiate into thymus (T) lymphocytes. When the T lymphocytes mature, they migrate to secondary lymph organs to become thymus-derived lymphocytes (T cells) that regulate immune responses. In mammals, the precursor cells destined to become B cells differentiate into lymphocytes in the hematopoietic tissue of the bone marrow and then migrate to secondary lymph organs to become B cells, which produce antibodies for the humoral immune responses. In birds, B cell precursors migrate to the bursa of Fabricius, where they differentiate into bursa lymphocytes. Many of these lymphocytes die, but others migrate to secondary lymph organs to become bursa-derived lymphocytes (B cells).

them. As well as antibody-producing B cells, there are also "memory" B cells, which remember which antibodies the body has created. B cells also produce interleukins, which play a role in immunoregulation.

13.1.4 Natural killer (NK) cells
NK cells are large lymphocytes that develop in the bone marrow. They are stored in the secondary lymph organs, from which they are released to circulate in the bloodstream. NK cells are able to spontaneously kill virus-infected cells and tumor cells. NK cells also release interleukins and interferon-γ.

13.1.5 Granulocytes
Granulocytes are mature granular leukocytes (white blood cells). There are three types of granulocytes: neutrophils, eosinophils and basophils (see Figure 13.7). Neutrophils destroy dead cells, bacteria and other foreign cells by phagocytosis. Eosinophils destroy parasites and modulate inflammatory responses in damaged tissues. Basophils destroy parasites by phagocytosis and are involved in immediate hypersensitivity (i.e. allergic) reactions through their secretion of histamine. They are also a source of interleukins. These cells are the primary

source of immune system cytokines, although they are only a small part of the total immune response system. It is beyond the scope of this chapter to discuss the immune system in more detail, and more detailed reviews of the cells involved in the immune response can be found elsewhere (Tonegawa 1985; Sompayrac 2012; Abbas and Lichtman 2010).

13.2 The thymus gland and its hormones

The thymus is a two-lobed gland that is located in the upper part of the chest cavity, behind the sternum and in front of the heart. It contains two important regions: one in which the T lymphocytes mature (see Figure 13.1), and one containing epithelial cells, which secrete thymic hormones. These hormones were introduced in Chapter 2 (section 2.2.5). The thymus gland produces at least 25 active peptides, collectively referred to as the thymic hormones or *thymosins* (Goldstein and Badamchian 2004). These include thymosin α1, thymosin β4, thymulin and thymic humoral factor-α.

The thymus has been called "the master gland of the immune system" because thymosins are essential for both cell-mediated and humoral immune responses. Thymosins facilitate the production of T cell precursors in the bone marrow, regulate the differentiation of precursor T cells into helper, cytotoxic and suppressor T cells in the thymus gland, and activate mature T cells in the spleen and lymph nodes (Figure 13.2). Thymosins, such as thymosin α1, increase cytokine production, NK cell activity and anti-tumor activity in animals, but also have important clinical effects in patients (Goldstein and Badamchian 2004).

The activity of the thymus gland is regulated by autonomic nerve pathways as well as by pituitary, adrenal, thyroid and gonadal hormones, as shown in Figure 13.2. The thymus gland is particularly sensitive to adrenal and gonadal steroids, which inhibit its growth and function (Brooks *et al.* 2005; Windmill and Lee 1998; Greenstein *et al.* 1986). In turn, thymosins act on both neural and endocrine cells to regulate their activity, as described in section 13.5.3.

The thymus gland is essential for the maturation and differentiation of T cells and the development of cell-mediated immune responses in newborn animals. If the thymus gland is absent at birth, T cells fail to develop and the animal lacks the ability to cope with viral and bacterial infections. In neonatal rats and mice, for example, thymus gland removal impairs growth and leads to "wasting syndrome." Such animals have reduced growth, a hunched posture and difficulty walking, suffer from diarrhea, are dirty and ungroomed and show delayed puberty. These symptoms are caused by overwhelming infections due to the absence of mature T cells (Asanuma *et al.* 1970). Children born without a thymus often die within the first few weeks of life because of the absence of T cells and consequent overpowering infections. However, they can be treated successfully with thymosin (Goldstein and Badamchian 2004).

In most mammals, the absolute size of the thymus gland increases until puberty, after which it begins to decline in both size and function (Tosi *et al.* 1982). However, in humans this is not the case. The thymus remains more or less constant in size, but the thymic epithelium – that part of the gland responsible for thymic cell production – declines dramatically as people age (Figure 13.3) (Hale 2004). This is paralleled by a loss of thymosin output, a decrease in thymus-dependent immunity and a corresponding increase in age-related diseases. In animals, the decline in thymus gland size after puberty is due to a developmental increase in circulating sex steroids. This inhibitory influence of sex steroids on the thymus is so powerful that castration reverses thymus atrophy in both male and female animals, while

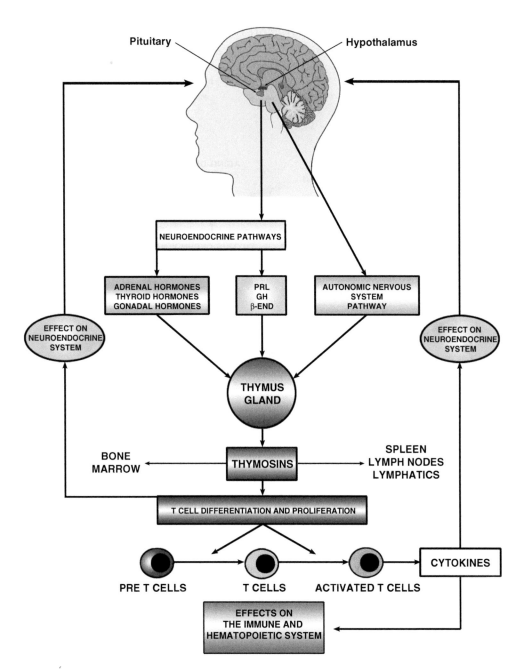

Figure 13.2 The interactions between the thymus gland and the neuroendocrine system
Thymus gland development, and thymic hormone (thymosin) secretion from the thymus, is regulated by hormones from the pituitary, adrenal gland and gonads, and by the autonomic nervous system. Thymosins regulate T cell precursor development in the bone marrow, the differentiation and proliferation of T cells in the thymus gland and the maturation of T cells in the secondary lymph organs. The thymosins and cytokines released from activated T cells have effects on the brain and neuroendocrine system, as well as on the immune and hematopoietic systems.

Abbreviations: β-END, β-endorphin; GH, growth hormone; PRL, prolactin.

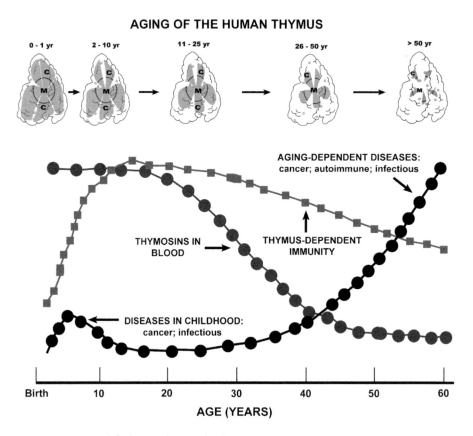

Figure 13.3 Aging of the human thymus gland

In most mammals, the size of the thymus gland declines after puberty. In humans this is not the case. The thymus remains essentially constant in size, but the thymic epithelium – responsible for thymic cell production (pink) – declines dramatically as people age. This is paralleled by a loss of thymosin output, a decrease in thymus-dependent immunity and a corresponding increase in age-related diseases. Note that the frequency of diseases associated with cellular immunity is high in newborns, as the T cells in the thymus must develop neonatally. Abbreviations: M, medulla; C, cortex. Images of the thymus reproduced with permission (Hale 2004).

the re-administration of sex steroids inhibits this thymus regeneration (Windmill and Lee 1998; Greenstein *et al.* 1986). A comparable effect is seen in elderly humans, where castration induces a resurgence of circulating T cells (Sutherland *et al.* 2005).

13.3 Cytokines: the messengers of the immune system

Cytokines are protein/peptide messengers that are produced and released by macrophages, T cells, B cells and other cells of the immune system when these cells are activated by antigen presenting cells, neural stimulation or hormones. Cytokines communicate between the cells of the immune system through paracrine and/or autocrine pathways and are essential for the activation of immune responses and the development of blood cells (hematopoiesis). In addition, cytokines may be considered as hormones (in the broadest definition of the term), since they are produced in one cell (e.g. T lymphocytes) and may travel through the blood to act at distant cells, as defined for endocrine communication in Chapter 1. Figure 1.2

illustrates how interleukin-1 (IL-1) enters the bloodstream and exerts effects in the brain. When first discovered, cytokines were named according to their function (e.g. T cell growth factor) and classified as cytokines or monokines (secreted from macrophages). However, it soon became clear that each cytokine had a number of different functions and that two or more cytokines often served the same function. In an attempt to reduce the confusion, individual cytokines were named *interleukins*, based on their ability to communicate between leukocytes (Aarden 1979). However, as the structure and function of individual cytokines was identified, it became necessary to reclassify them. Some of these, and their functions, are listed in Table 13.1 (Kindt *et al.* 2006).

Table 13.1 **Examples of some cytokines, their cellular source and functions**

Cytokine	Cellular source	Function
Interferon γ (IFNγ)	Activated T cells and NK cells	Activates macrophages Stimulates B cell proliferation and differentiation
Interleukin 1 (IL-1α and β)	Macrophages, B cells, NK cells, fibroblasts, etc.	Enhances T and B cell activation Promotes proliferation of T and B cells Stimulates prostaglandin synthesis Causes fever Stimulates synthesis of IFNγ, IL-2, IL-3 and IL-4
Interleukin 2 (IL-2)	Activated T cells	Stimulates T cell growth Stimulates B cell growth and proliferation Stimulates activity of NK cells
Interleukin 4 (IL-4)	Macrophages, activated T cells, NK cells	Promotes T cell growth Stimulates B cell proliferation and differentiation Activates macrophages
Interleukin 6 (IL-6)	Monocytes, activated T cells, fibroblasts, macrophages	Promotes B cell proliferation and differentiation Activates T cells Induces fever
Interleukin 15 (IL-15)	T cells, macrophages	Stimulates NK cells, B cell proliferation
Interleukin 17 (IL-17)	T cells, NK cells, epithelial cells	Stimulates macrophages and initiates inflammation
Interleukin 37 (IL-37)	Red blood cells	Inhibits macrophages and T cell numbers
Tumor necrosis factor α (TNFα)	Macrophages, T cells	Destruction of tumor cells Mediates endotoxic shock reaction Inhibits feeding Causes cachexia (wasting)
Colony stimulating factors (CSFs)	Fibroblasts, monocytes	Stimulate growth and differentiation of hematopoietic stem cells in bone marrow
Oncostatin M (OSM)	Macrophages, T cells	Inhibits tumor growth
Tranforming growth factor β (TGF-β)	Macrophages, lymphocytes	Induces IL-1 Stimulates B cells Inhibits inflammation

Each cytokine has its own specific receptor located on the surface of its target cells and the synthesis of these receptors is usually activated at the same time as the synthesis of the cytokines. Following the successful cloning of their genes, the structures of cytokine receptors are now known. Almost all of the receptors belong to a *superfamily* and are of the type shown in Figure 13.9; that is, cytokine receptors consist of a dimer of two polypeptide chains that penetrate the cell membrane. Cytokine receptors will be covered in more detail in section 13.5.1.

A detailed description of the multiple biological activities of cytokines (see Table 13.1) became possible following the development of a highly specific assay called *Enzyme-linked Immunosorbent Assay (ELISA)*, which is capable of quantifying levels of cytokines in many biological systems. This assay technique was described in detail in Chapter 8 (see Figure 8.1) and is dependent on the availability of highly specific monoclonal antibodies raised against individual cytokines.

13.3.1 Interferon γ (IFNγ)

Interferon γ is produced by T lymphocytes and NK cells following stimulation by specific antigens or mitogens (substances which induce DNA synthesis and cell proliferation). Interferon γ is an antiviral protein that activates macrophages, NK cells and stimulates B cell proliferation and differentiation. Interferon γ can also increase the synthesis of IL-2 and its receptors in helper T cells. Two other interferons (α and β) are produced in other cells of the immune system and act to inhibit the spread of viruses, bacteria, protozoa and other antigens (Schroder *et al.* 2004).

13.3.2 The interleukins

As of 2013, 38 interleukins have been identified. Some examples of their main functions, and their relationship to the neuroendocrine system, are outlined in this section (see also Table 13.1).

Interleukin 1 (IL-1)

Interleukin 1 exists in two forms (IL-1α and IL-1β) that are synthesized by monocytes, macrophages, T cells, fibroblasts and other immune cells which have been activated by endotoxins or other noxious factors. Once released, IL-1 has many roles in immunity and inflammation. IL-1 promotes T cell and B cell differentiation and the growth and activity of macrophages, neutrophils and NK cells during inflammatory responses. IL-1 promotes prostaglandin synthesis during the inflammatory response. IL-1 also stimulates the synthesis of other interleukins and IFNγ by helper T cells and the synthesis of IL-2 receptors on helper T cells (Kindt *et al.* 2006). IL-1 receptors also exist on non-immune cells, including liver, bone, muscle and brain cells (see Figure 13.4). During immune responses, IL-1 enters the blood and acts as an endocrine signal to the brain to induce fever and sleep, decrease locomotor activity and suppress food intake (Plata-Salaman 1991; 2001). IL-1 also regulates glial cell growth and proliferation, modulates the release of hypothalamic and pituitary hormones, and may modulate the activity of endogenous opioid peptide and noradrenergic neurotransmitter systems. The link between IL-1 and the neuroendocrine system is discussed in section 13.5. IL-1 activity may also play a significant role in autoimmune diseases such as rheumatoid arthritis (Goldbach-Mansky 2012). Corticosteroid hormones and some immunosuppressive drugs, such as cyclosporine, inhibit IL-1 production and may be used to treat autoimmune disorders.

ANTIGEN

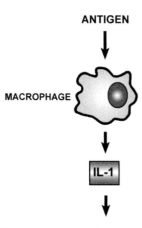

MACROPHAGE

IL-1

Figure 13.4 Some of the actions of interleukin 1 (IL-1)
As well as its actions on lymphocytes and bone marrow, IL-1 has receptors in the brain, liver, muscle, skin and bone and has a number of actions on these non-immune cells.

1. BRAIN: prostaglandin synthesis, sleep, anorexia, fever.
2. NEUROENDOCRINE SYSTEM: modulates release of hormones.
3. LYMPHOCYTES: promotes T and B cell proliferation and synthesis of cytokines.
4. BONE MARROW: hematopoiesis.
5. LIVER: acute phase proteins.
6. MUSCLE: protein synthesis.
7. BONE AND CARTILAGE: synthesis of prostaglandins.
8. ENDOTHELIUM AND EPITHELIUM: local inflammation and wound healing.

Interleukin 2 (IL-2)

IL-2 was originally named T cell growth factor. It is produced by helper T cells following activation by antigen presenting cells (macrophages) and acts to promote T cell growth and B cell differentiation, and activates NK cells (see Figure 13.5). IL-2 synthesis is promoted by IL-1 and a number of neuroendocrine factors (see Table 13.2). IL-2 reduces oxytocin and vasopressin secretion from magnocellular neurosecretory cells of the hypothalamus (Rettori *et al.* 2009). Immunosuppression, important in transplant recipients, can be achieved by blocking the IL-2 receptor (Webster *et al.* 2010).

Interleukin 4 (IL-4)

IL-4 is produced by activated helper T cells, NK cells and macrophages and is best known for stimulating the proliferation and differentiation of B cells. It also acts as a growth factor for T cells, mast cells and bone marrow hematopoietic cells and may stimulate other cells of the immune system (Kindt *et al.* 2006).

Interleukin 6 (IL-6)

IL-6 is produced in a number of cell types, including helper T cells, B cells, fibroblasts and macrophages. IL-6 has a variety of biological effects, including the up-regulation of IL-2 receptors on T cells and the differentiation of B cells (Kindt *et al.* 2006). It has also been implicated in human tumor formation (Sansone and Bromberg 2012). IL-6 often acts synergistically with IL-1 and has many functions in common with IL-1. IL-6 is also secreted by the basal hypothalamus and follicular stellate cells of the anterior pituitary gland (Tierney *et al.* 2005; Jankord *et al.* 2007).

Interleukin 15 (IL-15)

IL-15 has many effects in the immune system, including stimulation of T cell proliferation, the generation of cytotoxic T lymphocytes, stimulation of immunoglobulin synthesis by B cells, and the generation and persistence of NK cells. It also stimulates myocytes and muscle fibers to accumulate contractile protein and is able to slow muscle-wasting in rats with cancer-related cachexia (Steel *et al.* 2012). It is in clinical trials for treatment of human tumors (Steel *et al.* 2012).

Interleukin 17 (IL-17)

This cytokine is particularly active in non-immune cells such as fibroblasts and epithelial cells and acts to prevent fungal and bacterial infection. However, IL-17 may also be involved in pathogenesis of rheumatoid arthritis, multiple sclerosis and asthma (Pappu *et al.* 2011).

Interleukin 37 (IL-37)

IL-37 acts as an anti-inflammatory cytokine; that is, it reduces the formation of other cytokines, such as IL-1, and inhibits macrophage and T cell numbers (McNamee *et al.* 2011). In animal experiments, it reduced intestinal inflammation, and may be useful in treating human inflammatory bowel disease.

13.3.3 Other cytokines

Tumor necrosis factor α (TNFα)

TNFα is a proinflammatory cytokine produced chiefly by activated macrophages and T cells (Kindt *et al.* 2006). It is implicated in the cause of inflammation in rheumatoid arthritis, and TNFα antibodies are undergoing clinical trials for the treatment of this disease (Maini and Taylor 2000; Aggarwal *et al.* 2012). TNFα also induces tumor regression and has been implicated in a long list of diseases, including Alzheimer's, Parkinson's, multiple sclerosis and diabetes (Aggarwal *et al.* 2012). It stimulates the release of hypothalamic hormones (e.g. CRH) as well, and may act directly on the brain to suppress feeding (Plata-Salaman 2001). TNFα also mediates endotoxic shock, and induces fever, both by acting directly on the hypothalamus and by stimulating the release of IL-1 (Turrin and Plata-Salaman 2000).

Oncostatin M (OSM)

OSM is released from activated macrophages and T cells. This cytokine inhibits the growth of several types of tumor, including lung, breast and brain (Tanaka and Miyajima 2003). It may also act as an anti-inflammatory agent and promote wound healing and bone formation (Wahl and Wallace 2001). It reduces tissue destruction that occurs in rheumatoid arthritis.

13.4 The functions of cytokines in the immune and hematopoietic systems

Cytokines are important in the activation of T cells, B cells and macrophages, the production and maturation of red blood cells, the killing of infectious agents and tumor cells and in producing the inflammatory response initiated by infection and trauma (Sompayrac 2012; Kindt *et al.* 2006).

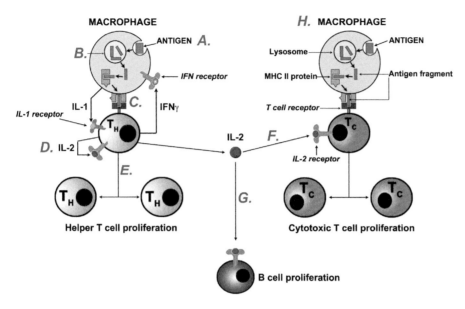

Figure 13.5 Response to antigens and the release of IL-1, IL-2 and IFNγ by helper T cells
Macrophages act as antigen presenting cells (APC). They phagocytose antigens (A) and process them into
smaller fragments within lysosomes (B). Helper T cells (T_H) are activated by APC cells that present antigen
fragments along with class II MHC proteins (major histocompatibility complex proteins) to the T cell receptors
(C). T_H cells respond by releasing IFNγ, which promotes MHC expression and stimulates IL-1 release from the
macrophages. IL-1 stimulates T_H cells to release IL-2, which up-regulates IL-2 receptors on the T_H cell (D). IL-2
also acts in an autocrine fashion to stimulate T_H cell proliferation (E), and stimulate the proliferation of cytotoxic
T cells (T_C) (F). IL-2 also activates B cells (G) to produce antibodies (immunoglobulins) (see Figure 13.6). *Note*:
T_C cells are also activated by macrophages (H) in the same way as T_H cells, via an MHC II protein.

13.4.1 T cell activation

T cells are stimulated to multiply when they are activated by antigen presenting cells and
several interleukins plus IFN (Sompayrac 2012; Kindt *et al.* 2006).

The mechanism involved in T cell activation is complex and may best be understood with
reference to Figure 13.5. In brief, since T cells cannot recognize antigens, antigen-
presenting cells (APC) break down the antigen and "present" the fragments to the T cells.
APC are normally monocytes and macrophages. They destroy the antigens by phagocytosis
and digestion by lysosomes and then present the resulting antigen fragments, bound to
major histocompatibility complex (MHC) proteins (which are self-recognition markers), to
receptors on helper T cells (T_H) and cytotoxic T cells (T_C). The antigen presenting macrophages release interleukins such as IL-1, which stimulate the helper T cell (Figure 13.5). The
helper T cell has receptors that recognize and bind the antigen-MHC receptor complex, as
well as interleukin receptors for IL-1 and IL-2. Stimulation of the T cell receptors by the
antigen-MHC complex, combined with the stimulation of interleukin receptors, activates
the synthesis of, for example, IL-2 and the up-regulation of IL-2 receptors on the surface of
the helper T cells. IL-2 stimulates the proliferation of helper T cells by autocrine action
and the proliferation of cytotoxic T cells by endocrine/paracrine action. IL-2, released from
T cells, also stimulates B cells (see below, section 13.4.2; and Figure 13.6). Helper T cells also
release interferon γ, which stimulates IL-1 synthesis and increased expression of the MHC

proteins in the APC, thus increasing the efficiency of antigen presentation in macrophages. When the antigen is destroyed, the T cells are no longer stimulated, so they stop producing IL-2, the IL-2 receptors are down-regulated, and T cell proliferation stops (Royer and Reinherz 1987).

13.4.2 B cell activation

Cytokines secreted from macrophages and especially activated helper T cells stimulate the activation, proliferation and differentiation of B cells as depicted in Figure 13.6 (Mitchison 2004; Kindt *et al.* 2006). IL-1 and IL-4 activate the resting B cell, causing the synthesis and release of IL-4 and the up-regulation of cytokine receptors. IL-4 also promotes B cell proliferation via autocrine action. The activated B cells then differentiate under the influence of several cytokines, including IL-2, IL-4, IL-5 and IL-6, as shown in Figure 13.6. The final step in this transformation of resting B cells is the appearance of plasma cells that are responsible for producing antibodies against the antigen fragments.

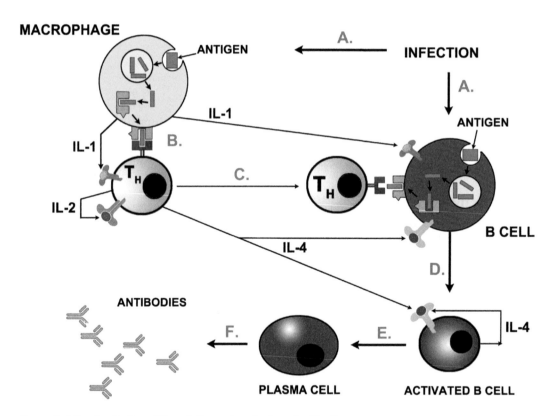

Figure 13.6 B cell activation, differentiation and production of antibodies
Resting B cells and macrophages are both activated by antigens (A; see also Figure 13.5). Helper T cells (T$_H$) are activated as shown (B) and then these activated T$_H$ cells bind to B cells through a class II MHC protein to activate them (C). This is known as an "immune synapse." Once activated (D), B cells produce their own IL-4, which has an autocrine action to up-regulate its own receptor. B cell proliferation (E) and differentiation into antibody-releasing plasma cells (F) is then regulated by a number of cytokines from T$_H$ cells, including IFNγ, IL-2, IL-4, IL-5 and IL-6.

13.4.3 Activation of macrophages

Resting macrophages are activated by antigens and release various cytokines, which stimulate the T_H cell to release interleukins and interferon γ. Interferon γ, and interleukins such as IL-4 and IL-10, then further activate the macrophage (Figure 13.5). Once activated, the macrophage begins to function in phagocytosis and tumor cell killing and releases a number of cytokines, including IL-1, IL-8 and TNFα (Sompayrac 2012; Kindt *et al.* 2006).

13.4.4 The production and maturation of blood cells (hematopoiesis)

An array of cytokines is required for the normal production and maturation of blood cells in the bone marrow (Sompayrac 2012; Kindt *et al.* 2006). Examples of these are shown in Figure 13.7. Multipotential stem cells in the bone marrow are the precursors for all of the different types of blood cells (i.e. they are hematopoietic precursor cells). When stimulated by colony stimulating factors (CSFs), these multipotential stem cells develop into colonies of specific blood cell types. Some of the interleukins, for example IL-3, IL-6 and IL-7, directly involved in hematopoiesis, are included in Figure 13.7. In addition, there are several CSFs: granulocyte-macrophage colony stimulating factor (GM-CSF); granulocyte colony stimulating factor (G-CSF) and macrophage colony stimulating factor (M-CSF). A complete description of cytokine involvement in hematopoiesis is beyond the scope of this book, but additional information may be found elsewhere (Sompayrac 2012; Kindt *et al.* 2006; Szilvassy 2003).

One of the responses of the immune system to antigens is the stimulation of bone marrow cells to increase their production of blood cells. In response to antigen stimulation, activated macrophages and helper T cells produce CSFs, IL-1 and TNF. These latter two cytokines then stimulate further production of CSFs from fibroblasts and other immune cells and these CSFs increase blood cell production.

13.4.5 Cytotoxicity

Some cytokines, such as TNF and lymphotoxin, kill infectious agents and tumor cells directly. Others act indirectly, by activating the cytotoxicity of other cells. For example, IFNγ activates cytotoxic T cells; IFNγ and IL-2 activate tumor killing in macrophages, and IFNγ, IL-1 and IL-2 facilitate the cytotoxic activity of NK cells.

13.4.6 Inflammatory response

Cytokines IL-1, IL-6 and TNFα are important mediators of the inflammatory responses of tissue to infection or trauma. IL-1, IL-6 and TNFα are produced by macrophages in response to a variety of toxic stimuli (e.g. bacterial endotoxins such as lipopolysaccharide) and are released into the circulation to activate target cells throughout the body and brain. Both cytokines facilitate local inflammation and wound healing and act on the brain to stimulate ACTH release, cause fever and induce sleep, all of which are adaptive responses to infection. TNFα also mediates inflammatory responses, but may do so indirectly, by stimulating the release of IL-1, prostaglandins or corticosteroids, which then act on their receptors to modulate inflammatory responses (Sompayrac 2012; Kindt *et al.* 2006).

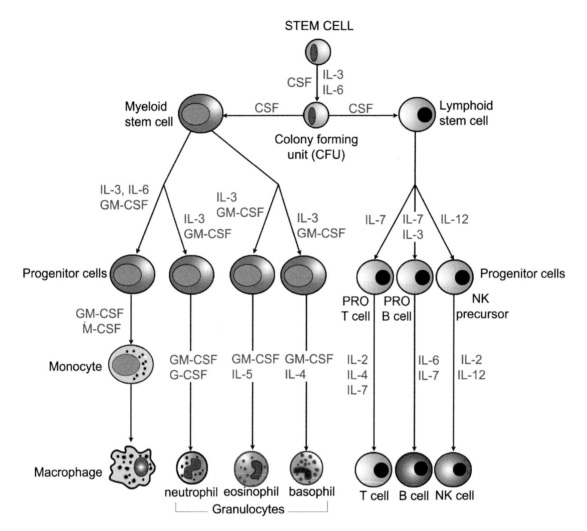

Figure 13.7 The role of cytokines in some stages of hematopoiesis
Multipotential (also called pluripotent) stem cells in the bone marrow respond to IL-3 and IL-6, plus colony stimulating factor (CSF), to produce more specific stem cells that give rise to myeloid cells (e.g. macrophages) and lymphoid cells (e.g. T cells). Myeloid stem cells receive additional instructions from the cytokines shown to give rise to specific progenitor cells that eventually produce an array of white blood cells. For example, monocytes/macrophages develop from a progenitor cell under the influence of macrophage colony stimulating factor (M-CSF) and granulocyte-macrophage colony stimulating factor (GM-CSF). Similarly, lymphoid stem cells respond to IL-7, IL-2 and IL-4 to produce T cells.

13.5 Effects of cytokines and other immunomodulators on the brain and neuroendocrine system

As shown in Figure 13.8 (see also Figure 1.2), there is bidirectional communication between the cells of the immune system and the brain and neuroendocrine system through the production of cytokines. Cytokines influence the brain and pituitary directly, and pituitary and target cell hormones (e.g. cortisol, estradiol, ACTH, LH, etc.)

affect immune cells via feedback (Petrovsky 2001). Immune cells also release several peptide hormones such as LH, GH, ACTH – normally seen as pituitary hormones – but these very likely influence immune cells (via autocrine/paracrine stimulation) rather then enter the bloodstream to affect other endocrine glands (Blalock 1989; Pallinger and Csaba 2008). The messengers of the immune system, including the cytokines, peptide hormones and thymosins, are conveniently termed *immunomodulators* (Plata-Salaman

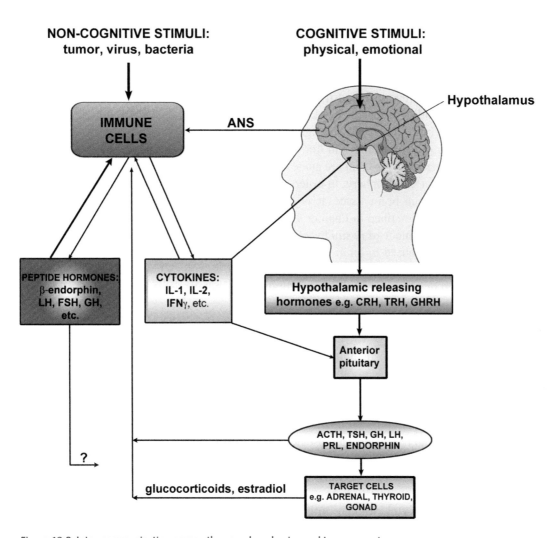

Figure 13.8 Inter-communication among the neural, endocrine and immune systems

The immune system is sensitive to non-cognitive stimuli, such as bacteria and viruses. Upon activation, cells of the immune system secrete cytokines and peptide hormones such as β-endorphin. The cytokines act on the brain and neuroendocrine system as well as on other immune cells. Peptide hormones released from the immune cells have autocrine and paracrine effects on immune cells. It remains unclear whether they enter the bloodstream in high enough concentration to affect endocrine cells. Perception of cognitive stimuli by the brain can also result in stimulation of the cells of the immune system by the peripheral nerves of the autonomic nervous system or through the activation of the neuroendocrine system. In this way, the brain can indirectly perceive the presence of bacteria and viruses. One of the functions of the immune system, in fact, may be to make the brain aware of these stimuli through the release of cytokines.

1989) because they have modulator rather than neurotransmitter actions on the brain and (neuro)endocrine system. Before examining the neuromodulatory effects of cytokines, the localization of cytokine receptors in the brain and pituitary gland will be discussed.

13.5.1 Localization of cytokines and their receptors in the brain

Previous sections have concentrated on discussing the production of cytokines in peripheral immune cells, such as lymphocytes, and the effects of these cytokines on the brain and neuroendocrine system. However, there is evidence that cytokines are also produced by brain cells as a result of pro-inflammatory processes (Srinivasan *et al.* 2004; Juttler *et al.* 2002). This possibility will be discussed briefly here before focusing on those cytokines that enter the brain from the bloodstream.

The availability of specific cytokine antibodies permitted the localization of individual cytokines within brain cells throughout the brain (e.g. see section 11.5 describing the localization of neuropeptides in brain tissue). Nonetheless, this technique cannot distinguish between cytokines that are made in the brain and those molecules that have entered the brain from the periphery. In order to prove conclusively that a particular cytokine is biosynthesized in brain tissue, it is necessary to localize and quantify cytokine gene expression. As described in Chapter 9 (section 9.2; Figure 9.5), this can be accomplished through the technique of *in situ* hybridization. For example, IL-6 mRNA and IL-6 receptor mRNA were localized to brain and pituitary tissue (Kurotani *et al.* 2001; Schobitz *et al.* 1993), firmly establishing that IL-6, and its receptor, is made by brain cells. A more detailed description of cytokine localization in brain and pituitary tissue can be found elsewhere (Turnbull and Rivier 1999).

Another example is the detection of TNFα and IL-1 mRNA following brain injury (Bruccoleri *et al.* 1998). The involvement of cytokines in mediating brain damage is an important field of investigation. In situations where brain tissue is injured, or when blood flow is compromised (e.g. during a stroke), cytokines such as IL-1, IL-6 and TNFα are released by brain cells (Skinner *et al.* 2009). This represents an inflammatory response analogous to that taking place in the peripheral immune system. There is evidence that blocking the effects of IL-1 following brain injury may be useful in the treatment of stroke, Alzheimer's and Parkinson's disease (Rothwell 2003).

Since cytokines are large polypeptides and proteins, it is unlikely that cytokines circulating in the blood could have effects in the brain. They would need to penetrate the blood-brain barrier that normally restricts access of such big molecules. However, like peptides such as insulin, cytokines may enter the brain through the *circumventricular organs* (see Figure 11.12), where there is no blood-brain barrier (Boron and Boulpaep 2005). In addition, Banks *et al.* (2002) showed that many cytokines, including IL-1, can be actively transported across the blood-brain barrier into and out of the brain. Thus, the brain is an accessible target for circulating cytokines.

Receptors for many cytokines are located throughout the brain (Turnbull and Rivier 1999). IL-1 receptors, for example, have their highest density in the olfactory bulbs, dentate gyrus, hippocampus, hypothalamus and choroid plexus of the rat brain. Moreover, these receptors are found on both neurons and astrocytes, as well as on brain tumor cells. Cytokine receptors are highly diverse and consist of multiple subtypes. For example, there are at least nine IL-1 receptor subtypes (Dinarello 2011), 13 for tumor necrosis factor (Croft *et al.* 2012) and three for the interferons (de Weerd and Nguyen 2012). The structures of cytokine receptors are of the general type shown in Figure 13.9. This receptor type was previously discussed in Chapter 10 for the leptin receptor (see Figure 10.6). Several examples of interleukin receptors are discussed in Rochman *et al.* (2009) and Javed *et al.* (2010) and interferon receptors by de Weerd and Nguyen (2012). IL-8 is an exception and it binds to a G-protein-coupled receptor (see, e.g., Figure 10.4). IL-8 therefore signals through the cAMP system, whereas most cytokine receptors use a variety of signaling molecules such as STAT1 and STAT5 as shown in Figure 13.9.

13.5.2 Neuromodulator effects of cytokines

There is growing evidence that cytokines, like neuropeptides, have a variety of neuromodulator functions. For example, IL-2 modulates neuronal firing rate, release of neurotransmitters, sleep, arousal and locomotion (Hanisch and Quirion 1995) and IL-1β is implicated in defensive-aggressive behavior (Pesce *et al.* 2011). Peripheral infection may lead directly to sickness behavior (e.g. reduced appetite; pain; fatigue; depression) via release of cytokines (Dantzer *et al.* 2008; Glaser and Kiecolt-Glaser 2005). Several specific examples of neuromodulator effects are given below.

Neuromodulator actions of IL-1

Peripheral and central administration of IL-1 induces norepinephrine release in the brain, particularly in the hypothalamus. Small changes in brain dopamine are also observed, but these effects are not regionally selective. IL-1 also increases brain concentrations of tryptophan, and the metabolism of serotonin (5-HT), throughout the brain (Dunn 2006). Part of this effect on serotonin neurons is via an inhibition of neuron firing rate (Brambilla *et al.* 2007). The effects of IL-1 modulation of hypothalamic function include induction of fever and slow wave sleep, the inhibition of feeding and the production of "sickness" behavior, with reduced exploration (Turrin and Plata-Salaman 2000; Krueger 2008).

Neuromodulator actions of IL-2

A detailed review by Hanisch and Quirion (1995) outlines the extensive literature on the central effects of IL-2, which include modulation of neurotransmitter release and behavior, growth of neurons and electrical activity of neurons. For example, IL-2 reduced locomotor behavior and increased sickness behavior in mice (Sudom *et al.* 2004). These effects seem to be mediated by changes in serotonin and norepinephrine turnover and the induction of IL-1 and TNFα in cerebral cortex and hippocampus (Anisman *et al.* 2008).

Neuromodulator action of IL-6

IL-6, which is also made in the brain, has neuroprotective properties and can prevent apoptosis (programmed cell death) of neurons (Juttler *et al.* 2002). IL-6 also inhibits glutamate release and prevents the spread of excitation in the cerebral cortex (D'Arcangelo *et al.* 2000). Inhibition of neuronal excitation could also be due to an up-regulation of adenosine receptors and a modulation of neuronal membrane potentials (Juttler *et al.* 2002).

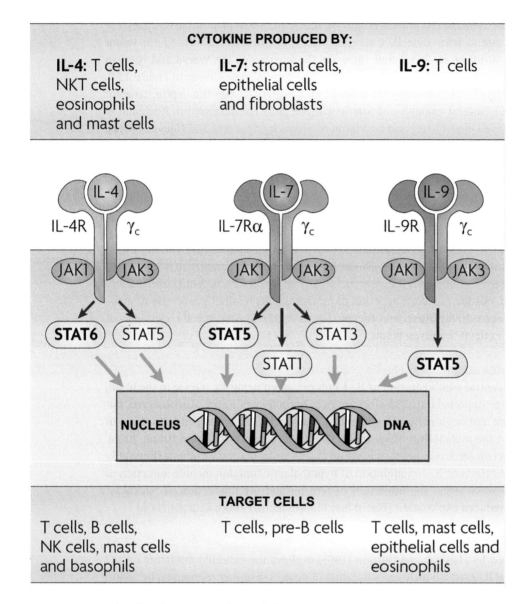

Figure 13.9 Examples of cytokine receptors for interleukins

Illustrated are the receptors for IL-4, IL-7 and IL-9. Each receptor consists of a dimer made up of a specific membrane receptor protein (e.g. IL-4R for IL-4), and a cytokine receptor γ chain (γc), common to all three receptor complexes. Binding of the interleukin to the receptor activates (phosphorylates) Janus kinases (JAK1 and JAK3), which in turn activate several *signal transducer and activator of transcription* (STAT) proteins. The phosphorylated STAT proteins enter the cell nucleus and induce changes in gene expression in the target cells (e.g. T cells) (see also Figure 10.6).

   Abbreviations: JAK, janus kinase; NKT, natural killer T cell; STAT, signal transducer and activator of transcription. Reproduced with permission (Rochman *et al.* 2009).

Neuromodulator effects of IL-10

Like IL-6, IL-10 also exerts neuroprotective effects in the brain. For example, the hippocampus is sensitive to hypoxia, causing cell death. This process is ameliorated by IL-10, probably by preventing the increase in intracellular calcium ion release (Turovskaya *et al.* 2012). IL-10 also prevents the hyperexcitability in hippocampal neurons that normally follows from a period of hypoxia (Levin and Godukhin 2011). In contrast to the known involvement of IL-1 in the production of fever (Lane and Lachmann 2011), IL-10 has antipyretic activity. Thus, injection of lipopolysaccharide (LPS) reliably induces fever in experimental animals and this can be prevented by IL-10, which reduces glutamate and prostaglandin levels in the hypothalamus (Kao *et al.* 2011).

13.5.3 Effects of cytokines in the neuroendocrine system

The endocrine significance of the cytokines is widely acknowledged (Turnbull and Rivier 1999; Padgett and Glaser 2003; Barrell 2007). Cytokines modulate the neuroendocrine system by influencing the hypothalamus, as well as by acting on the pituitary, adrenal glands and gonads to modulate hormone release. Cytokines also influence thyroid function, but probably via alterations in pituitary secretion of TSH. Insulin secretion from the pancreas is reduced by cytokines and this pathway may be important in the etiology of Type 2 diabetes (Larsen *et al.* 2007).

This section also summarizes some of the effects of cytokines and thymosins on the release of the hypothalamic and pituitary hormones. An abbreviated outline of this information is provided in Table 13.2. Some references refer to effects in the hypothalamus (e.g. on releasing hormones), whereas others focus on cytokine effects on the pituitary gland. It remains possible that cytokines sometimes have effects at both sites. It is also

Table 13.2 Effects of cytokines and thymosins on release of pituitary hormones

	GH	PRL	ACTH	TSH	LH
Cytokines:					
IFNγ	↑[4]	↑[2]	↑[3]	↑[12]	0[2]
IL-1	↑[6]	↑[8]	↑[3]	↓[10]	↓[1]
IL-2	0[7]	0[11]	↑[3]	↑[14]	↓[11]
IL-6	↑[5]	↑[15]	↑[3]	↓[13]	↓[9]
TNFα	0[7]	↓[16]	↑[3]	↓[10]	↓[1]
Thymosins:					
Thymulin[17]	↑	↑	↑	↑	↑
Tα1[18]	–	↓	↓	↓	–
Thymosinβ4	–	–	–	–	↑[18]

References: 1. Watanobe and Hayakawa 2003; 2. Cano *et al.* 2005; 3. Turnbull and Rivier 1999; 4. Then Bergh *et al.* 2007; 5. Nemet *et al.* 2006; 6. Gong *et al.* 2005; 7. Fry *et al.* 1998; 8. McCann *et al.* 2000; 9. Russell *et al.* 2001; 10. Wassen *et al.* 1996; 11. Umeuchi *et al.* 1994; 12. Nagai *et al.* 1996; 13. Torpy *et al.* 1998; 14. Karanth and McCann 1991; 15. Tsigos *et al.* 1997; 16. Harel *et al.* 1995; 17. Reggiani *et al.* 2011; 18. Goldstein and Badamchian 2004.

important to remember that many of the reports on effects of cytokines on the neuroendo-crine system are contradictory. This reflects the many different species, drug dosages, routes of administration, etc. used in experiments from different laboratories.

Interferon

Interferon, given by intracerebroventricular or peripheral injection, elevates CRH, ACTH and glucocorticoid release (Turnbull and Rivier 1999). Peripheral injection of interferon increases PRL secretion, but has no effect on LH (Cano *et al.* 2005). GH is also increased by intramuscular injection (Then Bergh *et al.* 2007) and TSH by intravenous injection (Nagai *et al.* 1996).

Interleukin 1

IL-1 stimulates secretion of GH, PRL and ACTH, but inhibits release of TSH and LH. The effect of IL-1 on GH is at the level of the pituitary (Gong *et al.* 2005), whereas PRL secretion is mediated via the hypothalamus (McCann *et al.* 2000). The regulation of ACTH secretion by IL-1 was investigated by injections into the brain as well as the periphery (Turnbull and Rivier 1999). This represents the most intensively studied of all the effects of interleukins on the neuroendocrine system. Figure 13.10 illustrates that IL-1 stimulates the release of CRH from neurons located in the paraventricular nucleus (PVN) of the hypothalamus. CRH, in turn, induces ACTH secretion from the anterior pituitary (Turnbull and Rivier 1999; Lightman and Conway-Campbell 2010). IL-1 regulates CRH release by entering the brain through three circumventricular organs (see pathways [1] in Figure 13.10); that is, median eminence, area postrema (AP) and the organum vasculosum of the lamina termi-nalis (OVLT). IL-1 entering through the OVLT induces the synthesis of prostaglandins (PGs; pathway [2]) which regulate CRH release. IL-1 entering through the median emi-nence acts specifically on CRH neuron terminals to enhance CRH secretion (pathway [3]). The AP and the nucleus tractus solitarius (NTS) respond to IL-1 by mobilizing brain catecholamines that stimulate PVN neurons (pathway [4]). There is also evidence that IL-1 exerts a direct effect on ACTH secretion from the anterior pituitary gland (pathway [5]; Turnbull and Rivier 1999). Whatever its mechanism of action, IL-1 signals to the brain that the immune system is being activated, by an infection for example, and triggers a neuroendocrine stress response. The elevation of glucocorticoid levels (pathway [6]) by IL-1 during this stress response provides a normal negative feedback mechanism to limit glucocorticoid output (see, e.g., Figure 4.10). However, elevations in glucocorticoid secre-tion inhibit the immune response and down-regulate IL-1 production (pathway [7]). This effect makes people open to opportunistic infections during stressful events, but on the other hand is a powerful approach to treat inflammatory disease and to prevent graft rejection (Franchimont 2004).

Interleukin 2

IL-2, like IL-1, enters the brain via the CVOs and also acts on the CRH system, elevating ACTH and β-endorphin release from corticotrophs in the anterior pituitary gland (Turnbull and Rivier 1999). IL-2 may do this via activation of norepinephrine systems in the PVN and median eminence (Lacosta *et al.* 2000). The inhibitory effects of IL-2 on LH secretion were revealed by injection of IL-2 directly into the brain, whereas the same molecule *stimulated*

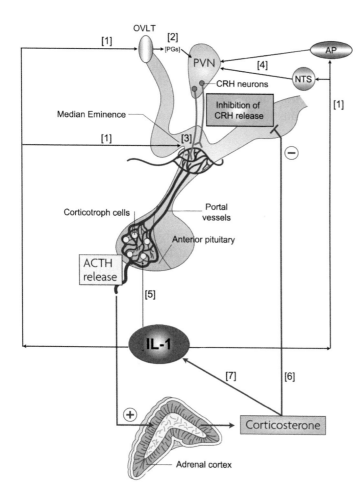

Figure 13.10 IL-1 induces CRH secretion
IL-1 enters the brain from the bloodstream through circumventricular organs (1); these are the OVLT, the area postrema and the median eminence. IL-1 stimulates prostaglandin (PG) synthesis to regulate CRH secretion (2). Prostaglandin also induces fever. The action of IL-1 at the median eminence (3) is to induce CRH secretion from the CRH neuron terminals. The action of IL-1 at the AP, and also the nucleus tractus solitarius (NTS), probably involves an intermediate step of catecholamine release (4). An additional stimulatory influence of IL-1 on ACTH secretion is exerted directly on cortico-troph cells in the anterior pituitary (5). Glucocorticoids, in this case corticosterone, exert negative feedback to the hypothalamus to inhi-bit CRH release (6), but also have an inhibitory effect on the immune system as well, reducing secretion of IL-1 from immune cells (7).

Abbreviations: ACTH, adrenocorticotropic hormone; CRH, corticotropin-releasing hormone; OVLT, organum vasculosum of the lamina terminalis; PVN, paraventricular nucleus. Data obtained form Turnbull and Rivier (1999) and figure reproduced with permission from Lightman and Conway-Campbell (2010).

LH secretion when incubated *in vitro* with pituitary cells (Umeuchi *et al.* 1994). Similar experiments showed that very low concentrations of IL-2 (10^{-15} M) increased TSH release from the pituitary gland (Karanth and McCann 1991). This result emphasizes that low circulating concentrations of IL-2, and probably other cytokines as well, can influence pituitary hormone secretion without entering the brain.

Interleukin 6

Like IL-1 and IL-2, IL-6 appears to act in the hypothalamus to modulate CRH release, and it also stimulates ACTH release directly from the anterior pituitary (Turnbull and Rivier 1999; Karanth and McCann 1991). In contrast, IL-6 inhibits LH secretion from the pituitary (Russell *et al.* 2001). Several studies in humans revealed that IL-6 releases GH and PRL following intravenous or subcutaneous injections (Nemet *et al.* 2006; Tsigos *et al.* 1997), whereas TSH is inhibited (Torpy *et al.* 1998). There is evidence that IL-6 may increase glucocorticoid secretion by a direct effect on the adrenal gland, although Turnbull and Rivier (1999) conclude that this may not be physiologically relevant.

Tumor necrosis factor

In keeping with the effects of other cytokines on the CRH/ACTH axis, TNF also elevates CRH, ACTH and β-endorphin secretion (Turnbull and Rivier 1999). However, it has no effect on GH release (Fry *et al.* 1998), but inhibits TSH, LH and PRL. TNF lowers LH secretion via an inhibitory influence on GnRH release from the preoptic area of the hypothalamus (Watanobe and Hayakawa 2003), but inhibits PRL secretion through a direct effect on the pituitary (Harel *et al.* 1995).

Thymosins

Thymulin is the best characterized of all the thymosins, especially in terms of its action on the pituitary. Thymulin stimulates release of ACTH, LH, GH, TSH and PRL (Reggiani *et al.* 2011). In contrast, thymosin β4 increases LH secretion when injected into the brain, but has no effect at the pituitary (Goldstein and Badamchian 2004). Also, a mixture of thymosins, called TF5 (thymosin fraction 5), increased blood levels of ACTH, β-endorphin and cortisol following peripheral injection in monkeys (Figure 13.11A) (Healy *et al.* 1983). There was no effect on GH, LH, PRL and TSH secretion. In further experiments, surgical removal of the thymus gland had the opposite effect to TF5 and *decreased* plasma ACTH, cortisol and β-endorphin levels in these monkeys (Figure 13.11B). This result suggests that the thymus contains a peptide, or peptides, that regulates secretion of ACTH. Mutant mice that lack a thymus (called the "nude" mouse) have unusually low levels of LH, GH and PRL (Savino *et al.* 1999). Whether this occurs via an influence on the brain or pituitary gland (or both) remains to be determined (Savino *et al.* 1999; Goya *et al.* 1999). Localization of thymosin receptors would be helpful, but there is little information available, although thymosin α1 binds to hypothalamic cells (Turrini and Aloe 1999).

Since thymosins are important regulators of pituitary hormone secretion, it follows that hormone release from target tissues such as the gonads and adrenal gland will also be affected. This is seen, for example, following removal of the thymus gland in experimental animals, and especially in animals such as the nude mouse where the thymus is absent (Reggiani *et al.* 2011; Hall *et al.* 1985).

13.5.4 The production of peptide hormones by cells of the immune system

Many cells of the immune system express the genes for peptide hormones that are identical to those produced in the hypothalamus and pituitary gland. For example, the thymus contains oxytocin and vasopressin (Blalock 1989) and human white blood cells contain ACTH and β-endorphin (Pallinger and Csaba 2008). Note that the detection of peptides in immune cells, for example by immunohistochemistry, might represent peptides made in the cell or may indicate uptake from the blood. For this reason, Table 13.3 includes references to the localization of both gene expression and peptide localization wherever possible.

Peptides are released from immune cells in response to immune stimulation from both antigens and hormones. For example, viral infections stimulate ACTH and β-endorphin release from lymphocytes, macrophages and spleen cells. Most cells of the immune system also have receptors for the peptides listed in Table 13.3. These receptors are discussed in many of the references provided in Table 13.3. The presence of receptors in immune cells suggests that immune cell peptides have autocrine/paracrine signaling activity.

Table 13.3 Peptide hormones in cells of the immune system

Cells	Hormones released	References
Thymus	Oxytocin, Vasopressin, Enkephalin, β-endorphin	1, 2, 3, 4
Lymphocytes	CRH, ACTH, β-endorphin, GH, Prolactin, GnRH, Ghrelin	5, 6, 7, 8, 9, 16
Monocytes	ACTH, β-endorphin, LH	7, 13, 15
Macrophages	Substance P, Enkephalin, β-endorphin	10, 11, 14
Leukocytes	β-endorphin, LH	12, 15
Spleen Cells	ACTH, GnRH, Enkephalin, CRH	3, 16, 17

References: 1. Blalock 1989; 2. Hansenne *et al.* 2005; 3. Assis *et al.* 2006; 4. Kvetnoy *et al.* 2003; 5. Muglia *et al.* 1994; 6. Sitte *et al.* 2007; 7. Csaba *et al.* 2009; 8. Hattori *et al.* 2001; 9. Gerlo *et al.* 2005; 10. Hino *et al.* 2009; 11. Machelska 2011; 12. Stein and Lang 2009; 13. Mousa *et al.* 2007; 14. Douglas *et al.* 2008; 15. Hotakainen *et al.* 2000; 16. Weesner *et al.* 1997; 17. Lyons and Blalock 1997.

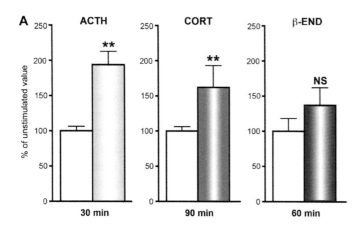

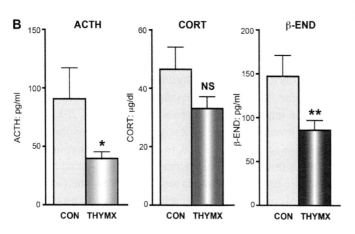

Figure 13.11 The regulation of the hypothalamic-pituitary-adrenal system by the thymus gland

A. Maximum increases in plasma ACTH, cortisol and β-endorphin levels in monkeys given an intravenous injection of either saline (white bars), or 10.0 mg/kg of thymosin fraction 5 (TF5; blue bars). Values are plotted as percentage changes compared to saline controls and are means ± SEM (n = 8). NS = not significant; **p < 0.02.
B. Plasma ACTH, cortisol and β-endorphin levels are reduced in monkeys that had their thymus surgically removed (THYMX; green bars) compared to control monkeys (CON; gray bars). Values are means ± SEM (n = 6). NS = not significant; *p < 0.05; **p < 0.02. Graphs were plotted using data from Healy *et al.* (1983).

13.6 Neural and endocrine regulation of the immune system

There is good evidence that brain injury adversely affects the immune response (Besedovsky and del Rey 1996; Quan and Banks 2007; Anthony *et al.* 2012). For example, lesions of the anterior hypothalamus decrease the proliferation of T cells and NK cells (Cross *et al.* 1984). Lesions of the cerebral cortex also alter immune function and, to complicate matters, lesions of the left cortex produce different effects on the immune system than lesions of the right cortex (Neveu 1988; Barnéoud *et al.* 1988). Brain damage in humans – for example, as a result of a stroke – causes immune depression and a high risk of infection, a leading cause of death in such patients (Klehmet *et al.* 2009).

This chapter has already outlined that the brain and immune systems communicate via cytokines and various neuroendocrine hormones. But, in addition, the CNS and the immune system are connected through innervation from sympathetic postganglionic neurons, although there is no neuroanatomical evidence for a parasympathetic or vagal nerve supply to any immune organ (see Figures 1.2 and 13.8; Nance and Sanders 2007). Thus, the primary pathway for the neural regulation of immune function is provided by the sympathetic nervous system (SNS) and its main neurotransmitter, norepinephrine. NE receptors are present on immune cells, together with those for 5-HT, substance P, VIP and histamine (Steinman 2004). For example, most white blood cells (e.g. T cells and B cells) possess serotonin receptors and serotonin can activate NK cells and induce proliferation of T cells (Ahern 2011). Changes in neurotransmitter levels in the hypothalamus and brain stem can also modify the responses of the immune system through changes in autonomic nervous system (ANS) activity, thymus gland stimulation and by altering neuroendocrine activity (Kordon and Bihoreau 1989). Drugs that alter neurotransmitter levels also alter immune responses. For example, serotonin agonist drugs suppress antibody responses, while serotonin antagonists facilitate antibody responses (Roszman and Brooks 1985). Serotonin may influence both T cells and B cells through its modulation of the neuroendocrine system; i.e. serotonin elevates ACTH levels (Locatelli *et al.* 2010) and ACTH/cortisol inhibits many immune system activities (see section 13.6.2.5, below). Serotonin also mediates the release of PRL, LH and GH (Table 6.1), and so may mediate immune responses by altering the release of these hormones (Table 13.3).

Catecholamines have wide-ranging effects on the immune system. Lymphocytes synthesize dopamine and also have DA receptor subtypes, suggesting that these cells may respond to DA released from the sympathetic nervous system or from lymphocytes (Buttarelli *et al.* 2011; Pacheco *et al.* 2009). DA has different effects depending on which receptor subtype is expressed. For example, D2-like receptors modulate T cell physiology by promoting enhanced production of interleukin-10, a cytokine that negatively regulates the function of effector T cells, whereas D1 receptor stimulation impairs T cell responses (Buttarelli *et al.* 2011). Since hypothalamic DA inhibits prolactin release, part of the action of DA on the immune system could also be mediated by changes in prolactin levels. NE and, to a smaller extent, epinephrine influence immune responses, by their action as hormones released from the adrenal medulla, by their action as neurotransmitters in the hypothalamus and brain stem, and following release from sympathetic nerves of the ANS. Intracerebroventricular injection of 6-hydroxy-dopamine (6-OHDA), a neurotoxin that decreases brain NE levels, impairs the production of antibodies in response to antigens (Roszman and Brooks 1985). As mentioned above, cholinergic innervation of the immune system appears to be absent.

Nevertheless, lymphocytes possess most components of the cholinergic system, including ACh, muscarinic and nicotinic ACh receptors and choline acetyltransferase (ChAT) for the biosynthesis of ACh, and ACh seems to regulate some aspects of the immune system (Kawashima and Fujii 2004). In other words, immune cells might release ACh to act as an autocrine/paracrine signal.

As outlined in Figure 13.8, the neural and endocrine systems can modulate the immune system in three different ways: (1) through the ANS; (2) through the release of hypothalamic and pituitary hormones; and (3) through the release of neuropeptides. Behavioral changes linked to the ANS (e.g. behaviors that alter body temperature or blood volume) or neuroendocrine system (e.g. changes in feeding behavior or sleep-wake cycles) may also alter immune responses.

13.6.1 ANS influences on the immune system

The ANS innervates all of the tissues of the immune system: bone marrow, thymus gland, spleen and lymph nodes (Nance and Sanders 2007). Cholinergic innervation is absent.

Adrenergic stimulation

Cells of the immune system use the $\beta2$-adrenergic receptor ($\beta2$-AR) to mediate the effects of NE, although $\beta1$-AR, $\alpha1$-AR and $\alpha2$-AR are also present (Marino and Cosentino 2013; Nance and Sanders 2007). Evidence for adrenergic sympathetic stimulation of cells in the immune system is substantial and there is also evidence that immune cells biosynthesize and release NE and epinephrine that would act as autocrine/paracrine signals (Marino and Cosentino 2013; Flierl *et al.* 2008). Extensive information is available on the effects of adrenergic stimulation of immune cells and a complete description is beyond the scope of this text (Marino and Cosentino 2013; Flierl *et al.* 2008). For example, T cells reduced their secretion of IL-2 when exposed to NE, suggesting that $\beta2$-AR stimulation affected the ability of activated T cells to expand in number (Nance and Sanders 2007). Stimulation of NK cells by NE/$\beta2$-AR inhibits their cytotoxic behavior, whereas in activated macrophages NE inhibits the production and secretion of TNF-α and IL-1. Adrenergic stimulation also regulates thymus function. For example, stimulation of α-AR reduces lymphocyte numbers in thymus and in spleen (Stevenson *et al.* 2001) and catecholamines are generally thought to be inhibitory to thymic lymphocyte production (Leposavic *et al.* 2008). Along with the catecholamines, the sympathetic branch of the SNS co-releases a number of neuropeptides, including VIP, CCK, NPY, somatostatin and others. These neuropeptides also regulate immune system function (see section 13.6.3).

Cholinergic stimulation

As noted, cholinergic innervation of the immune system appears to be absent, although lymphocytes synthesize and release ACh and possess receptors. It is therefore likely that ACh might regulate some aspects of the immune system (Kawashima and Fujii 2004). For example, ACh acts via nicotinic receptors to significantly attenuate the release of cytokines TNFα, IL-1, IL-6 and IL-10 from stimulated human macrophages (Borovikova *et al.* 2000; Wang *et al.* 2003). Because of the controversy surrounding the absence of cholinergic innervation, the origin of the ACh remains to be conclusively determined (Nizri and Brenner 2013;

Rosas-Ballina *et al.* 2011). Table 13.4 provides an overview of neural and hormonal effects of neurotransmitters and neuropeptides on the immune system (Petrovsky 2001).

13.6.2 Effects of hypothalamic and pituitary hormones on the immune system

The first indication that hypothalamic and pituitary hormones could modulate immune responses was the finding that lesions of the anterior hypothalamus, which interfered with

Table 13.4 Neuroendocrine and neurotransmitter / neuropeptide effects on immune function

Hormone	Cytokine / immune function
α-endorphin	Inhibits I_g production
α-MSH	Inhibits IL-1 and IL-2 production via inhibition of NF-κB
Acetylcholine	Stimulates T and NK cells and increases IFNγ production
ACTH	Inhibits IFNγ production and I_g production and blocks macrophage activation by IFNγ
Epinephrine	Inhibits IL-1 and IL-2 production
Angiotensin 2	Enhances IFNγ production
β-endorphin	Enhances IFNγ production and NK cell mediated cytotoxicity. Inhibits T cell proliferation
cAMP	Enhances IL-4 and IL-5 production. Inhibits IL-2 production
Calcitonin-gene-related peptide	Increases T cell adhesion and stimulates IL-2, IL-4 and IFNγ production
Catecholamines	Enhance Ig production. Decrease the number of T and NK cells in the peripheral circulation and inhibit NK cells
Cortisol	Inhibits IFNγ, IL-2, IL-6 and TNFα. Enhances IL-4 and TGF-β production. Enhances immune cell expression of IL-1, IL-2, IL-6 and IFNγ receptors
CRH	Activates macrophages. Inhibits IL-1 and IL-6 production
Growth hormone	Activates macrophages
Gonadotropin releasing hormone	Increases IL-2 R expression, T and B cell proliferation
Histamine	Inhibits IL-12, TNF and IFNγ and enhances IL-10 production
Inhibin	Inhibits IFNγ production
LH	Enhances IL-2 stimulated T cell proliferation
Macrophage inhibitory factor	Blocks glucocorticoid inhibition of T cell proliferation and cytokine production
Melatonin	Enhances IL-11, IL-2, IL-6 and IFNγ production
Met-enkephalin	Enhances antigen-specific proliferation
Nerve growth factor	Enhances B cell proliferation, IL-6 production, IL-2 receptor expression
Neuropeptide Y	Increases T cell adhesion and stimulates IL-2, IL-4 and IFNγ
Estradiol	Enhances T cell proliferation and activity of IFNγ promotor
Oxytocin	Enhances IFNγ production
PGE2	Inhibits IL-2 production

Table 13.4 (Cont.)

Hormone	Cytokine / immune function
Progesterone	Enhances IL-4 production
Prolactin	Enhances T cell proliferation, IFNγ, IL-2 receptor expression and macrophage function
Serotonin	Inhibits T cell proliferation and IFNγ induced MHC class II expression Enhances NK cytotoxicity
Somatostatin	Inhibits T cell proliferation and IFNγ production and macrophage action
Substance P	Enhances T cell proliferation and IL-1, IL-6, TNF
Testosterone	Enhances IL-10 production
TSH	Enhances IL-2, GM-CSF and I_g production
Thyroxine	Activates T cells
1,25 Vitamin D3	Inhibits IL-2 and IFNγ. Enhances IL-4 production
Vasopressin	Enhances IFNγ production
VIP	Inhibits T cell proliferation and IL-12. Enhances IL-5 and cAMP production

Abbreviations: ACTH, adrenocorticotropic hormone; cAMP, cyclic adenosine monophosphate; CRH, corticotropin-releasing hormone; IFN, interferon; I_g, immunoglobulin; IL, interleukin; MSH, melanocyte-stimulating hormone; NK, natural killer; PGE2, prostaglandin E2; TNF, tumor necrosis factor; TSH, thyroid-stimulating hormone; VIP, vasoactive intestinal polypeptide. The information in this table is adapted from that given in Petrovsky 2001.

regulation of the neuroendocrine system, modified the immune response (Cross *et al.* 1984; see above, section 13.6). Since this demonstration, all of the hypothalamic and pituitary hormones have been shown to modulate the immune system, as have the hormones of the adrenal cortex, thyroid gland and the gonads (Blalock 1989). The appropriate receptors are also found in the immune system (Weigent and Blalock 1987).

Neurohypophyseal hormones
Both oxytocin and vasopressin receptors are detectable in thymus tissue, and the peptides and their genes are also found there (Hansenne 2005). There is also evidence for vasopressin and oxytocin to have opposite autocrine/paracrine effects on lymphocyte number (Maccio *et al.* 2010). It remains unknown whether oxytocin or vasopressin could modulate the synthesis and secretion of thymosins. Oxytocin may also stimulate the release of interferon γ from cytotoxic T cells and act to regulate hormone production by the cells of the immune system (see section 13.7).

GH and prolactin
Growth hormone and prolactin have wide-ranging effects on the immune system (Meazza *et al.* 2004; Yu-Lee 2002). GH receptors are present in immune cells (e.g. T and B lymphocytes) and GH has several actions; for example, GH enhances thymic lymphocyte development, modulates cytokine production and stimulates B cell development (Hattori 2009). GH is also made and released from immune cells and has a direct effect on thymus function (Savino *et al.* 2002; Savino and Dardenne 2010). It enhances cytokine production, increases thymocyte proliferation and stimulates migration of lymphocytes from the gland

(Smaniotto *et al.* 2011). Dwarf mice, which lack GH, are also immunodeficient, and their immune system activity can be normalized by GH treatment (Duquesnoy and Pedersen 1981). Prolactin is made in T and B cells, and in thymocytes, which also possess receptors for PRL (Buckley 2001; De Mello-Coelho *et al.* 1998). PRL stimulates T cells, B cells, natural killer (NK) cells, macrophages and neutrophils, and increases the number of progenitors of other immune cell lineages of T cells, B cells and NK cells (see Figure 13.7) (Yu-Lee 2002). Prolactin and GH also restore thymic function and T cell activity in aged rats whose thymus gland has atrophied (Kelley *et al.* 1987; French *et al.* 2002). In contrast to its significance in the normal immune response, PRL has been implicated in the etiology of autoimmune diseases, which are more prevalent in women (Shelly *et al.* 2012).

The hypothalamic-pituitary-thyroid system

Thyroid hormones triiodothyronine (T3) and thyroxine (T4) modulate immune responses, including cell-mediated immunity, NK cell activity, the antiviral action of IFN, proliferation of T and B cells and production of cytokines (de Vito *et al.* 2011). TRH increases proliferation of thymocytes and splenocytes and may reverse the inhibitory effects of glucocorticoids on lymphocytes (Kruger 1996). This review also describes evidence that the immune system biosynthesizes and responds to TSH. TSH can enhance the effect of IL-2 on NK cells and regulate B cell activity depending on whether the cells are activated or not; that is, TSH has no effect on resting B cells, but stimulated proliferation of pre-B cells. TSH, TRH and thyroid hormone receptors have been identified in several immune cell types (Pallinger and Csaba 2008; Klein 2003; Raiden *et al.* 1995; Villa-Verde *et al.* 1992). The important role of thyroid hormones in thymus function is revealed following removal of the thyroid (thyroidectomy); that is, thymus weight decreases and the production of thymulin is inhibited. This is also seen in human hypothyroid patients whose thyroid hormone production is compromised (Kruger 1996).

The hypothalamic-pituitary-gonadal system

Although males and females have the same immune system in terms of specific cell populations and chemical signaling, the responses to toxins, pathogens and allergens are different. Females mount higher immune responses than do males, possibly producing enhanced susceptibility to inflammatory and autoimmune diseases in females when compared with males. One reason for this is because gonadal sex hormones, such as estradiol, progesterone and testosterone, regulate immune cell responses to immunological stimuli. Since males and females have different circulating sex hormones, the immune responses are therefore different in males and females (Klein 2012; Fragala *et al.* 2011). *Sex hormone receptors* are present in cells of the immune system (Munoz-Cruz *et al.* 2011; Kovats *et al.* 2010). For example, human and mouse T cells, B cells, NK cells, monocytes and macrophages possess ERα, ERβ, PR and AR receptors that bind estradiol, progesterone and testosterone, respectively. Hematopoietic progenitor cells (Figure 13.7) also possess ER and these cells are depleted by estradiol treatment (Kovats *et al.* 2010).

In the innate immune system (i.e. the immune response shown in Figure 13.5) where the host responds to pathogens such as bacteria or fungi, investigations in rodents, monkeys and humans reveal that macrophage, monocyte and NK cell numbers, and activity, are different in females compared to males (Melgert *et al.* 2010; Xia *et al.* 2009; Yovel *et al.*

2001). Detailed reviews of the effects of sex hormones on human immune cells are provided by Oertelt-Prigione (2012), by Kovats *et al.* (2010) and by Bouman *et al.* (2005). For example, T lymphocyte secretion of IL-2, TNFα and IFNγ is regulated in a biphasic manner by estradiol, and low doses are stimulatory. IL-4 and IL-10 are unaffected by estradiol, but IL-4 production was increased by progesterone. In B cells, the effect of estradiol is to increase the output of immunoglobulins (antibodies), whereas testosterone inhibits production. This is in keeping with the known increase in circulating antibodies in women, compared to men. Monocytes are also a target for sex hormones. Monocytes positive for IL-1, IL-12 and TNFα are found at higher levels in men than women, but in pregnant women monocyte IL-12 production is down-regulated, suggesting an inhibitory effect of estradiol and progesterone. NK cells are also sensitive to sex hormones. Numbers of NK cells are increased in situations where estradiol levels are low; for example, in post-menopausal women and in men. NK cell numbers were lower in fertile women whose sex hormone levels are expected to be high. Details of comparable sex hormone effects in cells from experimental animals are provided elsewhere (Ansar Ahmed *et al.* 1985; Munoz-Cruz *et al.* 2011; Klein 2012).

A potentially important aspect of estrogenic influence on the immune system relates to so-called *environmental estrogens* (also called *endocrine disrupting chemicals;* see Chapter 15). These substances are found in foods; for example, meat, eggs and dairy products from animals pretreated with relatively high concentrations of estrogens. There are *industrial estrogens* such as bisphenol A (BPA), found in plastic water bottles and food containers (Vandenberg *et al.* 2012). Estrogens are also ingested through medications (contraceptive pill, hormone replacement therapy). There is now much evidence in animal models, and in cell studies, that such compounds exert immunotoxic effects (Chighizola and Meroni 2012). Possible widespread effects of environmental estrogens on human health remain to be established (Clayton *et al.* 2011; Welshons *et al.* 2006; see also section 15.3).

Given the sex hormone sensitivity of the immune system, the human menstrual cycle provides a means to determine whether the natural, physiological variations in sex hormone levels influence immune responses (see Figure 8.2A for hormone levels through the menstrual cycle). Symptoms of diseases such as autoimmune diseases and asthma do worsen during the premenstrual period, probably due to the large fluctuations in blood levels of estradiol and progesterone that take place (Oertelt-Prigione 2012). A physiological variation in immune response could be important for embryo implantation and pregnancy to occur. This was suggested by Lee *et al.* (2010), who observed a reduction in circulating T cell levels in the luteal phase (high progesterone) compared to the follicular phase (high estradiol, low progesterone) of the cycle. They also saw variations in NK cell levels and cytotoxicity. This is an important field of study and interested students should consult more detailed reviews (see, e.g., Bouman *et al.* 2005).

As noted already (see section 13.2), sex steroids have a pronounced effect in the thymus and contribute to thymic involution and immune aging. For example, castration induces a complete reversal of age-related thymic atrophy and a significant improvement in bone marrow function that can be seen in animals and in men (Sutherland *et al.* 2005). The reverse of this is equally dramatic, in that injection of testosterone into male mice causes prompt thymic atrophy and aging of the immune system in terms of thymocyte function (Olsen *et al.* 1991). Such experiments have implications for the aging human immune system and immune systems that are compromised following, for example, radiation or chemotherapy treatments. Clinical studies are needed to examine the usefulness of sex hormone removal on immune system rejuvenation (Calder *et al.* 2011). Note, however, that a *physiological* role for thymic involution is seen during pregnancy. The high levels of estradiol and progesterone that occur during pregnancy induce a thymic atrophy that reverses itself after birth (Zoller *et al.* 2007). This compromised immune function may be necessary to prevent rejection of the fetus and to allow pregnancy to proceed.

The hypothalamic-pituitary-adrenal system

Stressors, or activators of the hypothalamic-pituitary-adrenal system, are known to make us more sensitive to infectious agents, to diminish our response to vaccines, to delay our recovery from tissue damage such as wounds and to induce depression (Cohen *et al.* 2007). Stressors may be psychological (e.g. academic examinations or unemployment) or caused by something more acute, such as a car crash or an earthquake. Whatever the cause, the brain instructs the adrenal gland to rapidly release glucocorticoids, and the adrenal medulla to secrete epinephrine and NE (see Figure 1.2). All of the hormones implicated in this response to stress – CRH, ACTH, glucocorticoids and catecholamines – have direct effects on cells of the immune system. Some of the effects are persistent and the resultant immune changes continue to be felt for some time (Glaser and Kiecolt-Glaser 2005). The influence of catecholamines on immune cells has been covered already (section 13.6.1) and the present section will primarily consider the role of glucocorticoids even though all three hormones of the hypothalamic-pituitary-adrenal system (CRH, ACTH and glucocorticoids) modulate immune system activity. For example, CRH is expressed in immune cells and induces proliferation of lymphocytes, stimulates macrophage activation, and stimulates cytokine secretion by monocytes and lymphocytes (Singh *et al.* 1990; Goetzl *et al.* 2008). ACTH receptors are present on T and B cells (Johnson *et al.* 2001) and ACTH has been identified in leukocytes (Pallinger and Csaba 2008). This neuropeptide enhances lymphocyte proliferation and increases IL-2 receptors (Gonsalkorale *et al.* 1995). B cells are also increased by ACTH (Blalock 1989) and IFNγ release from cytotoxic lymphocytes is elevated (Johnson *et al.* 2005).

Glucocorticoids have a wide-ranging influence on the immune system and this is mediated through glucocorticoid receptors. Recall that glucocorticoids bind to two distinct receptors, mineralocorticoid (MR) and glucocorticoid (GR) (section 9.7). The cells of the immune system bind glucocorticoids at the GR site (Padgett and Glaser 2003) and receptors are found in T and B cells, macrophages, monocytes and neutrophils (Glaser and Kiecolt-Glaser 2005). In general, glucocorticoids inhibit immune system function and suppress maturation, differentiation and proliferation of immune cells such as T and B cells. For example, in healthy volunteers injected with the glucocorticoid hydrocortisone, blood levels of IL-1, IL-6 and TNFα were severely reduced, even though white blood cell count increased at the same time as lymphocyte count decreased (Derijk *et al.* 1997). Similar changes were observed when volunteers were subjected to exercise stress, indicating that physiological release of glucocorticoids was as effective as injected hormone. In molecular terms, animal experiments revealed that glucocorticoids regulate cytokine gene expression. They suppress the pro-inflammatory cytokines IL-1, IL-2, IL-6, IL-8, IL-11, IL-12, TNFα, IFNγ and GM-CSF, but up-regulate the anti-inflammatory cytokines IL-4 and IL-10 (Webster *et al.* 2002). An example can be seen in patients with rheumatoid arthritis whose levels of TNFα are reduced by treatment with glucocorticoid (Steer *et al.* 1998). Glucocorticoids also suppress cellular immunity by inhibiting thymus gland development, inducing atrophy of the mature thymus and inhibiting the development and differentiation of T cells (Bauer *et al.* 2009). Prolonged corticosteroid release also causes apoptosis of T cells (Herold *et al.* 2006).

Many of the effects of corticosteroids on the immune system are biphasic. On the one hand, as we have seen, the immunosuppressive and anti-inflammatory action of glucocorticoids is the reason they are used to control inflammation and to prevent organ rejection after transplantation. In contrast, there are examples of glucocorticoids being used to *enhance* resistance to infection, usually by low dose treatment (Jefferies 1991; 1994; Sapolsky *et al.* 2000). One possible mechanism is that each response uses a different glucocorticoid receptor; that is, either MR or GR (Sapolsky *et al.* 2000). Another example is the ability of a glucocorticoid to enhance, rather than suppress, an immune response following an endotoxin challenge. Volunteers were given lipopolysaccharide to induce an immune response (e.g. fever). Glucocorticoid co-treatment prevented (suppressed) the stimulation of IL-6 and TNFα, as expected. However, *pretreatment* with cortisol significantly increased the IL-6 and TNFα response (Barber *et al.* 1993).

Glucocorticoids as negative feedback signals for the immune system

Chapter 8 described the importance of negative feedback systems for the homeostatic regulation of hormone levels in the body. Section 13.5 outlined how cells of the immune system produce cytokines (such as IL-1) that stimulate the hypothalamic-pituitary-adrenal system to increase glucocorticoid release. Further, it is known that the immediate response to an inflammatory or endotoxin insult is the secretion of cytokines such as IL-1, IL-6 and TNFα that precede the elevation of ACTH release (Turnbull and Rivier 1999). Thus, the immune response probably generates the CRH-ACTH-glucocorticoid activation. Since glucocorticoids inhibit the activity of the immune system, the immunosuppressive action of glucocorticoids can be seen as negative feedback acting on the immune system, as shown in Figure 13.12 (Besedovsky and del Ray 1996). Such a negative feedback effect of glucocorticoids could thus act to prevent an out-of-control immune system. For example, the immunosuppressive action of glucocorticoid negative feedback may prevent the development of autoimmune diseases such as rheumatoid arthritis, in which the immune system attacks and destroys self cells as well as foreign cells.

Summary

Pituitary and target tissue hormones have well-described effects on the immune system. Figure 13.13 summarizes the extent of this influence. In general, growth hormone, prolactin and the hormones of the pituitary-thyroid axis stimulate immune system function, whereas the pituitary-adrenal system, and especially glucocorticoids, is inhibitory. Note, however, that ACTH, which induces the release of glucocorticoids, has stimulatory effects on lymphocytes. The sex hormones of the pituitary-gonadal axis, estradiol, progesterone and testosterone have biphasic effects and these are seen in the sex differences in immune responses in males and females. A profound inhibitory influence is apparent in the ability of estradiol and testosterone to cause thymic involution and aging. The posterior pituitary hormones oxytocin and vasopressin also have biphasic effects on lymphocyte numbers, but oxytocin stimulates cytotoxic T cells to release IFNγ in the secondary lymphoid tissues.

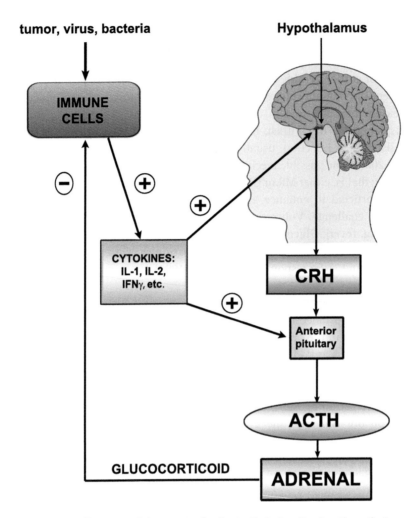

Figure 13.12 Illustration of the negative feedback effect of corticosteroids on the immune system
Cytokines released from immune cells such as macrophages following antigen presentation stimulate both immune and neuroendocrine target cells. The immune target cells release cytokines such as TNFα and IL-1. These cytokines release CRH from hypothalamic neurons and ACTH from the anterior pituitary gland. ACTH, in turn, stimulates corticosteroid release from the adrenal cortex that provides negative feedback to inhibit immune activity and the release of cytokines from cells of the immune system.

Abbreviations: ACTH, adrenocorticotropic hormone; CRH, corticotropin-releasing hormone; IFNγ, interferon γ; IL, interleukin.

13.6.3 Effects of neuropeptides on the immune system

In addition to hypothalamic and pituitary hormones, many neuropeptides modulate the activity of the immune system and the release of cytokines (Table 13.4). These include substance P, CRH, NPY, prolactin, GH, somatostatin, the endogenous opioid peptides and VIP. Some of these peptides are expressed and released from immune cells (see Table 13.3), but they also act as neurotransmitters/neuropeptides and are co-released from neurons in the brain and peripheral nerves of the ANS as discussed in Chapter 11. The effects of

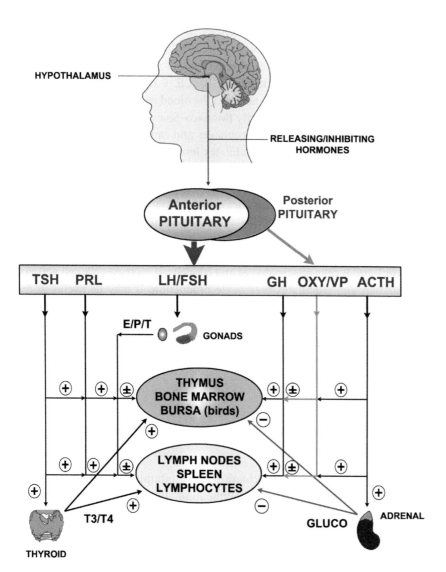

Figure 13.13 The regulation of lymphoid tissue and immune function by hormones of the pituitary gland
Pituitary and target tissue hormones have effects on the immune system. In general, growth hormone (GH), prolactin (PRL) and the hormones of the pituitary-thyroid axis (T3/T4, triiodothyronine / thyroxine; TSH, thyroid-stimulating hormone) stimulate immune system function, whereas the pituitary-adrenal system, and especially glucocorticoids (GLUCO), is inhibitory. Note, however, that adrenocorticotropin (ACTH), which induces the release of glucocorticoids, has stimulatory effects on lymphocytes. The sex hormones of the pituitary-gonadal axis, estradiol (E), progesterone (P) and testosterone (T), have biphasic effects and these are reflected in the sex differences in immune responses in males and females. A profound inhibitory influence is apparent in the ability of estradiol and testosterone to cause thymic involution and aging. The posterior pituitary hormones oxytocin (OXY) and vasopressin (VP) also have biphasic effects on lymphocyte numbers, but OXY stimulates cytotoxic T cells to release IFNγ in the secondary lymphoid tissues.

neuropeptides on the immune response are extensive and we will provide only a few selected examples (for further reading, see Souza-Moreira *et al.* 2011).

T and B cells, monocytes and macrophages have receptors for *Substance P* and the cells respond to the peptide by increasing cytokine release (IL-1, IL-6 and TNF). Substance P also increases trafficking of cells from the lymph nodes to the blood (Glaser and Kiecolt-Glaser 2005) and inhibits natural killer cell cytotoxicity (Monaco-Shawver *et al.* 2011). *NPY* has receptors on T and B cells, monocytes and macrophages and can down-regulate antibody production via an effect on B cells. It also modulates immune cell trafficking, cytokine production, T helper cell differentiation and natural killer cell activity (Dimitrijevic and Stanojevic 2013). *CRH* receptors are found on T cells and macrophages and CRH increases release of IL-1, IL-6 and TNFα (Goetzl *et al.* 2008). *Prolactin* receptors are widespread in the immune system (e.g. T and B cells, macrophages) and control progenitor cell proliferation. Prolactin also regulates the activation, differentiation and proliferation of T cells and increases the production of IFNγ and TNFα (Legorreta-Haquet *et al.* 2012). *Growth hormone* binds to receptors on T and B cells, macrophages and NK cells. It increases NK cell activity and antibody production and stimulates thymocyte production. It may be useful in reversing thymic involution (Savino and Dardenne 2010; Taub *et al.* 2010). *Somatostatin* receptors are present on human macrophages, reduce IL-8 secretion and may be immunosuppressive (Armani *et al.* 2007; Pinter *et al.* 2006). VIP receptors are found on lymphocytes and this peptide is anti-inflammatory, reducing the production of IL-2 and the number of lymphocytes (Ganea 1996; Ganea *et al.* 2006).

13.7 Hypothalamic integration of the neuroendocrine and immune systems

If the immune response is considered as a mechanism for maintaining homeostasis among the cells of the body in response to antigen stimulation, then cytokines provide the communication signals for maintaining immune system homeostasis. As described previously in this chapter, cytokines serve two functions: they inform the brain about the type of immune response being activated (an affector or sensory function); and they regulate the immune response (an effector function).

As shown in Figure 13.8, the immune system serves as a sensory system, responding to the presence of non-cognitive stimuli (such as bacteria, viruses, tumors and other antigens), which are not detected by the central or peripheral nervous systems. When these non-cognitive stimuli are detected, cell-mediated or humoral immune responses occur, depending on the type of antigen, and this information is sent to the brain via cytokine release. Peptide hormones are also released from immune cells, although they are more likely to exert autocrine/paracrine effects than to enter the blood and affect distant endocrine targets. Cytokines enter the brain through the circumventricular organs and stimulate hypothalamic neurons in the medial preoptic area, anterior and ventromedial hypothalamus, paraventricular nucleus and median eminence/arcuate nucleus. These hypothalamic nuclei then modulate the immune response by activating the ANS, the neuroendocrine system, and the cognitive and behavioral correlates of these systems, as described in Chapters 4 and 11.

Antigens can be regarded as stressful stimuli and the neuroendocrine response activated when the immune cells detect an antigen is a "stress response" with the same characteristics as the stress response to a cognitive stimulus; that is, activation of the hypothalamic-pituitary-adrenal axis. In order to coordinate adaptive responses to stressful stimuli, whether cognitive or non-cognitive, the brain and immune system communicate via the neuroendocrine-cytokine messenger systems and the information from these messenger systems is integrated in the hypothalamus. The neuroimmune stress response involves the activation of the defense mechanisms of the immune system (macrophages, T cells, B cells, NK cells, etc.) which are modulated by the neuroendocrine system to ensure the rapid production and proliferation of these cells in the thymus, spleen and lymph nodes and the development of their precursor cells in the bone marrow (e.g. see Figure 13.2).

Activation of the immune system alters neurotransmitter release and electrical activity in hypothalamic neurons and damage to the hypothalamus results in abnormal neuroimmune responses as well as abnormal neuroendocrine responses (Besedovsky and del Rey 1996; Anthony *et al*. 2012). The area of the hypothalamus which appears to have the greatest involvement in the neuroendocrine-immune response is the PVN. At least six lines of evidence support the central role of the PVN in the integration of the neuroendocrine-immune response.

First, the PVN has receptors for cytokines and thymic hormones, so these immunomodulators can directly influence PVN neural activity. Second, the PVN is involved in both the afferent and efferent sympathetic and parasympathetic pathways of the ANS. Activation of the ANS results in rapid stimulation of the bone marrow, thymus gland, spleen and lymph nodes, as well as the adrenal medulla. This causes the activation of both immune and endocrine cells by peripheral neurotransmitter, neuropeptide and hormone release. Third, the PVN contains magnocellular neurons which release oxytocin and vasopressin, both of which act on the thymus gland to influence T cell development. Fourth, the PVN contains parvicellular neurons that release TRH, CRH and other hypothalamic hormones. Activation of the neurosecretory cells of the hypothalamus by immunomodulators results in the release of hypothalamic and pituitary hormones (such as GH, PRL and TSH), which stimulate both endocrine and immune cells. As a result of neuroendocrine stimulation, there is an elevation of gonadal and adrenal steroid hormones that modulate the immune system. The glucocorticoids, in particular, provide negative feedback to inhibit immune responses. Fifth, the PVN receives input from circumventricular organs such as the OVLT and AP (Figure 13.10), structures that have no blood-brain barrier and thus allow large peptide molecules to pass between the circulation and the brain. Sixth, the PVN receives neural input from the neocortex, amygdala and hippocampus, areas of the brain that mediate cognitive functions, including emotional arousal and learning. Cognitive stressors can influence immune system activity through their action on the hypothalamic neuroendocrine-immune integrating mechanisms. Isolation rearing, crowding, low dominance status and social stress all influence the ability of animals to defend themselves against disease and infection (Reber *et al*. 2006). In humans, stressors such as depression, bereavement and examination stress all suppress immune responses (Glaser and Kiecolt-Glaser 2005; Dantzer *et al*. 2008). Both humoral and cell-mediated immune responses can also be conditioned to neutral stimuli (Cohen *et al.* 1994). Thus, the PVN may be able to integrate the cognitive responses to stressful psychosocial stimuli with the activation of the neuro-endocrine-immune system, resulting in a psycho-neuro-endocrine-immune system response (Daruna and Morgan 1990).

Other functions of the hypothalamus are also influenced by neuroimmune activity. When the immune system is activated by certain antigens, such as bacterial pyrogens, body temperature is elevated, causing fever. Some cytokines (e.g. IL-1) also cause fever through their receptors in the hypothalamus. The hypothalamus then initiates responses from the ANS (sweating or shivering), the neuroendocrine system (e.g. TSH and ACTH release) and behavioral responses to changes in body temperature (Dinarello 1999).

Cytokines, such as IL-1 and IL-6, also alter hunger and feeding behavior mediated by the ventromedial and lateral hypothalamic nuclei, and changes in food intake alter neuroendocrine functions (Guijarro *et al.* 2006; Plata-Salaman 2001). For example, fasting reduces ANS activity, while feeding activates the release of insulin and the gastrointestinal peptides (see Chapter 12). Cytokines also produce cognitive, psychiatric and behavioral side effects, such as fatigue, anorexia, sleep disruption and disturbances of emotional behavior (Dantzer *et al.* 2008).

Virtually all pituitary hormones show day/night or sleep/wake rhythms of activity. For example, GH (Figure 4.7) and PRL secretion peaks during sleep and ACTH and cortisol secretion is highest just before waking (Figure 6.5). Many of these hypothalamic cycles are regulated by a clock mechanism in the suprachiasmatic nuclei of the hypothalamus. Immune responses also show day/night cycles and these may be controlled by hypothalamic neuroendocrine rhythms. Disturbance or desynchronization of these rhythms, as a result of shift work, for example, may increase vulnerabilities to infection and disease (Logan and Sarkar 2012; Bechtold *et al.* 2010).

Neurotransmitters, peptide hormones and cytokines act via a variety of membrane receptors to activate second messenger systems in their target cells. These second messengers regulate ion channels and other receptors in the target cell membrane and the transcription of genomic information, via mRNA, from the cell nucleus (see Chapter 10). The interactions between neural, endocrine and immune systems are mediated through common receptor and second messenger mechanisms. In T lymphocytes, for example, cAMP mediates the effects of prostaglandins (PGs), adenosine, histamine, β-adrenergic agonists, neuropeptide hormones and β-endorphin (Mosenden and Tasken 2011). Since immune cells also possess sex steroid and glucocorticoid hormone receptors, this means that hormones, neurotransmitters and peptides may, therefore, all interact to regulate the responses of the cells of the immune system to antigens. Cytokines act through cell membrane receptors of the type shown in Figure 13.9. Since the cells of the immune system can produce neuropeptides, it is unsurprising to find that neurons can produce cytokines (see section 13.5.1). For example, IL-1, IL-3, interferons and thymosins are all produced in the brain and neuroendocrine system. Their target cells are the brain's resident set of macrophages, known as microglia (Wake *et al.* 2011).

13.8 Summary

This chapter examined the interaction of the neuroendocrine system and the immune system. The immune system consists of a number of specific cell types, including macrophages, T and B cells, and NK cells which control cell-mediated and humoral responses to antigens. T cells mature in the thymus gland, the so-called "master-gland of the immune system." The thymus gland secretes a number of thymic hormones (thymosins) which are important regulators of immune system functions. T lymphocytes and other cells of the immune system produce cytokines, which include interferon γ, the interleukins, tumor necrosis factor and colony stimulating factors. These cytokines regulate T cell activation, B cell activation, the production of blood cells (hematopoiesis), cytotoxicity and inflammatory responses. Cells of the immune system can regulate the brain and endocrine system through the actions of the cytokines on specific cytokine receptors in the brain. Cytokines and thymic hormones can enter the brain and stimulate neurotransmitter and neuropeptide release, thus regulating autonomic, neuroendocrine and behavioral actions. Neural regulation of the immune system occurs through ANS activation and release of hormones from the hypothalamic-pituitary-endocrine organ systems. Complex neuroendocrine-immune system interactions occur between the thymus gland and gonadal hormones and between cytokines and adrenal steroid hormones. Glucocorticoids provide negative feedback to inhibit the release of cytokines and suppress immune system activity, thus preventing some autoimmune diseases. Many neuropeptides, including opioid peptides, Substance P and somatostatin, also regulate the activity of the immune system. Neuroendocrine-immune system interactions are mediated through the hypothalamus, and especially the PVN, which plays a central role in the integration of these three systems. Immune responses to antigens can be compared with neural responses to stressful cognitive stimuli as both initiate neuroendocrine stress responses. The activation of neuroendocrine and autonomic nervous activity in response to cognitive and non-cognitive stimuli is an adaptive response that maintains the homeostasis of the body in response to these stressors. Through the integrative activity of the hypothalamus, cognitive stressors such as overcrowding, exam stress and depression can alter the activity of the immune system, while interleukins and other cytokines can have cognitive and behavioral side effects. Through the action of the brain, immune responses may also be conditioned to neutral (non-antigenic) stimuli. The integration of neural, endocrine and immune system activity occurs through common receptors and second messenger systems.

FURTHER READING

Abbas, A. K. and Lichtman, A. H. (2010). *Basic Immunology*, 3rd edn. (Philadelphia, PA: Saunders)

Anthony, D. C., Couch, Y., Losey, P. and Evans, M. C. (2012). "The systemic response to brain injury and disease," *Brain Behav Immun* 26, 534–540.

Blalock, J. E. (1989). "A molecular basis for bidirectional communication between the immune and neuroendocrine systems," *Physiolog Rev* 69, 1–32.

Bouman, A., Heineman, M. J. and Faas, M. M. (2005). "Sex hormones and the immune response in humans," *Hum Reprod Update* 11, 411–423.

Javed, M. J., Richmond, T. D. and Barber, D. L. (2010). "Cytokine receptor signaling" in R. A. Bradshaw and E. A. Dennis (eds.), *Handbook of Cell Signaling*, 2nd edn. (Waltham, MA: Academic Press), pp. 451–466.

Kindt, T. J., Goldsby, R. A. and Osborne, B. A. (2006). *Kuby Immunology* (New York: Freeman).

Locatelli, V., Bresciani, E., Tamiazzo, L. and Torsello, A. (2010). "Central nervous system-acting drugs influencing hypothalamic-pituitary-adrenal axis function," *Endocr Dev* 17, 108–120.

Sompayrac, L. (2012). *How the Immune System Works*, 4th edn. (Hoboken: Wiley-Blackwell).

Souza-Moreira, L., Campos-Salinas, J., Caro, M. and Gonzalez-Rey, E. (2011). "Neuropeptides as pleiotropic modulators of the immune response," *Neuroendocrinology* 94, 89–100.

Turnbull, A. V. and Rivier, C. L. (1999). "Regulation of the hypothalamic-pituitary-adrenal axis by cytokines: actions and mechanisms of action," *Physiol Rev* 79, 1–71.

REVIEW QUESTIONS

13.1 What is the difference between T cells and B cells in the immune system?

13.2 What two immune functions does the thymus gland perform?

13.3 Which two cells of the immune system are most prominent in cytokine production?

13.4 Which cytokine stimulates fever, sleep, and prostaglandin synthesis and inhibits eating?

13.5 What is the main function of TNF?

13.6 Which two neurohormones are produced in the thymus gland?

13.7 Name two of the hormones synthesized from POMC that are produced by lymphocytes.

13.8 Describe the steps involved in the immune-corticosteroid negative feedback system.

13.9 How do GH and prolactin influence immune system activity?

13.10 How do progesterone and corticosteroids influence immune system activity?

13.11 Which area of the hypothalamus appears to regulate the neuroendocrine-immune system interaction?

13.12 Why is the immune response considered a stress response?

ESSAY QUESTIONS

13.1 Discuss the interactions between the immune system and the hormones of the hypothalamic-pituitary-adrenal axis.

13.2 Discuss the interactions between the immune system and the hormones of the hypothalamic-pituitary-gonadal axis.

13.3 Discuss the role of the ANS in regulating immune system function.

13.4 Discuss the mechanisms by which infections can activate the neuroendocrine system through the release of cytokines.

13.5 Discuss the functions of the thymus hormones in the neural, endocrine and immune systems.

13.6 Discuss how the neuroendocrine system suppresses the immune system during pregnancy.

13.7 Discuss the function of the peptide hormones produced by the cells of the immune system.

13.8 Discuss the neural, endocrine and immune functions of IL-1.

13.9 Discuss the effects of psychological stress on the immune system.

REFERENCES

Aarden, L. A. (1979). "Revised nomenclature for antigen-nonspecific T cell proliferation and helper factors," *J Immunol* 123, 2928–2929.

Abbas, A. K. and Lichtman, A. H. (2010). *Basic Immunology*, 3rd edn. (Philadelphia, PA: Saunders).

Aggarwal, B. B., Gupta, S. C. and Kim, J. H. (2012). "Historical perspectives on tumor necrosis factor and its superfamily: 25 years later, a golden journey," *Blood* 119, 651–665.

Ahern, G. P. (2011). "5-HT and the immune system," *Curr Opin Pharmacol* 11, 29–33.

Anisman, H., Gibb, J. and Hayley, S. (2008). "Influence of continuous infusion of interleukin-1beta on depression-related processes in mice: corticosterone, circulating cytokines, brain monoamines, and cytokine mRNA expression," *Psychopharmacology (Berl)* 199, 231–244.

Ansar Ahmed, S., Penhale, W. J. and Talal, N. (1985). "Sex hormones, immune responses, and autoimmune diseases. Mechanisms of sex hormone action," *Am J Pathol* 121, 531–551.

Anthony, D. C., Couch, Y., Losey, P. and Evans, M. C. (2012). "The systemic response to brain injury and disease," *Brain Behav Immun* 26, 534–540.

Armani, C., Catalani, E., Balbarini, A., Bagnoli, P. and Cervia, D. (2007). "Expression, pharmacology, and functional role of somatostatin receptor subtypes 1 and 2 in human macrophages," *J Leukoc Biol* 81, 845–855.

Asanuma, Y., Goldstein, A. L. and White, A. (1970). "Reduction in the incidence of wasting disease in neonatally thymectomized CBA-W mice by the injection of thymosin," *Endocr* 86, 600–610.

Assis, M. A., Collino, C., Figuerola Mde, L., Sotomayor, C. and Cancela, L. M. (2006). "Amphetamine triggers an increase in met-enkephalin simultaneously in brain areas and immune cells," *J Neuroimmunol* 178, 62–75.

Banks, W. A., Farr, S. A. and Morley, J. E. (2002). "Entry of blood-borne cytokines into the central nervous system: effects on cognitive processes," *Neuroimmunomodulation* 10, 319–327.

Barber, A. E., Coyle, S. M., Marano, M. A., Fischer, E., Calvano, S. E., Fong, Y. *et al.* (1993). "Glucocorticoid therapy alters hormonal and cytokine responses to endotoxin in man," *J Immunol* 150, 1999–2006.

Barnéoud, P., Neveu, P. J., Vitiello, S., Mormede, P. and Le Moal, M. (1988). "Brain neocortex immunomodulation in rats," *Brain Res* 474, 394–398.

Barrell, G. K. (2007). "Immunological influences on reproductive neuroendocrinology," *Soc Reprod Fertil Suppl.* 64, 109–122.

Bauer, M. E., Jeckel, C. M. and Luz, C. (2009). "The role of stress factors during aging of the immune system," *Ann NY Acad Sci* 1153, 139–152.

Bechtold, D. A., Gibbs, J. E. and Loudon, A. S. (2010). "Circadian dysfunction in disease," *Trends Pharmacol Sci* 31, 191–198.

Besedovsky, H. O. and del Rey, A. (1996). "Immune-neuro-endocrine interactions: facts and hypotheses," *Endocr Rev* 17, 64–102.

Blalock, J. E. (1989). "A molecular basis for bidirectional communication between the immune and neuroendocrine systems," *Physiol Rev* 69, 1–32.

Boron, W. F. and Boulpaep, E. L. (2005). *Medical Physiology*, updated edn. (Philadelphia, PA: Elsevier Saunders).

Borovikova, L. V., Ivanova, S., Zhang, M., Yang, H., Botchkina, G. I., Watkins, L. R. *et al.* (2000). "Vagus nerve stimulation attenuates the systemic inflammatory response to endotoxin," *Nature* 405, 458–462.

Brambilla, D., Franciosi, S., Opp, M. R. and Imeri, L. (2007). "Interleukin-1 inhibits firing of serotonergic neurons in the dorsal raphe nucleus and enhances GABAergic inhibitory post-synaptic potentials," *Eur J Neurosci* 26, 1862–1869.

Brooks, K. J., Bunce, K. T., Haase, M. V., White, A., Changani, K. K., Bate, S. T. *et al.* (2005). "MRI quantification in vivo of corticosteroid induced thymus involution in mice: correlation with ex vivo measurements," *Steroids* 70, 267–272.

Bruccoleri, A., Brown, H. and Harry, G. J. (1998). "Cellular localization and temporal elevation of tumor necrosis factor-alpha, interleukin-1 alpha, and transforming growth factor-beta 1 mRNA in hippocampal injury response induced by trimethyltin," *J Neurochem* 71, 1577–1587.

Buckley, A. R. (2001). "Prolactin, a lymphocyte growth and survival factor," *Lupus* 10, 684–690.

Buttarelli, F. R., Fanciulli, A., Pellicano, C. and Pontieri, F. E. (2011). "The dopaminergic system in peripheral blood lymphocytes: from physiology to pharmacology and potential applications to neuropsychiatric disorders," *Curr Neuropharmacol* 9, 278–288.

Calder, A. E., Hince, M. N., Dudakov, J. A., Chidgey, A. P. and Boyd, R. L. (2011). "Thymic involution: where endocrinology meets immunology," *Neuroimmunomodulation* 18, 281–289.

Cano, P., Cardinali, D. P., Jimenez, V., Alvarez, M. P., Cutrera, R. A. and Esquifino, A. I. (2005). "Effect of interferon-gamma treatment on 24-hour variations in plasma ACTH, growth hormone, prolactin, luteinizing hormone and follicle-stimulating hormone of male rats," *Neuroimmunomodulation* 12, 146–151.

Chighizola, C. and Meroni, P. L. (2012). "The role of environmental estrogens and autoimmunity," *Autoimmun Rev* 11, A493–A501.

Clayton, E. M., Todd, M., Dowd, J. B. and Aiello, A. E. (2011). "The impact of bisphenol A and triclosan on immune parameters in the U.S. population, NHANES 2003–2006," *Environ Health Perspect* 119, 390–396.

Cohen, N., Moynihan, J. A. and Ader, R. (1994). "Pavlovian conditioning of the immune system," *Int Arch Allergy Immunol* 105, 101–106.

Cohen, S., Janicki-Deverts, D. and Miller, G. E. (2007). "Psychological stress and disease," *JAMA* 298, 1685–1687.

Croft, M., Duan, W., Choi, H., Eun, S. Y., Madireddi, S. and Mehta, A. (2012). "TNF superfamily in inflammatory disease: translating basic insights," *Trends Immunol* 33, 144–152.

Cross, R. J., Markesbery, W. R., Brooks, W. H. and Roszman, T. L. (1984). "Hypothalamic-immune interactions: neuromodulation of natural killer activity by lesioning of the anterior hypothalamus," *Immunology* 51, 399–405.

Csaba, G., Tekes, K. and Pallinger, E. (2009). "Influence of perinatal stress on the hormone content in immune cells of adult rats: dominance of ACTH," *Horm Metab Res* 41, 617–620.

D'Arcangelo, G., Tancredi, V., Onofri, F., D'Antuono, M., Giovedi, S. and Benfenati, F. (2000). "Interleukin-6 inhibits neurotransmitter release and the spread of excitation in the rat cerebral cortex," *Eur J Neurosci* 12, 1241–1252.

Dantzer, R., O'Connor, J. C., Freund, G. G., Johnson, R. W. and Kelley, K. W. (2008). "From inflammation to sickness and depression: when the immune system subjugates the brain," *Nat Rev Neurosci* 9, 46–56.

Daruna, J. H. and Morgan, J. E. (1990). "Psychosocial effects on immune function: neuroendocrine pathways," *Psychosomatics* 31, 4–12.

De Mello-Coelho, V., Savino, W., Postel-Vinay, M. C. and Dardenne, M. (1998). "Role of prolactin and growth hormone on thymus physiology," *Dev Immunol* 6, 317–323.

De Vito, P., Incerpi, S., Pedersen, J. Z., Luly, P., Davis, F. B. and Davis, P. J. (2011). "Thyroid hormones as modulators of immune activities at the cellular level," *Thyroid* 21, 879–890.

de Weerd, N. A. and Nguyen, T. (2012). "The interferons and their receptors – distribution and regulation," *Immunol Cell Biol* 90, 483–491.

DeRijk, R., Michelson, D., Karp, B., Petrides, J., Galliven, E., Deuster, P. *et al.* (1997). "Exercise and circadian rhythm-induced variations in plasma cortisol differentially regulate interleukin-1 beta (IL-1 beta), IL-6, and tumor necrosis factor-alpha (TNF alpha) production in humans: high sensitivity of TNF alpha and resistance of IL-6," *J Clin Endocrinol Metab* 82, 2182–2191.

Dimitrijevic, M. and Stanojevic, S. (2013). "The intriguing mission of neuropeptide Y in the immune system," *Amino Acids* 45, 41–53.

Dinarello, C. A. (1999). "Cytokines as endogenous pyrogens," *J Infect Dis* 179(Suppl. 2), S294–S304.

Dinarello, C. A. (2011). "Interleukin-1 in the pathogenesis and treatment of inflammatory diseases," *Blood* 117, 3720–3732.

Douglas, S. D., Lai, J. P., Tuluc, F., Schwartz, L. and Kilpatrick, L. E. (2008). "Neurokinin-1 receptor expression and function in human macrophages and brain: perspective on the role in HIV neuropathogenesis," *Ann NY Acad Sci* 1144, 90–96.

Dunn, A. J. (2006). "Effects of cytokines and infections on brain neurochemistry," *Clin Neurosci Res* 6, 52–68.

Duquesnoy, R. J. and Pedersen, G. M. (1981). "Immunologic and hematologic deficiencies of the hypopituitary dwarf mouse" in M. E. Gershwin and B. Merchant (eds.), *Immunologic Defects in Laboratory Animals* (New York: Plenum Press), pp. 309–324.

Flierl, M. A., Rittirsch, D., Huber-Lang, M. S., Sarma, J. V. and Ward, P. A. (2008). "Molecular events in the cardiomyopathy of sepsis" *Mol Med* 14, 327–336.

Fragala, M. S., Kraemer, W. J., Denegar, C. R., Maresh, C. M., Mastro, A. M. and Volek, J. S. (2011). "Neuroendocrine-immune interactions and responses to exercise," *Sports Med* 41, 621–639.

Franchimont, D. (2004). "Overview of the actions of glucocorticoids on the immune response: a good model to characterize new pathways of immunosuppression for new treatment strategies," *Ann NY Acad Sci* 1024, 124–137.

French, R. A., Broussard, S. R., Meier, W. A., Minshall, C., Arkins, S., Zachary, J. F. *et al.* (2002). "Age-associated loss of bone marrow hematopoietic cells is reversed by GH and accompanies thymic reconstitution," *Endocr* 143, 690–699.

Fry, C., Gunter, D. R., McMahon, C. D., Steele, B. and Sartin, J. L. (1998). "Cytokine-mediated growth hormone release from cultured ovine pituitary cells," *Neuroendocrinology* 68, 192–200.

Ganea, D. (1996). "Regulatory effects of vasoactive intestinal peptide on cytokine production in central and peripheral lymphoid organs," *Adv Neuroimmunol* 6, 61–74.

Ganea, D., Gonzalez-Rey, E. and Delgado, M. (2006). "A novel mechanism for immunosuppression: from neuropeptides to regulatory T cells," *J Neuroimmune Pharmacol* 1, 400–409.

Gerlo, S., Verdood, P., Hooghe-Peters, E. L. and Kooijman, R. (2005). "Modulation of prolactin expression in human T lymphocytes by cytokines," *J Neuroimmunol* 162, 190–193.

Glaser, R. and Kiecolt-Glaser, J. K. (2005). "Stress-induced immune dysfunction: implications for health," *Nat Rev Immunol* 5, 243–251.

Goetzl, E. J., Chan, R. C. and Yadav, M. (2008). "Diverse mechanisms and consequences of immunoadoption of neuromediator systems," *Ann NY Acad Sci* 1144, 56–60.

Goldbach-Mansky, R. (2012). "Immunology in clinic review series; focus on autoinflammatory diseases: update on monogenic autoinflammatory diseases: the role of interleukin (IL)-1 and an emerging role for cytokines beyond IL-1," *Clin Exp Immunol* 167, 391–404.

Goldstein, A. L. and Badamchian, M. (2004). "Thymosins: chemistry and biological properties in health and disease," *Expert Opin Biol Ther* 4, 559–573.

Gong, F. Y., Deng, J. Y. and Shi, Y. F. (2005). "Stimulatory effect of interleukin-1beta on growth hormone gene expression and growth hormone release from rat GH3 cells," *Neuroendocrinology* 81, 217–228.

Gonsalkorale, W. M., Dascombe, M. J. and Hutchinson, I. V. (1995). "Adrenocorticotropic hormone as a potential enhancer of T-lymphocyte function in the rat mixed lymphocyte reaction," *Int J Immunopharmacol* 17, 197–206.

Goya, R. G., Brown, O. A. and Bolognani, F. (1999). "The thymus-pituitary axis and its changes during aging," *Neuroimmunomodulation* 6, 137–142.

Greenstein, B. D., Fitzpatrick, F. T., Adcock, I. M., Kendall, M. D. and Wheeler, M. J. (1986). "Reappearance of the thymus in old rats after orchidectomy: inhibition of regeneration by testosterone," *J Endocrinol* 110, 417–422.

Guijarro, A., Laviano, A. and Meguid, M. M. (2006). "Hypothalamic integration of immune function and metabolism," *Prog Brain Res* 153, 367–405.

Hale, L. P. (2004). "Histologic and molecular assessment of human thymus," *Ann Diagn Pathol* 8, 50–60.

Hall, N. R., McGillis, J. P., Spangelo, B. L. and Goldstein, A. L. (1985). "Evidence that thymosins and other biologic response modifiers can function as neuroactive immunotransmitters," *J Immunol* 135, 806s–811s.

Hanisch, U. K. and Quirion, R. (1995). "Interleukin-2 as a neuroregulatory cytokine," *Brain Res Rev* 21, 246–284.

Hansenne, I. (2005). "Thymic transcription of neurohypophysial and insulin-related genes: impact upon T-cell differentiation and self-tolerance," *J Neuroendocrinol* 17, 321–327.

Hansenne, I., Rasier, G., Pequeux, C., Brilot, F., Renard, C., Breton, C. *et al*. (2005). "Ontogenesis and functional aspects of oxytocin and vasopressin gene expression in the thymus network," *J Neuroimmunol* 158, 67–75.

Harel, G., Shamoun, D. S., Kane, J. P., Magner, J. A. and Szabo, M. (1995). "Prolonged effects of tumor necrosis factor-alpha on anterior pituitary hormone release," *Peptides* 16, 641–645.

Hattori, N. (2009). "Expression, regulation and biological actions of growth hormone (GH) and ghrelin in the immune system," *Growth Horm IGF Res* 19, 187–197.

Hattori, N., Saito, T., Yagyu, T., Jiang, B. H., Kitagawa, K. and Inagaki, C. (2001). "GH, GH receptor, GH secretagogue receptor, and ghrelin expression in human T cells, B cells, and neutrophils," *J Clin Endocrinol Metab* 86, 4284–4291.

Healy, D. L., Hodgen, G. D., Schulte, H. M., Chrousos, G. P., Loriaux, D. L., Hall, N. R. *et al.* (1983). "The thymus-adrenal connection: thymosin has corticotropin-releasing activity in primates," *Science* 222, 1353–1355.

Herold, M. J., McPherson, K. G. and Reichardt, H. M. (2006). "Glucocorticoids in T cell apoptosis and function," *Cell Mol Life Sci* 63, 60–72.

Hino, M., Ogata, T., Morino, T., Horiuchi, H. and Yamamoto, H. (2009). "Intrathecal transplantation of autologous macrophages genetically modified to secrete proenkephalin ameliorated hyperalgesia and allodynia following peripheral nerve injury in rats," *Neurosci Res* 64, 56–62.

Hotakainen, P. K., Serlachius, E. M., Lintula, S. I., Alfthan, H. V., Schroder, J. P. and Stenman, U. E. (2000). "Expression of luteinising hormone and chorionic gonadotropin beta-subunit messenger-RNA and protein in human peripheral blood leukocytes," *Mol Cell Endocrinol* 162, 79–85.

Jankord, R., Turk, J. R., Schadt, J. C., Casati, J., Ganjam, V. K., Price, E. M. *et al.* (2007). "Sex difference in link between interleukin-6 and stress," *Endocr* 148, 3758–3764.

Jefferies, W. M. (1991). "Cortisol and immunity," *Med Hypotheses* 34, 198–208.

Jefferies, W. M. (1994). "Mild adrenocortical deficiency, chronic allergies, autoimmune disorders and the chronic fatigue syndrome: a continuation of the cortisone story," *Med Hypotheses* 42, 183–189.

Johnson, E. W., Hughes, T. K., Jr. and Smith, E. M. (2001). "ACTH receptor distribution and modulation among murine mononuclear leukocyte populations," *J Biol Regul Homeost Agents* 15, 156–162.

Johnson, E. W., Hughes, T. K., Jr. and Smith, E. M. (2005). "ACTH enhancement of T-lymphocyte cytotoxic responses," *Cell Mol Neurobiol* 25, 743–757.

Juttler, E., Tarabin, V. and Schwaninger, M. (2002). "Interleukin-6 (IL-6): a possible neuromodulator induced by neuronal activity," *Neuroscientist* 8, 268–275.

Kao, C. H., Huang, W. T., Lin, M. T. and Wu, W. S. (2011). "Central interleukin-10 attenuated lipopolysaccharide-induced changes in core temperature and hypothalamic glutamate, hydroxyl radicals and prostaglandin-E(2)," *Eur J Pharmacol* 654, 187–193.

Karanth, S. and McCann, S. M. (1991). "Anterior pituitary hormone control by interleukin 2," *Proc Natl Acad Sci USA* 88, 2961–2965.

Kawashima, K. and Fujii, T. (2004). "Expression of non-neuronal acetylcholine in lymphocytes and its contribution to the regulation of immune function," *Front Biosci* 9, 2063–2085.

Kelley, K. W., Brief, S., Westly, H. J., Novakofski, J., Bechtel, P. J., Simon, J. *et al.* (1987). "Hormonal regulation of the age-associated decline in immune function," *Ann NY Acad Sci* 496, 91–97.

Kindt, T. J., Goldsby, R. A. and Osborne, B. A. (2006). *Kuby Immunology* (New York: Freeman).

Klehmet, J., Harms, H., Richter, M., Prass, K., Volk, H. D., Dirnagl, U. *et al.* (2009). "Stroke-induced immunodepression and post-stroke infections: lessons from the preventive antibacterial therapy in stroke trial," *Neurosci* 158, 1184–1193.

Klein, J. R. (2003). "Physiological relevance of thyroid stimulating hormone and thyroid stimulating hormone receptor in tissues other than the thyroid," *Autoimmunity* 36, 417–421.

Klein, S. L. (2012). "Immune cells have sex and so should journal articles," *Endocr* 153, 2544–2550.

Kordon, C. and Bihoreau, C. (1989). "Integrated communication between the nervous, endocrine and immune systems," *Horm Res* 31, 100–104.

Kovats, S., Carreras, E. and Agrawal, H. (2010). "Sex steroid receptors in immune cells" in
    S. L. Klein and C. W. Roberts (eds.), *Sex Hormones and Immunity to Infection* (Berlin
    Heidelberg: Springer-Verlag), pp. 53–91.

Krueger, J. M. (2008). "The role of cytokines in sleep regulation," *Curr Pharm Des* 14, 3408–3416.

Kruger, T. E. (1996). "Immunomodulation of peripheral lymphocytes by hormones of the
    hypothalamus-pituitary-thyroid axis," *Adv Neuroimmunol* 6, 387–395.

Kurotani, R., Yasuda, M., Oyama, K., Egashira, N., Sugaya, M., Teramoto, A. *et al*. (2001).
    "Expression of interleukin-6, interleukin-6 receptor (gp80), and the receptor's signal-
    transducing subunit (gp130) in human normal pituitary glands and pituitary adenomas,"
    Mod Pathol 14, 791–797.

Kvetnoy, I. M., Polyakova, V. O., Trofimov, A. V., Yuzhakov, V. V., Yarilin, A. A., Kurilets, E. S. *et al*.
    (2003). "Hormonal function and proliferative activity of thymic cells in humans:
    immunocytochemical correlations," *Neuro Endocrinol Lett* 24, 263–268.

Lacosta, S., Merali, Z. and Anisman, H. (2000). "Central monoamine activity following acute
    and repeated systemic interleukin-2 administration," *Neuroimmunomodulation* 8, 83–90.

Lane, T. and Lachmann, H. J. (2011). "The emerging role of interleukin-1beta in autoinflammatory
    diseases," *Curr Allergy Asthma Rep* 11, 361–368.

Larsen, C. M., Faulenbach, M., Vaag, A., Volund, A., Ehses, J. A., Seifert, B. *et al*. (2007).
    "Interleukin-1-receptor antagonist in type 2 diabetes mellitus," *N Engl J Med* 356,
    1517–1526.

Lee, S., Kim, J., Jang, B., Hur, S., Jung, U., Kil, K. *et al*. (2010). "Fluctuation of peripheral blood T,
    B, and NK cells during a menstrual cycle of normal healthy women," *J Immunol* 185,
    756–762.

Legorreta-Haquet, M. V., Chavez-Rueda, K., Montoya-Diaz, E., Arriaga-Pizano, L., Silva-Garcia, R.,
    Chavez-Sanchez, L. *et al*. (2012). "Prolactin down-regulates CD4+CD25hiCD127low/−
    regulatory T cell function in humans," *J Mol Endocrinol* 48, 77–85.

Leposavic, G., Pilipovic, I., Radojevic, K., Pesic, V., Perisic, M. and Kosec, D. (2008).
    "Catecholamines as immunomodulators: a role for adrenoceptor-mediated mechanisms in
    fine tuning of T-cell development," *Auton Neurosci* 144, 1–12.

Levin, S. G. and Godukhin, O. V. (2011). "Anti-inflammatory cytokines, TGF-beta1 and IL-10,
    exert anti-hypoxic action and abolish posthypoxic hyperexcitability in hippocampal slice
    neurons: comparative aspects," *Exp Neurol* 232, 329–332.

Lightman, S. L. and Conway-Campbell, B. L. (2010). "The crucial role of pulsatile activity of the
    HPA axis for continuous dynamic equilibration," *Nat Rev Neurosci* 11, 710–718.

Logan, R. W. and Sarkar, D. K. (2012). "Circadian nature of immune function," *Mol Cell
    Endocrinol* 349, 82–90.

Lyons, P. D. and Blalock, J. E. (1997). "Pro-opiomelanocortin gene expression and protein
    processing in rat mononuclear leukocytes," *J Neuroimmunol* 78, 47–56.

Maccio, A., Madeddu, C., Chessa, P., Panzone, F., Lissoni, P. and Mantovani, G. (2010).
    "Oxytocin both increases proliferative response of peripheral blood lymphomonocytes
    to phytohemagglutinin and reverses immunosuppressive estrogen activity," *In Vivo* 24,
    157–163.

Machelska, H. (2011). "Control of neuropathic pain by immune cells and opioids," *CNS Neurol
    Disord Drug Targets* 10, 559–570.

Maini, R. N. and Taylor, P. C. (2000). "Anti-cytokine therapy for rheumatoid arthritis," *Annu Rev
    Med* 51, 207–229.

Marino, F. and Cosentino, M. (2013). "Adrenergic modulation of immune cells: an update,"
    Amino Acids 45, 55–71.

McCann, S. M., Kimura, M., Karanth, S., Yu, W. H., Mastronardi, C. A. and Rettori, V. (2000). "The mechanism of action of cytokines to control the release of hypothalamic and pituitary hormones in infection," *Ann NY Acad Sci* 917, 4–18.

McNamee, E. N., Masterson, J. C., Jedlicka, P., McManus, M., Grenz, A., Collins, C. B. *et al.* (2011). "Interleukin 37 expression protects mice from colitis," *Proc Natl Acad Sci USA* 108, 16711–16716.

Meazza, C., Pagani, S., Travaglino, P. and Bozzola, M. (2004). "Effect of growth hormone (GH) on the immune system," *Pediatr Endocrinol Rev* 1(Suppl. 3), 490–495.

Melgert, B. N., Oriss, T. B., Qi, Z., Dixon-McCarthy, B., Geerlings, M., Hylkema, M. N. *et al.* (2010). "Macrophages: regulators of sex differences in asthma?" *Am J Respir Cell Mol Biol* 42, 595–603.

Mitchison, N. A. (2004). "T-cell-B-cell cooperation," *Nat Rev Immunol* 4, 308–312.

Monaco-Shawver, L., Schwartz, L., Tuluc, F., Guo, C. J., Lai, J. P., Gunnam, S. M. (2011). "Substance P inhibits natural killer cell cytotoxicity through the neurokinin-1 receptor," *J Leukoc Biol* 89, 113–125.

Mosenden, R. and Tasken, K. (2011). "Cyclic AMP-mediated immune regulation – overview of mechanisms of action in T cells," *Cell Signal* 23, 1009–1016.

Mousa, S. A., Straub, R. H., Schafer, M. and Stein, C. (2007). "Beta-endorphin, Met-enkephalin and corresponding opioid receptors within synovium of patients with joint trauma, osteoarthritis and rheumatoid arthritis," *Ann Rheum Dis* 66, 871–879.

Muglia, L. J., Jenkins, N. A., Gilbert, D. J., Copeland, N. G. and Majzoub, J. A. (1994). "Expression of the mouse corticotropin-releasing hormone gene in vivo and targeted inactivation in embryonic stem cells," *J Clin Invest* 93, 2066–2072.

Munoz-Cruz, S., Togno-Pierce, C. and Morales-Montor, J. (2011). "Non-reproductive effects of sex steroids: their immunoregulatory role," *Curr Top Med Chem* 11, 1714–1727.

Nagai, Y., Ohsawa, K., Ieki, Y. and Kobayashi, K. (1996). "Effect of interferon-beta on thyroid function in patients of chronic hepatitis C without preexisting autoimmune thyroid disease," *Endocr J* 43, 545–549.

Nance, D. M. and Sanders, V. M. (2007). "Autonomic innervation and regulation of the immune system (1987–2007)," *Brain Behav Immun* 21, 736–745.

Nemet, D., Eliakim, A., Zaldivar, F. and Cooper, D. M. (2006). "Effect of rhIL-6 infusion on GH->IGF-I axis mediators in humans," *Am J Physiol Regul Integr Comp Physiol* 291, R1663–R1668.

Neveu, P. J. (1988). "Cerebral neocortex modulation of immune functions," *Life Sci* 42, 1917–1923.

Nizri, E. and Brenner, T. (2013). "Modulation of inflammatory pathways by the immune cholinergic system," *Amino Acids* 45, 55–71.

Oertelt-Prigione, S. (2012). "Immunology and the menstrual cycle," *Autoimmun Rev* 11, A486–A492.

Olsen, N. J., Watson, M. B., Henderson, G. S. and Kovacs, W. J. (1991). "Androgen deprivation induces phenotypic and functional changes in the thymus of adult male mice," *Endocr* 129, 2471–2476.

Pacheco, R., Prado, C. E., Barrientos, M. J. and Bernales, S. (2009). "Role of dopamine in the physiology of T-cells and dendritic cells," *J Neuroimmunol* 216, 8–19.

Padgett, D. A. and Glaser, R. (2003). "How stress influences the immune response," *Trends Immunol* 24, 444–448.

Pallinger, E. and Csaba, G. (2008). "A hormone map of human immune cells showing the presence of adrenocorticotropic hormone, triiodothyronine and endorphin in immunophenotyped white blood cells," *Immunology* 123, 584–589.

Pappu, R., Ramirez-Carrozzi, V. and Sambandam, A. (2011). "The interleukin-17 cytokine family: critical players in host defence and inflammatory diseases," *Immunology* 134, 8–16.

Pesce, M., Speranza, L., Franceschelli, S., Ialenti, V., Patruno, A., Febo, M. A. *et al.* (2011). "Biological role of interleukin-1beta in defensive-aggressive behaviour," *J Biol Regul Homeost Agents* 25, 323–329.

Petrovsky, N. (2001). "Towards a unified model of neuroendocrine-immune interaction," *Immunol Cell Biol* 79, 350–357.

Pinter, E., Helyes, Z. and Szolcsanyi, J. (2006). "Inhibitory effect of somatostatin on inflammation and nociception," *Pharmacol Ther* 112, 440–456.

Plata-Salaman, C. R. (1989). "Immunomodulators and feeding regulation: a humoral link between the immune and nervous systems," *Brain Behav Immun* 3, 193–213.

Plata-Salaman, C. R. (1991). "Immunoregulators in the nervous system," *Neurosci Biobehav Rev* 15, 185–215.

Plata-Salaman, C. R. (2001). "Cytokines and feeding," *Int J Obes Relat Metab Disord* 25(Suppl. 5), S48–S52.

Quan, N. and Banks, W. A. (2007). "Brain-immune communication pathways," *Brain Behav Immun* 21, 727–735.

Raiden, S., Polack, E., Nahmod, V., Labeur, M., Holsboer, F. and Arzt, E. (1995). "TRH receptor on immune cells: in vitro and in vivo stimulation of human lymphocyte and rat splenocyte DNA synthesis by TRH," *J Clin Immunol* 15, 242–249.

Reber, S. O., Obermeier, F., Straub, R. H., Falk, W. and Neumann, I. D. (2006). "Chronic intermittent psychosocial stress (social defeat/overcrowding) in mice increases the severity of an acute DSS-induced colitis and impairs regeneration," *Endocr* 147, 4968–4976.

Reggiani, P. C., Poch, B., Console, G. M., Rimoldi, O. J., Schwerdt, J. I., Tungler, V. *et al.* (2011). "Thymulin-based gene therapy and pituitary function in animal models of aging," *Neuroimmunomodulation* 18, 350–356.

Rettori, V., Fernandez-Solari, J., Mohn, C., Zorrilla Zubilete, M. A., de la Cal, C., Prestifilippo, J. P. *et al.* (2009). "Nitric oxide at the crossroad of immunoneuroendocrine interactions," *Ann NY Acad Sci* 1153, 35–47.

Rochman, Y., Spolski, R. and Leonard, W. J. (2009). "New insights into the regulation of T cells by gamma(c) family cytokines," *Nat Rev Immunol* 9, 480–490.

Rosas-Ballina, M., Olofsson, P. S., Ochani, M., Valdes-Ferrer, S. I., Levine, Y. A., Reardon, C. (2011). "Acetylcholine-synthesizing T cells relay neural signals in a vagus nerve circuit," *Science* 334, 98–101.

Roszman, T. L. and Brooks, W. H. (1985). "Neural modulation of immune function," *J Neuroimmunol* 10, 59–69.

Rothwell, N. (2003). "Interleukin-1 and neuronal injury: mechanisms, modification, and therapeutic potential," *Brain Behav Immun* 17, 152–157.

Royer, H. D. and Reinherz, E. L. (1987). "T lymphocytes: ontogeny, function, and relevance to clinical disorders," *N Engl J Med* 317, 1136–1142.

Russell, S. H., Small, C. J., Stanley, S. A., Franks, S., Ghatei, M. A. and Bloom, S. R. (2001). "The in vitro role of tumour necrosis factor-alpha and interleukin-6 in the hypothalamic-pituitary gonadal axis," *J Neuroendocrinol* 13, 296–301.

Sansone, P. and Bromberg, J. (2012). "Targeting the interleukin-6/Jak/stat pathway in human malignancies," *J Clin Oncol* 30, 1005–1014.

Sapolsky, R. M., Romero, L. M. and Munck, A. U. (2000). "How do glucocorticoids influence stress responses? Integrating permissive, suppressive, stimulatory, and preparative actions," *Endocr Rev* 21, 55–89.

Savino, W., Arzt, E. and Dardenne, M. (1999). "Immunoneuroendocrine connectivity: the paradigm of the thymus-hypothalamus/pituitary axis," *Neuroimmunomodulation* 6, 126–136.

Savino, W. and Dardenne, M. (2010). "Pleiotropic modulation of thymic functions by growth hormone: from physiology to therapy," *Curr Opin Pharmacol* 10, 434–442.

Savino, W., Postel-Vinay, M. C., Smaniotto, S. and Dardenne, M. (2002). "The thymus gland: a target organ for growth hormone," *Scand J Immunol* 55, 442–452.

Schobitz, B., de Kloet, E. R., Sutanto, W. and Holsboer, F. (1993). "Cellular localization of interleukin 6 mRNA and interleukin 6 receptor mRNA in rat brain," *Eur J Neurosci* 5, 1426–1435.

Schroder, K., Hertzog, P. J., Ravasi, T. and Hume, D. A. (2004). "Interferon-gamma: an overview of signals, mechanisms and functions," *J Leukoc Biol* 75, 163–189.

Shelly, S., Boaz, M. and Orbach, H. (2012). "Prolactin and autoimmunity," *Autoimmun Rev* 11, A465–A470.

Singh, V. K., Warren, R. P., White, E. D. and Leu, S. J. C. (1990). "Corticotropin-releasing factor-induced stimulation of immune functions," *Ann NY Acad Sci* 594, 416–419.

Sitte, N., Busch, M., Mousa, S. A., Labuz, D., Rittner, H., Gore, C. *et al.* (2007). "Lymphocytes upregulate signal sequence-encoding proopiomelanocortin mRNA and beta-endorphin during painful inflammation in vivo," *J Neuroimmunol* 183, 133–145.

Skinner, R., Georgiou, R., Thornton, P. and Rothwell, N. (2009). "Psychoneuroimmunology of stroke," *Immunol Allergy Clin North Am* 29, 359–379.

Smaniotto, S., Martins-Neto, A. A., Dardenne, M. and Savino, W. (2011). "Growth hormone is a modulator of lymphocyte migration," *Neuroimmunomodulation* 18, 309–313.

Sompayrac, L. (2012). *How the Immune System Works*, 4th edn. (Hoboken: Wiley-Blackwell).

Srinivasan, D., Yen, J. H., Joseph, D. J. and Friedman, W. (2004). "Cell type-specific interleukin-1beta signaling in the CNS," *J Neurosci* 24, 6482–6488.

Steel, J. C., Waldmann, T. A. and Morris, J. C. (2012). "Interleukin-15 biology and its therapeutic implications in cancer," *Trends Pharmacol Sci* 33, 35–41.

Steer, J. H., Ma, D. T., Dusci, L., Garas, G., Pedersen, K. E. and Joyce, D. A. (1998). "Altered leucocyte trafficking and suppressed tumour necrosis factor alpha release from peripheral blood monocytes after intra-articular glucocorticoid treatment," *Ann Rheum Dis* 57, 732–737.

Stein, C. and Lang, L. J. (2009). "Peripheral mechanisms of opioid analgesia," *Curr Opin Pharmacol* 9, 3–8.

Steinman, L. (2004). "Elaborate interactions between the immune and nervous systems," *Nat Immunol* 5, 575–581.

Stevenson, J. R., Westermann, J., Liebmann, P. M., Hortner, M., Rinner, I., Felsner, P. *et al.* (2001). "Prolonged alpha-adrenergic stimulation causes changes in leukocyte distribution and lymphocyte apoptosis in the rat," *J Neuroimmunol* 120, 50–57.

Sudom, K., Turrin, N. P., Hayley, S. and Anisman, H. (2004). "Influence of chronic interleukin-2 infusion and stressors on sickness behaviors and neurochemical change in mice," *Neuroimmunomodulation* 11, 341–350.

Sutherland, J. S., Goldberg, G. L., Hammett, M. V., Uldrich, A. P., Berzins, S. P., Heng, T. S. *et al.* (2005). "Activation of thymic regeneration in mice and humans following androgen blockade," *J Immunol* 175, 2741–2753.

Szilvassy, S. J. (2003). "The biology of hematopoietic stem cells," *Arch Med Res* 34, 446–460.

Tanaka, M. and Miyajima, A. (2003). "Oncostatin M, a multifunctional cytokine," *Rev Physiol Biochem Pharmacol* 149, 39–52.

Taub, D. D., Murphy, W. J. and Longo, D. L. (2010). "Rejuvenation of the aging thymus: growth hormone-mediated and ghrelin-mediated signaling pathways," *Curr Opin Pharmacol* 10, 408–424.

Then Bergh, F., Kumpfel, T., Yassouridis, A., Lechner, C., Holsboer, F. and Trenkwalder, C. (2007). "Acute and chronic neuroendocrine effects of interferon-beta 1a in multiple sclerosis," *Clin Endocrinol (Oxf)* 66, 295–303.

Tierney, T., Patel, R., Stead, C. A., Leng, L., Bucala, R. and Buckingham, J. C. (2005). "Macrophage migration inhibitory factor is released from pituitary folliculo-stellate-like cells by endotoxin and dexamethasone and attenuates the steroid-induced inhibition of interleukin 6 release," *Endocr* 146, 35–43.

Tonegawa, S. (1985). "The molecules of the immune system," *Sci Am* 253, 122–131.

Torpy, D. J., Tsigos, C., Lotsikas, A. J., Defensor, R., Chrousos, G. P. and Papanicolaou, D. A. (1998). "Acute and delayed effects of a single-dose injection of interleukin-6 on thyroid function in healthy humans," *Metabolism* 47, 1289–1293.

Tosi, P., Kraft, R., Luzi, P., Cintorino, M., Fankhauser, G., Hess, M. W. *et al.* (1982). "Involution patterns of the human thymus. I Size of the cortical area as a function of age," *Clin Exp Immunol* 47, 497–504.

Tsigos, C., Papanicolaou, D. A., Defensor, R., Mitsiadis, C. S., Kyrou, I. and Chrousos, G. P. (1997). "Dose effects of recombinant human interleukin-6 on pituitary hormone secretion and energy expenditure," *Neuroendocrinology* 66, 54–62.

Turnbull, A. V. and Rivier, C. L. (1999). "Regulation of the hypothalamic-pituitary-adrenal axis by cytokines: actions and mechanisms of action," *Physiol Rev* 79, 1–71.

Turovskaya, M. V., Turovsky, E. A., Zinchenko, V. P., Levin, S. G. and Godukhin, O. V. (2012). "Interleukin-10 modulates [Ca(2+)](i) response induced by repeated NMDA receptor activation with brief hypoxia through inhibition of InsP(3)-sensitive internal stores in hippocampal neurons," *Neurosci Lett* 516, 151–155.

Turrin, N. P. and Plata-Salaman, C. R. (2000). "Cytokine-cytokine interactions and the brain," *Brain Res Bull* 51, 3–9.

Turrini, P. and Aloe, L. (1999). "Evidence that endogenous thymosin alpha-1 is present in the rat central nervous system," *Neurochem Int* 35, 463–470.

Umeuchi, M., Makino, T., Arisawa, M., Izumi, S., Saito, S. and Nozawa, S. (1994). "The effect of interleukin-2 on the release of gonadotropin and prolactin in vivo and in vitro," *Endocr J* 41, 547–551.

Vandenberg, L. N., Colborn, T., Hayes, T. B., Heindel, J. J., Jacobs, D. R., Jr., Lee, D. H. *et al.* (2012). "Hormones and endocrine-disrupting chemicals: low-dose effects and nonmonotonic dose responses," *Endocr Rev* 33, 378–455.

Villa-Verde, D. M., Defresne, M. P., Vannier-dos-Santos, M. A., Dussault, J. H., Boniver, J. and Savino, W. (1992). "Identification of nuclear triiodothyronine receptors in the thymic epithelium," *Endocr* 131, 1313–1320.

Wahl, A. F. and Wallace, P. M. (2001). "Oncostatin M in the anti-inflammatory response," *Ann Rheum Dis* 60(Suppl. 3), iii75–iii80.

Wake, H., Moorhouse, A. J. and Nabekura, J. (2011). "Functions of microglia in the central nervous system – beyond the immune response," *Neuron Glia Biol* 7, 47–53.

Wang, H., Yu, M., Ochani, M., Amella, C. A., Tanovic, M., Susarla, S. *et al.* (2003). "Nicotinic acetylcholine receptor alpha7 subunit is an essential regulator of inflammation," *Nature* 421, 384–388.

Wassen, F. W., Moerings, E. P., Van Toor, H., De Vrey, E. A., Hennemann, G. and Everts, M. E. (1996). "Effects of interleukin-1 beta on thyrotropin secretion and thyroid hormone uptake in cultured rat anterior pituitary cells," *Endocr* 137, 1591–1598.

Watanobe, H. and Hayakawa, Y. (2003). "Hypothalamic interleukin-1 beta and tumor necrosis factor-alpha, but not interleukin-6, mediate the endotoxin-induced suppression of the reproductive axis in rats," *Endocr* 144, 4868–4875.

Webster, A. C., Ruster, L. P., McGee, R., Matheson, S. L., Higgins, G. Y., Willis, N. S. *et al.* (2010). "Interleukin 2 receptor antagonists for kidney transplant recipients," *Cochrane Database Syst Rev*, CD003897.

Webster, J. I., Tonelli, L. and Sternberg, E. M. (2002). "Neuroendocrine regulation of immunity," *Annu Rev Immunol* 20, 125–163.

Weesner, G. D., Becker, B. A. and Matteri, R. L. (1997). "Expression of luteinizing hormone-releasing hormone and its receptor in porcine immune tissues," *Life Sci* 61, 1643–1649.

Weigent, D. A. and Blalock, J. E. (1987). "Interactions between the neuroendocrine and immune systems: common hormones and receptors," *Immunol Rev* 100, 79–108.

Welshons, W. V., Nagel, S. C. and vom Saal, F. S. (2006). "Large effects from small exposures. III. Endocrine mechanisms mediating effects of bisphenol A at levels of human exposure," *Endocr* 147, S56–S69.

Windmill, K. F. and Lee, V. W. (1998). "Effects of castration on the lymphocytes of the thymus, spleen and lymph nodes," *Tissue Cell* 30, 104–111.

Xia, H. J., Zhang, G. H., Wang, R. R. and Zheng, Y. T. (2009). "The influence of age and sex on the cell counts of peripheral blood leukocyte subpopulations in Chinese rhesus macaques," *Cell Mol Immunol* 6, 433–440.

Yovel, G., Shakhar, K. and Ben-Eliyahu, S. (2001). "The effects of sex, menstrual cycle, and oral contraceptives on the number and activity of natural killer cells," *Gynecol Oncol* 81, 254–262.

Yu-Lee, L. Y. (2002). "Prolactin modulation of immune and inflammatory responses," *Recent Prog Horm Res* 57, 435–455.

Zoller, A. L., Schnell, F. J. and Kersh, G. J. (2007). "Murine pregnancy leads to reduced proliferation of maternal thymocytes and decreased thymic emigration," *Immunology* 121, 207–215.

Methods for the study of behavioral neuroendocrinology

Behavioral neuroendocrinology involves the study of the interactive effects of steroid and peptide hormones, neuropeptides, cytokines and neurotransmitters on behavior. Previous chapters have mentioned the role of hypothalamic nuclei in behavior (section 4.1), the behavioral effects of neurotransmitter agonists and antagonists (section 5.8.4), the neuroendocrine correlates of psychiatric disorders (section 6.8), the behavioral functions of steroid hormones (sections 9.9.3 and 9.9.4), the cognitive and behavioral effects of neuropeptides (section 12.7) and the effects of cytokines on the brain and behavior (section 13.5). The present chapter discusses behavioral methods used in the study of neuroendocrinology, the neural and genetic mechanisms mediating the effects of hormones on behavior, and some of the special problems involved in conducting behavioral neuroendocrinology research.

Neuroendocrine research utilizes several specific methods such as immunoradiometric assays for quantifying hormone levels (Chapter 8), immunohistochemistry (Chapter 9) and immunofluorescence techniques for localizing hormones (Chapters 11 and 12). The study of behavioral neuroendocrinology relies on specific behavioral methodologies or behavioral bioassays. As discussed in Chapter 8 (section 8.1.3), a bioassay measures physiological changes in an animal or cell culture to determine the concentration or potency of a hormone in the circulation. Thus, for example, the size of a cock's comb is a bioassay for blood testosterone level and in rats the size and weight of the adrenal glands is a bioassay for the level of circulating ACTH or corticosterone. A behavioral bioassay measures behavioral changes to estimate the concentration or potency of a hormone.

14.1 Behavioral bioassays

A behavioral bioassay requires precise qualitative (verbal) descriptions of the behaviors of interest and accurate quantitative (mathematical) measures of the latency, frequency and duration of these behaviors. Thus, the measurement of behavior involves two stages: the observation and description of units of behavior and the quantitative measurement of these behavior units. Before these procedures can begin, however, one must determine which behaviors to record.

14.1.1 What behavior is to be recorded?

Behavior can be classified into two broad groups: ethological (unconditioned) responses and conditioned responses, as shown in Table 14.1. Ethological studies involve the observation and

Table 14.1 Some examples of behavioral measures used in neuroendocrine research

A. Ethological (unconditioned) responses

- locomotion (walking, swimming)
- grooming
- exploratory behavior (open field test)
- feeding, food preferences
- drinking, taste preferences
- "emotional behavior" (defecation, freezing, escape, avoidance)
- scent-marking
- vocalizations (audible and ultrasonic)
- social behavior (huddling, play, sexual, parental, aggressive, defensive)

B. Conditioned responses

- habituation/dishabituation
- conditioned suppression (conditioned emotional response)
- passive and active avoidance conditioning
- taste/odor aversion tests
- maze learning for food or water reward
- operant conditioning for food or water reward
- brain stimulation reward learning

C. Developmental parameters

- developmental reflexes (self-righting, auditory startle, negative geotaxis)
- spontaneous movements (crawling, standing, climbing, swimming)
- homing to nest
- vocalizations
- exploration
- learning

D. Abnormal behaviors

- hyperactivity
- "illness"-related behavior (vomiting, lethargy)
- feeding disorders (anorexia, obesity)
- affective disorders (anxiety, depression)
- neuromuscular disorders (dyskinesias)
- social disorders (extreme aggression or passivity)
- cognitive disorders
- developmental delays

recording of "natural" behaviors, such as grooming, vocalizations, sexual, parental and aggressive behavior in a laboratory or field setting (Kelley 1989; Krsiak 1991). Studies involving conditioned responses usually involve training and testing procedures in the laboratory (Adams 1986). Special categories of behavioral tests have been developed to measure the development of behavior (Zbinden 1981) and the display of abnormal behavior (Baumeister and Sevin 1990).

14.1.2 Observation and description

Before any neuroendocrine manipulation is carried out, the normal sequence of behavior must be observed and described in terms of the behavior units involved. These units include identifiable postures, motor acts, vocalizations, etc., which are stereotyped in form, duration or orientation and can be easily recognized by the observer. A description of sexual behavior of the male rat, for example, can be divided into a number of behavioral units such as: approach female, investigate (sniff female), mount female without intromission, mount with intromission, mount with ejaculation and ultrasonic vocalization (Brown and McFarland 1979). Using such behavior units, the sequence of male rat copulatory behavior can be described as shown in Figure 14.1. Other uncondi-tioned behaviors such as grooming, aggression and parental behavior can be described using similar qualitative methods.

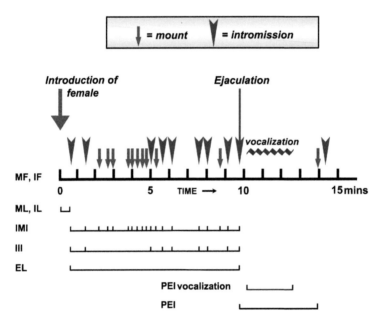

Figure 14.1 Example of temporal patterns and units of sexual behavior in a male rat

Behavioral units recorded: mount = mount female without intromission; intromission = mount female with intromission; ejaculation = mount female with intromission and ejaculation; post-ejaculatory vocalization = 22-kHz ultrasonic vocalizations of the male. Quantitative measures recorded: mount frequency (MF) = total number of mounts which occur in an ejaculatory series; intromission frequency (IF) = number of mounts with intromission in each ejaculatory series, including the ejaculatory thrust; mount latency (ML) = time from the introduction of the female until the first mount; intromission latency (IL) = time from the introduction of the female until the first intromission; inter-mount interval (IMI) = time between each mount in an ejaculatory series; inter-intromission interval (III) = time between each intromission in an ejaculatory series; ejaculation latency (EL) = time from the first mount to an ejaculation; this is called an ejaculatory series; post-ejaculatory interval (PEI) = time from the ejaculation to the first mount of the next ejaculatory series. There is a 22-kHz post-ejaculatory vocalization during the PEI that divides it into an absolute refractory period (time from ejaculation to the end of the song) and a relative refractory period (time from the end of the post-ejaculatory song to the first mount of the next ejaculatory series). Figure redrawn from Brown and McFarland, 1979.

14.1.3 Quantitative measurement of behavior units

To measure changes in behavior associated with neuroendocrine activity, each behavior unit must be quantified by measuring its latency, frequency or duration within a specific test session (see Figure 14.1). The accurate quantitative measurement of behavior units is a skill which requires a knowledge of experimental design, sampling procedures and recording methods, and usually requires the use of some apparatus such as check sheets, event recorders, videotapes, audiotapes, sound analyzers or automatic recording devices such as photocells or microswitches. Computer programs are now available for recording behavioral observations. If behavioral bioassays are to be reliable and valid measures of hormonal concentration or potency, considerable effort must be made to ensure the accuracy of the behavioral measures used (see Lehner 1979; Martin and Bateson 1986; and Donat 1991).

14.2 Correlational studies of hormonal and behavioral changes

Behavioral changes can be correlated with naturally occurring hormonal fluctuations or the hormonal changes associated with endocrine tumors or diseases (Beach 1974; 1975). This type of study measures hormone levels and behavioral variables simultaneously and looks for a relationship between them, without any experimental manipulation of the hormones or the behavior.

14.2.1 Natural fluctuations in hormone levels and behavior

Many behavioral changes occur in conjunction with normal fluctuations in hormone secretion. In female mammals, for example, characteristic changes in the circulating levels of estradiol and progesterone occur at puberty, during the reproductive cycle and during the mating season in seasonally breeding animals and at menopause when human cycles cease. Behavioral changes can be correlated with the fluctuations in gonadal hormone levels. In women, for example, sleep patterns are modified by variations in levels of estradiol and progesterone such as occur during the menstrual cycle, pregnancy and menopause. Sleep patterns also respond to oral contraceptive use in young women and hormone replacement therapy in post-menopausal women (Shechter and Boivin 2010; Hachul *et al.* 2008). Figure 14.2 shows the fluctuations in aggressive behavior in female mice correlated with estradiol and progesterone levels over their estrous cycles. In this experiment, female mice from each stage of the estrous cycle were confronted with an intruder mouse. The time taken for the intruder to be attacked (latency) was dependent on hormone levels during the cycle (Davis and Marler 2003). The attack took place significantly faster when estradiol levels were at their lowest in diestrus 3 (D3) compared to proestrus (PRO) and estrus (see also Hyde and Sawyer 1977).

Hormone levels can also be correlated with daily (circadian) or seasonal rhythms of daylight or temperature. Sex differences, as well as gonadal hormones, modify circadian rhythms and sleep in humans (Mong *et al.* 2011). The timing of gonadal hormone release and reproductive behavior in seasonally breeding animals, such as domestic ducks, can also be correlated with seasonal changes in day length, as shown in Figure 14.3. Sexual behavior, such as mounts and copulations, are associated with the rise in plasma testosterone levels. However, social displays, a form of courtship behavior, occur earlier than the peak in testosterone and may therefore be responsive to the gradual increase in testosterone. An alternative possibility is that social displays are dependent on additional hormones such as FSH (Balthazart and Hendrick 1976).

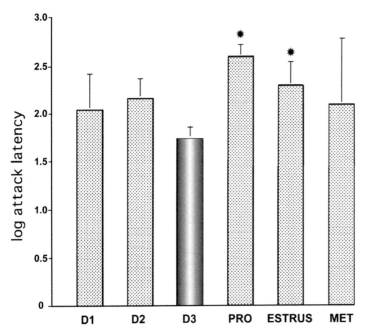

Figure 14.2 Attack latencies of female mice exposed to an intruder mouse across the estrous cycle

Attack latency (measured in seconds) varied through the estrous cycle, with D3 females (red) attacking significantly faster than either PRO or Estrus females. Values are mean ± SEM
(n = 3–12 per group; *p < 0.05).

Abbreviations: D, diestrus; PRO, proestrus; MET, metestrus. Reproduced with permission (Davis and Marler 2003).

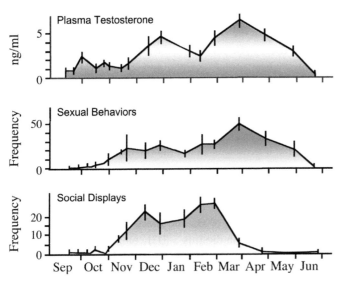

Figure 14.3 Seasonal variation in plasma testosterone (T) levels correlated with annual changes in sexual behavior and social displays in male domestic ducks

Data (means ± standard errors; n = 5 per group) reveal a high correlation between changes in plasma T levels and in the frequencies of sexual behaviors (mounts and copulations). The annual peak in social displays, a form of courtship behavior, occurred *before* the maximal levels of plasma T were reached, indicating that either these behaviors were maximally activated by rising levels of T (in November and December) or their activation depends at least in part on other hormones, such as FSH. Reproduced with permission (Ball and Balthazart 2004).

14.2.2 Endocrine and behavioral abnormalities

Diseases or malfunctions of the endocrine system, whether occurring in adulthood or during development, can result in changes in behavior. For example, chronic stress is accompanied by high cortisol secretion and is a risk factor for depression (de Kloet *et al.* 2005). Similarly, patients with Cushing's Syndrome have excessive cortisol secretion and a number of cognitive and behavioral symptoms, including depression, irritability and anxiety (Pereira *et al.* 2010). These symptoms improve when cortisol secretion is normalized. Abnormal hormone secretion in the prenatal period may lead to behavioral changes later in development. For example, disturbances of gonadal hormone secretion during sexual differentiation may influence later social and sexual behavior (Balthazart 2011; Wallen and Hassett 2009). Diminished levels of thyroid hormones during early development leads to stunted bodily growth, and neuroendocrine and cognitive deficits (Carreon-Rodriguez and Perez-Martinez 2012). Depression and schizophrenia can also be correlated with alterations in the neuroendocrine system (de Kloet *et al.* 2005; Walker *et al.* 2008).

14.2.3 The problem of hormone sampling during behavioral studies

In order to correlate hormonal and behavioral changes, hormone levels need to be measured at the same time as behavior is observed. This can be done by direct measures of the hormones in the circulation or by indirect measures.

Direct (invasive) measures

Direct measures of hormone levels depend on taking samples of blood or cerebrospinal fluid and then using bioassays or ELISA to measure hormones of interest in these bodily fluids (see section 8.1). While this procedure provides an accurate measure of the circulating level of the hormone, it has the disadvantage that such direct sampling methods are likely to disrupt the behavior being observed. For example, to correlate testosterone level with sexual behavior in male rats, one might take blood samples every 10 minutes. But taking these blood samples using a syringe or tail catheter involves restraining the animal and such restraint is stressful for the rat and disrupts the behavior being observed. The stress hormones released during the restraint period may confound the effects of testosterone (see Figure 14.11). To allow stress-free blood sampling during behavioral studies, many techniques have been developed for implanting chronic indwelling cannulae. Figure 14.4 shows an intravenous (jugular) catheterization which is used for taking blood samples from freely moving mice and rats during behavioral studies (Kelley *et al.* 1997). Animals can be easily detached, or reattached, to the tether during experiments.

Indirect (non-invasive) measures

Because it is not always possible to collect blood samples by direct means during behavioral studies, a number of indirect methods of determining hormone levels may be used. These include the analysis of hormones or their metabolites in the urine, saliva and feces (see section 8.1) and the use of bioassays. Because many hormones or their metabolites are eliminated in the urine (Table 7.3), urine samples can be used as an assay for the levels of certain hormones. Methods for the analysis of urinary hormones or their metabolites are well established (McWhinney *et al.* 2010) and provide an accurate estimate of the level of the hormone in the blood. Urinary hormone analysis is a practical, non-stressful method for monitoring hormone levels in children, adults, wild animals or animals in zoos (Anestis 2006; Andelman *et al.* 1985). Steroid hormone levels may be analyzed from saliva samples, as discussed in section 8.1, and the saliva can be collected using cotton swabs or chewing gum (Dabbs 1991; Higashi 2012).

Bioassays are also useful in behavioral studies where hormone levels are changing. For example, the vaginal secretions of female rodents are used as a bioassay for estradiol

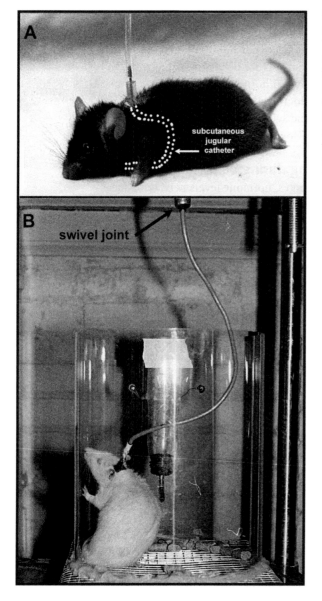

subcutaneous jugular catheter

B

swivel joint

Figure 14.4 A. **A mouse with a chronic intravenous cannula inserted**
The dotted outline shows the path of the indwelling catheter, under the skin, leading from the jugular vein in the neck to the plastic housing behind the head. The housing can be attached to plastic tubing (as shown) either for blood withdrawal or for drug injections.
B. Illustrates how the experimental animal (in this case a rat) is unrestrained and free to move around. The plastic tubing is connected to a swivel that permits unrestricted movement. Reproduced with permission (Kelley *et al.* 1997; Bertram *et al.* 1997).

and progesterone levels (Johnson and Everett 1988; Everett 1989). This is a simple technique able to determine the stage of the estrous cycle, and therefore the hormonal status of the animal (see Figure 14.5). Conscious rodents are gently picked up during daily swabbing of the vagina, and the epithelial cells obtained are then examined on a microscope slide (Marcondes *et al.* 2002). This method is invaluable for behavioral studies, enabling the phase of the estrous cycle of female rodents to be determined as shown in Figure 14.5. This approach dispenses with the need to collect and analyze blood samples. Changes in behavior can thus be correlated with vaginal cell type and, indirectly, hormone levels, for example as shown for aggressive behavior in Figure 14.2.

A. RODENT ESTROUS CYCLE

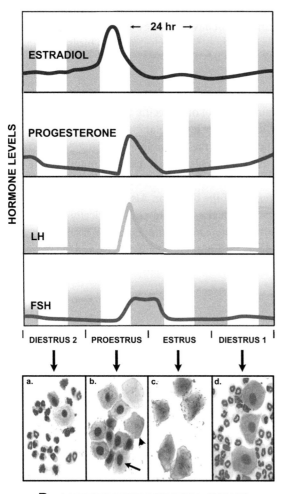

B. CORRESPONDING VAGINAL SMEARS

Figure 14.5 Association of hormonal changes and vaginal smears in the rodent estrous cycle

A. Illustration of the hormonal changes that take place through a typical four-day rodent estrous cycle. Each cycle is divided into light/dark daily periods (shaded bars; 12 hours) and is designated as sequential days: diestrus 1, proestrus, estrus and diestrus 2. The fluctuations in hormone levels in blood occur in a precise way: proestrus day is characterized by a large peak in estradiol secretion, followed, later in the day, by progesterone and luteinizing hormone (LH), and then follicle-stimulating hormone (FSH). Estrus is the day when the female is most responsive to the male, and the female will adopt a lordosis posture in response to the male (see Figure 9.16). B. Illustrates photomicrographs showing the characteristic cytology of vaginal smears obtained from rats in diestrus 2 (a), proestrus (b), estrus (c) and diestrus 1 (d). Note that when steroid hormone levels are low, i.e. in diestrus 1 and diestrus 2, the smears show predominantly small leukocytes. *In contrast*, when estradiol levels are high, as in proestrus, the smear is characterized by large round nucleated epithelial cells (arrow) and occasional larger, non-nucleated cells (arrowhead). In estrus, the day of maximal sexual receptivity, the vaginal smear consists of large, flat, non-nucleated cells ("cornified" cells). Reproduced with permission (A. Staley and Scharfman 2005; B. Brack *et al.* 2006).

14.3 Experimental studies I: behavioral responses to neuroendocrine manipulation

Correlational studies provide an assessment of the relationship between behavioral and neuroendocrine variables, and, if the correlation is high enough, allow prediction of the behavioral changes associated with changes in levels of particular hormones. It is not, however, possible to infer that the hormonal changes cause the correlated change in behavior. In order to show that a change in hormone level causes a change in behavior, or vice versa, systematic manipulation of the hormone or the behavior in a controlled experiment is necessary (Leshner 1978). The level of a hormone can be manipulated as the independent variable in an experiment, and behavior measured as the dependent variable during each phase of the hormonal manipulation, or behavior can be manipulated as the independent variable, and hormonal responses measured as the dependent variable. While the majority of the experiments discussed in this chapter concern the effects of steroid hormones on mammalian behavior, similar problems are faced in studying the effects of neuropeptides, such as opioids and cytokines, on behavior (Kelley 1989; Kelley *et al.* 2005) and in studying behavioral neuroendocrinology in other species, such as birds (Balthazart 1983; Balthazart *et al.* 2009).

14.3.1 The standard hormone removal and replacement experiment

Three criteria must be met before the hormonal control of a specific behavior can be demonstrated. First, removal of the hormone or its source (e.g. the gland secreting that hormone) should modify the behavior. Second, replacement of the hormone by grafting a gland from another animal or, more specifically, by hormone replacement therapy should restore the behavior to its original level. Third, variations in hormone concentration should result in correlated changes in the frequency or intensity of the behavior. Also, the hormone replacement therapy must activate the behavior at dose levels that are physiologically meaningful (Balthazart 1983). Experiments involving the systematic manipulation of hormone levels, therefore, have three phases: the baseline phase, the hormone removal phase and the hormone replacement phase (Figure 14.6), and behavioral tests can be applied in each phase.

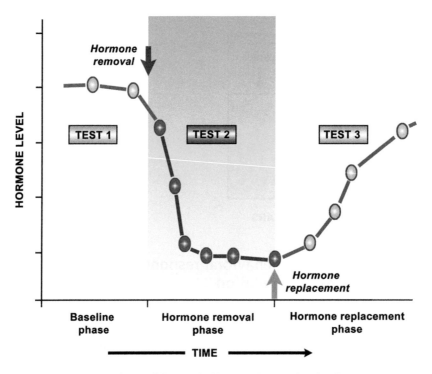

Figure 14.6 The three phases of the standard hormonal removal and replacement experiment
Hormones are normally removed from the test animals by surgery; for example, blood levels of testosterone are lowered by castration, and glucocorticoid levels are lowered by adrenalectomy. The baseline phase is therefore the time before surgery, followed by the hormone removal phase when hormone levels fall. The hormone replacement phase occurs when animals are injected with the hormone (e.g. testosterone) in order to restore normal circulating levels. Behavioral tests (bioassays) are conducted one or more times during the baseline, hormone removal and hormone replacement phases. A typical experiment is shown in Figure 14.7.

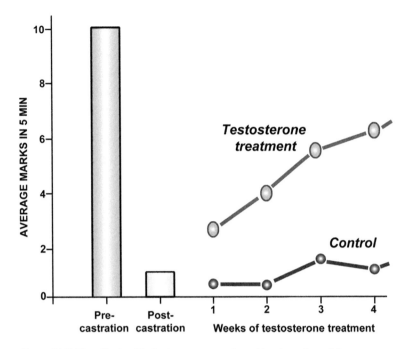

Figure 14.7 The effects of testosterone on scent-marking in male gerbils
Vertical bars indicate the frequency of ventral gland scent-marking in a 5-minute test period by adult male gerbils both before (yellow) and after (blue) castration. The frequency of scent-marking is therefore dependent on testosterone. This is confirmed by injecting the castrate gerbils with testosterone. Such treatment restores the high frequency of scent-marking over several weeks compared to the effects of injecting the vehicle by itself (control). This figure is based on data originally published by Yahr and Thiessen (1972).

An example of such an experiment is shown in Figure 14.7. Male gerbils use their ventral sebaceous gland to scent-mark their environment. To investigate the relationship between scent-marking frequency and testosterone levels, testosterone was removed by castration. Males were tested in the baseline (pre-castration) period and were tested again after removal of the testes (post-castration). Testosterone replacement injections were then given for four weeks and scent-marking was recorded each week. Castration reduced the frequency of scent-marking almost to zero and testosterone replacement increased scent-marking frequency. The control group of castrated males given only the vehicle and no testosterone injections did not show a significant increase in scent-marking frequency (Yahr and Thiessen 1972). Conducting this type of experiment requires careful consideration of the methods to be used for hormone removal and replacement.

14.3.2 Methods of hormone removal

There are three general methods of hormone removal: surgical, pharmacological and immunological. Strictly speaking, the latter two approaches remove the *effects* of hormones – for example, by blocking specific receptors. A fourth approach is to prevent the action of the hormone by removing the receptor protein.

Surgical methods (gland removal or ablation)

The traditional type of experimental procedure in behavioral neuroendocrinology is to surgically remove a gland (e.g. testes or thyroid) and observe physiological or behavioral changes in the subject. Thus, the testes of a rooster can be removed (castration) and the size of the comb measured (bioassay) or the amount of crowing, courtship, mating and aggressive behavior recorded (behavioral bioassay). If these are all found to decrease after castration, testes implants or purified hormone (testosterone) can be given in replacement therapy to see if this reverses the effect of the gland removal. Such an experiment was first conducted by Berthold in 1849 (Beach 1981). In the male rat, for example, the sex behavior shown in Figure 14.1 is completely absent when the testes are removed, but can be reinstated following testosterone treatment.

Surgical removal is possible for many endocrine glands (testes, ovaries, adrenal glands, thyroid, etc.), but some surgical procedures are complex (e.g. surgical removal of the pituitary gland) and, since many glands produce two or more hormones, elimination of the other hormones produced by the gland may confound the experiment. For example, surgical removal of the adrenal glands will eliminate both glucocorticoids and the mineralocorticoids (from the adrenal cortex; see Figure 2.5). Also, since it is difficult to dissect the adrenal cortex without also removing the adrenal medulla (see Figure 2.5), the source of epinephrine and norepinephrine is also eliminated by adrenalectomy. Thus, the controlled study of the effects of individual hormones on behavior may require more selective approaches than simply the removal of glands.

Pharmacological methods

It is possible to use drugs to prevent specific hormone synthesis and release from endocrine glands. For example, since we know that the effects of testosterone are often mediated via its conversion to estradiol (see Figure 9.14), it is feasible to block the production of estradiol by inhibiting the enzyme *aromatase* (i.e. the enzyme that converts testosterone to estradiol). Other drugs act as hormone antagonists at target cell receptors. Flutamide, for example, is an androgen antagonist that blocks testosterone receptors, thus preventing androgen-dependent male sexual behavior (see Figure 14.1; Gladue and Clemens 1980; Harding and McGinnis 2004). Likewise, tamoxifen, an estradiol antagonist, can inhibit sexual behavior in female rats (Patisaul *et al.* 2004). Many drugs that modify neurotransmitter function in the brain also alter hormone release and sexual behavior. For example, serotonin (5-HT) plays a role in both male and female sexual behavior. Thus, reduction of 5-HT function, by blocking serotonin receptors, facilitates, whereas stimulation of receptors inhibits sexual behavior (Olivier *et al.* 2011). Since modification of brain neurotransmitter or neuropeptide function also affects hormone secretion (see Chapters 6 and 12), hormone-dependent behavior is also changed.

Immunological methods

It is impractical to obtain drugs that are totally specific for hormone receptors, so immunological methods were developed to produce antibodies that are more specific, not only for receptors, but for the hormones as well. It is therefore theoretically possible to block receptors, or the action of hormones, by binding them to antibodies. In male pigs, for example, testosterone levels and sex behavior are significantly reduced by vaccination against GnRH (gonadotropin-releasing hormone). The production of antibodies to GnRH effectively blocks the release of LH from the anterior pituitary (see Figure 4.3) and therefore the production of testosterone by the testes (Zamaratskaia *et al.* 2008). Since highly specific antibodies are available for the assay of many hormones and peptides (see Chapter 8), these make possible the investigation of their role in behavior following direct systemic injection into animals or following injection into specific areas of the brain. For example, brain injection of an antibody to oxytocin delays the onset of maternal behavior in rats, revealing the crucial importance of brain oxytocin in this behavior (see section 12.7.2; Pedersen *et al.* 1985).

Molecular biological methods
Hormonal effects on behavior can also be investigated by eliminating, rather than blocking, the hormone receptor site. Hormone receptor genes can be inactivated in experimental animals, thus largely eliminating the receptor proteins. For example, reproductive behaviors in female rats are dependent on sex hormones such as estradiol acting at estrogen receptors (ER). The specific inactivation of the ERα gene, particularly in the hypothalamus, reduces sexual incentive motivation, receptivity and copulatory behavior (Spiteri *et al.* 2010; Tetel and Pfaff 2010; see also section 9.6.1). Similar studies show that brain androgen receptors constitute a critical component of male sexual behavior (Raskin *et al.* 2009).

14.3.3 Methods of hormone replacement
Hormone replacement can be achieved by replacing an entire gland, such as an ovary, by injecting purified hormones or by injecting synthetic hormones.

Gland replacement
In studies where an entire gland is surgically ablated, that gland or one from another animal can surgically replace it and, as long as it can be reattached to the circulatory system, hormone release can be reinstated. Thus, testes or ovaries can be removed and replaced beneath the skin or in the body cavity where they become revascularized and continue to secrete their hormones. The pituitary gland can be surgically removed from the head and transplanted beneath the skin or under the kidney capsule. There, free from the inhibitory action of hypothalamic inhibiting factors such as dopamine (see Figure 6.8), these transplanted pituitaries release high levels of prolactin into the bloodstream.

Injection of purified hormones
Rather than replacing an entire gland, purified hormones can be injected. Many purified hormones are obtained from the endocrine glands of cattle (bovine), sheep (ovine), pigs (porcine) or horses (equine). Thus, purified ovine prolactin, porcine growth hormone or pregnant mare's serum gonadotropin (equine) are available for medical or scientific use. Some hormones, however, are species specific and in these cases hormones cannot be used for replacement therapy in other species. For example, ovine and bovine growth hormones have only about 60 percent identity with human growth hormone in their amino acid structure and thus are inactive in humans. Human patients requiring growth hormone or insulin injections, for example, require human hormones, which can be purified from human pituitary glands or pancreas taken from organ donors. However, such treatment

was revolutionized through the availability of recombinant (genetically engineered) hormones. Thus, children with short stature and adults with growth hormone deficiency are now treated with recombinant growth hormone (Richmond and Rogol 2010).

Injection of synthetic hormones

Knowledge of the chemical structure of hormones, such as steroids (see Figure 7.3), enabled the production of synthetic hormones to serve as hormone replacement therapy in animals and humans. A wide variety of synthetic hormones are available, some of which mimic exactly the effects of the natural hormones (full agonists) and some of which are only partial agonists, having some, but not all, of the effects of the natural hormone, due to differences in their chemical structure. Some synthetic steroid hormones used in behavioral studies include testosterone propionate (a synthetic androgen), ethinyl estradiol (a synthetic estrogen), norethindrone (a synthetic progestin) and dexamethasone (a synthetic glucocorticoid). In the case of peptide and neuropeptide hormones, gene sequencing permits the production of recombinant hormones that act identically to the native peptides. Alternatively, some shorter peptides, such as kisspeptin (see Figure 6.7 and section 12.2.1), can be synthesized from their constituent amino acids (Jayasena *et al.* 2009).

14.3.4 Factors to consider when giving hormone replacement therapy

Although the term "hormone replacement" appears quite simple, there are a number of factors to be considered before giving hormone replacement therapy. These include: (1) chemical preparation of the hormone; (2) the dose to be used; (3) the vehicle in which the hormone is given; (4) the route of administration; and (5) the timing of the injections.

Chemical preparation

In addition to the natural hormones, there are many synthetic preparations of each hormone, some of which differ in their potency and therefore they may have different effects on behavior. Birth control pills, for example, utilize a number of different synthetic estrogens and progestins, some of which may have unwanted side effects. Different hormone preparations may be substituted to avoid these side effects. Likewise, some synthetic androgens are aromatizable to estradiol, while others are not. These can be used to determine whether the behavioral effects of androgens are due to their action at androgen or estradiol receptors (see Chapter 9).

Dose

Once the type of hormone to be used is chosen, the dose to be injected must be estimated. Since different doses of a hormone may have different effects on behavior, studies are required to determine dose-response relationships. Figure 14.8, for example, shows that fighting behavior is almost eliminated in castrated male mice, but can be reinstated via increasing doses of injected testosterone. Numerous behaviors respond to hormone replacement in this dose-dependent fashion (see also Figure 14.7).

Vehicle

Hormones are administered in a vehicle such as distilled water, physiological saline or oil. The selection of the vehicle depends on the hormone to be administered. For example,

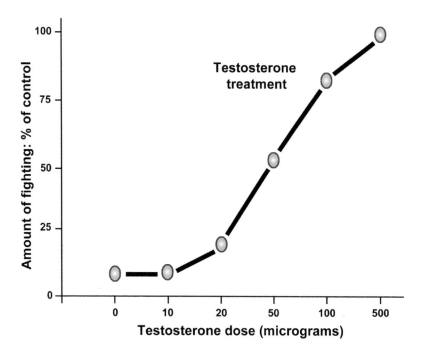

Figure 14.8 **Effects of testosterone treatment on aggression**
The percentage of castrated adult male mice fighting, compared to castrate controls, at each dose of
testosterone. Based on data obtained from Edwards (1969).

peptide, protein and neuropeptide hormones are soluble in physiological saline. On the
other hand, most steroid hormones are insoluble in saline and they have to be prepared as
suspensions in corn oil or peanut oil. The control groups of animals, such as those in
Figure 14.8, are injected with the vehicle alone without any hormone. Some steroids may be
incorporated into pellets implanted subcutaneously (see below).

Route of administration

Some hormones, such as steroids, can be given orally; for example, women taking birth
control pills. However, subjecting experimental animals to oral treatments is stressful.
Rats and mice, for example, need to be restrained so that the hormone solution can be
forcibly placed in the mouth. Nevertheless, oral treatments are valuable when the
hormone can be included in the drinking water. For example, this non-stressful, non-
invasive technique revealed that the hormone melatonin (section 2.2.1), included in
drinking water, significantly increased nocturnal activity in rats (Terron *et al.* 2013;
Figure 14.9). This method can be readily be applied to many drugs and hormones, such as
corticosterone (Olausson *et al.* 2013).

Other drugs, such as insulin, are normally injected. Injections can be subcutaneous (s.c.),
intramuscular (i.m.) or intraperitoneal (i.p.). Hormones can also be injected into discrete
brain regions by intracerebral injections using stereotaxic surgery. Such an approach is
described in section 13.3.2 (Spiteri *et al.* 2010) and a good illustration of this brain injection
technique can be found in Ikemoto and Wise (2004). This technique can be modified so that

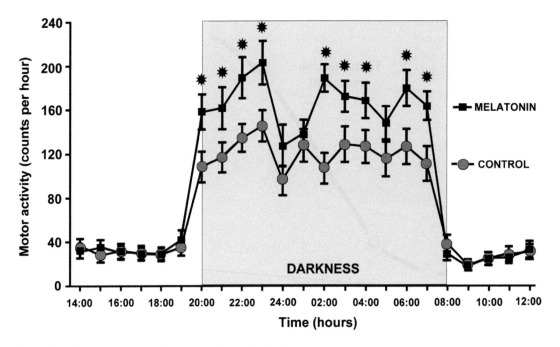

Figure 14.9 Treatment of rats with oral melatonin via drinking water
The data show the diurnal and nocturnal activity of Wistar rats, untreated (control), and treated with melatonin in tap water (20 µg per ml). Activity of the rats, in counts per hour, was measured electronically by an infrared activimetry system. Note the increase in activity that normally occurs in darkness (lights off, 20:00; lights on, 08:00) in control rats. Exposure of rats to melatonin in the drinking water, over a period of two weeks, induced a significant increase in activity during darkness, but not in the light. Each value represents the mean ± SEM of the values obtained from groups of ten rats. * $p < 0.05$ with regard to the corresponding control values. Reproduced with permission (Terron *et al.* 2013).

the experimental animal is unrestrained and freely moving, thus providing a non-stressful treatment period (see, e.g., Ludvig *et al.* 2002). Similarly, hormones can be microinjected into the cerebral spinal fluid (CSF) in the ventricles by intracerebroventricular injections.

One advantage of the drinking water approach to hormone treatments is that experiments can easily be continued for days or weeks. An alternative to this involves hormones that can be implanted, usually subcutaneously, in pellets or controlled release capsules made from silastic tubing. Because the silastic tubing is porous and allows the hormone to diffuse into the circulation at a constant rate, these capsules are often used for steroid hormone implants. Different lengths of tubing can be used for different hormone doses and these capsules may continue to secrete hormones at a constant rate, producing physiological blood levels, for several weeks (Jones *et al.* 2012; Isaksson *et al.* 2011). This approach has been standardized and commercialized and many different types of hormone are now available as small pellets made from a biodegradable matrix that effectively and continuously releases the active product in the animal (Innovative Research of America; www. innovrsrch.com/pellets.asp). A wide variety of dose levels and release rates allow experiments to be performed over periods up to three months (see, e.g., Brown *et al.* 2012). However, there is evidence that silastic pellets, at least for estradiol, are superior to commercial pellets and injections (Strom *et al.* 2008). Finally, an alternative, and also commercially available, approach to hormone treatments in laboratory animals is the use of osmotic minipumps. ALZET osmotic pumps are small (1 to 2 centimeters long), implantable pumps that deliver hormones at continuous and controlled rates over periods up to six weeks, without the need for frequent handling (www.alzet.com/). They can be used for a wide variety of hormone exposures, including leptin and estradiol (Cui *et al.* 2011; Sarvari *et al.* 2010).

Timing of hormone replacement and behavioral testing

The study of behavioral effects of hormones in experimental animals is best accomplished when the source of endogenous hormone has been removed – for example, when estradiol is eliminated by ovariectomy. The timing of the hormone replacement and the temporal patterning of the injections with regard to hormone removal and behavioral observations must also be considered, especially for sex steroid hormones such as estradiol and testosterone. Hormone replacement can be given immediately after hormone removal, in which case there is little time for the circulating levels of hormones to decline before replacement is given. On the other hand, hormone replacement can be delayed for some days or weeks to allow baseline hormone levels to decline after gland removal.

When glands are surgically removed, and steroid hormone levels fall, compensatory changes in receptor numbers, with up- or down-regulation of receptors, occur (discussed in sections 9.5 and 10.1.4). In order to restore hormone responsiveness once receptor numbers are down-regulated following gland removal, it may require one or more hormone treatments before the receptors are restored to their former levels. Thus, some hormone pretreatment may be necessary before behavioral changes occur in response to the hormone replacement regimen. Figure 14.7, for example, shows that the frequency of scent-marking by castrated male gerbils in response to testosterone injection is low on the first week of replacement therapy and reaches an asymptote by the fourth week. How frequently the hormone injections are given depends on the active life of the hormone in the circulation. Some hormones can be injected once per day, others once every two days and others twice per day. Hormone release from pellets or pumps overcomes these problems. Figure 14.10 shows the level of estradiol in the blood of ovariectomized female rats from 30 minutes to 24 hours after subcutaneous estradiol injection. The level of circulating estradiol quickly reaches a peak 30 minutes after the injection and then gradually declines to within the physiological range by 4 hours (Isaksson *et al.* 2011). Other workers, also working with ovariectomized rats, reported a successful attempt to reproduce the normal variation in estradiol levels, characteristic of the estrous cycle, following injection of estradiol benzoate. These injections mimicked the rapid rise and subsequent decline of blood estradiol that occurs on the day of proestrus (Schwartz and Mong 2013; see Figure 14.5 for the normal variation in estradiol levels that occurs through the estrous cycle in intact female rats).

Finally, the timing of the behavioral tests with regard to the hormone injections must be determined. Should tests be given immediately after the hormone injection or some hours later? While Figure 14.10 suggests that testing could be done at 1 to 2 hours after estradiol injection (i.e. following peak levels), it could be argued that 4 to 5 hours might be more appropriate, when levels are once more within the physiological range. In contrast, other

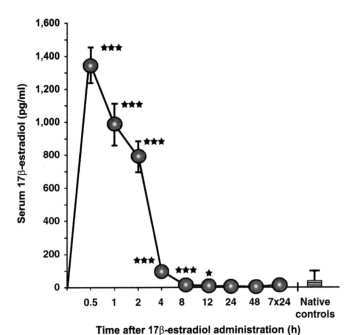

Figure 14.10 Blood levels of estradiol following a single injection in ovariectomized rats
Female rats were ovariectomized (14 days) and then given a single subcutaneous injection of estradiol (10 μg/kg in sesame oil). Blood samples were obtained by venipuncture from a hind limb at the times shown and assayed for estradiol content by radioimmunoassay. The native control group consisted of ten ovary-intact rats sacrificed during estrus and proestrus to provide an estradiol value for the normal physiological range. Values are means ± SEM (ten rats per group). Note that blood levels of estradiol rapidly reach peak values (approximately nine times the physiological level) at 30 minutes and fall to within the physiological range within 4 hours. *** $p < 0.001$, and * $p < 0.05$ vs. pre-injection levels. Data were generously provided by Dr. J. O. Strom (see Isaksson *et al.* 2011).

hormones that have very short half-lives should be tested sooner after the injection. It is therefore important that blood levels of each hormone be determined as shown in Figure 14.10.

As noted above, when hormone capsules are implanted, the level of hormone released may remain constant for up to three months and testing can be done over an extended period following implantation without repeated injections being necessary. While this has many advantages, in some circumstances there are disadvantages in the use of sustained release capsules. For example, some hormones, such as GnRH, are active only when released in pulses (see Figure 7.4). Constant release of GnRH, as occurs with sustained release capsules, causes GnRH receptor down-regulation, so that the high constant levels of GnRH inhibit rather than stimulate their pituitary target cells. Since GnRH pulses occur naturally about once every 90 minutes (see Figure 4.4; section 4.3.3), injections given this frequently are impractical. To solve this problem, experimental animals can be stimulated with pulsatile GnRH delivered from programmable infusion pumps such that a specific dose of GnRH can be delivered at various frequencies, thus providing synthetic pulsatile hormone release (Wildt *et al.* 1981).

14.4 Experimental studies II: neuroendocrine responses to environmental, behavioral and cognitive stimuli

14.4.1 Neuroendocrine responses to environmental changes

The neuroendocrine system of an individual cannot easily be isolated from environmental cues. For example, under physiological conditions, neuroendocrine rhythms are correlated with daily light-dark cycles and changes in day length over the seasons of the year (e.g. see Figure 14.3), and on a daily basis the light-dark cycle regulates the secretion of melatonin from the pineal gland (Figure 8.5). Sleep is also a trigger for hormone secretion such as growth hormone (Figure 4.7) and cortisol (Figure 6.5), whereas the diurnal increase in TSH secretion is inhibited by sleep (Roelfsema and Veldhuis 2013). The neuroendocrine system also responds to environmental stimuli such as stress (see Figure 1.2). Figure 14.11 shows

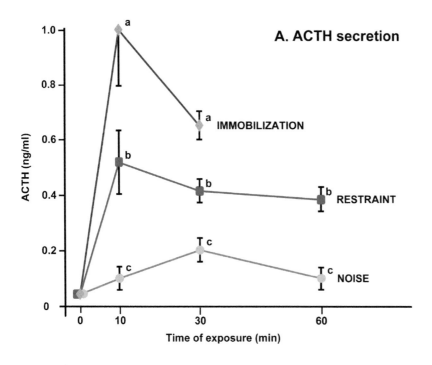

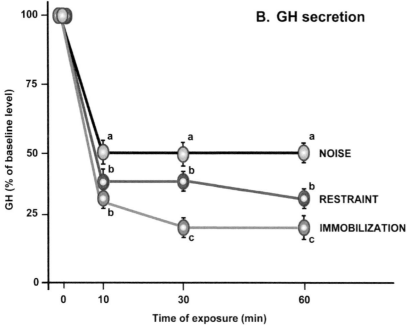

Figure 14.11 Effects of various stressors on secretion of GH and ACTH in male rats
Adult male rats (n = five to seven per group) were subjected to three different stressors: noise, restraint (confinement in tubes) and immobilization. Blood levels of ACTH and GH were determined at various times after the stress episode and values are plotted as means ± standard error. Group means that differed statistically are labeled with different letters.
A. Immobilization was much more effective than restraint or noise in elevating ACTH secretion, although all three stressors significantly increased ACTH release above background levels (significance not indicated). B. In contrast to the effects on ACTH, all three stressors *reduced* GH levels, with immobilization causing the greatest decrease. The GH response to stress was always significant. The figure is redrawn from the original data with the kind permission of Dr. A. Armario (Armario and Jolin 1989).

the changes in ACTH and GH secretion in rats following exposure to three different stressors: noise, restraint and immobilization. Immobilization stimulates greater ACTH release than restraint, which, in turn, stimulates more ACTH release than noise. All three stimuli inhibit GH release, with immobilization producing the greatest inhibition and noise the least (Armario and Jolin 1989). Likewise, the ingestion of food stimulates the secretion of neuropeptides from the gastrointestinal tract (see section 12.6.2). For example, when we are hungry, the stomach releases ghrelin, which then stimulates hypothalamic neurons to increase appetite and induce feeding behavior (see Figure 12.13). In contrast, when food is withheld (fasting), there is a profound inhibition of GH secretion (Steyn *et al.* 2011). In these examples, external stimuli alter the neuroendocrine system, which then influences behavior. Social interactions with other animals can have similar effects.

14.4.2 Neuroendocrine responses to social interactions

Hormone levels can be altered in response to social, including sexual, interactions. For example, access to females results in transient increases in LH and testosterone levels in males, while aggressive interactions result in a decrease in testosterone levels and an increase in corticosteroid levels (Harding 1981). Figure 14.12 shows testosterone secretion in a single adult male rhesus monkey over a seven-month period. This male had *increased* testosterone secretion on the two occasions when he had access to receptive females, but showed prolonged *decreases* in testosterone secretion following defeat in aggressive encounters with dominant males (Rose *et al.* 1975). In male rats, exposure to receptive females also elevates testosterone levels, thus influencing their sexual and aggressive behavior (Flannelly and Lore 1977; Nyby 2008).

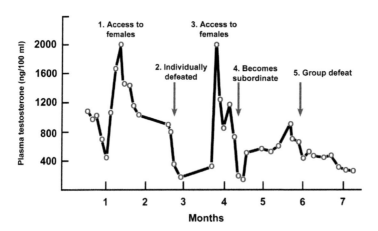

Figure 14.12 Plasma testosterone levels assayed in a single male rhesus monkey over a seven-month period
This monkey showed increases in testosterone secretion on two occasions (1 and 3) coincident with ready access to receptive female monkeys. In contrast, a marked decline in testosterone secretion was observed following defeat in three different circumstances. On the first occasion (2), he was defeated after a brief exposure to an all-male group of 34 animals. On the second occasion (4), he became the subordinate male in a newly formed group consisting of three other adult males and 13 females. On the third occasion (5), he was defeated, along with the three other males, by a large, well-established breeding group. Reproduced with permission and redrawn (Rose *et al.* 1975).

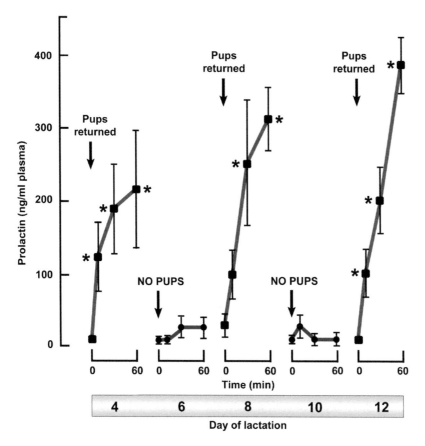

Figure 14.13 Suckling-induced prolactin secretion in nursing mother rats
Plasma prolactin levels (mean ± SEM) in lactating female rats (n = 5–7 per group) which had their pups removed for 4 hours and then returned for 60 minutes on postnatal days 4, 8 and 12. On days 6 and 10, pups were removed, but not returned. Blood samples were taken before pups were returned (time 0) and then 10, 30 and 60 minutes after pup return. * $p < 0.05$ to 0.001 compared to prolactin level at time 0. The figure is redrawn from the original data with the kind permission of Dr. R. S. Bridges (Samuels and Bridges 1983).

Prolactin secretion in lactating female rats is stimulated by the suckling behavior of her pups. Figure 14.13 shows prolactin levels in nursing female rats from day 4 to 12 of lactation. Pups were separated from their mother for 4 hours on days 4, 8 and 12, and then replaced. Blood samples were then taken from the mother four times during the first hour after the pups were returned. The suckling behavior of the pups stimulated the secretion of prolactin in the mother. In contrast, on days 6 and 10, when no pups were returned, no increase in prolactin release occurred in the mother (Samuels and Bridges 1983). A similar suckling-induced release of prolactin in nursing women was illustrated in Figure 8.4 (Tay *et al.* 1996).

14.4.3 Neuroendocrine reflexes
Many sensory stimuli trigger the release of hormones in a reflex fashion, representing a special form of social interaction as described above (section 14.4.2). Chemical signals

from conspecifics (pheromones) can prime the reproductive system by triggering the release of gonadotropins and can thus influence the timing of puberty, reproductive behavior and pregnancy in many mammals. Odors of females also stimulate the release of LH and testosterone in males (Brown 1985b; Ziegler *et al.* 2005). Vaginal stimulation during mating induces release of prolactin and oxytocin in the female rat (Komisaruk and Steinman 1986). Females of many species of mammals, such as cats and rabbits, are called *induced ovulators*. These animals require tactile stimulation of the vagina during sexual behavior to trigger the LH release that stimulates ovulation. Without this external stimulation, they will not ovulate. Tactile stimulation of the nipple when pups are suckling from a lactating mother rat induces prolactin and oxytocin release (Figure 14.13). Likewise, in ring doves, the act of placing a breeding pair together rapidly activates the reproductive axis in both birds (see below; section 14.4.4).

14.4.4 Chaining of neuroendocrine and behavioral responses

In some cases, a series of behavioral and neuroendocrine events can be viewed as a chain reaction (Harding 1981). Such hormone-behavior chains involve three factors: (1) environmental stimulation of hormonal change; (2) hormonal stimulation of behavioral change; and (3) behavioral feedback on hormone levels – which lead to the next level of behavioral change (Leshner 1978; Beach 1975). A good example of such a hormone-behavior chain is the reproductive behavior of the female ring dove, in which social interactions with a male are essential for the timing of the transition from one phase of the female's reproductive cycle to the next and the stimulation of ovulation (Lehrman *et al.* 1961; Lehrman 1965).

When a female ring dove is paired with a male, the male responds with an increase in hypothalamic GnRH synthesis within 2 hours (Mantei *et al.* 2008) and blood testosterone levels are doubled within 4 hours (Feder *et al.* 1977). The visual, auditory and tactile stimuli from the male then trigger the release of gonadotropins in the female ring dove, leading to follicular production of estradiol and then to ovulation (e.g. see Figure 4.6). The temporal sequence of behavior, follicular maturation and ovulation is illustrated in Figure 14.14 (Cheng 1986).

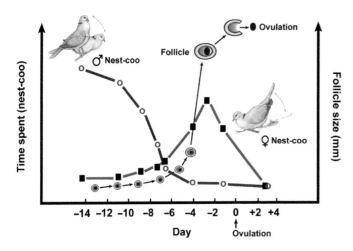

Figure 14.14 Courtship behavior of ring doves induces follicular development and ovulation in female ring doves

As shown, courtship behavior includes nest-cooing (nest-coo; bowing behavior) which continues over several days. As noted in the text, this behavior induces rapid neuroendocrine changes in the male (increase in GnRH, followed by increased testosterone secretion), which in turn induce follicular maturation in the female. A schematic representation of the sequence of the male (blue line) and female (red line) ring dove's nest-cooing behavior in relation to the pattern of follicular development is plotted with reference to follicle rupture and release of an ovum (ovulation; Day 0). Reproduced with permission, and redrawn (Cheng 1986; de Kort and ten Kate 2004).

The nest-coo call of the female ring dove to the male during this courtship phase may also facilitate her follicular development through self-feedback (Cheng 1986). The increase in estradiol levels also stimulates female courtship and copulatory behavior and initiates nest building. After ovulation occurs, progesterone levels rise and estradiol levels fall, resulting in a decline in sexual behavior. In the presence of the nest and eggs, these hormonal changes stimulate the female to sit on the nest and incubate the eggs. The tactile stimulation received by sitting on the eggs stimulates prolactin release. Prolactin stimulates growth of the crop sac and secretion of "crop milk," which is fed to the young doves when they hatch. Prolactin also induces parental behavior towards the hatchlings and their presence stimulates continued prolactin release. As the young birds grow and begin to feed on their own, they no longer stimulate prolactin release from the female. As her prolactin levels decline, there is a rise in FSH, which stimulates the growth of new follicles. In the presence of the male, more FSH and estradiol are secreted, stimulating follicular growth and another cycle of reproductive behavior begins. Similar neuroendocrine-behavior chains occur in the reproductive behavior of other birds, fish, amphibia and mammals (Brown 1985a; Baggerman 1968; Crews 1980).

14.4.5 Neuroendocrine responses to cognitive stimuli

Because the neuroendocrine system is primarily regulated by the central nervous system, the way in which external stimuli are perceived can influence neuroendocrine responses. For example, cognitive factors in the emotional response to an external stimulus may determine the pattern of neuroendocrine responses to that stimulus. Emotional arousal involves a physiological response to a stimulus and the cognitive appraisal of that stimulus. The physiological response involves activation of the autonomic nervous system by the hypothalamus and limbic system, resulting in increased heart rate, blood pressure, temperature, perspiration, etc., as described in section 12.7 (Figure 12.20). Cognitive appraisal involves assessing the stimulus situation as positive, in which case the physiological arousal may be associated with joy or happiness; or negative, in which case the physiological arousal may be associated with fear or anger (Plutchik 1962). Different emotional states have been associated with different patterns of neuroendocrine response. The hypothalamic-pituitary-adrenal stress system (H-P-A; see Figures 1.2 and 6.4) is especially sensitive. For example, listening to music is used for stress release in several medical contexts (Nelson *et al.* 2008) and is reported to reduce cortisol secretion in intensive care patients (Beaulieu-Boire *et al.* 2013). On the other hand, excessive stimulation of the H-P-A axis (i.e. high cortisol levels) is seen in a substantial proportion of adults and children with depressive and anxiety disorders (Guerry and Hastings 2011) and patients with major depression have high cortisol levels associated with deficits in verbal memory, visuospatial memory, working memory and selective attention (Hinkelmann *et al.* 2009). In contrast, cortisol levels are abnormally low in post-traumatic stress disorder (PTSD; Daskalakis *et al.* 2013). However, a marked response of the H-P-A axis is also seen in healthy people. For example, competitive ballroom dancers exhibit high cortisol secretion, independent of the physical strain of dancing (Rohleder *et al.* 2007), and students facing mental challenges, such as an examination, also release high levels of cortisol (Johansson *et al.* 1983). Figure 14.15 shows the levels of epinephrine (released from the adrenal medulla) and cortisol (released

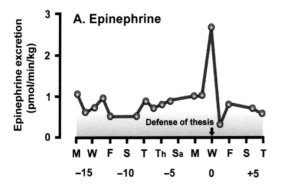

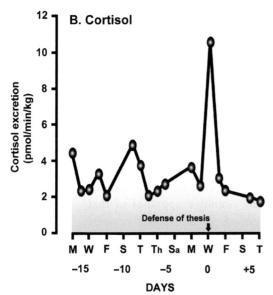

Figure 14.15 Adrenal response to the stress of a Ph.D. examination
The male subject (age 30 years) provided a urine sample at 11.00 each day. Urine was analyzed for epinephrine and cortisol content. The graphs show: (A) epinephrine excretion and (B) cortisol excretion on the days preceding, following and coinciding with the public defense of the subject's Ph.D. thesis. Reproduced with permission (Johansson *et al.* 1983).

from the adrenal cortex) excreted in the urine in a 30-year-old male student on the days before, during and after the oral examination of his Ph.D. thesis. Note the pronounced peak in the release of these adrenal hormones on the day of his examination and their return to baseline afterwards. Given that adrenal hormones are released in stressful situations, one can infer that the changes in ACTH secretion in rats in response to immobilization, restraint and noise shown in Figure 14.11 represent stress responses to these acute stimuli. Corticosterone secretion was also increased in mice by the perception of even very distant events (330 kilometers), such as an earthquake (Yanai *et al.* 2012).

Note that growth hormone secretion is also affected by stress (Figure 14.11) and other pituitary and pineal hormones are influenced by different forms of emotional arousal. For example, as described in section 8.2.1, oxytocin secretion is rapidly increased in nursing mothers, not just by suckling, but also on hearing the baby cry (Figure 8.4). In other examples, prolactin levels are elevated, and melatonin decreased, in panic patients, whereas the hypothalamic-pituitary-gonadal axis can be suppressed in anxious patients (Cameron and Nesse 1988). The hypothalamic-pituitary-thyroid system also appears to be compromised in depressed patients (Gary *et al.* 2003).

14.4.6 Conditioned hormone release

Since environmental stimuli and cognitive factors influence the release of neurotransmitters, hormones, neuropeptides and cytokines (Chapters 6, 12 and 13; and see Larson *et al.* 2001), it is not surprising to learn that hormone release can be conditioned to environmental stimuli. Ader (1976) showed that corticosterone release is conditioned in an illness-induced taste aversion test, in which saccharin taste is paired with drug-induced illness. Rats learned to avoid saccharin-flavored water and also produced an elevated corticosterone secretion to the taste of saccharin. Similar conditioned release of corticosterone was seen in rats when nicotine treatment was paired with a distinct environment (Davis *et al.* 2005). Human babies can also respond to an anticipated stress (e.g. disruption in parent-baby interaction) with a significant release of cortisol (Haley *et al.* 2011). *Decreases* in corticosterone release were conditioned to stimuli associated with feeding and drinking in rats (Coover *et al.* 1977); increased LH and testosterone secretion were conditioned to stimuli associated with sexual behavior in male rats (Graham and Desjardins 1980) and in sheep (Gelez *et al.* 2004); and histamine release was conditioned to odor stimuli in the guinea pig (Russell *et al.* 1984). Immune responses can also be conditioned, indicating that the neuroimmune system can respond to conditioned stimuli as well as the neuroendocrine system (Ader 2003).

14.4.7 Long- and short-term neuroendocrine changes in response to external stimuli

The changes in hormone levels in response to environmental, behavioral or cognitive stimuli may be long-term changes in the secretion of a hormone, as occurred following defeat in aggressive encounters in the rhesus monkey (Figure 14.12), or short-term surges in hormone release, as occurs during examination stress in humans (Figure 14.15). The neuroendocrine responses to external stimuli may also trigger other physiological responses as occur when pup-induced prolactin release stimulates milk production in the female rat (Figure 14.13).

Hormonal surges represent short-term neuroendocrine responses to external stimuli and are often associated with particular behavioral responses such as sexual behavior, maternal behavior or stress reactions. Chronic changes in the baseline secretion of hormones, on the other hand, are often associated with long-term changes in behavioral responsiveness. The attenuated levels of testosterone in the subordinate male rhesus monkey, for example, are associated with submissive behavior in a number of different situations, which results in the animal being treated differently by other monkeys in the group. There are two important distinctions to make between short-term, stimulus-induced changes in hormone release and long-term changes in baseline hormonal levels with regard to their effects on behavior. First, an altered baseline level of hormone release means that the hormone level is changed before the animal or person enters a particular stimulus situation, whereas short-term changes in hormone release occur in response to a particular stimulus. Second, short-term hormonal changes occur within minutes of stimulus onset and quickly return to baseline, whereas changes in the baseline hormone levels develop slowly, taking hours or days to occur, and result in new baseline hormone levels, which remain for days or weeks (Leshner 1979). This difference can be seen by comparing the neuroendocrine stress responses of the male student in Figure 14.15 with those of the male rhesus monkey in

Figure 14.12. The student's epinephrine and cortisol levels increased just before he defended his thesis, and when this stress was over, the hormone levels returned to baseline. When the monkey was placed with other monkeys, however, he remained with them and was under prolonged stress for over three months and his testosterone level remained depressed for this entire period.

14.4.8 Summary

Timing of many neuroendocrine events is regulated by external stimuli. These can be environmental changes, social interactions or cognitive processes. Some of these stimuli are cyclic, such as day/night and seasonal light cycle rhythms, some are short-term stimuli, such as noise, restraint or immobilization, and others are more chronic, such as long-term changes in social status. Often, the release of hormones does not occur unless triggered by external stimuli, as occurs with induced ovulation and lactation. The changes in neuroendocrine activity caused by these external stimuli alter the behavior of the animal, resulting in hormone-behavior chains, as occurs in the mating behavior of the ring dove. Because neuroendocrine activity can be conditioned to external stimuli, the animal or person may have conditioned hormone release in certain stimulus situations. These learned neuroendocrine responses may lead to chronic or acute changes in hormone secretion, depending on the conditioning stimulus. The result of these neuroendocrine responses to external stimuli is that the timing of certain neuroendocrine events, such as puberty, may be regulated by a host of environmental stimuli, such as nutrition, day-night cycles, stress, social interactions and pheromones released by adult animals (see Brown 1985b).

14.5 Neural and genomic mechanisms mediating neuroendocrine-behavior interactions

Hormones, neuropeptides and cytokines do not stimulate behavior directly: they act as neuromodulators to alter the activity of their target neurons. This neuromodulatory action regulates membrane permeability and genomic activity in the target neuron, as discussed in sections 9.4, 9.8, 10.7, 12.3 and 13.5. Changes in membrane permeability alter the electrophysiological activity of neurons and the release of neurotransmitters. Genomic action, particularly by steroids, regulates gene transcription (mRNA production) and translation (protein synthesis). Endocrine stimulation of target neurons activates behavioral changes and hormone synthesis in adults (Nugent *et al.* 2012). However, during embryonic and neonatal development, sex steroid hormones such as estradiol and testosterone regulate cell death, neuronal migration, neurogenesis and neurotransmitter plasticity. In addition, these hormones direct the formation of sexually dimorphic circuits through axonal guidance and synaptogenesis (Simerly 2002; Morris *et al.* 2004). Although the embryonic and neonatal periods are generally considered to be especially sensitive to gonadal steroid action, there is also evidence that steroid-dependent organization of behavior also occurs during adolescence (Schulz *et al.* 2009; Vigil *et al.* 2011). These effects of steroids therefore may have a profound influence on the regulation of behavior. This section discusses the neural and genomic mediators of hormonal modulation of behavior in adult and developing animals.

14.5.1 Neural mechanisms

Hormones can influence behavior by modulating three different aspects of neural activity. These involve the hormonal modulation of sensory receptors and sensory input to the brain, the motor pathways controlling behavior and the central mechanisms responsible for the integration of sensory information with the organization of behavioral response patterns. These neural mechanisms have been studied using a variety of techniques, including electrical stimulation of brain pathways, electrophysiological recording of neural activity, lesion techniques, intracerebroventricular injection of hormones (Beach 1974; 1975; Leshner 1978), localization of sex hormone receptors and gene expression (Ball and Balthazart 2004).

Sensory receptors and sensory input

Gonadal, thyroid and adrenocortical hormones modulate gustatory (taste), olfactory, tactile and other sensory systems by acting directly on the peripheral sense organs and by affecting the processing of sensory stimuli in the brain (Gandelman 1983; Henkin 1975). Steroid hormones influence taste and smell sensitivity by altering the oral and nasal mucosa, as well as by modulating the neural pathways processing olfactory and taste stimuli (Gandelman 1983). Estrogen-associated brain pathways are implicated in sensory experience in the mouse auditory cortex and estrogenic modulation is an important component of the operational framework of central auditory networks (Tremere *et al.* 2011; Caras 2013). In a specific example, hearing thresholds of women are modified through the menstrual cycle, especially by estradiol (Walburger *et al.* 2004). Disorders of thyroid and adrenocortical hormone secretion alter taste, smell and auditory perception (Henkin 1975). For example, thyroid hormone deficiency causes irreversible damage to peripheral and central auditory systems (Knipper *et al.* 2000). Also in humans, acute stress (i.e. high cortisol secretion) attenuates taste intensity (Al'Absi *et al.* 2012), whereas olfactory perception is modified through the menstrual cycle (Watanabe *et al.* 2002). Estradiol increases the size of the tactile sensory field of the pudendal nerve, increasing the sensitivity of the skin receptors around the genital area of the female rat during estrus (Komisaruk *et al.* 1972; Adler *et al.* 1977). There is good evidence that women are at greater risk for some clinical pain conditions, and post-operative pain may be more severe among women than men (Fillingim *et al.* 2009; Rezaii *et al.* 2012; Traub and Yaping 2013). Estradiol is a critical hormone for this difference, but the neurochemical pathways and biochemical mechanisms involved are complex (Amandusson and Blomqvist 2013).

Motor pathways

Hormones influence the motor pathways controlling a number of behavioral acts, including locomotion, sexual reflexes and grooming behavior. Excessive and prolonged grooming behavior in rats is stimulated by intracranial injections of ACTH and MSH-like peptides (Mul *et al.* 2013). This effect is mediated through melanocortin MC4 receptors, which are also responsible for inhibition of feeding behavior (see Figure 12.13). Other neuropeptides, such as orexin, neuromedin and somatostatin, increase motor activity (Nixon *et al.* 2012; Teske *et al.* 2008; Viollet *et al.* 2008; see also Table 12.6). Increased locomotor activity, stimulated by estradiol secretion, is characteristic of estrus in rodents and many other mammals, including cows, and can be used to predict sexual receptivity in females (Burke and Broadhurst 1966; Basterfield *et al.* 2009; Rorie *et al.* 2002). However, in non-human primates (Rhesus monkeys), the opposite seems to be true; that is, the *absence* of estradiol increases physical activity (Sullivan *et al.* 2012). The spinal nucleus of the bulbocavernosus (SNB), a cluster of motor neurons located in the spinal cord, innervates the perineal muscles bulbocavernosus and levator ani (BC/LA), which participate in copulatory behavior. The size of this spinal nucleus and the number of its dendritic connections, as well as the development of the penile muscles innervated by these nerves, are regulated by testosterone, which is essential for penile erection and other penile reflexes in the male rat (Sengelaub and Forger 2008) and in men (Morales 2011).

Central integrative functions

For some hormone-dependent behaviors, it is difficult to distinguish the effects of hormones on the sensory and motor pathways from those on the motivational or arousal mechanisms in the brain. Hormones seem to perform an integrative function, enabling each component of the behavioral system to interact with the others to produce functional sequences of behavior. The central integrative function of the neuroendocrine system in (1) the control of bird vocalizations and (2) in the arousal of sexual behavior in male and female rats will be discussed in this section. Male sexual behavior has been studied in birds since the nineteenth century. Birds are especially useful for the study of the neuroendocrine regulation of sexual behavior since it can be studied in the context of much field data, and known neural circuits related to reproductive behavior have been described (Ball and Balthazar 2004). For example, in song birds, such as zebra finches, only the males sing, and the development of seasonal singing is dependent on rising levels of testosterone (Ball *et al.* 2004). Figure 14.16 illustrates the androgen-sensitive brain areas of song birds, such as zebra finches, which are involved in the arousal and motor control of singing (Arnold 1981; Wade and Arnold 2004). Although it is difficult to separate the motivational and motor effects of androgens in stimulating singing in these particular birds, it is possible to do this with the vocalizations of ring doves (Cohen 1983). Testosterone implants into the septum and anterior hypothalamic-medial preoptic area of the brain activate reproductive motivational systems and stimulate increased courtship and vocal behavior in ring doves, while testosterone implants into the nucleus intercollicularis (ICo; see Figure 14.16) stimulate vocalizations, but not courtship behavior. Lesions of the vocal control system (area RA) and the motor neurons controlling the vocal cords (nXIIts) alter vocal behavior in a number of behavioral contexts. Thus, testosterone activates singing as a component of courtship behavior when it stimulates receptors in the hypothalamus and limbic system, whereas singing alone is induced by acting at receptors in the motor control neurons. A good overall description of the action of sex steroids on the birdsong system is provided by Schlinger (1997).

The sexual behavior of the *male* rat has both arousal (motivational) and performance (motor) components and testosterone sensitive areas of the anterior hypothalamic-medial preoptic area (mPOA) of the brain integrate these components into a functional sequence of behavior (Davidson and Trupin 1975; Soulairac and Soulairac 1978; Everitt 1990). For example, lesions of the mPOA impair copulation in male rats, whereas stereotaxic implants of testosterone in the mPOA restore most aspects of male copulatory behavior in castrated

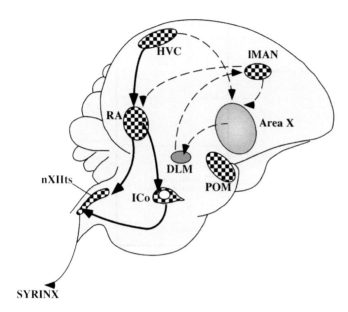

Figure 14.16 Vocal control system in zebra finch and canary brain illustrating the schematic distribution of androgen receptors

Black dots indicate the presence of cells containing androgen receptors. The nuclei of the forebrain pathway involved in song production (HVC, RA, ICo and nXIIts) are connected by *black arrows*. Nuclei of the anterior forebrain pathway involved in song learning and song maintenance in adulthood (HVC, Area X, IMAN and RA) are connected by *dashed arrows*. The medial preoptic nucleus (POM) of the hypothalamus also participates in the activation of singing behavior, by integrating sensory inputs and initiating motor output.

Abbreviations: DLM, nucleus dorsolateralis anterior, pars medialis; HVC, hyperstriatum ventrale pars caudale; ICo, nucleus intercollicularis; IMAN, lateral magnocellular nucleus of the anterior nidopallium; nXIIts, tracheosyringeal portion of the nucleus hypoglossus; RA, robust nucleus of the arcopallium; Area X, song nucleus of the medial striatum. Reproduced with permission (Ball *et al.* 2004).

rats (Balthazart and Ball 2007). Full integration of male sexual behavior depends on testosterone activating receptors in the mPOA and limbic system of the brain, as well as in the spinal neurons of the SNB and the penile striated muscles (Arnold and Jordan 1988; Sengelaub and Forger 2008). There is also good evidence that testosterone action in the mPOA is mediated by conversion to estradiol (Balthazart *et al.* 2004; see also Figure 9.14). A number of neuropeptides, such as oxytocin, are also involved in the activation of male sexual behavior (Argiolas and Melis 2004; and see Table 12.7).

The sexual behavior of the *female* rat involves the action of estradiol and progesterone on olfactory pathways, tactile receptors of the pudendal nerves of the genital area, and on neurons in the AH-MPOA, ventromedial hypothalamus (VMH) and other limbic structures in the brain (Pfaff 1989; Pfaff *et al.* 2008). These effects of sex hormones on the neural components underlying female sexual behavior were discussed in section 9.9.3 (see Figure 9.16). Neuropeptides are also implicated in the regulation of female sexual behavior. The classical view provides a model of the interactions among estradiol, progesterone, GnRH, endogenous opioid peptides and the catecholamine neurotransmitters in the integration of the sensory, motivational and motor components of sexual behavior in the

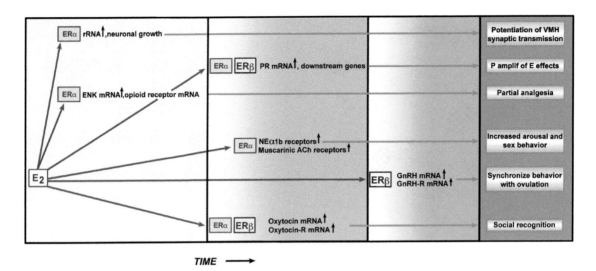

TIME →

Figure 14.17 Outline of estradiol-mediated influence on various genes, receptors and neuropeptides implicated in sex behavior

All of these pathways foster female reproductive behavior. ERα-dependent pathways are signaled in red, ERβ are labeled in blue. Functional consequences are in green, and time reads from left to right.

Abbreviations: ACh, acetylcholine; E_2, estradiol; ENK, enkephalin; ER, estradiol receptor; GnRH, gonadotropin-releasing hormone; P, progesterone; PR, progesterone receptor; mRNA, messenger ribonucleic acid; NE, norepinephrine; rRNA, ribosomal ribonucleic acid; VMH, ventromedial nucleus of hypothalamus. Reproduced with permission (Mong and Pfaff 2004).

female rat (Crowley 1986). However, this interpretation is further clarified through application of more sophisticated techniques, including an assessment of estradiol-induced gene expression, and this is presented in Figure 14.17.

The classical view remains at the heart of this outline, but specific receptors, such as the estradiol receptor subtypes ERα and ERβ (see section 9.1.1) and neuropeptides, such as oxytocin, are now known to be involved (Mong and Pfaff 2004). For example, GnRH facilitates sex behavior in rodents (Moss and McCann 1973; Schiml and Rissman 2000) and synchronizes this behavior with release of LH and induction of ovulation. Oxytocin is also a critical component of social recognition ensuring that aggression and anxiety, for example, are attenuated in order to promote mating (Figure 14.18; Shelley *et al.* 2006). This process is facilitated by estradiol, acting through ERβ, to increase oxytocin secretion from PVN neurons that send axons to the amygdala. The oxytocin that is released in the amygdala binds to oxytocin receptors whose synthesis is also increased by estradiol acting at ERα (Figure 14.18).

14.5.2 Genomic mechanisms

As described in Chapter 9, steroid hormones such as estradiol and progesterone modulate mRNA and protein synthesis through direct actions on nuclear receptors in neurons. These effects are called *genomic* because the steroid-receptor complex acts as a *transcription factor* in the nucleus to modify target gene activity (see Figure 9.1). However, as outlined in section 9.4.1, steroid hormone receptors can also be located in the cell membrane. These

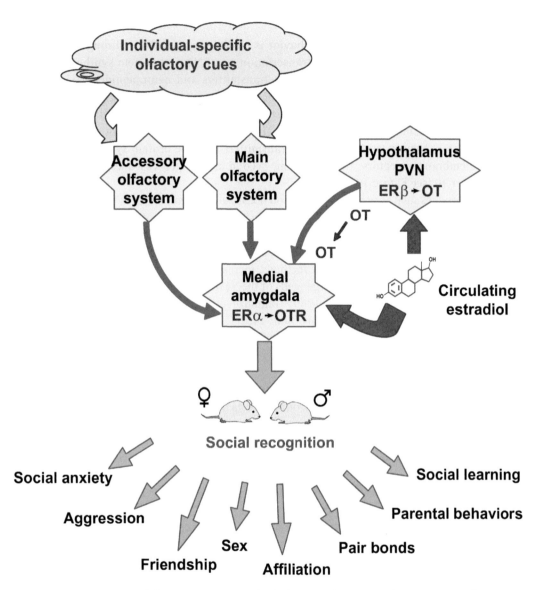

Figure 14.18 Outline of estradiol-oxytocin interactions that facilitate social recognition during mating in rodents Circulating ovarian estradiol binds to ERβ in neurons of the PVN that project axons to the amygdala. Oxytocin production is increased and release of oxytocin from PVN terminals in the amygdala is elevated. Estradiol, acting via ERα, regulates the expression of oxytocin receptors in the amygdala. Oxytocin binds to oxytocin receptors, ultimately facilitating social recognition.

Abbreviations: ER, estradiol receptor; OT, oxytocin; OTR, oxytocin receptor; PVN, paraventricular nucleus of the hypothalamus. Reproduced with permission (Shelley *et al.* 2006).

receptors also affect genomic responses, but do so indirectly via intracellular processes such as cAMP (Figure 9.9). Thus, estradiol (Figure 9.9) and progesterone (Mani and Oyola 2012) can modify neuronal activity, and behavior, by simultaneously binding to two distinct types of receptor.

Genomic modification of behavior

Steroid hormonal activation of behavior is mediated genomically through stimulation of gene expression via increased transcription of mRNA and protein synthesis in target cells, as well as by the release of neurotransmitters and neuropeptides. For example, Figure 14.17 shows that several genes, including those for GnRH, oxytocin and progesterone receptor (PR), are regulated by estradiol to alter behavior. In addition to intracellular mechanisms, the cell membrane is also responsible for mediating changes in gene expression. Neuropeptides and neurotransmitters regulate protein synthesis by binding to membrane receptors and activating second messenger cascades, as discussed in Chapter 10 (see Figure 10.1).

Techniques such as *in situ* hybridization (localization of mRNA; see Figures 9.5 and 9.13; see also Simerly *et al.* 1990), Northern analysis (quantification of mRNA) and the reverse transcriptase polymerase chain reaction (RT-PCR; quantification of mRNA) are commonly used to assess genomic responses to neuroendocrine activation (Pfaff 1989; Wilcox 1986; Gagnidze *et al.* 2013). Using these methods, increased levels of PR mRNA and PR protein are detectable in the estradiol target cells of the VMH of the female rat brain 6 to 24 hours after estradiol injection (Pfaff 1989; Gagnidze *et al.* 2013). Other proteins produced via estradiol-induced gene expression include muscarinic acetylcholine receptors, α1b-noradrenergic receptors, GnRH and oxytocin (Figure 14.17), all of which are implicated in the induction of female sexual behavior.

14.5.3 Neuroendocrine organization of neural development

Chapter 9 outlined some of the profound effects of steroid hormones in the mammalian brain. Acting via genomic as well as non-genomic mechanisms, steroid hormones such as estradiol, progesterone and cortisol can modify neurotransmitter release, neurotransmitter receptors and steroid receptors (section 9.8). In functional terms, these effects induce changes in, for example, neuronal firing (Figure 9.6), neuropeptide secretion (Figure 9.7), overall brain activation (Figure 9.8) and the development of neurotransmitter pathways (Herlenius and Lagercrantz 2004). They also influence emotional, motivational, sensory and behavioral systems. The developing brain is also a steroid target organ and contains a variety of steroid hormone receptors and the modulation of genomic activity in estradiol target neurons during the prenatal and neonatal periods regulates the growth and development of these cells (Balazs 1976; Balazs *et al.* 1975; Lauder 1983). Estradiol induces sex differences in regional volume, cell number, connectivity, morphology, physiology, neurotransmitter phenotype and molecular signaling (Lenz and McCarthy 2010; see also section 9.9.5). Figure 14.19 illustrates in general terms the complex interplay of hormones, neurotransmitters and neuropeptides that regulate and control the many maturational steps necessary to produce a mature nervous system.

Estradiol not only affects the number of neurons in various nuclei, but also regulates many developmental changes, including neurogenesis, neural connectivity, synaptogenesis, cell death and neurotransmitter plasticity (Simerly 2002; Tsutsui *et al.* 2011). Figure 14.20 illustrates that determination of neuron size, dendrite growth, synapse formation, myelination and enzyme activity can all be used to investigate the quantitative effects of hormonal (estradiol) stimulation during cell development.

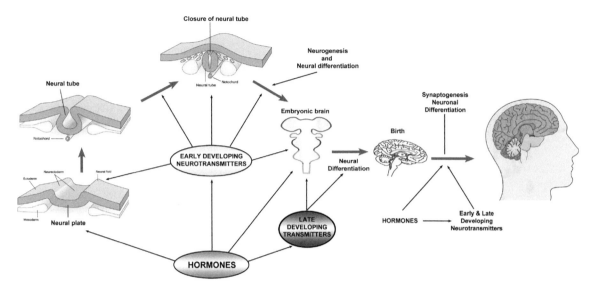

Figure 14.19 The role of the neuroendocrine system in neural development

Hormones, neurotransmitters and neuropeptides assist in promoting neural growth and differentiation at each stage of brain development and may have different effects, depending on when and where they appear in the developing nervous system. This figure is based on information provided in the paper by Lauder and Krebs (1986). The images of the developing neural tube are reproduced with permission (Gammill and Bronner-Fraser 2003).

The image of the embryonic brain was obtained from Wikipedia Commons (EmbryonicBrain.svg); author, Nrets.

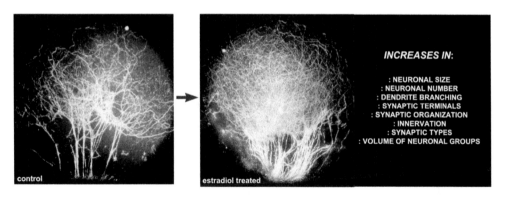

Figure 14.20 The effects of estradiol stimulation on neural circuits

This figure outlines the profound effect of neuroendocrine (estrogenic) stimulation on various aspects of neural maturation, including axonal growth, dendritic branching, synapse formation and synaptic organization (Toran-Allerand 1991). Reproduced with permission (Toran-Allerand 1980).

As outlined in Figure 14.19, hormones, neuropeptides and neurotransmitters interact to modulate brain development during critical periods of neural growth and differentiation (Lauder and Krebs 1986; Weiss *et al.* 1998). These neuroregulators assist the intracellular gene expression program to organize the development of the neural circuits that mediate sensory input, motor activity and the central integrative functions of the brain. As a result,

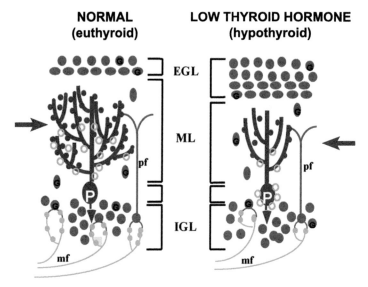

NORMAL (euthyroid)

LOW THYROID HORMONE (hypothyroid)

EGL

ML

pf

IGL

mf

Figure 14.21 Effect of thyroid hormone deficiency (hypothyroidism) on neural development in the cerebellum
Purkinje cells are profoundly affected by the absence of thyroid hormone. Among the many changes, note the striking reduction in the number of dendrites (blue; blue arrows) in the molecular layer (ML). The number of synaptic connections (purple dots) between these dendrites and the parallel fibers (pf) are also severely depleted.

Abbreviations: EGL, external granule cell layer; G, granule cells; IGL, internal granule cell layer; mf, mossy fibers; ML, molecular layer; P, Purkinje cells; pf, parallel fibers. Reproduced with permission (Koibuchi *et al.* 2003).

the general metabolic state of the animal and the functioning of the neuroendocrine, autonomic and central nervous systems are all "shaped" by the modulatory actions of hormone-neurotransmitter-neuropeptide interactions during neural development (Weiss *et al.* 1998; Herlenius and Lagerkrantz 2001). Because these neuroregulators contribute to the development of the brain, any factors which disrupt their secretion during the embryonic and perinatal periods may alter neural development and thus have permanent effects on the brain and behavior. For example, a lack of thyroid hormones during human embryonic development results in impaired neurological development, severe mental retardation and impaired postnatal growth (Bernal 2005). The effect on one specific part of the brain – the cerebellum – will serve to illustrate the critical importance of thyroid hormones in neuronal development (Koibuchi *et al.* 2003). Figure 14.21 shows how the neuronal structure of the cerebellum is markedly changed when thyroid hormone is absent. For example, in the molecular layer (ML), there is a large reduction in Purkinje cell dendrites and in the number of synaptic connections with parallel fibers (pf).

Thyroid hormone deficiency is not the only factor capable of affecting neural development. If pregnant females suffer from nutritional deficiencies, take drugs which alter neurotransmitter or neuropeptide levels, are subjected to environmental toxins or stimuli which alter their steroid hormone release, such as severe stress, the neuroendocrine development of the fetus can be disrupted, resulting in perturbations in the modulation of brain development (Crinnion 2009; Frederick and Stanwood 2009; Ferguson *et al.* 2013).

14.5.4 Integration of neuroendocrine-behavior-environmental interactions

The interactions between the neuroendocrine system, the external environment and behavior involve a number of levels of integrative mechanisms as shown in Figure 14.22 (Leshner 1978). During embryonic development, hormones and other neuroregulators

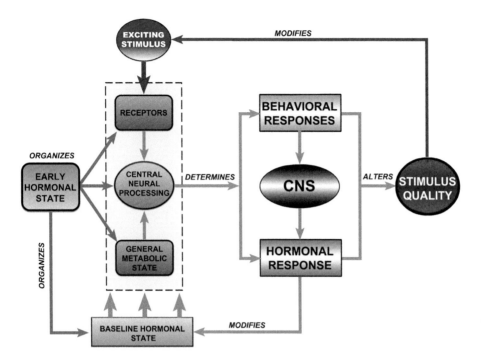

Figure 14.22 A model of neuroendocrine-behavior interactions
The *organizational* and *activational* effects of hormones on behavior, and the feedback effects of behavior on hormone responses and the baseline hormonal state are illustrated. Blue arrows denote the organizational effects of hormones present early in development; note that the baseline hormonal state also influences the *general metabolic state, receptors* and *central neural processing* (broken line box); red arrows denote relationships within an adult individual and gray arrows denote interactions between individuals. The information in this figure is adapted from the paper by Leshner (1978).

organize the development of the baseline hormonal state, the general metabolic state, the central nervous system pathways and the sensitivity of the organism's sensory receptors. In adulthood, perception of an external stimulus by sensory receptors is modulated by baseline hormone levels, as is the neural response to this stimulus. Through the neural mediating mechanism, the stimulus can elicit a behavioral and/or a neuroendocrine response and the hormones released can feed back to modulate both the ongoing behavioral response and the baseline hormonal state of the animal. The hormonal state and behavioral activity of the animal can, in turn, influence its quality as a stimulus in social interactions with other animals. Since the initial behavioral and hormonal responses of animals to external stimuli will depend on the organizational effects of the neuroendocrine system on baseline hormone levels, there are three levels of neuroendocrine-behavior interactions to consider: (1) the neuroendocrine organization of behavior during development (organizational effects); (2) the neuroendocrine modulation of behavior during adulthood (activational effects); and (3) the behavioral modulation of the neuroendocrine system (behavioral feedback).

Table 14.2 **Confounding variables that influence hormone-behavior interactions**
1. Genetic differences between species
2. Individual differences Genetic and other biological differences Organizational effects Uterine position
3. Conditioning and experiential factors Learning/conditioning Social experiences Dominance status
4. Environmental factors Day-night and annual cycles of hormone release Social stimuli and pheromones
5. The stimulus situation Test apparatus Time of day Subject-experimenter interaction
6. The present state of the subject Presence of other hormones or drugs Psychological expectancies Emotional and motivational state Nutritional variables-feeding cycles Sleep-wake cycles
7. Interactions among hormones, neurotransmitters, cytokines, environmental stimuli and behavior

14.6 Confounding variables in behavioral neuroendocrinology research

Any research on the interactions between the neuroendocrine system and behavior must contend with a host of extraneous or confounding variables, some of which are listed in Table 14.2 (Beach 1974; 1975).

14.6.1 Species differences

Not all species show the same behavioral responses to hormonal changes. There are, for example, phylogenetic differences in the effects of gonadal hormones on sexual and parental behavior (Aronson 1959; Brown 1985a). Even closely related species may show different behavioral responses to hormonal manipulation. For example, genetically different inbred strains of mice differ in the level of sexual and aggressive behavior activated by gonadal hormones (Whalen 1986). There are also phylogenetic differences in the perinatal organizational effects of hormones, particularly with reference to sexual differentiation. In mammals, for example, the neutral sex is thought to be female and androgens are necessary for masculinization, while in birds, the neutral sex is male and estrogens are necessary for feminization (Adkins-Regan 1981). Thus, the genotype of the animal may determine its sensitivity to hormonal stimulation and this makes it difficult to generalize the effects of hormones on the behavior of one species to that of another. The discipline of comparative behavioral neuroendocrinology has developed to study evolutionary and ecological differences in the effects of hormones on behavior (Crews 1986).

14.6.2 Individual differences

Just as there are genetic differences between species, there are biological and experiential differences between individuals of the same species and the effects of hormones on behavior may differ between individuals. There are, for example, large individual differences in the decline of sexual behavior after castration in male cats (Aronson 1959). Genetic differences may account for some of these effects, but even animals from genetically identical inbred strains show differences in behavioral responses to hormones. One explanation for these differences is that the intrauterine environment may not be the same for every embryo, even if they are all members of the same litter (Ryan and Vandenbergh 2002; Kawata 2013). For example, individual mouse embryos are exposed to different levels of gonadal hormones during prenatal development, depending on their uterine position. From 6 to 16 mice may be born in the same litter, half of which develop in each uterine horn, as shown in Figure 14.23.

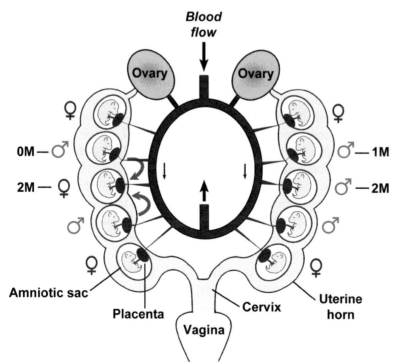

Figure 14.23 Representation of the distribution of pups in a pregnant mouse uterus

The diagram shows the arrangement of the uterine horns, the fetuses and the blood flow in a pregnant mouse (21-day-old embryos). Pups in each horn are classified according to how they are flanked by other fetuses; for example, 0M refers to an embryo flanked by two females; 1M refers to an embryo flanked by only one male; and 2M refers to an embryo flanked by two males. This scheme is therefore applicable to identify both male and female fetuses. Blue arrows indicate a 2M female exposed to testosterone derived from the two adjacent male embryos. Reproduced with permission, and redrawn (Kawata 2013).

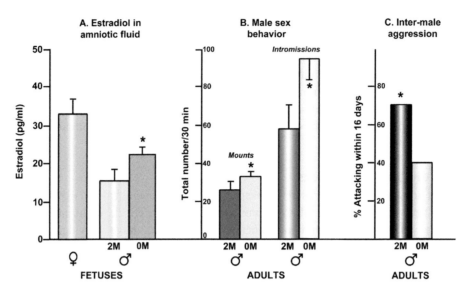

Figure 14.24 Effect of uterine environment on male mouse estradiol and behavior
(A) Estradiol levels in amniotic fluid of female, 0M and 2M male mouse fetuses on day 17 of gestation;
(B) number of mounts and intromissions during a 30-minute test with a sexually receptive 1M female for
gonadally intact, 90-day-old 0M and 2M male mice (20 per group); (C) the percentage of neonatally castrated,
90-day-old 0M and 2M male mice (20 per group) that exhibited a 5-second biting attack towards a 1M male
intruder within 16 days after the 0M and 2M males were implanted with a silastic capsule containing
testosterone. 10-minute tests for aggression were conducted every other day after implantation of the
testosterone. * p < 0.05 for t-tests comparing 0M values with 2M values (A and B) and Chi Square tests (C; 0M
vs. 2M). Values are means ± SEM. The figure is constructed from data published in vom Saal (1983).

The figure illustrates that mice may occupy a position between two males (2M), between
two females (0M) or between a male and a female (1M). The level of sex steroids to which
each fetus is exposed has been shown to differ, depending on the uterine position. A male
fetus flanked by two female fetuses (0M) is exposed to a significantly higher-than-normal
level of estradiol originating from the females. In mice, this occurs by diffusion through the
amniotic fluid (Ryan and Vandenbergh 2002). Conversely, a 2M *female* mouse fetus
(between two males) has a higher level of testosterone in both her blood and her amniotic
fluid than does a 0M female (vom Saal and Bronson 1980). These slight differences in sex
hormone levels can alter the sexual differentiation of the mouse and its adult physiology
and behavior. Thus, in the case of male mouse fetuses exposed to estradiol, there are
alterations in amniotic fluid estradiol levels (Figure 14.24A), male sex behavior
(Figure 14.24B) and in inter-male aggression (Figure 14.24C). These prenatal effects can
also differ between species (Ryan and Vandenbergh 2002) (see Table 14.3). This would
account for differences in the behavior of *genetically identical* mice, reared in the same
environment, due to their intrauterine position during embryonic development (vom Saal
1983).

The human fetus might be similarly affected. Because twins occupy the same womb, it is
possible that in opposite sex twins the female would be exposed to higher-than-normal
levels of testosterone, originating from the male twin. The majority of studies reported so

Table 14.3 Some effects of intrauterine position affecting physiology and behavior in mice and gerbils

Sex	0M (flanked by two females)	2M (flanked by two males)
Female mice	Lower fetal T	Higher fetal T
Female mice	Fewer male offspring	More male offspring
Female mice	Mated/impregnated earlier	Mated/impregnated later
Male and female mice	Less sensitive to T	More sensitive to T
Female mice	Less likely to mount other females	More likely to mount other females
Male mice	Less parental behavior	More parental behavior
Male and female mice	Less aggressive	More aggressive
Male and female mice	Smaller home range	Larger home range
Male gerbil	Lower adult T	Higher adult T
Male gerbil	Less attractive to females	More attractive to females
Male gerbil	Lower impregnation rate	Higher impregnation rate
Male gerbil	Scent-mark less often	Scent-mark more often
Male gerbil	More parental behavior	Less parental behavior
Male and female gerbil	Receive less parental attention	Receive more parental attention

Abbreviation: T, testosterone. Information adapted from Ryan and Vandenbergh (2002).

far have been carefully reviewed (Tapp *et al.* 2011; Ryan and Vandenbergh 2002). Two reports suggest that females with a male, rather than a female, co-twin exhibit increased sensation-seeking behavior when tested as teenagers. Otherwise, there is little evidence for behavioral changes due to opposite-sex twins. However, there is more consistency in studies of perception (otoacoustic emissions) and cognition (in particular vocabulary acquisition and visuo-spatial ability). The evidence supports the induction of masculinized performance in female twins with a male co-twin compared to twins with a female co-twin. An effect of a male co-twin on female attitudes to disordered eating has also been detected (Culbert *et al.* 2013). The prevalence of eating disorders is much higher in young girls than in boys (varying from 3:1 to 10:1; Hudson *et al.* 2007) and exposure to a male twin *in utero* reduces (masculinizes) this incidence (Culbert *et al.* 2013).

14.6.3 Learning and experiential factors

Since hormone release may be conditioned (see section 14.4.6), specific experiences may be associated with neuroendocrine responses in individual subjects. Thus, individual differences in social interactions or stressful experiences may result in different patterns of neuroendocrine reactivity. For example, animals reared in isolation have different neuroendocrine and behavioral responses to novel stimuli than animals reared in social groups (Brain and Benton 1979; Ruscio *et al.* 2007), sexually experienced animals respond differently to opposite sex conspecifics than sexually naive animals, and dominant animals respond differently than subordinates. Since hormone release can be conditioned to environmental cues, conditioned hormone release may occur independently of the stimuli being manipulated by the experimenter. If, for example, a male rat is conditioned to show an LH release to the chamber in which it has been tested with an estrous female, that male will show the LH release when placed in the test chamber, independently of the stimulus animal placed into the chamber by the experimenter. In this case, erroneous conclusions may be drawn about the effects of the stimulus animal on the hormone response of the test male. Likewise, sexually experienced male dogs may still show sexual responses to estrous females after castration, while sexually naive males require hormonal stimulation before responding to the female. Thus, sexual experience may alter behavior independently of hormone levels.

14.6.4 Environmental factors

As discussed in section 14.4, neuroendocrine activity is synchronized with environmental variables such as day/night cycles and seasonal rhythms and can be influenced by subtle social stimuli, such as pheromones. The effects of light cycles and pheromones on neuroendocrine activity means that the housing conditions of the animals may influence their baseline hormonal levels and behavior independently of the variables of interest to the experimenter. For example, seasonally breeding animals which mate in the long days of the spring will not mate during short days (8 to 12 hrs of light per day) and may require 16 hours of light per day. Even when such seasonally breeding animals are held in constant laboratory conditions, they may show a breeding depression in the winter (Bermant and Davidson 1974; Dawson *et al.* 2001; Clarke and Caraty 2013). Thus, experiments conducted at different times of the year may result in different behavioral responses because of the underlying seasonal change in hormone levels (e.g. see Figure 14.3).

An animal's sleep-wake cycle is usually related to the environmental light-dark cycle. Rodents, for example, are active in darkness and sleep during the light phase. In humans also, the pattern of sleep dictates the timing of hormone release; for example, growth hormone secretion (Figure 4.7) is triggered by sleep and ACTH/cortisol (Figure 6.5) secretion occurs at the end of sleep. Changing the light/dark cycle induces changes in neuroendocrine rhythms and behavioral responsiveness, such as occurs in "jet-lag," which requires the traveller's physiological activity to shift to accommodate a change in the day-night cycle following a change in time zones (Sack 2009; Leatherwood and Dragoo 2013). Housing animals under constant (24-hour) light alters neuroendocrine rhythms, and affects the timing of puberty, the estrous cycle of adult females and the sexual behavior of male rats (Fantie *et al.* 1984; Schwartz 1982). Constant light during pregnancy is also effective at changing the neuroendocrine system. Exposure of pregnant non-human primates (*Cebus apella*) to constant light disrupts the cortisol and body temperature rhythms measured in the newborn (Serón-Ferré *et al.* 2002; 2013).

As a result of daily activity and neuroendocrine rhythms, hormone treatment during the active, waking phase may have different behavioral effects than if given during the inactive sleep phase of the day-night cycle. Likewise, stimuli presented at different times of the day may have different neuroendocrine effects. For example, corticosteroid levels in rats are much higher at the start of the active (dark) phase of the light-dark cycle than they are at the start of the light phase. This is also true in humans (Figure 6.5). If rats are handled or placed into a novel environment during the light phase of the cycle, their corticosteroid response is much lower than if this same stimulation occurs during the dark phase of the cycle, as shown in Figure 14.25 (Brown and Martin 1974). Thus, the effects of the light-dark cycle

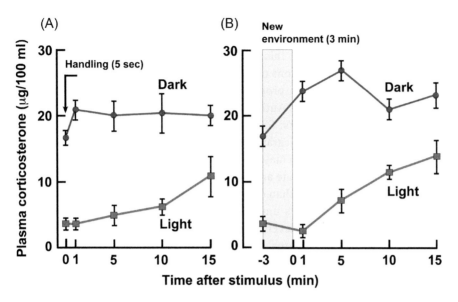

Figure 14.25 Effect of environmental factors on stress responses
Levels of plasma corticosterone in male rats in response to (A) 5 seconds of handling and (B) 3 minutes in a novel environment; comparison of light and dark phases. Values are means ± SEM. The responses are different during the light and dark phases of the day/night cycle and light appears to blunt the stress response. Reproduced with permission and redrawn (Brown and Martin 1974).

may have effects on the neuroendocrine control of behavior that are independent of the stimuli introduced in the experiment. Likewise, the social conditions under which the subjects are housed can influence neuroendocrine activity, as discussed in detail in section 14.4 (see Figure 14.11).

14.6.5 The stimulus situation
Behavioral responses to hormonal stimulation may very depending on the specific stimuli present during the experiment. Behavior in a novel environment may differ from behavior in a familiar environment and the type of test cage, the method of stimulus presentation, the novelty of the stimuli and the design of the experiment can all influence the results (Johnston 1981). Likewise, the time in the light-dark cycle that the animal is tested (Fantie *et al.* 1984), the type of test used, and the behavior of the experimenter may all influence the animal's response. Finally, in tests of social behavior, the familiarity of the animals tested may influence the results. Male rats, for example, will mate with both familiar and unfamiliar females, but male gerbils are more likely to mate with a familiar estrous female (their mate) and will be aggressive towards an unfamiliar estrous female (Swanson 1974).

14.6.6 The present state of the subject
Many variables that could alter the state of the subject must be considered in the study of behavioral neuroendocrinology. When one hormone is of interest, the effects of other hormones or drugs that the subject may be taking must be considered. Women taking birth control pills, for example, may show different responses to certain stimuli than

women not taking birth control pills. Similarly, consumption of alcohol, caffeine, nicotine and antibiotic drugs may alter hormone-behavior interactions. Males given anti-psychotic drugs, such as the major tranquilizer, thioridazine, may experience a number of sexual dysfunctions. The anticholinergic actions of these drugs interfere with both erection and ejaculation and their antidopaminergic properties may inhibit sexual motivation (Mitchell and Popkin 1983; La Torre *et al.* 2013). Antidepressant drugs, especially of the Prozac type (selective serotonin reuptake inhibitors; SSRIs; see Figure 5.16) also induce a high incidence of sexual dysfunction in men (Segraves and Balon 2013).

Baseline levels of hormone secretion may also be determined by the social and cognitive situation of the subject, with subordinate animals or chronically depressed subjects showing different neuroendocrine activity than other subjects. Thus, the emotional state of the subject should be considered. Subjects under stress have high ACTH and corticosteroid levels and low LH and testosterone levels (e.g. see Figures 14.11 and 14.12) and may respond to hormone treatment or external stimuli differently from non-stressed animals. Likewise, animals may respond differently when they are hungry or thirsty than when they are satiated after a meal.

14.6.7 Neuroendocrine, environmental and behavioral interactions

Unless hormones are given under controlled conditions, it may be difficult to distinguish between their effects and the influence of other confounding variables on behavior. It is, for example, difficult to demonstrate a one-to-one correspondence between neurotransmitters and neuropeptides on hormone release (Chapters 6 and 12) and behavioral changes because of the complex interactions among hormones, neurotransmitters, neuropeptides, environmental stimuli and behavior (Beach 1974; 1975). Hormone release can also be synchronized with environmental or behavioral changes so that hormone-behavior interactions form a feedback chain. Hormones do not act alone. They interact with, and depend on, the presence of receptor proteins, enzymes, second messengers and essential precursors, such as cholesterol and amino acids. They interact with neurotransmitters, neuropeptides, cytokines and other hormones. They also form part of feedback loops, so manipulation of one hormone can stimulate or inhibit the release of other hormones.

The neural control of hormone release is a complex problem. Each of the pituitary hormones can be regulated by a multitude of neurotransmitters (Table 6.1), neuropeptides (Table 12.2) and cytokines (Table 13.2). In addition, some neurotransmitters or neuropeptides can have opposite effects on the secretion of a hormone, depending on the mode of administration or on the levels of other hormones (see sections 6.6 and 6.7). Thus, to say that there is a neural mediation of hormone-behavior interactions does not imply that the nature of this neural mediation is easy to identify.

As with neurotransmitter-hormone interactions, there are many cases in which two or more hormones regulate the release of a third hormone. As shown in Figure 4.9, both TRH and PrRP stimulate prolactin secretion. Likewise, LH, FSH and prolactin can all regulate testosterone synthesis and secretion (Bartke *et al.* 1978). Hormones may also act in pairs, with one hormone priming a target cell to respond to a second hormone. For example, estradiol primes its target neurons for progesterone stimulation by increasing the synthesis of progesterone receptors (see section 9.5). Hormones can also have antagonistic actions on each other, so that an increase in one hormone may inhibit the secretion of others. High prolactin levels, for example, inhibit the release of gonadotropins. These complex neuroendocrine interactions make it difficult to determine exactly what combination of neuroendocrine changes is associated with behavioral events. Many different neuroendocrine changes may cause the same behavioral change and a single environmental stimulus may cause different neuroendocrine responses in different individuals or in the same individual at different times.

14.6.8 Special problems with human subjects

All of the confounding variables listed in Table 14.2 apply to research on humans as well as animals, but there are other variables operating which further confound human behavioral neuroendocrinology research. If these confounding variables are not controlled, erroneous conclusions will be drawn about the hormone-behavior interactions. Research studies on animal subjects can control each of these variables to ensure that hormone levels are the only variables influencing behavioral changes. In research with human subjects, however, it is not possible to control all of these extraneous variables, so careful experimental designs are required to ensure that effects ascribed to hormones are not caused by some other factors. The special problems in doing research on human subjects include the choice of subjects, the design of experiments, the subject-experimenter interaction, and the social and political implications of neuroendocrine research. There are also special ethical principles to consider when conducting research on human subjects (Barber 1976; American Psychological Association 1981; World Medical Association 2013).

The choice of subjects

How are subjects selected for research in behavioral neuroendocrinology? Studies can be conducted on groups of volunteers, medical patients or randomly selected groups. Each group of subjects has special considerations with regard to both experimental and ethical treatments. The ethics of psychological and medical experimentation now preclude many types of research that have been carried out in the past on populations of prisoners, racial minorities and unsuspecting medical patients (Rutstein 1969; Edsall 1969; Rodriguez and Garcia 2013). No treatment may be given to human subjects that may cause unknown or detrimental effects.

Experimental design

The effects of hormones on behavior can be studied by three methods: (1) direct observation and experimentation; (2) indirect observation using questionnaires or subjective reports; and (3) by using clinical medical reports. While it is relatively easy to conduct an experiment that presents different stimuli to subjects and involves taking blood samples to measure hormonal responses to these stimuli, it is difficult to design an experiment to alter hormonal levels in humans and look for behavioral changes. This is especially true when the effects of hormonal disruptions during prenatal development are of interest (Reinisch and Gandelman 1978). However, there are "natural experiments" – for example, where the human fetus is inadvertently exposed to sex hormones – which can advantageously be used for investigational purposes. Children with *congenital adrenal hyperplasia* (*CAH*) have a genetic disease that exposes them to high levels of testosterone during gestation. Although both sexes are affected by CAH, girls exposed to excess androgens have been the subject of many studies that show masculinization of play behavior in childhood and changes in sexual orientation in adulthood (Berenbaum and Beltz 2011; Hines 2011). Research also suggests that sex steroid hormone exposure may contribute to sex differences in the risk of autism, affective disorders and schizophrenia, although the results are inconsistent for aggression and spatial ability. Thus, for many studies on hormones and behavior in human subjects, medical patients with endocrine abnormalities are used as subjects. This poses problems with identifying, selecting and locating enough potential subjects from medical records and having them agree to take part in the study.

Once subjects have been selected, they must be matched in treatment and control conditions on factors such as age, sex, body weight, reproductive status, race and medical condition. Patients on drug treatments, radiation therapy or undergoing psychiatric treatment may not be suitable for some research. If the patients being studied are involved in a treatment program, different physicians may be prescribing different treatment regimes for the same disorder, so standardizing the independent variables may not be possible. Randomization of patients into control and treatment conditions may be impossible as it may be unethical to withhold treatment from patients in the control group. As a result, techniques must be developed to find control groups matched to the patients with endocrine disorders. These may be sibling controls, in which the brother or sister of the patient serves as the untreated control. This allows for control of race, parental environment, sex, etc. and for partial control of heredity, but there may still be individual differences in biological, medical and psychological variables.

Statistical methods such as the analysis of covariance, multivariate procedures or correlational methods can be used to control covariates that may confound the hormonal effects. Factors attributable to age, sex, type of hormone given, dose levels and treatment duration may be factored out using statistical methods rather than control groups (Reinisch and Gandelman 1978). The problem of developing randomized clinical trials in medical research is discussed by Silverman (1985) and by Umscheid *et al.* (2011). It may be difficult to study behavior following hormone treatment, so subjects are often asked to complete questionnaires on their moods, thoughts, or their aggressive and sexual behaviors or feelings. How the results of these pencil and paper tests correlate with actual performance of sexual, aggressive and other behaviors is largely unknown.

Subject and experimenter bias

Subjects and experimenters both enter into experimental situations with expectancies that may bias the results. The experimenters have specific hypotheses to test and the subjects have expectancies of being cured of a disease or showing some altered behavior or psychological state. Many procedures exist to control for these expectancy effects. Experiments can be run in single- and double-blind designs and subjects in some groups may be given placebo treatments (Rosenthal 1966). But the mood, temperament and personality variables of the subjects are not controllable, nor are the social factors involved in the subject-experimenter interaction. The experimenter's sex, mannerisms and general friendliness, as perceived by the subjects, can influence the behavior and the willingness of the subject to continue in the experiment. Also, the instructions given to the subject and what the subject thinks they are really "supposed to do" can influence the results (Rosenthal 1966).

Animal models

One way to study the effects of hormones on human behavior is to perform experimentally controlled studies with animals. Such animal models of human studies can follow the traditional experimental design discussed in section 14.3 that looks at the effects of specific hormone parameters (chemistry, dose, timing, etc.) on behavior. On the other hand, one can attempt to mimic a human disorder in an animal and then test different methods of inducing and treating the disorder. This "human analog model" of hormonal research is designed to examine the etiology and treatment of specific biomedical disorders (Reinisch and Gandelman 1978).

Social and political implications

Biomedical and sociobiological studies of human behavior have been the subject of criticism for fostering a biologically deterministic model of human behavior which is opposed by those who see behavior as determined by social learning and socio-cultural factors. Various critics have claimed that biological theories of human behavior support the capitalist system, are racist and sexist. These arguments against biological theories of human behavior are discussed by Ruse (1979). The interpretation of results from neuroendocrine research on human behavior has also been questioned. A biological interpretation is that behavioral changes are determined by neuroendocrine changes. A psychological interpretation is that emotional or cognitive responses to the neuroendocrine change alter the behavior, and a socio-cultural explanation is that socially acceptable responses to the neuroendocrine change are learned and, if this cultural attitude was changed, the response would be different. These arguments occur most often in sensitive areas such as the influence of hormones on sex differences in behavior and studies of premenstrual disorders. The fact that there may be sex and racial differences in the physiological responses to drugs and other neuroregulators does not seem to be a political issue.

Finally, the application of knowledge about the neuroendocrine control of behavior has produced a biotechnology with wide-ranging social and political implications (Austin 1972; Jaffe 1973). How, for example, should hormonal birth control be regulated? Should there be social, parental and religious control over who can use hormonal methods of birth control, especially for the "morning after pill" (Ceva and Moratti 2013)? Should birth control be forced upon people in over-populated countries? Should mentally disabled people be forced to use birth control? How should fertility drugs be used? Should hormonal treatments for inducing abortion be used? These are some of the questions raised by our ability to hormonally alter reproductive processes. Likewise, the social control of male sex hormones is also controversial (Campo-Engelstein 2012). How, for example, can the use of performance-enhancing anabolic steroid hormones and growth hormone in sports be controlled (Hoffman et al. 2009)? Should sex offenders and extremely aggressive men be treated by castration or anti-androgen therapy? Many other issues in the use of neuroendocrine regulation for social and political purposes will develop as neuroendocrine technology becomes more widely applied to everyday life.

Summary

Research on the neuroendocrine control of human behavior is difficult because of the problems in designing carefully controlled experiments that eliminate confounding variables. Such considerations are extremely important because of the high level of neuroendocrine biotechnology in modern society. Taking drugs or hormonal treatment during pregnancy, birth control pills, hormone supplements in old age, medication for psychological or neurological disorders, cancer treatments and other medical treatments may all alter the neuroendocrine system, but how can the neuroendocrine effects of these treatments be determined? Much of the research on the neuroendocrine regulation of human behavior is from correlational studies, anecdotal case study reports, studies using medical patients or from studies using questionnaires and self reports rather than behavioral data. As a consequence, much of the basic information in behavioral neuroendocrinology is from controlled experiments with animals or experiments with animal analogs of human disorders. It is usually assumed that the results of these animal studies can be generalized to humans, but this is often difficult because of species and other differences between animals and humans.

14.7 Summary

This chapter has introduced some of the methods used to study behavioral neuroendocrinology. First, the behaviors to be recorded must be determined and careful observational and descriptive methods used to define the specific behavior units to be recorded. Accurate quantitative methods are then necessary for measuring the frequency, duration and patterning of these behavior units. Correlational studies can then be conducted to record the behavioral changes associated with natural fluctuations in hormone levels or disorders of the neuroendocrine system. Hormone levels for these correlational studies can be measured directly from blood samples or indirectly from urine, saliva or by using bioassays. Experimental studies can be of two types: (1) hormones can be manipulated as the independent variable and behavioral responses recorded; or (2) external stimuli can be manipulated and hormonal responses recorded. When neuroendocrine variables are manipulated, the experiments include a baseline, hormone removal and hormone replacement phase. These are required to demonstrate that it is the hormone that alters behavior and not some non-hormonal confounding factor. In these studies, careful consideration must be given to the methods of hormone removal and replacement. Consideration must also be given to the chemical composition of synthetic hormones used, the dose level, vehicle, route of administration and the timing of hormone replacement therapy with regard to behavioral testing. Neuroendocrine changes can also be measured in response to environmental stimuli, behavioral interactions and cognitive functions. These stimuli can cause brief hormone surges, as occur in a neuroendocrine reflex, or long-term changes in baseline hormonal secretion. There can also be chains of hormone-behavior responses, in which neuroendocrine and behavioral stimuli interact to produce a sequence of coordinated behaviors, as occurs during reproduction in the ring dove. Neuroendocrine responses can also be conditioned to environmental stimuli.

The effects of hormones on behavior are mediated by neural and genomic mechanisms. Through their neuromodulatory actions on neurotransmitter release, hormones alter neural sensitivity to sensory input, regulate the motor pathways controlling behavior and modulate the central neural mechanisms that integrate complex behavioral responses. Hormones act through genomic mechanisms to stimulate gene expression and protein synthesis, resulting in cell growth and differentiation, as well as long-term changes in behavior. The effects of hormones on neural and genomic mechanisms may occur in adulthood, but especially during sensitive embryonic and postnatal development, during which time they can alter the organization of the developing nervous system and have permanent effects on brain and behavior. Research in behavioral neuroendocrinology requires control of many confounding variables, such as species and individual differences, previous experience of the subjects, environmental factors, the stimulus situation and the state of the subject. There are also complex interactions among hormones, neurotransmitters, neuropeptides, cytokines and environmental variables that may influence behavior. As a result, hormone secretion may be altered by changes in other hormones or neurotransmitters, synchronized with environmental changes and modified by feedback effects from behavior.

Behavioral neuroendocrinology research with human subjects involves special considerations of subject selection, experimental design, and experimenter and subject bias. Much of the research on hormones and behavior that is relevant to humans is done using animal models of human disorders. There are also a number of social and political questions that arise with regard to the use of the biotechnology resulting from knowledge about the neuroendocrine system and the ability to manipulate human hormones and behavior.

FURTHER READING

Adams, J. (1986). "Methods in behavioural teratology" in E. P. Riley and C. V. Vorhees (eds.), *Handbook of Behavioral Teratology* (New York: Plenum Press), pp. 67–97.

Argiolas, A. and Melis, M. R. (2013). "Neuropeptides and central control of sexual behaviour from the past to the present: a review," *Prog Neurobiol* 108, 80–107.

Arnold, A. P. and Breedlove, S. M. (1985). "Organizational and activational effects of sex steroids on brain and behavior: a reanalysis," *Horm Behav* 19, 469–498.

Beach, F. A. (1974). "Behavioral endocrinology and the study of reproduction," *Biol Reprod* 10, 2–18.

Henry, J. P. (1986). "Neuroendocrine patterns of emotional response" in R. Plutchik and H. Kellerman (eds.), *Emotion: Theory, Research, and Experience*, vol. 3: *Biological Foundations of Emotion* (Orlando, FL: Academic Press), pp. 37–60.

Komisaruk, B. R., Siegel, H. I., Cheng, M.-F. and Feder, H. H. (eds.) (1986). "Reproduction: a behavioral and neuroendocrine perspective," *Ann NY Acad Sci* 474, 1–465.

Lauder, J. M. (1983). "Hormonal and humoral influences on brain development," *Psychoneuroendocr* 8, 121–155.

REVIEW QUESTIONS

14.1   What is a behavioral bioassay?

14.2   What is the difference between the qualitative and quantitative description of behavior?

14.3   What are two non-invasive or indirect measures of hormone levels which can be used in behavioral studies.

14.4   What are the three phases of the standard hormone and behavior experiment?

14.5   If one did not want to use surgical gland ablation to remove a hormone, what two other methods could be used?

14.6   What is the difference between subcutaneous and intracerebral hormone injection?

14.7   What are the advantages and disadvantages of using sustained release (silastic) capsules for hormone replacement?

14.8   What is the difference between long-term and short-term feedback effects of hormones in response to environmental stimuli?

14.9   What is the difference between neural and genomic mechanisms mediating the effects of hormones on behavior?

14.10 How can the intrauterine environment lead to individual differences in the behavior of genetically identical mice?

14.11 Why are hormone-hormone interactions important problems for the study of behavior?

14.12 What is a "human analog model"?

ESSAY QUESTIONS

14.1   Discuss the advantages, disadvantages and confounding variables in the use of sexual behavior in male rats as a behavioral bioassay for androgen levels.

14.2   Of what value is the study of behavioral changes during the menstrual cycle of adult human females for understanding the role of gonadal hormone changes in controlling behavior?

14.3   Discuss a way of adapting the standard hormone removal and replacement experiment for the study of how: (a) anabolic steroids influence exercise performance in human males; or (b) exercise level affects gonadal hormone secretion in human females.

14.4   Discuss the variables that must be controlled for a study on the hormonal control of the onset of parental (maternal) behavior in sheep (or rats).

14.5   Starting with Figure 14.12, discuss the effects of social stimuli on gonadal and adrenal steroid levels in monkeys.

14.6   Can pituitary hormone release be conditioned? Discuss, for example, conditioned release of LH in males or oxytocin in females.

14.7   Describe the neuroendocrine control of the lordosis reflex in the female rat.

14.8   Describe the hormonal control of the neuroanatomical changes underlying birdsong.

14.9   Discuss the role of androgens/estrogens as neural growth factors during fetal development.

14.10 Discuss the phenomenon of "uterine position" as it relates to individual differences in sexual differentiation and subsequent sexually dimorphic behaviors.

14.11 Discuss the problems underlying the controlled study of hormones and human sexual behavior.

REFERENCES

Adams, J. (1986). "Methods in behavioral teratology" in E. P. Riley and C. V. Vorhees (eds.), *Handbook of Behavioral Teratology* (New York: Plenum Press), pp. 67–97.

Ader, R. (1976). "Conditioned adrenocortical steroid elevations in the rat," *J Comp Physiol Psychol* 90, 1156–1163.

Ader, R. (2003). "Conditioned immunomodulation; research needs and directions," *Brain Behav Immun* 17(Suppl. 1), S51–S57.

Adkins-Regan, E. (1981). "Early organizational effects of hormones: an evolutionary perspective" in N. T. Adler (ed.), *Neuroendocrinology of Reproduction: Physiology and Behavior* (New York: Plenum Press), pp. 159–228.

Adler, N. T., Davis, P. G. and Komisaruk, B. R. (1977). "Variation in the size and sensitivity of a genital sensory field in relation to the estrous cycle in rats," *Horm Behav* 9, 334–344.

Al'Absi, M., Nakajima, M., Hooker, S., Wittmers, L. and Cragin, T. (2012). "Exposure to acute stress is associated with attenuated sweet taste," *Psychophysiol* 49, 96–103.

American Psychological Association. (1981). "Ethical principles of psychologists," *Amer Psychol* 36, 633–638.

Amandusson, A. and Blomqvist, A. (2013). "Estrogenic influences in pain processing," *Front Neuroendocr* 34, 329–349.

Andelman, S. J., Else, J. G., Hearn, J. P. and Hodges, J. K. (1985). "The non-invasive monitoring of reproductive events in wild Vervet monkeys (*Cercopithecus aethiops*) using urinary pregnanediol-3a-glucuronide and its correlation with behavioural observations," *J Zool London (A)* 205, 467–477.

Anestis, S. F. (2006). "Testosterone in juvenile and adolescent male chimpanzees (*Pan troglodytes*): effects of dominance rank, aggression and behavioral style," *Am J Phys Anthropol* 130, 536–545.

Argiolas, A. and Melis, M. R. (2004). "The role of oxytocin and the paraventricular nucleus in the sexual behavior of male mammals," *Physiol Behav* 83, 309–317.

Armario, A. and Jolin, T. (1989). "Influence of intensity and duration of exposure to various stressors on serum TSH and GH levels in adult male rats," *Life Sci* 44, 215–221.

Arnold, A. P. (1981). "Logical levels of steroid hormone action in the control of vertebrate behavior," *Amer Zool* 21, 233–242.

Arnold, A. P. and Jordan, C. L. (1988). "Hormonal organization of neural circuits," *Front Neuroendocr* 10, 185–214.

Aronson, L. A. (1959). "Hormones and reproductive behavior: some phylogenetic considerations" in A. Gorbman (ed.), *Comparative Endocrinology* (New York: John Wiley), pp. 98–120.

Austin, C. R. (1972). "The ethics of manipulating human reproduction" in C. R. Austin and R. V. Short (eds.), *Reproduction in Mammals. 5. Artificial Control of Reproduction* (Cambridge University Press), pp. 141–152.

Baggerman, B. (1968). "Hormonal control of reproductive and parental behaviour in fishes" in E. J. W. Barrington and C. B. Jorgensen (eds.), *Perspectives in Endocrinology* (London: Academic Press), pp. 351–404.

Balazs, R. (1976). "Hormones and brain development," *Prog Brain Res* 45, 139–159.

Balazs, R., Patel, A. J. and Hajos, F. (1975). "Factors affecting the biochemical maturation of the brain: effects of hormones during early life," *Psychoneuroendocr* 1, 25–36.

Ball, G. F. and Balthazart, J. (2004). "Hormonal regulation of brain circuits mediating male sexual behavior in birds," *Physiol Behav* 83, 329–346.

Ball, G. F., Auger, C. J., Bernard, D. J., Charlier, T. D., Sartor, J. J., Riters, L. V. *et al.* (2004). "Seasonal plasticity in the song control system: multiple brain sites of steroid hormone action and the importance of variation in song behavior," *Ann NY Acad Sci* 1016, 586–610.

Balthazart, J. (1983). "Hormonal correlates of behavior" in D. S. Farner, J. R. King and K. C. Parkes (eds.), *Avian Biology*, vol. VII (New York: Academic Press), pp. 221–335.

Balthazart, J. (2011). "Minireview: hormones and human sexual orientation," *Endocr* 152, 2937–2947.

Balthazart, J. and Ball, G. F. (2007). "Topography in the preoptic region: differential regulation of appetitive and consummatory male sexual behaviors," *Front Neuroendocr* 28, 161–178.

Balthazart, J. and Hendrick, J. (1976). "Annual variation in reproductive behavior, testosterone and plasma FSH levels in the Rouen duck, *Anas platyrhynchos*," *Gen Comp Endocrinol* 28, 171–183.

Balthazart, J., Baillien, M., Cornil, C. A. and Ball, G. F. (2004). "Preoptic aromatase modulates male sexual behavior: slow and fast mechanisms of action," *Physiol Behav* 83, 247–270.

Balthazart, J., Taziaux, M., Holloway, K., Ball, G. F. and Cornil, C. A. (2009). "Behavioral effects of brain-derived estrogens in birds," *Ann NY Acad Sci* 1163, 31–48.

Barber, B. (1976). "The ethics of experimentation with human subjects," *Sci Amer* 234, 25–31.

Bartke, A., Hafiez, A. A., Bex, F. J. and Dalterio, S. (1978). "Hormonal interactions in regulation of androgen secretion," *Biol Reprod* 18, 44–54.

Basterfield, L., Lumley, L. K. and Mathers, J. C. (2009). "Wheel running in female C57BL/6 J mice: impact of oestrus and dietary fat and effects on sleep and body mass," *Int J Obes* 33, 212–218.

Baumeister, A. A. and Sevin, J. A. (1990). "Pharmacologic control of aberrant behavior in the mentally retarded: toward a more rational approach," *Neurosci Biobehav Rev* 14, 253–262.

Beach, F. A. (1974). "Behavioral endocrinology and the study of reproduction," *Biol Reprod* 10, 2–18.

Beach, F. A. (1975). "Behavioral endocrinology: an emerging discipline," *Amer Sci* 63, 178–187.

Beaulieu-Boire, G., Bourque, S., Chagnon, F., Chouinard, L., Gallo-Payet, N. and Lesur, O. (2013). "Music and biological stress dampening in mechanically-ventilated patients at the intensive care unit ward – a prospective interventional randomized crossover trial," *J Crit Care* 28, 442–450.

Berenbaum, S. A. and Beltz, A. M. (2011). "Sexual differentiation of human behavior: effects of prenatal and pubertal organizational hormones," *Front Neuroendocr* 32, 183–200.

Bermant, G. and Davidson, J. M. (1974). *Biological Bases of Sexual Behavior* (New York: Harper & Row).

Bernal, J. (2005). "Thyroid hormones and brain development," *Vitam Horm* 71, 95–122.

Bertram, E. H., Williamson, J. M., Cornett, J. F., Spradlin, S. and Chen, Z. F. (1997). "Design and construction of a long-term continuous video-EEG monitoring unit for simultaneous recording of multiple small animals," *Brain Res Protocols* 2, 85–97.

Brack, K. E., Jeffrey, S. M. T. and Lovick, T. A. (2006). "Cardiovascular and respiratory responses to a panicogenic agent in anaesthetised female Wistar rats at different stages of the oestrous cycle," *Europ J Neurosci* 23, 3309–3318.

Brain, P. and Benton, D. (1979). "The interpretation of physiological correlates of differential housing in laboratory rats," *Life Sci* 24, 99–116.

Brown, G. M. and Martin, J. B. (1974). "Corticosterone, prolactin, and growth hormone responses to handling and new environment in the rat," *Psychosom Med* 36, 241–247.

Brown, R. E. (1985a). "Hormones and paternal behavior in vertebrates," *Amer Zool* 25, 895–910.

Brown, R. E. (1985b). "The rodents I: effects of odours on reproductive physiology (primer effects)" in R. E. Brown and D. W. Macdonald (eds.), *Social Odours in Mammals* (Oxford: Clarendon Press), vol. 1, pp. 245–344.

Brown, R. E. and McFarland, D. J. (1979). "Interaction of hunger and sexual motivation in the male rat: a time-sharing approach," *Animal Behav* 27, 887–896.

Brown, R. E., Wilkinson, D. A., Imran, S. A., Caraty, A. and Wilkinson, M. (2012). "Hypothalamic *kiss1* mRNA and kisspeptin immunoreactivity are reduced in a rat model of polycystic ovary syndrome (PCOS)," *Brain Res* 1467, 1–9.

Burke, A. W. and Broadhurst, P. L. (1966). "Behavioural correlates of the oestrous cycle in the rat," *Nature* 209, 223–224.

Campo-Engelstein, L. (2012). "Contraceptive justice: why we need a male pill," *Virtual Mentor* 14, 146–151.

Cameron, O. G. and Nesse, R. M. (1988). "Systemic hormonal and physiological abnormalities in anxiety disorders," *Psychoneuroendocr* 13, 287–307.

Caras, M. L. (2013). "Estrogenic modulation of auditory processing: a vertebrate comparison," *Front Neuroendocr* 34, 285–299.

Carreon-Rodriguez, A. and Perez-Martinez, L. (2012). "Clinical implications of thyroid hormones effects on nervous system development," *Pediatr Endocr Rev* 9, 644–649.

Ceva, E. and Moratti, S. (2013). "Whose self-determination? Barriers to access to emergency hormonal contraception in Italy," *Kennedy Inst Ethics* 23, 139–167.

Cheng, M. F. (1986). "Individual behavioral response mediates endocrine changes induced by social interaction," *Ann New York Acad Sci* 474, 4–12.

Clarke, I. J. and Caraty, A. (2013). "Kisspeptin and seasonality of reproduction," *Adv Exp Med Biol* 784, 411–430.

Cohen, J. (1983). "Hormones and brain mechanisms of vocal behavior in non-vocal learning birds" in J. Balthazart, E. Pröve and R. Gilles (eds.), *Hormones and Behaviour in Higher Vertebrates* (Berlin: Springer-Verlag), pp. 422–436.

Coover, G. D., Sutton, B. R. and Heybach, J. P. (1977). "Conditioning decreases in plasma corticosterone level in rats by pairing stimuli with daily feedings," *J Comp Physiol Psychol* 91, 716–726.

Crews, D. (1980). "Interrelationships among ecological, behavioral and neuroendocrine processes in the reproductive cycle of *Anolis carolensis* and other reptiles," *Adv Study Behav* 11, 1–74.

Crews, D. (1986). "Comparative behavioral endocrinology," *Ann NY Acad Sci* 474, 187–198.

Crinnion, W. J. (2009). "Maternal levels of xenobiotics that affect fetal development and childhood health," *Alt Med Rev* 14, 212–222.

Crowley, W. R. (1986). "Reproductive neuroendocrine regulation in the female rat by central catecholamine-neuropeptide interactions: a local control hypothesis," *Ann NY Acad Sci* 474, 423–436.

Cui, J. G., Tang, G. B., Wang, D. H. and Speakman, J. R. (2011). "Effects of leptin infusion during peak lactation on food intake, body composition, litter growth and maternal neuroendocrine status in female Brandt's voles (*Lasiopodomys brandtii*)," *Am J Physiol Integ Comp Physiol* 300, R447–R459.

Culbert, K. M., Breedlove, S. M., Sisk, C. L., Burt, S. A. and Klump, K. L. (2013). "The emergence of sex differences in risk for disordered eating attitudes during puberty: a role for prenatal testosterone exposure," *J Ab Psychol* 122, 420–432.

Dabbs, J. M. (1991). "Salivary testosterone measurements: collecting, storing, and mailing saliva samples," *Physiol Behav* 49, 815–817.

Daskalakis, N. P., Lehrner, A. and Yehuda, R. (2013). "Endocrine aspects of post-traumatic stress disorder and implications for diagnosis and treatment," *Endocrinol Metab Clin North Am* 42, 503–513.

Davidson, J. M. and Trupin, S. (1975). "Neural mediation of steroid-induced sexual behavior in rats" in M. Sandler and G. L. Gessa (eds.), *Sexual Behavior: Pharmacology and Biochemistry* (New York: Raven Press), pp. 13–20.

Davis, E. S. and Marler, C. A. (2003). "The progesterone challenge: steroid hormone changes following a simulated territorial intrusion in female *Peromyscus californicus,*" *Horm Behav* 44, 185–198.

Davis, K. W., Cepeda-Benito, A., Harraid, J. H. and Wellman, P. J. (2005). "Plasma corticosterone in the rat in response to nicotine and saline injections in a context previously paired or unpaired with nicotine," *Psychopharmacol* 180, 466–472.

Dawson, A., King, V. M., Bentley, G. E. and Ball, G. F. (2001). "Photoperiodic control of seasonality in birds," *J Biol Rhythms* 16, 365–380.

de Kloet, E. R., Joels, M. and Holsboer, F. (2005). "Stress and the brain: from adaptation to disease," *Nat Rev Neurosci* 6, 463–475.

De Kort, S. R. and ten Kate, C. (2004). "Repeated decrease in vocal repertoire size in *Streptopelia* doves," *Anim Behav* 67, 549–557.

Donat, P. (1991). "Measuring behaviour: the tools and the strategies," *Neurosci Biobehav Rev* 15, 447–454.

Edsall, G. (1969). "A positive approach to the problem of human experimentation," *Daedalus* 98, 463–479.

Edwards, D. A. (1969). "Early androgen stimulation and aggressive behavior in male and female mice," *Physiol Behav* 4, 333–338.

Everitt, B. J. (1990). "Sexual motivation: a neural and behavioural analysis of the mechanisms underlying appetitive and copulatory responses of male rats," *Neurosci Biobehav Rev* 14, 217–232.

Everett, J. W. (1989). *Neurobiology of Reproduction in the Female Rat: A Fifty Year Perspective* (Berlin: Springer-Verlag).

Fantie, B. D., Brown, R. E. and Moger, W. H. (1984). "Constant lighting conditions affect sexual behaviour and hormone levels in adult male rats," *J Reprod Fert* 72, 435–441.

Feder, H. H., Storey, A., Goodwin, D., Reboulleau, C. and Silver, R. (1977). "Testosterone and 5α-dihydrotestosterone levels in peripheral plasma of male and female ring doves (*Streptopelia risoria*) during the reproductive cycle," *Biol Reprod* 16, 666–677.

Ferguson, K. T., Cassells, R. C., MacAllister, J. W. and Evans, G. W. (2013). "The physical environment and child development: an international review," *Int J Psychol* 48, 437–468.

Fillingim, R. B., King, C. D., Ribeiro-Dasilva, M. C., Rahim-Williams, B. and Riley, J. L. (2009). "Sex, gender, and pain: a review of recent clinical and experimental findings," *J Pain* 10, 447–485.

Flannelly, K. and Lore, R. (1977). "The influence of females upon aggression in domesticated male rats (*Rattus norvegicus*)," *Anim Behav* 25, 654–659.

Frederick, A. L. and Stanwood, G. D. (2009). "Drugs, biogenic amine targets and the developing brain," *Dev Neurosci* 31, 7–22.

Gagnidze, K., Weil, Z. M., Faustino, L. C., Schaafsma, S. M. and Pfaff, D. W. (2013). "Early histone modifications in the ventromedial hypothalamus and preoptic area following oestradiol administration," *J Neuroendocr* 25, 939–955.

Gammill, L. S. and Bronner-Fraser, M. (2003). "Neural crest specification: migrating into genomics," *Nat Rev Neurosci* 4, 795–805.

Gandelman, R. (1983). "Gonadal hormones and sensory function," *Neurosci Biobehav Rev* 7, 1–17.

Gandelman, R. (1984). "Relative contributions of aggression and reproduction to behavioral endocrinology," *Aggress Behav* 10, 123–133.

Gary, K. A., Sevarino, K. A., Yarbrough, G. G., Prange, A. J. and Winokur, A. (2003). "The thyrotropin releasing hormone (TRH) hypothesis of homeostatic regulation: implications for TRH-based therapeutics," *J Pharmacol Exp Therap* 305, 410–416.

Gelez, H., Archer, E., Chesneau, D., Campan, R. and Fabre-Nys, C. (2004). "Importance of learning in the response of ewes to male odor," *Chem Senses* 29, 555–563.

Gladue, B. A. and Clemens, L. G. (1980). "Flutamide inhibits testosterone-induced masculine sexual behavior in male and female rats," *Endocr* 106, 1917–1922.

Graham, J. M. and Desjardins, C. (1980). "Classical conditioning: induction of luteinizing hormone and testosterone secretion in anticipation of sexual activity," *Science* 210, 1039–1040.

Guerry, J. D. and Hastings, P. D. (2011). "In search of HPA axis dysregulation in child and adolescent depression," *Clin Child Fam Psychol Rev* 14, 135–160.

Hachul, H., Bittencourt, L. R., Andersen, M. L., Haidar, M. A., Baracat, E. C. and Tufic, S. (2008). "Effects of hormone therapy with estrogen and/or progesterone on sleep patterns in post-menopausal women," *Int J Gynaecol Obstet* 103, 207–212.

Haley, D. W., Cordick, J., Mackrell, S., Antony, I. and Ryan-Harrison, M. (2011). "Infant anticipatory stress," *Biol Lett* 23, 136–138.

Harding, C. F. (1981). "Social modulation of circulating hormone levels in the male," *Amer Zool* 21, 223–231.

Harding, S. M. and McGinnis, M. Y. (2004). "Androgen blockade in the MPOA or VMN: effects on male sociosexual behavior," *Physiol Behav* 81, 671–680.

Henkin, R. I. (1975). "Effects of ACTH, adrenocorticosteriods and thyroid hormone on sensory function" in W. E. Stumpf and L. D. Grant (eds.), *Anatomical Neuroendocrinology* (Basel: Karger), pp. 298–316.

Herlenius, E. and Lagercrantz, H. (2001). "Neurotransmitters and neuromodulators during early human development," *Early Hum Dev* 65, 21–37.

Herlenius, E. and Lagercrantz, H. (2004). "Development of neurotransmitter systems during critical periods," *Exp Neurol* 190, S8–S21.

Higashi, T. (2012). "Salivary hormone measurement using LC/MS/MS: specific and patient-friendly tool for assessment of endocrine function," *Biol Pharm Bull* 35, 1401–1408.

Hines, M. (2011). "Prenatal endocrine effects on sexual orientation and on sexually differentiated childhood behavior," *Front Neuroendocr* 32, 170–182.

Hinkelmann, K., Moritz, S., Botzenhardt, J., Riedesel, K., Wiedemann, K., Kellner, M. *et al.* (2009). "Cognitive impairment in major depression: association with salivary cortisol," *Biol Psych* 66, 879–885.

Hoffman, J. R., Kraemer, W. J., Bhasin, S., Storer, T., Ratamess, N. A., Haff, G. G. *et al.* (2009). "Position stand on androgen and growth hormone use," *J Strength Cond Res* 23(Suppl. 5), S1–S59.

Hudson, J. I., Hiripi, E., Pope, H. G. and Kessler, R. C. (2007). "The prevalence and correlates of eating disorders in the National Comorbidity Survey Replication," *Biol Psych* 61, 348–358.

Hyde, J. S. and Sawyer, T. F. (1977). "Estrous cycle fluctuations in aggressiveness of house mice," *Horm Behav* 9, 290–295.

Ikemoto, S. and Wise, R. A. (2004). "Mapping of chemical trigger zones for reward," *Neuropharmacol* 47, 190–201.

Isaksson, I. M., Theodorsson, A., Theodorsson, E. and Strom, J. O. (2011). "Methods for 17β-oestradiol administration to rats," *Scand J Clin Lab Invest* 71, 583–592.

Jaffe, F. S. (1973). "Public policy on fertility control," *Sci Amer* 229, 17–23.

Jayasena, C. N., Nijher, G. M. K., Chaudhri, O. B., Murphy, K. G., Ranger, A., Lim, A. *et al.* (2009). "Subcutaneous injection of kisspeptin-54 stimulates gonadotrophin secretion in women with hypothalamic amenorrhea, but chronic administration causes tachyphylaxis," *J Clin Endocr Metab* 94, 4315–4323.

Johansson, G., Collins, A. and Collins, V. P. (1983). "Male and female psychoneuroendocrine response to examination stress: a case report," *Motivat Emotion* 7, 1–9.

Johnson, M. and Everett, B. (1988). *Essential Reproduction*, 3rd edn. (Oxford: Blackwell).

Johnston, R. E. (1981). "Attraction to odors in hamsters: an evaluation of methods," *J Comp Physiol Psychol* 95, 951–960.

Jones, S. L., Ismail, N., King, L. and Pfaus, J. G. (2012). "The effects of chronic administration of testosterone propionate with or without estradiol on the sexual behavior and plasma steroid levels of aged female rats," *Endocr* 153, 5928–5939.

Kawata, M. (2013). "Nurture: effects of intrauterine position on behaviour," *J Neuroendocr* 25, 422–423.

Kelley, A. E. (1989). "Behavioural models of neuropeptide action" in G. Fink and A. J. Harmer (eds.), *Neuropeptides: A Methodology* (Chichester: Wiley), pp. 301–331.

Kelley, A. E., Baldo, B. A. and Pratt, W. E. (2005). "A proposed hypothalamic-thalamic-striatal axis for the integration of energy balance, arousal and food reward," *J Comp Neurol* 493, 72–85.

Kelley, B. M., Bandy, A.-L. E. and Middaugh, L. D. (1997). "A novel and detachable indwelling jugular catheterization procedure for mice," *Physiol Behav* 62, 163–167.

Knipper, M., Zinn, C., Maier, H., Praetorious, M., Rohbock, K., Köpschall, I. *et al.* (2000). "Thyroid hormone deficiency before the onset of hearing causes irreversible damage to peripheral and central auditory systems," *J Neurophysiol* 83, 3101–3112.

Koibuchi, N., Jingu, H., Iwasaki, T. and Chin, W. W. (2003). "Current perspectives on the role of thyroid hormone in growth and development of cerebellum," *Cerebellum* 2, 279–289.

Komisaruk, B. R., Adler, N. T. and Hutchison, J. (1972). "Genital sensory field: enlargement by estrogen treatment in female rats," *Science* 178, 1295–1298.

Komisaruk, B. R. and Steinman, J. L. (1986). "Genital stimulation as a trigger for neuroendocrine and behavioral control of reproduction," *Ann NY Acad Sci* 474, 64–75.

Krsiak, M. (1991). "Ethopharmacology: a historical perspective," *Neurosci Biobehav Rev* 15, 439–445.

La Torre, A., Conca, A., Duffy, D., Giupponi, G., Pompili, M. and Grozinger, M. (2013). "Sexual dysfunction related to psychotropic drugs: a critical review Part II: antipsychotics," *Pharmacopsych* 46, 201–208.

Larson, M. R., Ader, R. and Moynihan, J. A. (2001). "Heart rate, neuroendocrine and immunological reactivity in response to an acute laboratory stressor," *Psychosom Med* 63, 493–501.

Lauder, J. M. (1983). "Hormonal and humoral influences on brain development," *Psychoneuroendocr* 8, 121–155.

Lauder, J. M. and Krebs, H. (1986). "Do neurotransmitters, neurohumors, and hormones specify critical periods?" in W. T. Greenough and J. M. Juraska (eds.), *Developmental Neuropsychobiology* (Orlando, FL: Academic Press), pp. 119–174.

Leatherwood, W. E. and Dragoo, J. L. (2013). "Effect of airline travel on performance: a review of the literature," *Br J Sports Med* 47, 561–567.

Lehner, P. N. (1979). *Handbook of Ethological Methods* (New York: Garland Press).

Lehrman, D. S. (1965). "Interaction between internal and external environments in the regulation of the reproductive cycle" in F. A. Beach (ed.) *Sex and Behavior* (New York: Wiley), pp. 355–380.

Lehrman, D. S., Brody, P. N. and Wortis, R. P. (1961). "The presence of the mate and of nesting material as stimuli for the development of incubation behavior and for gonadotropin secretion in the ring dove (*Streptopelia risoria*)," *Endocrinol* 68, 507–516.

Lenz, K. M. and McCarthy, M. M. (2010). "Organized for sex – steroid hormones and the developing hypothalamus," *Europ J Neurosci* 32, 2096–2104.

Leshner, A. I. (1978). *An Introduction to Behavioral Endocrinology* (New York: Oxford University Press).

Leshner, A. I. (1979). "Kinds of hormonal effects on behavior: a new view," *Neurosci Biobehav Rev* 3, 69–73.

Ludvig, N., Kovacs, L., Kando, L., Medveczky, G., Tang, H. M., Eberle, L. P. *et al*. (2002). "The use of a remote-controlled minivalve, carried by freely moving animals on their heads, to achieve instant pharmacological effects in intracerebral drug-perfusion studies," *Brain Res Protocols* 9, 23–31.

Mani, S. K. and Oyola, M. G. (2012). "Progesterone signaling mechanisms in brain and behavior," *Front Endocr* 3, 1–8.

Mantei, K. E., Ramakrishnan, S., Sharp, P. J. and Buntin, J. D. (2008). "Courtship interactions stimulate rapid changes in GnRH synthesis in male ring doves," *Horm Behav* 54, 669–675.

Marcondes, F. K., Bianchi, F. J. and Tanno, A. P. (2002). "Determination of the estrous cycle phases of rats: some helpful considerations," *Braz J Biol* 62, 609–614.

Martin, P. and Bateson, P. (1986). *Measuring Behaviour* (Cambridge University Press).

McWhinney, B. C., Briscoe, S. E., Ungerer, J. P. and Pretorius, C. J. (2010). "Measurement of cortisol, cortisone, prednisolone, dexamethasone and 11-deoxycortisol with ultra high performance liquid chromatography-tandem mass spectrometry: application for plasma, plasma ultrafiltrate, urine and saliva in a routine laboratory," *J Chromatogr B Analyt Technol Biomed Life Sci* 878, 2863–2869.

Mitchell, J. and Popkin, M. (1983). "The pathophysiology of sexual dysfunction associated with antipsychotic drug therapy in males: a review," *Arch Sexual Behav* 12, 173–183.

Mong, J. A. and Pfaff, D. W. (2004). "Hormonal symphony: steroid orchestration of gene modules for socio-sexual behaviors," *Mol Psych* 9, 550–556.

Mong, J. A., Baker, F. C., Mahoney, M. H., Paul, K. N., Schwartz, M. D., Semba, K. *et al*. (2011). "Sleep, rhythms, and the endocrine brain: influence of sex and gonadal hormones," *J Neurosci* 31, 16107–16116.

Morales, A. (2011). "Androgens are fundamental in the maintenance of male sexual health," *Curr Urol Rep* 12, 453–460.

Morris, J. A., Jordan, C. L. and Breedlove, S. M. (2004). "Sexual differentiation of the vertebrate nervous system," *Nat Neurosci* 7, 1034–1039.

Moss, R. and McCann, S. (1973). "Induction of mating behavior in rats by luteinizing hormone releasing factor," *Science* 181, 177–179.

Mul, J. D., Spruijt, B. M., Brakkee, J. H. and Adan, R. A. H. (2013). "Melanocortin MC4 receptor-mediated feeding and grooming in rodents," *Europ J Pharmacol* 719, 192–201.

Nelson, A., Hartl, W., Jauch, K. W., Fricchione, G. L., Benson, H., Warshaw, A. L. *et al*. (2008). "The impact of music on hypermetabolism in critical illness," *Curr Opin Clin Nutr Metab Care* 11, 790–794.

Nixon, J. P., Kotz, C. M., Novak, C. M., Billington, C. J. and Teske, J. A. (2012). "Neuropeptides controlling energy balance: orexins and neuromedins," *Handbk Exp Pharmacol* 209, 77–109.

Nugent, B. M., Tobet, S. A., Lara, H. E., Lucion, A. B., Wilsom, M. E., Recabarren, S. E. *et al.* (2012). "Hormonal programming across the lifespan," *Horm Metab Res* 44, 577–586.

Nyby, J. G. (2008). "Reflexive testosterone release: a model system for studying the non-genomic effects of testosterone upon male behavior," *Front Neuroendocr* 29, 199–210.

Olausson, P., Kiraly, D. D., Gourley, S. L. and Taylor, J. R. (2013). "Persistent effects of prior exposure to corticosterone on reward-related learning and motivation in rodents," *Psychopharmacol* 225, 569–577.

Olivier, B., Chan, J. S., Snoeren, E. M., Olivier, J. D., Veening, J. G., Vinkers, C. H. *et al.* (2011). "Differences in sexual behavior in male and female rodents: role of serotonin," *Curr Top Behav Neurosci* 8, 15–36.

Patisaul, H. B., Luskin, J. B. and Wilson, M. E. (2004). "A soy supplement and tamoxifen inhibit sexual behavior in female rats," *Horm Behav* 45, 270–277.

Pedersen, C. A., Caldwell, J. D., Johnson, M. F., Fort, S. A. and Prange, A. J., Jr. (1985). "Oxytocin antiserum delays onset of ovarian steroid-induced maternal behavior," *Neuropeptides* 6, 175–182.

Pereira, A. M., Tiemensma, J. and Romijn, J. A. (2010). "Neuropsychiatric disorders in Cushing's syndrome," *Neuroendocr* 92(Suppl. 1), 65–70.

Pfaff, D. W. (1989). "Features of a hormone-driven defined neural circuit for a mammalian behavior," *Ann NY Acad Sci* 563, 131–147.

Pfaff, D. W., Kow, L. M., Loose, M. D. and Flanagan-Cato, L. M. (2008). "Reverse engineering the lordosis behavior circuit," *Horm Behav* 54, 347–354.

Plutchik, R. (1962). *The Emotions: Facts, Theories and a New Model* (New York: Random House).

Raskin, K., de Gendt, K., Duittoz, A., Liere, P., Verhoeven, G., Tronche, F. *et al.* (2009). "Conditional inactivation of androgen receptor gene in the nervous system: effects on male behavioral and neuroendocrine responses," *J Neurosci* 29, 4461–4470.

Reinisch, J. M. and Gandelman, R. (1978). "Human research in behavioral endocrinology: methodological and theoretical considerations" in G. Dorner and M. Kawakami (eds.), *Hormones and Brain Development* (Amsterdam: Elsevier), pp. 77–86.

Rezaii, T., Hirschberg, A. L., Carlström, K. and Emberg, M. (2012). "The influence of menstrual phases on pain modulation in healthy women," *J Pain* 13, 646–655.

Richmond, E. and Rogol, A. D. (2010). "Current indications for growth hormone therapy for children and adolescents," *Endocr Dev* 18, 92–108.

Rodriguez, M. A. and Garcia, R. (2013). "First, do no harm: the US sexually transmitted disease experiments in Guatemala," *Am J Public Health* 103, 2122–2126.

Roelfsema, F. and Veldhuis, J. D. (2013). "Thyrotropin secretion patterns in health and disease," *Endocr Revs* 34, 619–657.

Rohleder, N., Beulen, S. E., Chen, E., Wolf, J. M. and Kirschbaum, C. (2007). "Stress on the floor: the cortisol stress response to social-evaluative threat in competitive ballroom dancers," *Pers Soc Psychol Bull* 33, 69–84.

Rorie, R. W., Bilby, T. R. and Lester, T. D. (2002). "Application of electronic estrus detection technologies to reproductive management of cattle," *Theriogenol* 57, 137–148.

Rose, R. M., Bernstein, I. S. and Gordon, T. P. (1975). "Consequences of social conflict on plasma testosterone levels in rhesus monkeys," *Psychosom Med* 37, 50–61.

Rosenthal, R. (1966). *Experimenter Effects in Behavioral Research* (New York: Appleton-Century-Crofts).

Ruscio, M. G., Sweeny, T., Hazelton, J., Suppatkul, P. and Carter, C. S. (2007). "Social environment regulates corticotropin releasing factor, corticosterone and vasopressin in juvenile prairie voles," *Horm Behav* 51, 54–61.

Ruse, M. (1979). *Sociobiology: Sense or Nonsense?* (Boston, MA: D. Reidel).

Russell, M., Dark, K. A., Cummins, R. W., Ellman, G., Callaway, E. and Peeke, H. V. S. (1984). "Learned histamine release," *Science* 225, 733–734.

Rutstein, D. D. (1969). "The ethical design of human experiments," *Daedalus* 98, 523–541.

Ryan, B. C. and Vandenbergh, J. G. (2002). "Intrauterine position effects," *Neurosci Biobehav Revs* 26, 665–678.

Sack, R. L. (2009). "The pathophysiology of jet lag," *Travel Med Infect Dis* 7, 102–110.

Samuels, M. H. and Bridges, R. S. (1983). "Plasma prolactin concentration in parental male and female rats: effects of exposure to rat young," *Endocr* 113, 1647–1654.

Sarvari, M., Kallo, I., Hrabovszky, E., Solymosi, N., Toth, K., Liko, I. *et al.* (2010). "Estradiol replacement alters gene expression of genes related to neurotransmission and immune surveillance in the frontal cortex of middle-aged ovariectomized rats," *Endocr* 151, 3847–3862.

Schiml, P. A. and Rissman, E. F. (2000). "Effects of gonadotropin-releasing hormones, corticotropin-releasing hormone, and vasopressin on female sexual behavior," *Horm Behav* 37, 212–220.

Schlinger, B. A. (1997). "Sex steroids and their actions on the birdsong system," *J Neurobiol* 33, 619–631.

Schulz, K. M., Molenda-Figueira, H. A. and Sisk, C. L. (2009). "Back to the future: the organizational-activational hypothesis adapted to puberty and adolescence," *Horm Behav* 55, 597–604.

Schwartz, M. D. and Mong, J. A. (2013). "Estradiol modulates recovery of REM sleep in a time-of-day-dependent manner," *Am J Physiol* 305, R271–R280.

Schwartz, S. M. (1982). "Effects of constant bright illumination on reproductive processes in the female rat," *Neurosci Biobehav Rev* 6, 391–406.

Segraves, R. T. and Balon, R. (2014). "Antidepressant-induced sexual dysfunction in men," *Pharmacol Biochem Behav* 121, 132–137.

Sengelaub, D. R. and Forger, N. G. (2008). "The spinal nucleus of the bulbocavernosus: firsts in androgen-dependent neural sex differences," *Horm Behav* 53, 596–612.

Serón-Ferré, M., Torres, C., Parraguez, V. H., Vergara, M., Valladares, L., Forcelledo, M. L. *et al.* (2002). "Perinatal neuroendocrine regulation. Development of the circadian time-keeping system," *Mol Cell Endocr* 186, 169–173.

Serón-Ferré, M., Forcelledo, M. L., Torres-Farfan, C., Valenzuela, F. J., Rojas, A., Vergara, M. *et al.* (2013). "Impact of chronodisruption during primate pregnancy on the maternal and newborn temperature rhythms," *PLoS One* 8, e57710.

Shechter, A. and Boivin, D. B. (2010). "Sleep, hormones and circadian rhythms throughout the menstrual cycle in healthy women and women with premenstrual dysphoric disorder," *Int J Endocrinol* 2010, 1–17.

Shelley, D. N., Choleris, E., Kavaliers, M. and Pfaff, D. W. (2006). "Mechanisms underlying sexual and affiliative behaviors of mice: relation to generalized CNS arousal," *Soc Cog Affect Neurosci* 1, 260–270.

Silverman, W. A. (1985). *Human Experimentation: A Guided Step into the Unknown* (Oxford University Press).

Simerly, R. B. (2002). "Wired for reproduction: organization and development of sexually dimorphic circuits in the mammalian brain," *Annu Rev Neurosci* 25, 507–536.

Simerly, R. B., Chang, C., Muramatsu, M. and Swanson, L. W. (1990). "Distribution of androgen and estrogen receptor mRNA-containing cells in the rat brain: an in situ hybridization study," *J Comp Neurol* 294, 76–95.

Soulairac, A. and Soulairac, M. L. (1978). "Relationships between the nervous and endocrine regulation of sexual behavior in male rats," *Psychoneuroendocr* 3, 17–29.

Spiteri, T., Musatov, S., Ogawa, S., Ribeiro, A., Pfaff, D. W. and Agmo, A. (2010). "Estrogen-induced sexual incentive motivation, proceptivity and receptivity depend on a functional estrogen receptor α in the ventromedial nucleus of the hypothalamus but not in the amygdala," *Neuroendocr* 91, 142–154.

Staley, K. and Scharfman, H. (2005). "A woman's prerogative," *Nat Neurosci* 8, 697–699.

Steyn, F. J., Huang, L., Ngo, S. T., Leong, J. W., Tan, H. Y., Xie, T. Y. *et al.* (2011). "Development of a method for the determination of pulsatile growth hormone secretion in mice," *Endocr* 152, 3165–3171.

Strom, J. O., Theodorsson, E. and Theodorsson, A. (2008). "Order of magnitude differences between methods for maintaining physiological 17β-oestradiol concentrations in ovariectomized rats," *Scand J Clin Lab Invest* 68, 814–822.

Sullivan, E. L., Shearin, J., Koegler, F. H. and Cameron, J. L. (2012). "Selective estrogen receptor modulator promotes weight loss in ovariectomized female rhesus monkeys (*Macaca mulatta*) by decreasing food intake and increasing activity," *Am J Physiol Endocr Metab* 302, E759–E767.

Swanson, H. H. (1974). "Sex differences in behaviour of the Mongolian gerbil (*Meriones unguiculatus*) in encounters between pairs of same or opposite sex," *Anim Behav* 22, 638–644.

Tapp, A. L., Maybery, M. T. and Whitehouse, A. J. O. (2011). "Evaluating the twin testosterone transfer hypothesis: a review of the empirical evidence," *Horm Behav* 60, 713–722.

Tay, C. C. K., Glasier, A. F. and McNeilly, A. S. (1996). "Twenty-four hour patterns of prolactin secretion during lactation and the relationship to suckling and the resumption of fertility in breast-feeding women," *Hum Reprod* 11, 950–955.

Terron, M. P., Delgado-Adamez, J., Pariente, J. A., Barriga, C., Paredes, S. D. and Rodriguez, A. B. (2013). "Melatonin reduces body weight gain and increases nocturnal activity in male Wistar rats," *Physiol Behav* 118, 8–13.

Teske, J. A., Billington, C. J. and Kotz, C. M. (2008). "Neuropeptidergic mediators of spontaneous physical activity and non-exercise thermogenesis," *Neuroendocr* 87, 71–90.

Tetel, M. J. and Pfaff, D. W. (2010). "Contributions of estrogen receptor-alpha and estrogen receptor-β to the regulation of behavior," *Biochim Biophys Acta* 1800, 1084–1089.

Toran-Allerand, C. D. (1980). "Sex steroids and the development of the newborn mouse hypothalamus and preoptic area in vitro. II. Morphological correlates and hormonal specificity," *Brain Res* 189, 413–427.

Toran-Allerand, C. D. (1991). "Organotypic culture of the developing cerebral cortex and hypothalamus: relevance to sexual differentiation," *Psychoneuroendocr*, 16, 7–24.

Traub, R. J. and Yaping, J. (2013). "Sex differences and hormonal modulation of deep tissue pain," *Front Neuroendocr* 34, 350–366.

Tremere, L. A., Burrows, K., Jeong, J.-K. and Pinaud, R. (2011). "Organization of estrogen-associated circuits in the mouse primary auditory cortex," *J Exp Neurosci* 2011, 45–60.

Tsutsui, K., Ukena, K., Sakamoto, H., Okuyama, S.-I. and Haraguchi, S. (2011). "Biosynthesis, mode of action, and functional significance of neurosteroids in the Purkinje cell," *Front Endocr* 2, 1–9.

Umscheid, C. A., Margolis, D. J. and Grossman, C. E. (2011). "Key concepts of clinical trials: a narrative review," *Postgrad Med* 123, 194–204.

Vigil, P., Orellana, R. F., Cortés, M. E., Molina, C. T., Switzer, B. E. and Klaus, H. (2011). "Endocrine modulation of the adolescent brain: a review," *J Ped Adolesc Gynecol* 24, 330–337.

Viollet, C., Lepousez, G., Loudes, C., Videau, C., Simon, A. and Epelbaum, J. (2008). "Somatostatinergic systems in brain: networks and function," *Mol Cell Endocr* 286, 75–87.

Vom Saal, F. S. (1983). "The interaction of circulating oestrogens and androgens in regulating mammalian sexual differentiation" in J. Balthazart, E. Prove and R. Gilles (eds.), *Hormones and Behaviour in Higher Vetebrates* (Berlin: Springer-Verlag), pp. 159–177.

Vom Saal, F. S. and Bronson, F. H. (1980). "Sexual characteristics of adult female mice are correlated with their blood testosterone levels during prenatal development," *Science* 208, 597–599.

Wade, J. and Arnold, A. P. (2004). "Sexual differentiation of the zebra finch song system," *Ann NY Acad Sci* 1016, 540–559.

Walburger, V., Pietrowsky, R., Kirschbaum, C. and Wolf, O. T. (2004). "Effects of the menstrual cycle on auditory event-related potentials," *Horm Behav* 46, 600–606.

Wallen, K. and Hassett, J. M. (2009). "Sexual differentiation of behavior in monkeys: role of prenatal hormones," *J Neuroendocr* 21, 421–426.

Walker, E., Mittal, V. and Tessner, K. (2008). "Stress and the hypothalamic pituitary adrenal axis in the developmental course of schizophrenia," *Ann Rev Clin Psychol* 4, 189–216.

Watanabe, K., Umezu, K. and Kurahashi, T. (2002). "Human olfactory contrast changes during the menstrual cycle," *Japan J Physiol* 52, 353–359.

Weiss, E. R., Maness, P. and Lauder, J. M. (1998). "Why do neurotransmitters behave like growth factors?" *Perspect Dev Neurobiol* 5, 323–335.

Whalen, R. E. (1986). "Hormonal control of behavior – a cautionary note," *Ann NY Acad Sci* 474, 354–361.

Wilcox, J. N. (1986). "Analysis of steroid action on gene expression in the brain," *Ann NY Acad Sci* 474, 453–460.

Wildt, L., Hausler, A., Marshall, G., Hutchison, J. S., Plant, T. M., Belchetz, P. E. *et al.* (1981). "Frequency and amplitude of gonadotropin-releasing hormone stimulation and gonadotropin secretion in the Rhesus monkey," *Endocr* 109, 376–385.

World Medical Association (2013). "World Medical Association Declaration of Helsinki: ethical principles for medical research involving human subjects," *JAMA* 310, 2191–2194.

Yahr, P. and Thiessen, D. D. (1972). "Steroid regulation of territorial scent marking in the Mongolian gerbil (*Meriones unguiculatus*)," *Horm Behav* 3, 359–368.

Yanai, S., Semba, Y. and Endo, S. (2012). "Remarkable changes in behavior and physiology of laboratory mice after the massive 2011 Tohoku earthquake in Japan," *PLoS One* 7, e44475.

Zamaratskaia, G., Rydhmer, L., Andersson, H. K., Chen, G., Lowagie, S., Andersson, K. *et al.* (2008). "Long-term effect of vaccination against gonadotropin-releasing hormone, using Improvac, on hormonal profile and behavior of male pigs," *Anim Reprod Sci* 108, 37–48.

Zbinden, G. (1981). "Experimental methods in behavioral teratology," *ArchToxicol* 48, 69–88.

Ziegler, T. E., Schultz-Darken, N. J., Scott, J. J., Snowdon, C. T. and Ferris, C. F. (2005). "Neuroendocrine response to female ovulatory odors depends upon social conditions in male common marmosets, *Callithrix jacchus*," *Horm Behav* 47, 56–64.

15 An overview of behavioral neuroendocrinology: present, past and future

15.1 The aim of this book

The aim of this book is to introduce students to the language and concepts of neuroendocrinology, including how the neuroendocrine system influences behavior. It began with a consideration of the many chemical messengers in the body, classified as "true" hormones, neurohormones, neurotransmitters, pheromones, parahormones, prohormones, growth factors, cytokines, adipokines, vitamins and neuropeptides. As more became known about the neuroendocrine system, it became clear that these classifications are not unambiguous and a single chemical might fit into two or more classes of messenger and perform different functions. This means that while the classification of chemical messengers is useful to begin the study of neuroendocrinology, by the end it provides little help in understanding the different actions of peptides, steroids, neuropeptides and neurotransmitters on different target cells, even within the same tissue.

The hormones of the endocrine and pituitary glands are generally accepted as the major components of the neuroendocrine system. However, the traditional endocrine function of hormones being released into the bloodstream to act on peripheral target cells represents only a small part of the neuroendocrine activity of hormones such as testosterone, cholecystokinin or somatostatin. These hormones also have significant effects in the brain, via specific receptors, and can alter neural regulation of autonomic reflexes, behavior and emotional states. The hypothalamus provides the link between the brain and the traditional endocrine system and provides the locus for external factors, such as environmental influences, to regulate endocrine target organs. Thus, while the endocrine system consists of a number of closed-loop feedback systems that maintain homeostatic control over the synthesis, storage, release and deactivation of hormones, external stimuli can alter these systems. For example, environmental stimuli, social interactions and cognitive factors can greatly alter the functioning of the endocrine system by altering the neurotransmitter/neuropeptide pathways that regulate the release of hypothalamic hormones. Drugs, food, changes in light cycles, emotional arousal and other changes in the environment which alter hypothalamic neurotransmission can disrupt the hormonal control of feeding and drinking, arousal, sexual, aggressive and parental behavior, and responses to environmental stressors. Bacteria, viruses and other toxins also activate the neuroendocrine system by stimulating the release of cytokines from the cells of the immune system.

While the "classical" neurotransmitters, such as acetylcholine and norepinephrine, were originally considered to be the most important neurochemicals, it is now clear that the neuropeptides, steroid hormones and cytokines, which also modulate neural activity, may prove to be equally, if not more, important in determining brain function. As more was learned about the wide-ranging effects of neuropeptides on physiological and behavioral responses, it became obvious that these hormones have significantly different functions in the brain than they do elsewhere in the body, yet they are often oriented towards the same goals. For example, the gastrointestinal peptides help to digest food in the gut, but also play critical roles in the brain to regulate feeding and drinking behavior, body temperature and blood pressure. Another peptide, oxytocin, is secreted from the posterior pituitary gland to control milk let-down and uterine contractions, but exerts critical effects in the brain as a regulator of emotional, maternal and social behavior. Steroid hormones also have multiple functions. Sex hormones such as estradiol prepare the genitals for puberty and mating and develop the breasts for lactation, and act in the brain to regulate sexual arousal and sexual and parental behavior. The adrenal cortical and medullary hormones that increase heart rate and respiratory rate and facilitate muscular energy also prepare the brain to deal with stressful situations and to respond in the most adaptive way.

All of these effects result from the actions of the chemical messengers at receptors located either on, or inside, their target cells. The biochemical changes in the target cells which result from the activation of receptors, and the resulting changes in ion channel permeability, second messenger cascades, control of secretion and regulation of gene expression, involve chains of enzyme reactions, gene transcription, protein synthesis, cell growth and differentiation, resulting in neuromuscular changes in behavior and modified neurotransmission leading to emotional arousal and memory. Hormones, neurotransmitters, neuropeptides and cytokines act through a series of complex interactions, whereby a neurotransmitter controls hormone release. On the other hand, neuropeptides and other hormones control the release of neurotransmitters, as well as the response of the postsynaptic cell to the neurotransmitter. Cytokines are regulated by the neuroendocrine system and can act at receptors in these systems to influence both neural and hormonal responses.

The intention of this book was to introduce students to the complexity of the neuroendocrine system by presenting the necessary terminology and concepts in a gradual way. The peripheral endocrine glands, pituitary gland, hypothalamic hormones, neurotransmitters, neuropeptides and cytokines are the building blocks of the neuroendocrine system. Knowing about the synthesis, storage, transport and release of each of these chemical messengers provides a mechanism for understanding the nature of each of the individual elements in the neuroendocrine system, while knowing about receptor mechanisms and second messenger cascades leads to an understanding of how these elements are interconnected. Knowing how the steroid hormones, neuropeptides and cytokines modulate the electrophysiological and genomic activity of their neural target cells leads to an understanding of how these neuroregulators can modulate the visceral, behavioral and cognitive functions of the brain.

This book has not covered in any detail the functions of the hormones in medical, behavioral or psychiatric practice, although many examples have been provided. The function of this book is to provide the framework for more advanced study. The essay questions, for example, require extra reading, if they are to be answered with the most up-to-date information available. If, on finishing this book, you are able to read the references in scientific and medical journals without a dictionary, and are able to "see" in your mind's eye: the relationships among, for example, the hypothalamic-pituitary-gonadal hormones; the behavioral effects of oxytocin; the action of a dopamine agonist drug on the release of prolactin; and the ability of interleukin 1 to alter the hypothalamic-pituitary-adrenal feedback system, then this book has served its function and you are ready for more current and advanced topics in neuroendocrinology. Some of these are briefly outlined in section 15.3, with particular reference to the future of behavioral neuroendocrinology.

15.2 The history of endocrinology and behavioral neuroendocrinology

Endocrinology has a long history (Medvei 1982; Wilson 2005) and many discoveries of relevance to the endocrine system are important enough to have been awarded a Nobel Prize (Table 15.1).

The biographies of the many scientists who made these discoveries are available in a number of books (McCann 1988; Meites *et al.* 1975; 1978). Brief histories of the study of endocrinology are given in Turner and Bagnara (1976), Hadley (1992) and Wilson (2005). Gilman *et al.* (1985) and Tausk (1975) include several historical summaries of the discoveries of neuroendocrine phenomena and the drugs that influence them. The important history of the hypothalamic-releasing hormones is particularly well described, from the seminal work of G. W. Harris (Raisman 1997) to the race between the Guillemin and Schally laboratories that culminated in the isolation of the first two releasing hormones (TRH and LHRH) and ultimately the award of a Nobel Prize (Wade 1981). In the present era of molecular biology and the human genome, it is worth remembering that the crucial discovery of releasing hormones was made possible by a semi-industrial scale extraction of brain tissue. In the case of TRH, Guillemin's laboratory dissected and processed 5 million sheep brains that produced 500 tons of brain tissue (7 tons of hypothalamus) (Guillemin and Burgess 1972). Finally, after four years of work, 1 milligram of pure TRH was obtained.

The history of the study of hormonal influences on behavior, which is less well known, is described by Beach (1981) and Gandelman (1984). Table 15.2 gives some of the important milestones in the study of hormones and behavior from 1849 to the present. The study of hormones and behavior was unsystematic until 1948, when Frank Beach established the field of hormones and behavior with the publication of the book *Hormones and Behavior*. The history of research on hormones and behavior began in 1849, but was not developed into a scientific discipline until 1948. Since then, the field has grown rapidly (see, e.g., Argiolas and Melis 2013).

15.3 The future of behavioral neuroendocrinology

This book has outlined some of the important fundamentals that underpin our understanding of endocrine and neuroendocrine phenomena. Most importantly, (neuro)endocrinology is no

Table 15.1 Nobel prizes in endocrine-relevant fields

Recipients	Year	Citation
E. T. Kocher	1909	Physiology and surgery of the thyroid
F. G. Banting and J. J. R. Macleod	1923	Discovery of insulin
H. H. Dale and O. Loewi	1936	Chemical transmission of nerve impulses
A. F. J. Butenandt	1939	Discovery and synthesis of sex steroids
B. A. Houssay	1947	Pituitary hormone control of sugar metabolism
E. C. Kendall, T. Reichstein and P. S. Hench	1950	Hormones of the adrenal cortex
V. du Vigneaud	1955	Isolation and structure of oxytocin
F. Sanger	1958	Structure of insulin
C. B. Huggins	1966	Hormonal treatment of prostate cancer
B. Katz, U. von Euler and J. Axelrod	1970	Humoral transmitters in nerve terminals and the mechanism for their storage, release and inactivation
E. W. Sutherland	1971	Mechanisms of action of hormones
R. Guillemin and A. V. Schally	1977	Peptide hormone production in brain
R. Yalow	1977	Radioimmunoassay of peptide hormones
S. K. Bergstrom, B. I. Samuelsson and J. R. Vane	1982	Discoveries concerning prostaglandins
S. Cohen and R. Levi-Montalcini	1986	Discovery of growth factors
K. B. Mullis	1993	The polymerase chain reaction (PCR) method
A. R. Gilman and M. Rodbell	1994	G proteins in signal transduction
A. Carlsson, P. Greengard and E. R. Kandel	2000	Signal transduction in the nervous system
R. Axel and L. B. Buck	2004	Odorant receptors and the organization of the olfactory system
R. G. Edwards	2010	Development of *in vitro* fertilization
R. J. Lefkowitz and B. K. Kobilka	2012	G-protein-coupled receptors
J. E. Rothman, R. W. Schekman and T. C. Südhof	2013	Discoveries of machinery regulating vesicle traffic, a major transport system in our cells

longer focused solely on the role of hormones, but is best understood in terms of neuroscience, cell and molecular biology, immunology and genetics. Hormonal cell signaling systems are highly sophisticated and are supported by paracrine, autocrine and intracrine signals. It is inevitable that the continued acquisition of new knowledge, aided by development of novel technologies, will expand this understanding, especially with regard to endocrinology in general (Dhillo *et al.* 2006) and to behavioral neuroendocrinology in particular. Several new directions that will certainly influence studies in behavioral neuroendocrinology are briefly described below. In addition, although this book has focused on the physiology of hormone systems, the impact of exogenous, environmental endocrine disruptors on behavioral neuroendocrinology is an important future focus of particular relevance to human behavior.

Table 15.2 Historic milestones with implications for the study of behavioral neuroendocrinology (from Beach 1981; Gandelman 1984; and other sources)

1849	Arnold A. Berthold castrated roosters and replaced their testes; observed changes in crowing, copulation and aggression.
1889–1894	Charles Edouard Brown-Sequard developed "organo-therapy" by producing extracts of testes, thyroid and adrenal glands for treatment of human disorders, and became infamous for his rejuvenation experiments, injecting testes extracts into elderly men.
1894–1910	Eugene Steinach experimented on the effects of testes removal and replacement on the sexual behavior of amphibia, birds and mammals, and examined the effects of the gonads on sexual differentiation and the timing of puberty.
1901	J. Takamine isolated and purified the hormone adrenaline from animal glands.
1902–1905	E. H. Starling and W. M. Bayliss discover the hormone secretin and used the term "hormone," and N. Pende introduced the term "endocrinology" a few years later.
1910	H. Cushing *et al.* discover the anterior pituitary-gonadal link.
1914	H. H. Dale discovers acetylcholine's function as a neurotransmitter in the parasympathetic nervous system. He received the Nobel Prize in 1936.
1914	E. C. Kendall isolated thyroxine from thyroid gland extracts.
1917	Frank R. Lillie describes "free martins" as the female twin of a normal male calf who is masculinized by androgens from the male's testes. Between 1917 and 1922, Lillie outlined a number of experiments to determine the role of perinatal gonadal hormones in sexual differentiation.
1917	C. R. Stockard and G. N. Papanicolau describe the estrous cycle of the female guinea pig and correlate the stages of development of the ovarian follicle with changes in cell types of the vaginal mucosa. In 1922, J. A. Long and H. M. Evans described a similar estrous cycle in rats and used vaginal smears to correlate vaginal estrus with mating behavior. In 1923, G. H. Wang showed a correlation between spontaneous activity and the estrous cycle of the rat.
1921	F. G. Banting, C. Best, J. R. Macleod and J. B. Collip discover insulin. Banting and Macleod were awarded the Nobel Prize in 1923.
1923	Edgar Allen and Edward A. Doisey purify estrogen from the ovaries of mice and rats and show that estrogen injections into spayed females induce estrous behavior.
1925	F. H. Marshall and J. Hammond show that ovulation in rabbits is not spontaneous, but is induced by vaginal stimulation.
1930	C. Pfeiffer shows that neonatally castrated male rats have a feminine hypothalamic-pituitary-gonadal feedback system in adulthood.
1930	G. T. Popa and U. Fielding discover the hypophyseal portal veins, the vascular connection between the hypothalamus and pituitary gland.
1933	B. P. Weisner and N. M. Sheard demonstrate that maternal behavior in rats is dependent on hormones from the pituitary gland and stimuli from the young.
1933–1941	Curt Richter shows the effects of hormones on locomotor activity, dietary selection and motivated behavior.
1934	Walter Hohlweg discovers the positive feedback of estradiol on LH release and, in 1938, he develops the orally acting estrogen, ethinyl estradiol, an essential component of the modern birth control pill.
1935	Oscar Riddle shows that prolactin is important in the control of maternal behavior.

Table 15.2 (Cont.)

1935	K. G. David isolates testosterone from the testes.
1936	Hans Selye begins his studies on the neuroendocrine response to stress and develops the concept of the "general adaptation syndrome" of response to stress.
1937–1952	E. C. Kendall, T. Reichstein and P. S. Hench isolate adrenal corticosteroids. They win the Nobel Prize in 1950.
1937	Geoffrey W. Harris shows that electrical stimulation of the hypothalamus alters pituitary hormone secretion. In 1948, Harris conclusively demonstrates the hypothalamic control of the pituitary. He publishes his book on *Neural Control of the Pituitary Gland* in 1955.
1939	Philip Bard discovers the hypothalamic control of ovulation, emotional and sexual behavior in cats.
1940	J. G. Wilson, W. C. Young and J. B. Hamilton show that neonatal androgen injections masculinize female rats.
1940–1945	C. H. Li isolates LH, ACTH and GH.
1948	F. A. Beach publishes *Hormones and Behavior*, the first book of behavioral endocrinology.
1949	Wolfgang Bargmann discovers the neural connections between the hypothalamus and posterior pituitary.
1951	Ernst and Berta Scharrer describe hypothalamic neurosecretory cells in the brains of vertebrates and invertebrates, leading to the concept of hypothalamic releasing factors (see below, 1955).
1952	The Hodgkin–Huxley model of the axonal conduction of nerve impulses developed. They win the Nobel Prize in 1963.
1955	CRF-like activity, the first evidence for hypothalamic-releasing hormones, discovered simultaneously by M. Saffron and A. Schally and by R. Guillemin and B. Rosenberg. Subsequent work by the Guillemin and Schally groups led to the isolation and characterization of GnRH and TRH (see below).
1955	Gregory Pincus gives his report on the first successful clinical trials of the birth control pill.
1956	A. E. Fisher shows that intracranial injections of hormones could stimulate sexual and maternal behavior in rats.
1958	Geoffrey W. Harris and colleagues show that estradiol implanted directly into the anterior hypothalamus elicited sexual behavior in ovariectomized cats.
1959	Daniel Lehrman publishes his finding on the effects of external stimuli on hormones in birds.
1959	C. Phoenix and colleagues describe the organizational effects of prenatal androgens on the sexual behavior of the guinea pig. This paper established the importance of perinatal hormones on adult behavior.
1960	R. Yalow and S. Berson develop the radioimmunoassay method for measuring peptide hormone levels in the blood. Yalow wins the 1977 Nobel Prize for Physiology or Medicine (shared with Guillemin and Schally; see above).
1963	A. Soulairac demonstrates the hormone-neurotransmitter interaction in the control of rat sexual behavior.
1964	David de Wied discovers the effects of pituitary hormones, such as vasopressin, on learning in rats.
1971	A. Schally *et al.* isolate LH-RH and R. Guillemin *et al.* isolate TRH (see below).

Table 15.2 **(Cont.)**	
1972	J. Terkel and J. Rosenblatt show that blood transfusions from a lactating female rat will induce maternal behavior in a virgin female rat.
1971–1975	J. Hughes, H. Kosterlitz *et al.* discover brain enkephalins, the first endogenous opioid peptides.
1977	Work by the Guillemin and Schally groups on the isolation and characterization of GnRH and TRH is awarded a Nobel Prize in Physiology or Medicine in 1977.
1972–1978	A period of discovery of numerous brain neuropeptides such as somatostatin, neuropeptide Y, β-endorphin, etc.
1978	A. I. Leshner writes first textbook on hormones and behavior: *An Introduction to Behavioural Endocrinology*.
1981	Robert Ader edits first book on *Psychoneuroimmunology*.
1986	Rita Levi-Montalcini and Stanley Cohen win Nobel Prize in recognition of their discovery of nerve growth factor.
1994	J. M. Friedman and colleagues discover leptin, the first adipokine.

15.3.1 New peptide hormones

The Human Genome Project discovered that humans have approximately 20,000 to 25,000 genes; about the same number as in the mouse. However, individual genes, particularly in the endocrine system, can code for more than one signaling peptide or protein. For example, as described in Chapter 11, the pro-opiomelanocortin (*pomc*) gene codes for a large peptide molecule that can be processed to a variety of other signaling (neuro)peptides (see Figure 11.1). Samson and co-workers have used information from the Human Genome Project to predict previously unidentified, secreted, highly conserved peptide hormones such as phoenixin and neuronostatin (Yosten *et al.* 2013; Samson *et al.* 2008). *Phoenixin* is a hypothalamic neuropeptide that regulates gonadotropin secretion from the anterior pituitary by interfering with GnRH stimulation. However, its localization to other hypothalamic nuclei associated with the autonomic nervous system suggests additional behavioral roles. *Neuronostatin* is a peptide hormone derived from the pro-somatostatin polypeptide and is found in gut tissues, as well as in the brain. Injection of neuronostatin into rat brain caused a reduction in food intake and water drinking (Samson *et al.* 2008) and interfered with memory retention (Carlini *et al.* 2011). In contrast to somatostatin, it had no effect on GH secretion. These studies indicate that this genomic approach will identify many more peptides that possess behavioral neuroendocrine properties.

15.3.2 Gene knockouts

Investigation of the neural and hormonal signaling in behavioral neuroendocrinology is now facilitated relatively simply by generating mouse mutants that lack individual receptor types (Hewitt *et al.* 2005). For example, the role of individual estradiol receptors (ERα and ERβ) in sexual behavior was elucidated by generating ERα and ERβ knockout mice (ERαKO and ERβKO); that is, mice lacking functional ERα or ERβ (Tetel and Pfaff 2010). In this way, the ERα gene product was shown to be indispensable for the normal performance of female sex behavior, whereas mice lacking ERβ had normal sexual receptivity. This powerful technology can theoretically be applied to any receptor type. It is also possible to produce gene knockouts in specific areas of the brain, a further important refinement. For example, deletion of glucocorticoid receptors (GR) specifically in the mouse forebrain produce different behavioral responses than GR knockouts confined to the amygdala (Arnett *et al.* 2011). Such knockout studies are not confined to the mouse and other, non-traditional, species will prove to be valuable (Smale *et al.* 2005).

15.3.3 Epigenetics

The importance of gene expression in regulating the neuroendocrine system was emphasized in several chapters. Figures 9.1 and 10.1 outline how steroid hormones, or neurotransmitters, bind to their receptors to induce changes in gene expression in specific tissues or cell types. In this way, for example, estradiol induces increased expression of progesterone receptor and oxytocin genes in the brain (Figure 14.17). In all the examples provided in this book, gene expression does not involve changing the nucleotide sequence of the DNA; that is, specific genes are switched on, or off, by signals from hormones, neuropeptides, etc. It is now known that some types of stimuli (e.g. life experience) can alter DNA structure, not by changing the genetic code, but by the placement of chemical "marks" such as methyl or acetyl groups at specific sites on the DNA (Buchen 2010; Williams 2013; West and Orlando 2014). Such marks are called *epigenetic modifications*. In the case of *methylation*, its function is to silence nearby genes, whereas *acetylation* will activate gene transcription. Environmental events can directly influence the epigenetic state of DNA and thus the interaction between environmental signals and the epigenome determines differences in behavior and physiology (Zhang and Meaney 2010). Sex hormones also induce epigenetic changes and such studies will help in understanding the known sex differences in brain and behavior (McCarthy *et al.* 2009).

15.3.4 Endocrine disrupting chemicals

A momentous landmark in endocrine studies was the discovery that certain synthetic chemical pollutants, named *endocrine disrupting chemicals* (EDCs), can mimic the effects of natural hormones such as estradiol and testosterone. EDCs are produced industrially in large amounts, are pervasive in the environment and are essentially inescapable. We are therefore exposed to a cocktail of EDCs every day, from conception to death. Thus, many of the pathways discussed in this book – including estrogenic, androgenic, thyroid, neurotransmitter receptors and signaling systems – particularly in the brain, are potentially influenced by these chemicals. They leach from plastic water bottles (*bisphenol A*), are present in tap water, house dust, toys, fabrics, furniture, cookware and food containers. They influence male and female reproduction, breast development, neuroendocrinology, thyroid physiology, body weight, immune response and cardiovascular endocrinology. They are of great concern not only for human health, but also for their impact on animals, birds and fish (Diamanti-Kandarakis *et al.* 2009; Gore and Patisaul 2010). For example, embryonic exposure to bisphenol A induced enduring effects on adult social and anxiety-like behavior, and learning/memory behavior in mice (Kundakovic *et al.* 2013). The possibility that human endocrine systems might be dysregulated by such chemicals, especially during development, is a serious issue that is attracting increasing attention (Birnbaum 2013; Marques-Pinto and Carvalho 2013; Andersson *et al.* 2014).

ESSAY QUESTIONS

15.1 Discuss the contributions of one of the following scientists (all deceased) to the study of behavioral neuroendocrinology:
- (a) Hans Selye;
- (b) Frank Beach;
- (c) Curt Richter;
- (d) Geoffrey W. Harris; or
- (e) Daniel Lehrman.

15.2 Discuss the discovery of one of the following neuroendocrine phenomena:
- (a) endogenous opioid peptides;
- (b) thymus hormones;
- (c) insulin;
- (d) neurohypophyseal hormone system;
- (e) secretin;
- (f) "Dale's Principle"; or
- (g) hypothalamic-releasing hormones.

15.3 Discuss the importance of the following scientists in the history of neuroendocrinology:
- (a) Arnold A. Berthold;
- (b) Charles Edouard Brown-Sequard;
- (c) Eugene Steinach;
- (d) F. H. Marshall;
- (e) W. C. Young;
- (f) C. H. Li;
- (g) Ernst and Berta Scharrer;
- (h) A. Schally and R. Guillemin.

REFERENCES

Andersson, A.-M., Frederiksen, H., Grigor, K. M., Toppari, J. and Skakkebaek, N. E. (2014). "Special issue on the impact of endocrine disrupters on reproductive health," *Reprod* 147, E1.

Argiolas, A. and Melis, M. R. (2013). "Neuropeptides and central control of sexual behavior from the past to the present: a review," *Prog Neurobiol* 108, 80–107.

Arnett, M. G., Kolber, B. J., Boyle, M. P. and Muglia, L. J. (2011). "Behavioral insights from mouse models of forebrain- and amygdala-specific glucocorticoid receptor genetic disruption," *Mol Cell Endocr* 336, 2–5.

Beach, F. A. (1981). "Historical origins of modern research on hormones and behavior," *Horm Behav* 15, 325–376.

Birnbaum, L. S. (2013). "When environmental chemicals act like uncontrolled medicine," *Trends Endocr Metab* 24, 321–323.

Buchen, L. (2010). "In their nurture," *Nature* 467, 146–148.

Carlini, V. P., Ghersi, M., Gabach, L., Schioth, H. B., Perez, M. F., Ramirez, O. A. *et al.* (2011). "Hippocampal effects of neuronostatin on memory, anxiety-like behavior and food intake in rats," *Neurosci* 197, 145–152.

Diamanti-Kandarakis, E., Bourgignon, J.-P., Giudice, L. C., Hauser, R., Prins, G. S., Soto, A. M. *et al.* (2009). "Endocrine-disrupting chemicals: an endocrine society scientific statement," *Endocr Rev* 30, 293–342.

Dhillo, W. S., Murphy, K. G. and Bloom, S. (2006). "Endocrinology: the next 60 years," *J Endocrinol* 190, 7–10.

Gandelman, R. (1984). "Relative contributions of aggression and reproduction to behavioral endocrinology," *Aggress Behav* 10, 123–133.

Gilman, A. G., Goodman, L. S., Rall, T. W. and Murad, S. (eds). (1985). *Goodman & Gilman's The Pharmacological Basis of Therapeutics*, 7th edn. (New York: Macmillan).

Gore, A. C. and Patisaul, H. B. (2010). "Neuroendocrine disruption: historical roots, current progress, questions for the future," *Front Neuroendocr* 31, 395–399.

Guillemin, R. and Burgess, R. (1972). "The hormones of the hypothalamus," *Sci Amer* 227, 24–33.

Hadley, M. E. (1992). *Endocrinology*, 3rd edn. (Engelwood Cliffs, NJ: Prentice-Hall).

Hewitt, S. C., Harrell, J. C. and Korach, K. S. (2005). "Lessons in estrogen biology from knockout and transgenic animals," *Annu Rev Physiol* 67, 285–308.

Kundakovic, M., Gudsnuk, K., Franks, B., Madrid, J., Miller, R. L., Perera, F. P. *et al.* (2013). "Sex-specific epigenetic disruption and behavioral changes following low-dose in utero bisphenol A exposure," *Proc Natl Acad Sci USA* 110, 9956–9961.

Marques-Pinto, A. and Carvalho, D. (2013). "Human infertility: are endocrine disruptors to blame?" *Endocr Connect* 2, R15–R29.

McCann, S. M. (ed.) (1988). *Endocrinology. People and Ideas* (Bethesda, MD: American Physiological Society).

McCarthy, M. M., Auger, A. P., Bale, T. L., deVries, G. J., Dunn, G. A., Forger, N. G. *et al.* (2009). "The epigenetics of sex differences in the brain," *J Neurosci* 29, 12815–12823.

Medvei, V. C. (1982). *A History of Endocrinology* (Lancaster, PA: MTP Press).

Meites, J., Donovan, B. T. and McCann, S. M. (1975) *Pioneers in Neuroendocrinology* (New York: Plenum Press).

Meites, J., Donovan, B. T. and McCann, S. M. (1978) *Pioneers in Neuroendocrinology II* (New York: Plenum Press).

Raisman, G. (1997). "An urge to explain the incomprehensible: Geoffrey Harris and the discovery of the neural control of the pituitary gland," *Ann Rev Neurosci* 20, 533–566.

Samson, W. K., Zhang, J. V., Avsian-Kretchmer, O., Cui, K., Yosten, G. L. C., Klein, C. *et al.* (2008). "Neuronostatin encoded by the somatostatin gene regulates neuronal, cardiovascular and metabolic functions," *J Biol Chem* 283, 31949–31959.

Smale, L., Heideman, P. D. and French, J. A. (2005). "Behavioral neuroendocrinology in nontraditional species of mammals: things the 'knockout' mouse cannot tell us," *Horm Behav* 48, 474–483.

Tausk, M. (1975). *Pharmacology of Hormones* (Chicago, IL: Yearbook Medical Publishers).

Tetel, M. J. and Pfaff, D. W. (2010). "Contributions of estrogen receptor-alpha and estrogen receptor-β to the regulation of behavior," *Biochim Biophys Acta* 1800, 1084–1089.

Turner, C. D. and Bagnara, J. T. (1976). *General Endocrinology*, 6th edn. (Philadelphia, PA: W. B. Saunders).

Wade, N. (1981). *The Nobel Duel: Two Scientist's 21-Year Race to Win the World's Most Coveted Research Prize* (New York: Anchor Press).

West, A. E. and Orlando, V. (2014). "Epigenetics in brain function," *Neurosci* 264, 1–3.

Williams, S. C. P. (2013). "Epigenetics," *Proc Natl Acad Sci USA* 110, 3209.

Wilson, J. D. (2005). "The evolution of endocrinology," *Clin Endocr* 62, 389–396.

Yosten, G. L. C., Lyu, R.-M., Hsueh, A. J. W., Avsian-Kretchmer, O., Chang, J.-K., Tullock, C. W. *et al.* (2013). "A novel reproductive peptide, Phoenixin," *J Neuroendocr* 25, 206–215.

Zhang, T. Y. and Meaney, M. J. (2010). "Epigenetics and the environmental regulation of the genome and its function," *Annu Rev Psychol* 61, 439–466.

INDEX

acetylcholine (ACh) 82
  biosynthesis 89
  brain pathways 97
action potentials 92
activin
  testes 32
adipokine 13, 34, 307
  leptin 35
  resistin 36
adrenal glands 29
  cortex 29
    and steroid production 29
  medulla 29
    epinephrine 29
    norepinephrine 29
    and autonomic nervous
      system 175
adrenocorticotropic hormone
    (ACTH) 50
  and stress 124, 418
  and circadian rhythm 124
  in Cushing's Syndrome/Disease
    186
  neurotransmitter regulation of
    124
  release by CRH/ADH 124
  in brain 328
  effects on behavior
    inhibits food intake 328
    stimulates grooming 328
    reduces social interaction 328
    increased sex behavior 328
    effect on memory 328
agouti-related protein (AgRP)
    303
aldosterone 29
Alzheimer's Disease
  and estradiol 208, 217
amino acids as neurotransmitters
    80
  inhibitory (GABA) 81
  excitatory (GLU) 81

amplification (second messengers)
    249
amylin
  receptors in brain 332
  reduces motor activity 332
  inhibits sexual activity 332
anandamide 86
androgen receptor localization in
    brain 209
antidiuretic hormone (ADH) see
    vasopressin
arachidonic acid second messen-
    ger system 248
aromatase 210
astrocyte
  and hypothalamic function 143
  in human brain 143
  and tripartite synapse 144
  and D-serine 87
  and oxytocin release 145
  and vasopressin release 145
  retraction of processes 146
  and GnRH release 147
  and PGE2 147
  and estradiol receptors 149
ATP (adenosine triphosphate) as
    neurotransmitter 88
atrial natriuretic factor/peptide
    (ANP) 24
autocrine communication 7
autonomic nervous system 110
  central regulation of 113
  and hypothalamus 113
  control of hormone levels 175
  regulation of immune system
    375
  sympathetic nervous system 112

B lymphocytes (B cells) see
    immune system
B-type natriuretic peptide (BNP)
    24

behavior
  integration of hormonal, envir-
    onmental interaction
    432
  hormonal effects on neural
    activity 425
  hormones and sensory
    systems 425
  hormones and motor systems
    425
  aggression and vasopressin
    320
  emotional and social behavior
    and oxytocin 319
  sex behavior 217
    and opioids 298
    and oxytocin 319
    and GnRH 325
    and testosterone 427
    in female rats 427
  mating behavior in ring doves
    420
  singing in song birds 426
  parental behavior 217, 319
    and prolactin 327
  food craving 312
behavioral neuroendocrinology
    400
  and genomic mechanisms 424
  history 460
behavioral studies
  bioassays 400
    copulatory behavior 402
    vaginal cells 405
  confounding variables 434;
    Table 14.2
    species differences 434
    individual differences 435
    position of fetus in uterus
      435; Table 14.3
    human fetus 436
    environmental factors 438

behavioral studies (cont.)
  measures used in 400;
      Table 14.1
  correlation of behavior and
      hormonal changes 403
  effects of hormone removal/
      replacement 408
  methods of hormonal
      removal 409
  methods of hormone
      replacement 411
    drinking water 414
    implanted pellets/capsules
      414
    timing of 415
  abnormal endocrine hormones
      affect behavior 404
  genomic mechanisms influence
      behavior 424
  problems of hormone sampling
      405
  human studies 441
    animal models 443
    social and political
      implications 443
  human subjects, experimental
      studies, confounding
      variables
    choice of subjects 441
    experimental design 441
    bias 442
  ring doves 420
  song birds 426
bioassay 174
  in vitro: cell culture 174
  behavioral 400
bisphenol A 217, 465
  endocrine disrupting chemicals
      465
blood-brain barrier (BBB) 275
  and circumventricular organs
      277
  and cytokines 366
body weight and neuropeptides
      300
bone, as endocrine organ 38
  osteocalcin 39
brain development

critical periods 431
  and hormones 431
  and thyroid hormones 432

calcitonin 23
calcium
  blood level regulation 23
  as second messenger system
      246
  and see second messenger
      system
cAMP response element binding
      protein (CREB) 242
cAMP second messenger system
      244
  and see second messenger
      system
CART (cocaine- and
      amphetamine-regulated
      transcript)
  and TRH 63
catechol o-methyl transferase
      (COMT) 97
catecholamines 84
  and biosynthesis 88
  and synaptic vesicles 88
  (see also individual
      catecholamines)
cGMP second messenger system
      244
  and see second messenger
      system
chemical messenger
  adipokines 13
  cascade 120
  cytokines 13
  growth factors 12
  hormones 8
  neurohormones 10
  neurotransmitters 11
  parahormones 12
  pheromones 11
  phytohormones 7
  prohormones 12
  vitamins 14
  evolutionary pathway 262
childbirth (parturition) and
      oxytocin release 296

cholecystokinin (CCK) 26
  biosynthesis in brain 329
  and multiple behaviors 329
circadian rhythm
  and ACTH 124
  and pituitary hormones 386
  and immune response
      disturbances 386
  pineal gland and melatonin 175
circumventricular organs (CVO)
      277
co-localization
  neuropeptides and neurotrans-
      mitters 260, 266;
      Table 11.3
  two neuropeptides 260
  co-release/co-transmission 261
colony stimulating factors 363
conditioned release of hormones
      423
corticosterone 29
  rapid effect in brain 200
corticotropin releasing hormone
      (CRH) 64
  neurotransmitter regulation of
      64
  and ACTH release 184
  localization 322
  and interleukins 369
  behavioral effects 322
    reduces food intake 322
cortisol 29
  neurotransmitter control of 124
  control by CRH 184
  and Cushing's disease 186
Cushing's Disease 186; Table 8.3
cyclic adenosine monophosphate
      see cAMP
cyclic guanosine monophosphate
      (cGMP) see cGMP
cytokines 356; Table 13.1
  as immunomodulators 365
  interferon γ (IFNγ) 358
    stimulation of hormone
      secretion 370
  interleukins 358
    IL-1 induces fever and sleep
      367

IL-1 and hormone secretion 370

IL-2 and hormone secretion 370

neuroendocrine effects 365; Table 13.1

in hypothalamus feeding behavior 386

tumor necrosis factor α (TNFα) 360

stimulates CRH release 360

and hormone release 372

oncostatin M (OSM) 360

activation of T cells 361

activation of B cells 362

activation of macrophages 363

and maturation of blood cells 363

and cytotoxicity 363

and blood-brain barrier 366

and steroid hormone release 369

and peptide hormone release 369; Table 13.2

biosynthesis in brain 366

localization by *in situ* hybridization 366

receptors in brain 366

structure of 367

Dale's principle (Dale's law) 267

diacylglycerol (DAG) 245

L-Dopa and dopamine 102

dopamine

biosynthesis 90

and L-Dopa 102

brain pathways 97

controlling PRL release 69

and αMSH release 133

dynorphin (DYN) 259

ELISA (enzyme-linked immuno-sorbent assay) 170

and cytokines 358

endocannabinoids 85

and anandamide 86

brain localization 101

and oxytocin 134

endocrine communication 5

endocrine-disrupting chemicals 465

bisphenol A 217, 465

endogenous opioids 259

endorphin (β-endorphin) (β-END) 51, 259

"runner's high" 52

enkephalin (ENK) 259

enteric nervous system 112

epigenetics 465

estradiol (E)

and tanycytes 149

and astrocytes 149

and immune system 378

blood levels after injection 415

positive feedback 67, 181, 216

negative feedback 181

genomic effect 196, 198

and behavior 430

non-genomic effect 196, 200

stimulation of action potentials 200

stimulation of GnRH secretion 200, 216

and neuroprotection 208

induction of sexual behavior 215, 217

lordosis reflex 219

induction of parental behavior 219

effect on fetus 436

effect on neuron structure 215, 430

effect on sensory systems 425

effect on motor systems 425

and neuroprotection 217

and Alzheimer's Disease 208, 217

synthesis in brain 222

estradiol receptor

subtypes ERα and ERβ 196

membrane site 196

localization of ERα mRNA 198, 206

ERα knockouts (ERα KO) 206

ERα, ERβ and social recognition 428

ERα and ERβ and sex behavior 428

and gene expression 428

estrogen response element (ERE) 203

estrous cycle 406

follicle stimulating hormone (FSH) 51

folliculostellate cells (FS) 53

gamma-aminobutyric acid (GABA) 81

biosynthesis 88

brain pathways 97

gastrin 26

gastrointestinal hormones 24

cholecystokinin 26

gastrin 26

ghrelin 26

glucagon-like peptide (GLP-1) 26

peptide YY 26

secretin 26

gene knockouts 206, 464

genomic effects of steroids 214

and neuronal structure 215

ghrelin 26

and GH release 68, 130

and hunger 304

brain ghrelin

and circadian rhythms 330

ghrelin antagonist

reduces food intake 330

reduces alcohol and narcotic intake 330

ghrelin receptor

in brain 330

influence on sleep and drug addiction 331

glial cells

and hypothalamic hormone release 142

gliotransmitter

and astrocyte 144

and D-serine 144

and PGE2 147

glucagon 28

glucagon-like peptide-1 (GLP-1) 26, 306
　released after meals 306
　reduces appetite 306
　and compensatory mechanisms 307
　biosynthesis in brain 329
　effects in brain
　　reduces food intake 329
　　elevates blood pressure 329
glucocorticoid
　and adaptive behaviors 219
　suppression of immune system 29, 380, 381
　increased activity of neurons 200
　receptor types in brain 211
　and allostasis 216
　and stress 2, 219, 380
　and learning and memory 219
　effect on neonatal brain 221
glucocorticoid receptor (GR) 211
　brain localization 211
　GR and negative feedback on ACTH 211
　and immune system 380
　and stress 213
　two types, GR and MR 213
　repression of noradrenergic neurotransmission 214
　regulation of gene transcription 253
glucocorticoid response element (GRE) 253 (Figure 10.9)
glucose, blood regulation 28
glutamate (glutamic acid) 81
　and zinc ions 87
　and D-serine 87
　receptor structure 81
　biosynthesis 88
　brain pathways 97
gonadotropin hormones 126
　and blood-brain barrier 325
gonadotropin inhibitory hormone (GnIH) 127
　and action potentials 294

gonadotropin releasing hormone (GnRH) 64
　and sex behavior 325, 428
　and calcium ion permeability 250
　and kisspeptin 65
　and ovulation 67
　and tanycytes 147
　and TGFα 147
　and TGFβ1 148
　and PGE2 147
　receptor down-regulation 251
　location of neurons 64, 324; Table 12.6
　　and green fluorescent protein 295
　extra-hypothalamic release 65, 324
　regulation by estradiol 430
　regulation of secretion 64
　pulsatile secretion 66, 128
　　and starvation 67
　　and puberty 128
　　and electrical activity of neurons 135
　　and multiple unit activity (MUA) 135
　　inhibition of 163
　　continuous infusion 163
　　pulse generator 185
G-protein-coupled receptors 237
　subunits, Gα, Gβ and Gγ 237; Table 10.2
　Gq subunit 245
　therapeutic value of 237; Table 10.1
　and PRL secretion 241
　and steroid hormones 241
　and second messengers 242; Table 10.3
　and GnRH 250
　mutations 251
green fluorescent protein (GFP) 295
growth factors
　as chemical messengers 12
　TGFα and GnRH 147
　TGFβ1 and GnRH 148

growth hormone (GH) 50, 130
　expression in non-pituitary tissues 327
　and ghrelin 68, 130
　and sleep 68
　and stress 418
　neurotransmitter regulation of 131
　and somatostatin 130
　long-loop negative feedback 181
　short-loop negative feedback 182
　and acromegaly 186
　and immune system 377
　and thymus 377
growth hormone releasing hormone (GH-RH) 67, 130
　circadian rhythm 67
　control of secretion 67
　and behavior
　　stimulates feeding 322
　　effect on sleep 322
　　effect on circadian rhythm 322
growth hormone release inhibiting hormone (somatostatin) 67
　see also somatostatin
gut-brain axis 26
gut peptides, see peptide hormones and Table 2.2

Harris, G. W. 60
heart, as endocrine gland 24
　secretion of atrial natriuretic peptide (ANP) 24
　secretion of B-type natriuretic peptide (BNP) 24
heat shock protein 90 (HSP90) 192
hepatokines 40
histamine 84
history
　of endocrinology 460
　of behavioral neuroendocrinology 460; Table 15.2
　Nobel prizes 60, 460
hormones

and behavior
  neural mechanism of 425
structures 157; Table 7.1
steroid hormones 157
amine hormones 157
peptide hormones 159
biosynthesis
  pre-propeptides 159
  proprotein convertases 159
  steroid hormones 161
  peptides 159
vesicle storage 162
deactivation 166
metabolites 242; Table 7.3
methods of removal 409
methods of replacement 411
"free" (unbound) 164
levels
  analysis
    blood 170
    saliva 172
    urine 172
    feces 172
    hair 172
  regulation of 174
feedback
  negative feedback 179
  long-loop 181
  short-loop 182
  positive feedback 67, 181,
    216
assay
  ELISA (enzyme-linked immu-
    nosorbent assay) 170
  mass spectrometry 172
release by non-hormonal stimuli
  176; Table 8.1
release by social interactions
  418
modulation
  of neurotransmitter release
    183
  of sensory stimuli 425
  of motor pathways 425
  of pain 425
  of hearing threshold 425
conditioned release of
  hormones 423

evolutionary pathways 262
hormone response element (HRE)
  194
human chorionic gonadotropin
  (HCG) 34
human placental lactogen (HPL)
  34
hypophysiotropic hormones
  (releasing hormones) 60,
  62
  complications in study of 138
    effects of anesthesia 139
    species differences 140
    sex differences 140
  glial cells and 142
hypophyseal portal blood 46, 58,
  122
hypothalamus
  and neurotransmitters 122
  functions of 57
  integration functions of 183;
    Table 8.2
  and neuropeptides 271
  neurosecretory cells 58
  neuroanatomy 57
  hypophysiotropic hormones
    complexities of 71
    levels in portal blood 122
  magnocellular system 58, 120
  nuclei, functions 57
  parvicellular system 121

immune system
  and shift work 386
  sex differences 351, 378, 378
  effect of hormone circadian
    rhythms 386
  biosynthesis of neuropeptides in
    382
  cells of 351
    T lymphocytes (T cells) 352
    B lymphocytes (B cells) 352
    macrophages 352
    natural killer cells 353
    granulocytes 353
  cholinergic stimulation 375
  hypothalamic integration of
    384

key role of paraventricular
  nucleus (PVN) 385
humoral system 352
cell-mediated system 352
thymus gland
  source of T cells 354
regulation through sympathetic
  nervous system 374, 375
  adrenergic stimulation 375
control by hypothalamic/
  pituitary hormones
  358; Table 13.4
control by neuroendocrine
  system Table 13.4
control by steroid hormones
  377, 378
  toxic effects of environmental
    estrogens 379
  and menstrual cycle 379
  growth hormone, effects of 377
  neuropeptides, effects of 382;
    Table 13.4
  autocrine/paracrine effects
    384
  prolactin, effects of 377
  and thyroid hormones 378
  stress and 380
    inhibition by glucocorticoids
      380, 381
immunohistochemistry
  localization of steroid receptors
    196
in situ hybridization
  technique 198
  steroid receptor mRNA 196
  kiss1 gene expression 272
  ERα mRNA 198
indoleamines 84
inhibin
  and testes 32
inositol phospholipid system 245
  and see second messenger
    system
insulin
  and glucose 179
  and pancreas 28, 179
insulin receptor 246
  in brain 330

insulin receptor (cont.)
  in hypothalamus
    inhibition of food intake 330
  in hippocampus
    effects on memory 330
    and Alzheimer's Disease 330
interferon γ (IFNγ) see *cytokines*
interleukins see *cytokines*
intermediate lobe of pituitary (pars
    intermedia) 52; see
      pituitary gland
intracrine signaling 7

Janus kinase (JAK) 247

kidney 31
  and calcitriol 32
  and erythropoietin 32
  and vitamin D 31
kiss1 gene 65
  sex difference brain localization
    272
kisspeptin 65
  structure of pre-propeptide 300
  LH secretion 127, 302
  GnRH release 216, 302
  activity as a neuropeptide 289
  and action potentials 294
  and reproductive system 300

leptin 35, 307
  leptin deficiency and obesity
    308
  reduces body weight 308
  hypothalamic feedback 308
  resistance 310
  and reward 312
  food craving 312
  therapeutic uses 312; see
    Table 12.3
  treatment of obesity 312
leptin receptor 246, 309
  and leptin resistance 311
  and SOCS3 311
  and STAT3 311
leukotrienes 248
β-lipotropin (β-LPH or β lipotropic
    hormone) 51

liver, as endocrine organ 40
  hepatokines 40
lung as endocrine tissue 25
luteinizing hormone (LH) 51
  neurotransmitter regulation of
    126
  and kisspeptin 127
  and gonadotropin inhibitory
    hormone (GnIH)
    127
  inhibition of secretion by GnRH
    163
luteinizing hormone releasing
    hormone (LH-RH) see
      GnRH
lymphocytes see *immune system*

macrophages see *immune system*
magnocellular neurosecretory
    cells 120
mass spectrometry 172
melanocyte stimulating hormone
    (αMSH) 132
  derived from POMC 132
  neurotransmitter regulation of
    secretion 71, 133
melatonin 22, 84
  pineal gland 175
  and nocturnal activity 413
  in drinking water 413
membrane receptors 94, 236
  internalization 239
  regulation 238
  for steroid hormones 196, 201,
    241
  for neuropeptides 236, 237
menstrual cycle 173
  and immune responses 379
mineralocorticoid receptor
    feedback on CRH/ACTH 212
  brain localization 211
  MR binds aldosterone and
    cortisol 211
monoamine neurotransmitters 83
    (see *catecholamines*)
monoamine oxidase (MAO) 97
multiple unit activity (MUA)
    and GnRH/LH release 135

muscle
  skeletal 36
    and myokines 37
  heart (cardiac) 38

natural killer (NK) cells 353
negative feedback, hormonal 179
  and anterior pituitary 180
  and reproductive system 181
  testosterone 181
  estradiol 181
  cortisol 185
nervous system, divisions of 109
  autonomic 110
  enteric 112
  somatic 109
neural development
  neuroendocrine 220
  organization of 220
  perinatal development 220
neurocrine communication 5
neuroendocrine communication 7
neuroendocrine disruptors 217,
    465
  and bisphenol A 217
  plastic water bottles 217
neuroendocrine reflexes 175
  and oxytocin 175
  and melatonin 175
  and prolactin 420
  and oxytocin 420
  and LH 420
  induced ovulators 420
neuroendocrine research metho-
    dology 166; Table 7.4
neuroendocrine responses
  chaining with behavioral
    responses 420
  to cognitive stimuli 421
  to environmental stimuli 416
  to social interactions 418
neuroendocrine-environmental
    interactions 416
neuroendocrine-immune system
    364, 369
  hypothalamic integration of
    384
neuroendocrine transducer 60

neurohypophysis see *posterior
        pituitary*
neuromodulators 2, 15, 286
    criteria for definition 286
    long-range influence: neuro-
        modulation 287
    comparison with neurotrans-
        mitters 287
neurons
    and synapses 78
    and dendrites 78
neuropeptides
    classification of 257; Table 11.1
    comparison with neurotrans-
        mitters 287; Table 12.1
    criteria for action as neuro-
        transmitter 289
    evolution 262
    embryological origins 264;
        Table 11.4
    effects on immune system 382;
        Table 13.4
    and human genome 258
    and pre-propeptides 258, 262
    and presynaptic inhibition
        298
    opioids see *opioid peptides*
    as neurotransmitters 84
    as neuromodulators 293
    electrophysiological responses
        294
    biosynthesis 60, 84, 88
    localization/brain pathways
        269
    control of pituitary hormone
        secretion 120
    control of body weight 300
    control of food intake 302
    receptors 236, 251
        brain localization 274
        membrane location 236
    regulation of postsynaptic
        receptors 293
    storage 260
    co-localization with neuro-
        transmitters 260, 266;
        Table 11.3
    deactivation 261

co-release/co-transmission 261
    and Dale's Law 267
    in pituitary gland 269
and hypothalamus 271
sex differences 271
    vasopressin 272
    kiss1 272
and second messengers 272
and clinical use 273; Table 11.5
and blood-brain barrier 275
neuroendocrine effects 299;
        Table 12.2
behavioral effects of 313, 314
cognitive effects of 313
visceral effects of 313
learning and memory 314
neurophysin 50
    neurophysin I (oxytocin
        binding) 165
    neurophysin II (vasopressin
        binding) 165
neuroprotection and steroid
        hormones 216
    and allostasis 216
    and estradiol 217
neurosteroids 222
neurotransmitter 78, 79
    biosynthesis 88
    comparison with neuropeptides
        287; Table 12.1
    effect on endocrine glands 135
    evolution 262
    agonists and antagonists 103
        and Table 5.5
    amino acids 80
    categories of 80; Table 5.2
    non-classical 84
    criteria for 79
    release 92
    and action potentials 92
    inactivation 97
        enzymatic 97
        by re-uptake 97
    transporters 97
        and drugs 97
    brain pathways 97
    effect of drugs on 102;
        Table 5.4

receptors 236
    membrane location 236
    receptor subtypes, Table 5.3
    reuptake, blockade of 102
    nutrients and 107
    control of anterior pituitary
        hormone secretion 124;
        Table 6.1
    control of posterior pituitary
        hormone secretion 124;
        Table 6.1
    hormonal modulation of 183
    co-localization with neuropep-
        tides 260, 266;
        Table 11.3
    and Dale's Law 267
nitric oxide (NO) 84
    brain localization 101
    and glutamate 85
N-methyl-D-aspartate (NMDA)
        receptor 81
Nobel prizes 60, 460
norepinephrine
    biosynthesis 90
    storage and reuptake 91
    inactivation by reuptake 97
    tyrosine hydroxylase 91
    brain pathways 97
    and regulation of pituitary
        hormone secretion;
        Table 6.1
nuclear receptor superfamily 193

obesity 13; and see *leptin*
opioid peptides 259, 316;
        Table 11.2;
    analgesic effects 316
    behavioral effects 317
        binge eating 317
        alcoholism 318
        sex behavior 318
    cognitive effects 317
    gene families 259
        endorphin 259
        enkephalin 259
        dynorphin 259
        deltorphin 259
        endomorphin 259

opioid peptides (cont.)
 and autonomic nervous system
  316
 and propeptides 259
 and regulation of oxytocin
  release 298
 and neuroendocrine system
  316; Table 12.4
 decrease sex behavior 298
ovaries 32
 estradiol 32
 progesterone 33
 relaxin 34
oxyntomodulin (OXM) 307
 released after food intake 307
 decreases body weight 307
oxytocin
 and behavior 313
  emotional and social behavior
   319
  sex behavior 319, 428
  in humans 320
  maternal behavior 319
  social memory/recognition
   319, 428
  Bruce effect 320
 and romantic love 320
 and suckling stimulus 121
 neurotransmitter regulation of
  122
 and endocannabinoids 134
 and neuronal firing 135, 251
 and astrocyte retraction 146
 time scale of secretion 251
 and thymus gland 377
 regulation by estradiol 430
oxytocin neurons
 and secretion during childbirth
  296

pancreas 28, 179
 and β cells 28
 and glucagon 28
 and glucose 28
 and insulin 28
 and Islets of Langerhans 28
 and somatostatin 28
 and pancreatic polypeptide 28

pancreatic polypeptide (PP) 28,
  307
 elevated after meals 307
 reduces food intake 307
paracrine communication 5
parathyroid glands 23
parathyroid hormone (PTH) 23
 and calcium 23
paraventricular nucleus (PVN) 58
 and immune system 385
parvicellular neurosecretory cells
  58, 121
peptide hormones
 as immunomodulators 365
 peptide biosynthesis 159, 162
 biosynthesis in immune system
  372; Table 13.3
 vesicle storage 162
 secretion 162
 receptors 236
  membrane location 236
 gut peptides and appetite
  303
  ghrelin 304
  PYY$_{3-36}$ 306
  glucagon-like peptide (GLP-1)
   306
  oxyntomodulin 307
  pancreatic polypeptide (PP)
   307
  leptin 307
  new peptides 464
peptide YY$_{3-36}$ 26
 behavioral effects 331
pheromones, 11
 and vomeronasal organ 11
phosphatidylinositol diphosphate
  (PIP2) 245
phospholipase C (PLC) 245
pineal gland 175
 and melatonin 22, 84
pituitary gland 45
 anterior pituitary (adenohypophysis) 46
  and hypophyseal portal system 46 hormones of;
   Table 3.1
  folliculostellate cells 53

 paracrine relationship with
  anterior pituitary cells 53
 posterior pituitary (neurohypophysis) 45
  oxytocin 49, 133
  vasopressin 50, 133
  neurophysin 50
  neurotransmitter regulation
   126; Table 6.1
 intermediate lobe 48
  α-melanocyte stimulating
   hormone (α-MSH) 52
 pituitary hormones in brain 54
placenta 34; Table 2.3
 human chorionic gonadotropin
  (HCG) 34
 human placental lactogen (HPL)
  34
 human chorionic somatomammotropin (HCS) 34
polymerase chain reaction (PCR)
  166
positive feedback, estradiol 67,
  181
posterior pituitary
 neurotransmitter regulation
  133, Table 6.1
 and oxytocin 133
 and vasopressin 133
pre-propeptides 159
 processing 162
proenkephalin see opioid peptides
progesterone 33
progesterone receptor
 subtypes PRA and PRB 208
 membrane site 204
 and GABA receptor 204
 localization in brain 208
 induction by estradiol 208,
  430
prolactin (PRL) 50
 circadian rhythm 69, 327
 neurotransmitter regulation of
  secretion 69, 128
 sex differences in secretion 69
 and ghrelin 129
 secretion and G proteins 241
 biosynthesis in brain 327

behavioral effects
  maternal behavior 327
  food intake 327
and immune system 377, 378
release during suckling 419
prolactin inhibiting factor
      (dopamine) 69, 128
prolactin releasing peptide (PrRP)
      69, 129
pro-opiomelanocortin (POMC)
      132, 160
  and β-endorphin 132
  and α-MSH 132
proprotein convertases 51, 159
prostaglandin E2 (PGE2)
  and GnRH release 147
protein kinase 242
  protein kinase C (PKC) 246
psychiatric disorders
  neuroendocrine correlates 142
  neuorendocrine disorders 142
psychotropic drugs
  and neurohormone release
      141
pulsatile secretion 163
  GnRH/LH 66, 128
  GHRH 67
  PRL 69
  CRH/ACTH 124
  steroids 163
  and negative feedback 185
PYY$_{3-36}$ 306
  inhibits food intake 306

radioimmunoassay (RIA) 170
receptor
  steroid 192
  neurotransmitter 94, 236
  peptide hormone 236
  neuropeptide 236
  G-protein-coupled 95, 237
  ionotropic 94
  metabotropic 95, 237
  tyrosine receptor kinase (trk)
      95, 237
  subtypes 96; Table 5.3
  regulation 238
    down-regulation 238

  up-regulation 239
  regulation by neuropeptides
      293
  internalization 239
relaxin 34
reuptake mechanism, nerve term-
      inals 91

second messenger systems 242
  cAMP system 244; Table 10.3
  cGMP system 244;
      Table 10.3
  inositol phospholipid system
      245
    and phospholipase C 245
    and G-protein Gq 245
    and diacylglycerol (DAG) 245
    and protein kinase C (PKC)
      246
  calcium ion 246, 250
  tyrosine kinase (trk) system 246
    and insulin receptor 246
    and leptin receptor 246
    and STAT (signal transduc-
      tion and transcription)
      247
    and SOCS (suppressor of
      cytokine signaling) 248
  and neuropeptides 272
  arachidonic acid system 248
    and prostaglandins (PG) 248
    and leukotrienes 248
  signal amplification 249
  and protein synthesis 251
  time scale of neuronal response
      251
secretin 26
  biosynthesis in brain 331
  inhibits food intake 331
serine
  and tripartite synapse 144
  synergy with glutamate 87
  and oxytocin release 146
serotonin 84
  biosynthesis 90
  brain pathways 97
sex behavior and opioids 318
sex differences

  brain structure 220
  human behavior 220
  neurological diseases 220
  hypothalamic nuclei 220
  neurochemistry 220
  vasopressin 272
  kiss1 272
signal amplification (second
      messengers) 249
signaling mechanisms
  autocrine 7
  endocrine 5
  intracrine 7
  neurocrine 5
  neuroendocrine 7
  paracrine 5
skeletal muscle and myokines 36
  myostatin 37
  irisin 37
SOCS (suppressor of cytokine
      signaling) 248
  and leptin resistance 311
somatic nervous system 109
somatostatin 28, 68
  behavioral effects 321
  control of secretion 68
  and GH release 68, 130
  localization in brain 68, 321
somatostatin receptors 321
STAT (signal transduction and
      transcription) 247
steroid hormone
  and sensory systems 425
  and pain 425
  modulation of neuronal activity
      215
    and hypothalamic/pituitary
      activity 215
  biosynthesis 161
  structure 157, 161
  neurosteroids 222
  release 163
  blood levels 165; Table 7.2
  transport in blood 164
    sex hormone binding globulin
      164
    thyroid binding globulin 164
  and neuroprotection 216

steroid hormone (cont.)
  and behavior 217
  and endocrine-disrupting
    chemicals 465
steroid receptor 192
  and heat shock protein 90
    (HSP90) 192
  and hormone response ele-
    ments (HRE) 194
  dimerization 192
  superfamily 193
  structure 193
  localization methods 196
  localization in brain 206
  complex 198
  membrane site 201
  and G proteins 201
  and changes in neurotransmit-
    ter release 214
  and protein synthesis 214
  genomic and non-genomic
    action 198
  electrophysiological response of
    neurons 200
stress
  glucocorticoid release in 29
  effect on ACTH 124
  chronic, effects of 213
  and depression 213
  and neuronal damage 217
  and Ph.D. examination 422
  and environmental effects 438
Substance P
  brain localization 270
suckling stimulus 122
suppressor of cytokine signaling
  (SOCS) 248
suprachiasmatic nucleus (SCN)
  175
supraoptic nucleus (SON) 58
sympathetic nervous system 112
synapse 78
  tripartite synapse 144
synaptic vesicles 88
  and membrane fusion 92

tanycytes 147
  and GnRH release 147

and TGFα 147
and TGFβ1 148
testes 32
  and androgens 32
  and dihydrotestosterone 32
testosterone
  and aggression 412
  and scent-marking 409
  effect on fetus 436
  stimulates singing in birds 426
  and male copulatory behavior
    427
  negative feedback 181, 185
  as a prohormone 210
  aromatization to estradiol 210
  acts via estradiol receptor 210
  levels affected by social inter-
    action 418
tetrahydrocannabinol (THC) 86
thymosin see thymus gland
thymus gland 24
  age-related changes in 354
  regulation by autonomic
    nervous system 354
  regulation by endocrine
    hormones 354
  regulation by oxytocin/
    vasopressin 377
  effects of sex steroids 354, 379
  inhibition by glucocorticoids
    380
  thymosins 24, 354
    and neuroendocrine system
      369, 354; Table 13.2
    and pituitary hormone
      release 372
      effect of thymectomy 372
    as immunomodulators 365
    age-related decline 354
  in pregnancy 380
  peptides 354
  source of T cells 354
  and growth hormone 377
  and prolactin 378
  and thyroid hormones 378
thyroid gland 22
thyroid hormones
  and body temperature 126

and brain development 432
and neuronal structure 432
and growth 223
and metabolism 223
thyroxine (T4) 22
triiodothyronine (T3) 22
and thymus 378
thyroid hormone receptor
  in brain 223
  localization 223
thyroid stimulating hormone
  (TSH) 51
  neurotransmitter regulation of
    126
  TSH mRNA in brain 328
  TSH receptors in brain
    328
thyrotropin (thyroid hormone)
  releasing hormone
  (TRH) 63
  localization 323
  and TSH release 126
  multiple brain effects 323
  behavioral effects 323
  and circadian system 324
  and therapeutic applications
    324; Table 12.5
tripartite synapse
  and hypothalamus 144
  and glutamate 144
  and D-serine 144
tyrosine hydroxylase 91
tyrosine kinase receptors (trk) 237;
  see second messenger
  systems

urine
  hormone analysis in 172
  hormone metabolites in
    166

vasoactive intestinal polypeptide
  (VIP)
  biosynthesis in brain 329
  receptors in brain 329
  role in memory 330
  transmission of circadian
    information 330

role in Alzheimer's Disease
330
vasopressin (antidiuretic
hormone, ADH)
synergy with CRH
124

sex differences, brain localization 272
and behavior 313, 320
aggression 320
learning and memory
320

and thymus gland
377
vasopressin neurons
localization with green
fluorescent protein
295